TITANIUM SCIENCE AND TECHNOLOGY

The Metallurgical Society of AIME Proceedings published by Plenum Press

1968–Refractory Metal Alloys: Metallurgy and Technology
Edited by I. Machlin, R. T. Begley, and E. D. Weisert

1969–Research in Dental and Medical Materials
Edited by Edward Korostoff

1969–Developments in the Structural Chemistry of Alloy Phases
Edited by B. C. Giessen

1970–Corrosion by Liquid Metals
Edited by J. E. Draley and J. R. Weeks

1971–Metal Forming: Interaction Between Theory and Practice
Edited by A. L. Hoffmanner

1973–Titanium Science and Technology (4 Volumes)
Edited by R. I. Jaffee and H. M. Burte

A Publication of The Metallurgical Society of AIME

TITANIUM SCIENCE AND TECHNOLOGY

Proceedings of the Second International Conference organized by The Metallurgical Society of AIME, the American Society for Metals, and the Institute of Metals in association with the Academy of Sciences of the USSR and the Japan Institute of Metals, held at the Kresge Auditorium, Massachusetts Institute of Technology, Cambridge, Massachusetts, May 2-5, 1972

Edited by

R. I. Jaffee
Battelle-Columbus Laboratories
Columbus, Ohio

and

H. M. Burte
Air Force Materials Laboratory
Wright-Patterson Air Force Base, Ohio

VOLUME 2

PLENUM PRESS • NEW YORK – LONDON • 1973

Library of Congress Catalog Card Number 73-76701

ISBN 0-306-30728-6

Plenum Press is a division of Plenum Publishing Corporation
227 West 17th Street, New York, N.Y. 10011

United Kingdom edition published by Plenum Press, London
A Division of Plenum Publishing Company, Ltd.
Davis House (4th Floor), 8 Scrubs Lane, Harlesden, London,
NW10 6SE, England

Printed in the United States of America

VOLUME 2

SECTION VI

PHYSICAL PROPERTIES, ELECTRONIC STRUCTURE, PHASE STABILITY, PHASE EQUILIBRIA

CRITICAL REVIEW

PHASE STABILITY AND EQUILIBRIA AS AFFECTED BY THE PHYSICAL PROPERTIES AND ELECTRONIC STRUCTURE OF TITANIUM ALLOYS*

Larry Kaufman and Harvey Nesor
ManLabs, Inc.
Cambridge, Massachusetts 02139, U.S.A.

I. INTRODUCTION

Measurements of electronic specific heat coefficients, magnetic susceptibility, optical absorptivity, soft X-ray spectra, lattice parameters, elastic constants, Hall coefficient and conductivity are valuable in characterizing the physical and electronic properties of metals and alloys and can supplement conventional thermal analysis and metallographic techniques to reveal the phase constitution. However, in spite of many well-meaning efforts, utilization of these characteristics to define the electronic structure for the purpose of predicting phase stability has been unsuccessful. The main reason for failure is that the electronic structure is generally characterized by $n(E_F)$, the electronic density of states at the Fermi level, which is supposed to undergo a singular variation with electron/atom ratio as the phase boundary is approached yielding an energy-composition curve with similar characteristics (1).[±] The fallacy in this argument is that it completely ignores phase competition which lies at the basis of all phase diagrams (2). Indeed, the postulated inflection in the energy vs. e/a curve as the phase boundary is approached leads to serious inconsistancies (3), which are not as widely recognized (4) as they should be (5).

---

*This work has been sponsored by the Air Force Materials Laboratory under Contract F33615-71-C-1471.

±Underlined numbers in parentheses denote references.

The main shortcoming of the $n(E_F)$ vs. e/a description is that it treats stability as a property of a given phase rather than the outcome of competition between different structures and/or compositions (2, 6-9). Recent attempts at first principle calculations of the energy differences between competing structures of pure metals (10-13) have yielded encouraging results (14) and should be pursued. However, calculations which examine competing phases have not yet been applied to alloys.

Despite the difficulties in characterizing phase stability by the $n(E_F)$ vs. e/a relations (1, 5), this basis has been presented as a method for discerning the behavior of titanium alloys (15-20). Although critical e/a or group numbers may be useful as empirical guides for correlating data (17, 20, 21), assigning fundamental significance to such critical e/a values has little basis. This is demonstrated conclusively by a recent application of PHACOMP techniques to sigma phase formation in Ni-Cr-Mo alloys where the group number of Mo was changed from 4.66 to 9.66 (retaining a value of 4.66 for Cr) in order to conform with the phase boundary between the sigma and fcc phases (22).

By contrast, methods based on phase competition were first applied to titanium alloys (23). This technique, characterized by definition of the lattice stability of pure metals, has been applied to a variety of alloy systems (24, 2, 6, 14) in order to describe the energetics of stable, metastable, and unstable phases. The temperature-composition range for bcc$\rightarrow$hcp martensitic reactions in titanium alloys has been predicted (25-29) as well as the formation of miscibility gaps in bcc alloys. Moreover, this description considers metastable products such as the fcc structure, reported (30-33) to occur in a variety of alloys. This computer based technique (6), incorporates physical and electronic structure within the framework of phase competition, thus integrating thermochemical and structural data (34). Similar activities are now being pursued all over the world (35-40). These phase compeition analyses are completely different from the "solute enthalpy of transformation" approaches which have been applied to iron, titanium (19), and rare-earth systems (41). The basis for the latter analysis is questionable to say the least (6, p. 209; 2, p. 89)!

Including the physical and electronic factors within the thermochemical framework prior to examination of competative stability permits independent evaluation of diverse data and identification of the roles of individual factors (23, 34, 2, 6, 24). For example, Table 1 compares recent evaluation of the Debye temperatures of bcc and hcp titanium deduced from physical measurements without considering thermochemical factors (15) with earlier estimates which considered such factors (23, 29). The results indicate relatively good agreement on the electronic γ for hcp titanium and the Debye temperature for bcc titanium. However, substantial differences exist for the Debye temperature of hcp titanium and the electronic γ of bcc

Table 1. DEBYE TEMPERATURES AND ELECTRONIC SPECIFIC HEAT COEFFICIENTS FOR BCC AND HCP TITANIUM

H.C.P.	DEBYE TEMPERATURE ^{o}K	ELECTRONIC γ (cal/g.at.^{o}K^2)
Kaufman (1959)(15)	365	8.25×10^{-4}
Collings and Ho (1968)(23)	420	8.03×10^{-4}
BCC		
Kaufman (1959)(15)	299	5.7×10^{-4}
Collings and Ho (1968)(23)	295	17.4×10^{-4}

BCC MINUS HCP ENTROPY COMPONENTS at 1155°K	Vibrational (cal/g.at.^{o}K)	Electronic (cal/g.at.^{o}K)
Kaufman (1959)(15)	+1 180	-0.294
Collings and Ho (1968)(23)	+2.110	+1.082

ENTROPY DIFFERENCE (at 1155°K, cal./g.at.^{o}K)(BCC-HCP)	
Kaufman (1959)(15)	+0.89 (1.05 at 300°K)
Collings and Ho (1968)(23)	+3.19 (2.25 at 300°K)
OBSERVED	+0.90

titanium. If the recent values (15) are employed to compute the entropy difference between the phases at 1155°K (where the transition takes place at one atmosphere) and at 300°K, values near 3 and 2 entropy units result, compared with an observed value near one entropy unit. Although the recent values (15) were determined near 4°K, the entropy difference inferred departs so substantially from the thermochemical value that they should be questioned. Paradoxically, Collings and Ho make no mention of the entropy difference between bcc and hcp titanium even though their measurements of physical properties are employed to "deduce" characteristic $n(E_F)$ vs. e/a behavior presented as the basis for titanium alloy phase stability (15).

Although description of the physical components for the competing phases is limited by the paucity of thermochemical and physical data, application of computational methods (2, 6-9, 14, 26) permits consideration of a variety of practical problems including calculation of stable and metastable equilibria in binary and ternary systems, solubility limits and martensitic transformations. Incorporation of the characteristics of the important omega phase (6, p. 13; 42) has been considered. Moreover, expansion of the bcc/omega/hcp relations in titanium alloys can be viewed as a direct analogue of the current description of fcc/hcp/bcc relation in iron alloys (43). Examples of these calculations are presented below.

II. CALCULATION OF BINARY SYSTEMS

Calculation of the T_0-composition curve for martensitic transformations illustrates application of the above-mentioned techniques (2, 6-9, 14, 26). This approach was first applied more than ten years ago (23) and has been successfully employed (25, 27, 29) to a variety of titanium alloy systems. Application of this method (which requires data on the beta transus) by Jepson et. al. (27) yielded good results for Ti-Al but poor results for Ti-Cb. As a consequence, it is worthwhile re-examining the beta transus data employed in the T_0-x calculations performed by Jepson et. al. (27). Figure 1 compares these results (27) with independent calculations (26) indicating a more extensive two phase field. The latter (26) predict a T_0-x curve lying midway between the experimental findings of Jepson et.al. (27) and Huang (28), suggesting that beta transus curves (particularly those determined at low temperatures) must by viewed with caution unless equilibrium has been approached from different directions. Additional examples of this behavior will be presented below.

Calculation of phase stability and solubility limits in ternary titanium alloy systems is a more complex problem. A practical problem of interest is definition of the field of stability of the hcp structure to determine the temperature-composition range where improvement of mechanical properties due to alloying will not be impaired by trans formation or precipitation. The ternary systems of interest are

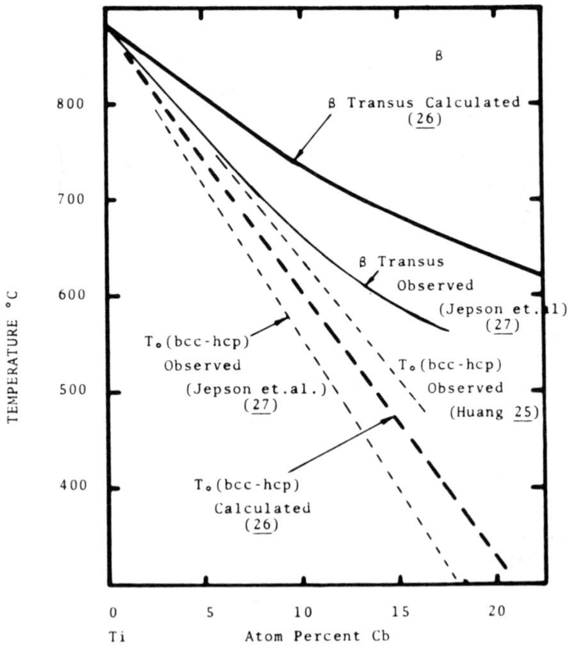

Fig. 1. Observed and calculated transformation temperatures in the Ti-Cb system.

Ti-Si-Zr, Ti-Be-Zr, Ti-Ga-Al and Ti-Sn-Al. This presents a challenge since some of the constituent binary systems have not been detailed experimentally (44-46) and limited thermochemical data have been reported (47, 48). Nevertheless, the present techniques provide an adequate description as detailed below and displayed in Figures 2-5 and Tables 2-11.

It is instructive to discuss the salient features of the calculations so that their scope becomes apparent. The basis for the calculation has been detailed earlier (26, 6) so that only phase notation need be discussed.* The designation of solutions phases bcc $=\beta$, hcp$=\varepsilon$, fcc $=\alpha$, and liquid= L will be maintained. Lattice stability parameters for the elements employed in the previous computations are also maintained (26, 6). In addition to the latter, similar descriptions are required for gallium, tin and silicon which are stable in the orthorhomic (ρ), body centered tetragonal (τ), and diamond cubic (δ) forms respectively as indicated in Tables 2, 6, and 9. Heats of fusion for the stable forms of these elements are taken from Hultgren's compilation (47, 48) while lattice stability parameters have been estimated using Miodownik's discussion of entropy changes as a guide (35). The only additional departure from earlier calculations (26, 6) is that interaction and compound parameters are treated as temperature dependent in some cases in order to obtain a more accurate description of the system under consideration.

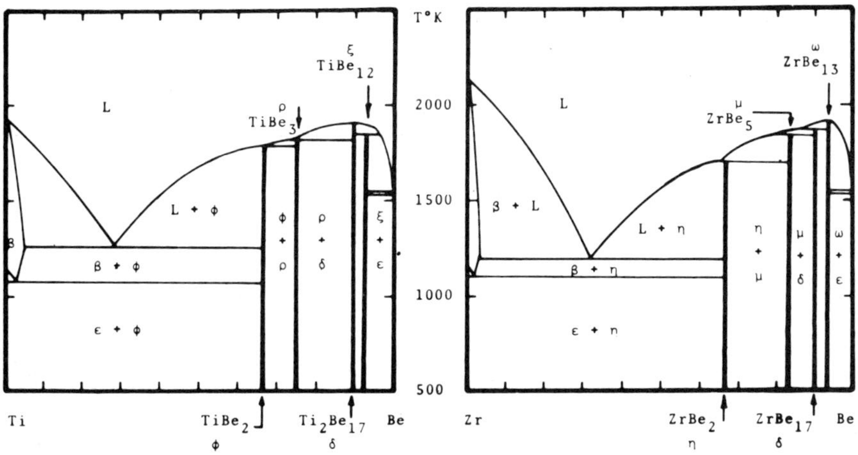

Fig. 2. Calculated Ti-Be and Zr-Be phase diagrams.

---

* Two typographical errors in reference 26 should be noted which have been corrected in reference 6. First, the melting point of hcp titanium in Table 2 of reference 26 should be 1697 not 1967°K. Second, Equation (5) of reference 26 is composed of three separate terms. The last term is $x_*(1-x_*)(\overline{L-C})$ as in Equation (91) of reference 6.

Table 2. SUMMARY OF FREE ENERGY DIFFERENCES FOR SILICON, GALLIUM AND TIN USED IN PHASE DIAGRAM CALCULATIONS (All Units in cal/g.at.)

$\Delta F_{Ga}^{\rho\to L}=1335-4.41T$	$\Delta F_{Sn}^{\tau\to L}=1720-3.41T$	$\Delta F_{Si}^{\delta\to L}=12100-7.17T$
$\Delta F_{Ga}^{\rho\to\beta}=1335-2.41T$	$\Delta F_{Sn}^{\tau\to\beta}=1720-1.91T$	$\Delta F_{Si}^{\delta\to\beta}=10600-4.67T$
$\Delta F_{Ga}^{\rho\to\varepsilon}=835-1.69T^{*}$	$\Delta F_{Sn}^{\tau\to\varepsilon}=1620-1.81T$	$\Delta F_{Si}^{\delta\to\varepsilon}=12100-4.27T$

Table 3. SOLUTION AND COMPOUND PARAMETERS FOR THE Ti-Be and Zr-Be SYSTEMS

Solution Phases

Phase	Symbol	Regular Solution Interaction Parameter (Cal/g.at.) Ti-Be	Zr-Be
Liquid	L	-9000	-10,000
bcc	β	0	+ 2000
hcp	ε	+6000(or +10,000)	+10,000

Compound Phases

Phase	Symbol ψ	x_*(at.fr.Be)	Base Phase θ	Compound Parameter (cal/g.at.)
$Ti_{1/3}Be_{2/3}$	ϕ	0.667	α	17,230
$Ti_{1/4}Be_{3/4}$	ρ	0.750	α	26,000-3.4T
$Ti_{2/19}Be_{17/19}$	δ	0.895	β	70,300-27.0T
$Ti_{1/13}Be_{12/13}$	ξ	0.923	β	86,900-35.7T
$Zr_{1/3}Be_{2/3}$	η	0.667	β	12,800-2.2T
$Zr_{1/6}Be_{5/6}$	μ	0.833	β	35,800-11.9T
$Zr_{2/19}Be_{17/19}$	δ	0.895	β	62,800-23.7T
$Zr_{1/14}Be_{13/14}$	ω	0.929	β	96,800-38.9T

*B. Predel and D. W. Stein, Acta Met. (1972) 20 515 estimate

$$\Delta H_{Ga}^{\rho\to\alpha} = 2000 \pm 1000 \text{ cal/g.at.}$$

Table 4. CALCULATION OF FREE ENERGY OF FORMATION OF TITANIUM-BERYLLIUM AND ZIRCONIUM-BERYLLIUM COMPOUNDS

$$F^{\psi} = (1-x_*)\, F_{Me} + x_*\, F_{Be} + x_*\, (1-x_*)(L-C^{\psi})$$

$$\Delta F_f{}^{\psi} = F^{\psi} - (1-x_*)\, F^{\circ}_{Me} - x_*\, F^{\circ}_{Be}$$

$F^{\circ}_{Zr}=F^{\varepsilon}_{Zr}$ for $T \leq 1155°K$; $F^{\circ}_{Ti}=F^{\varepsilon}_{Ti}$ for $T \leq 1155°K$

$F^{\circ}_{Zr}=F^{\beta}_{Zr}$ for $1155°K \leq T \leq 2125°K$; $F^{\circ}_{Ti}=F^{\beta}_{Ti}$ for $1155°K \leq T \leq 1940°K$

$F^{\circ}_{Zr}=F^{L}_{Zr}$ for $T \geq 2125°K$; $F^{\circ}_{Ti}=F^{L}_{Ti}$ for $T \geq 1940°K$

$F^{\circ}_{Be}=F^{\varepsilon}_{Be}$ for $T \leq 1528°K$

$F^{\circ}_{Be}=F^{\beta}_{Be}$ for $1528°K \leq T \leq 1556°K$

$F^{\circ}_{Be}=F^{L}_{Be}$ for $T \geq 1566°K$

Compound	x_* (at fr. Be)	ΔF_f (at 300°K) (cal/g.at)	ΔF_f (at 1300°K) (cal/g.at.)
$Ti_{1/3}Be_{2/3}$	(ϕ)(0.667)	$-5240 + 0.29T$	$-5590 + 0.59T$
$Ti_{1/4}Be_{3/4}$	(ρ)(0.750)	$-6015 + 0.96T$	$-6275 + 1.19T$
$Ti_{2/19}Be_{17/19}$	(δ)(0.895)	$-6360 + 1.81T$	$-6470 + 1.90T$
$Ti_{1/13}Be_{12/13}$	(ξ)(0.923)	$-5730 + 1.81T$	$-5810 + 1.88T$
$Zr_{1/3}Be_{2/3}$	(η)(0.667)	$-3980 - 0.29T$	$-4330 + 0.01T$
$Zr_{1/6}Be_{5/6}$	(μ)(0.833)	$-5280 + 0.90T$	$-5450 + 1.05T$
$Zr_{2/19}Be_{17/19}$	(δ)(0.895)	$-5750 + 1.05T$	$-5860 + 1.59T$
$Zr_{1/14}Be_{13/14}$	(ω)(0.929)	$-5950 + 1.84T$	$-6030 + 1.90T$

Table 5. SUMMARY OF PARAMETERS EMPLOYED FOR DESCRIBING THE Ti-Si and Zr-Si SYSTEMS

Solution Phases

Phase	Symbol	Regular Solution Interaction Parameter (cal/g.at.)	
		Ti-Si	Zr-Si
Liquid	L	-52,000	-50,000
bcc	β	-45,000	-40,000
hcp	ε	-60,000+13.2T	-40,000

Compound Phases
(Base Phase θ is the bcc β phase)

Phase	Symbol	Compound Parameter C^{ψ} (Cal/g.at.)
$Ti_{5/8}Si_{3/8}$	ξ	47,000-10.0T
$Ti_{1/2}Si_{1/2}$	ω	34,000-7.2T
$Ti_{1/3}Si_{2/3}$	ϕ	30,000-6.2T
$Zr_{3/4}Si_{1/4}$	τ	37,900-9.5T
$Zr_{2/3}Si_{1/3}$	π	46,900-10.4T
$Zr_{5/8}Si_{3/8}$	ξ	48,400-10.9T
$Zr_{3/5}Si_{2/5}$	σ	48,700-10.6T
$Zr_{6/11}Si_{5/11}$	μ	48,200-9.7T
$Zr_{1/2}Si_{1/2}$	ω	49,000-10.8T
$Zr_{1/3}Si_{2/3}$	ρ	43,200-13.7T

Table 6. CALCULATION OF FREE ENERGY OF FORMATION OF Ti-Si and Zr-Si COMPOUNDS

$$F^{\psi} = (1-x_*)\, F^{\theta}_{Me} + x_*\, F^{\theta}_{Si} + x_*\, (1-x_*)(L-C^{\psi})$$

$$\Delta F^{\psi}_{f} = F^{\psi} - (1-x_*)\, F^{\circ}_{Me} - x_*\, F^{\circ}_{Si}$$

$$F^{\circ}_{Si} = F^{\delta}_{Si} \text{ for } T \leq 1688^{\circ}K$$

$$F^{\circ}_{Si} = F^{L}_{Si} \text{ for } T \geq 1688^{\circ}K$$

Compound	x_* (at.fr.Si)	ΔF_f(at 300°K) (cal/g.at.)	ΔF_f(at 1300°K) (cal/g.at.)
$Ti_{5/8}Si_{3/8}$	ξ(0.375)	-19,100+0.25T	-19,500+0.59T
$Ti_{1/2}Si_{1/2}$	ω(0.500)	-15,700-0.99T	-16,200-0.54T
$Ti_{1/3}Si_{2/3}$	ϕ(0.667)	-10,600-2.33T	-11,300-1.73T
$Zr_{3/4}Si_{1/4}$	τ(0.250)	-13,500-0.07T	-13,800+0.61T
$Zr_{2/3}Si_{1/3}$	π(0.333)	-17,300+0.15T	-18,000+0.75T
$Zr_{5/8}Si_{3/8}$	ξ(0.375)	-18,400+0.24T	-19,000+0.80T
$Zr_{3/5}Si_{2/5}$	σ(0.400)	-18,800+0.13T	-19,500+0.67T
$Zr_{6/11}Si_{5/11}$	μ(0.455)	-19,000-0.20T	-19,500+0.29T
$Zr_{1/2}Si_{1/2}$	ω(0.500)	-18,900-0.07T	-19,500+0.36T
$Zr_{1/3}Si_{2/3}$	ρ(0.667)	-13,300-0.37T	-13,600-0.07T

Table 7. EQUATIONS FOR SOLUBILITY OF SILICON AND BERYLLIUM IN THE HCP AND BCC FORMS OF TITANIUM AND ZIRCONIUM

(Equations apply for small values of solute concentrations below 1.5% or 0.015 at.fr.)

Be in hcp Ti :

$$4.575\ T \log x = -13{,}869 + 0.43\ T$$

or

$$4.575\ T \log x = -17{,}869 + 0.43\ T$$

Be in hcp Zr :

$$4.575\ T \log x = -15{,}980 - 0.45\ T$$

Si in hcp Ti :

$$4.575\ T \log x = -1411 - 8.79\ T$$

Si in hcp Zr :

$$4.575\ T \log x = -24{,}300 + 4.03\ T$$

Si in bcc Ti :

$$4.575\ T \log x = -16{,}875 + 6.25\ T$$

Si in bcc Zr :

$$4.575\ T \log x = -25{,}925 + 7.13\ T$$

Table 8. SUMMARY OF PARAMETERS EMPLOYED FOR DESCRIBING THE Ti-Al, Ti-Ga and Ti-Sn SYSTEMS

Solution Phases

Phase	Symbol	Regular Solution Interaction Parameter (cal/g.at.)		
		Ti-Al	Ti-Ga	Ti-Sn
Liquid	L	-24,100+10.0T	-10,000	-18,000
bcc	β	-28,000+10.0T	-13,000	-21,000
hcp	ε	-29,000+10.0T	-13,800	-21,000
fcc	α	-24,200+10.0T	-	-

Compound Phases

Phase	Symbol	x_* at.fr.	Base Phase θ	Compound Parameter C^{ψ} (cal/g.at.)
$Ti_{3/4}Al_{1/4}$	ξ	0.250 (Al)	ε	7800 + 4.0 T
$Ti_{3/4}Ga_{1/4}$	ξ	0.250 (Ga)	ε	7000 + 4.5 T
$Ti_{3/4}Sn_{1/4}$	ξ	0.250 (Sn)	ε	6600 + 6.0 T
$Ti_{2/3}Ga_{1/3}$	η	0.333 (Ga)	ε	8900 + 4.2 T
$Ti_{2/3}Sn_{1/3}$	η	0.333 (Sn)	ε	9300 + 3.64 T
$Ti_{1/2}Al_{1/2}$	γ	0.500 (Al)	α	20,000 - 2.28 T
$Ti_{1/4}Al_{3/4}$	δ	0.750 (Al)	ε	25,000 - 4.50 T

Table 9. CALCULATION OF FREE ENERGY OF FORMATION OF Ti-Al, Ti-Ga and Ti-Sn COMPOUNDS

$$F^{\psi} = (1-x_*)\, F^{\theta}_{Ti} + x_*\, F^{\theta}_{Me} + x_*(1-x_*)(L-C^{\psi})$$

$$\Delta F^{\psi}_{f} = F^{\psi} - (1-x_*)\, F^{\circ}_{Ti} - x_*\, F^{\circ}_{Me}$$

Ti-Aℓ Solid Solution (hcp)

$$F^{\varepsilon} = (1-x)F^{\varepsilon}_{Ti} + xF^{\varepsilon}_{A\ell} + RT(x\ell nx + (1-x)\ell n(1-x)) +$$
$$(-29{,}000 + 10.0T)\ x\ (1-x)$$

$$\Delta F_{f} = F^{\varepsilon} - (1-x)\ F^{\circ}_{Ti} - xF^{\circ}_{A\ell}$$

$F^{\circ}_{Ga} = F^{\rho}_{Ga}$ for $T \leq 302°K$; $F^{\circ}_{Sn} = F^{\tau}_{Sn}$ for $286 \leq T \leq 504°K$

$F^{\circ}_{Ga} = F^{L}_{Ga}$ for $T \geq 302°K$; $F^{\circ}_{Sn} = F^{L}_{Sn}$ for $T \geq 504°K$

$F^{\circ}_{A\ell} = F^{\alpha}_{A\ell}$ for $T \leq 930°K$; $F^{\circ}_{A\ell} = F^{L}_{A\ell}$ for $T \geq 930°K$

Compound	x_* (at.fr.Me)	$\Delta F_f(300°K)$ (cal/g.at.)	$\Delta F_f(1300°K)$ (cal/g.at.)
hcp solid solution	ε (0.120)	-2900 + 0.38T	-2670 + 0.29T
Ti_3Al	ξ (0.250)	-5660 + 1.02T	-7400 + 2.49T
TiAl	γ (0.500)	-10,600 + 3.07T	-12,400 + 4.90T
$TiAl_3$	δ (0.750)	-8200 + 3.04T	-10,400 + 5.00T
Ti_3Ga	ξ (0.250)	-3000 - 1.26T	-3900 + 0.10T
Ti_2Ga	η (0.333)	-3900 - 1.48T	-4800 + 0.02T
Ti_3Sn	ξ (0.250)	-4200 - 1.58T	-5000 - 0.50T
Ti_2Sn	η (0.333)	-5510 - 1.41T	-6210 - 0.28T

MEASURED HEAT OF FORMATION AT 300°K

hcp solid solution	ε(0.120 Al)	-3400 ± 500
Ti_3Al	ξ(0.250 Al)	-6000 ± 500
TiAl	γ(0.500 Al)	-9000 ± 500
$TiAl_3$	δ(0.750 Al)	-8700 ± 500

Table 10. REGULAR SOLUTION PARAMETERS FOR THE Liquid(L), HCP(E) and BCC(B) PHASES

System	L	E	B	$T_c^{\beta}=B/2R$
		(cal/g.at.)		^{o}K
Ti-Zr	-746	+2090	+894	225
Ti-V	+871	+2659	+2659	669
Ti-Cb	+3125	+3125	+3125	786
Ti-Ta	+2779	+2790	+2790	702
Ti-Mo	+1560	+3671	+1241	312
Mo-Cb	-7100	-5500	-5500	
Al-Sn	2800	2800	2800	
Al-Ga	400	400	400	

Table 11. CALCULATED SUMMIT CONDITIONS FOR THE BCC ISOLATED MISCIBILITY GAPS IN THE Ti-Cb-Mo, Ti-Mo-V and Ti-Cb-V SYSTEMS

System	Interaction Parameters	Summit Conditions
Ti-Cb-Mo	3125, 1241, -5500	935°K, Ti=0.500, Mo=0.164
Ti-Mo-V	2659, 1241, -5500	859°K, Ti=0.500, Mo=0.185
Ti-Cb-V	3125, 2659, -5500	1076°K, Ti=0.500, V=0.229

The description of Ti-Be and Zr-Be shown in Figure 2 and Tables 3, 4 and 7 is based on a consideration of the fcc(C15) or Cu_2Mg type $TiBe_2$ phase (ϕ), the ρ(rhombohedral, Be_3Nb type $Ti_{1/4}Be_{3/4}$), δ(rhombohedral, $Be_{17}Nb_2$ type $Ti_{2/19}Be_{17/19}$) and ξ(body centered tetragonal, $Mn_{12}Th$ type $Ti_{1/13}Be_{12/13}$) compound phases in the Ti-Be system. Only one of the compound phases in the Zr-Be system has a counterpart in Ti-Be. This is the Zr_2Be_{17} δ phase which like its Ti_2Be_{17} counterpart δ phase exhibits the rhombohedral Nb_2Be_{17} structure. The remaining compounds including the hexagonal AlB_2 type $ZrBe_2$, the hexagonal $CuZn_5$ type $ZrBe_5$, and the fcc, $NaZn_{13}$ type $ZrBe_{13}$ do not have stable counterparts in the Ti-Be system. Parameters for these compounds designated η, μ, and ω respectively are given in Table 3. These parameters in combination with the regular solution interaction parameters for liquid, bcc, and hcp phases in Table 3 and the lattice stability parameters for Ti, Zr, and Be (2) permit calculation of Figure 2 and the free energy of formation of these compound phases. Limited constitution and thermochemical data are available for comparison. Recent studies of Ti-Be (49) suggest a bcc/L/$TiBe_2$ eutectic at 1302°K with 5.1 a/o Be in the bcc phase and a hcp/bcc/$TiBe_2$ eutectoid at 1087°K with 2.5% Be in the bcc phase. These features are reproduced by the calculations in Figure 2. This study (49) suggested that the eutectoid composition of the hcp phase is less than 0.28 a/o Be in equilibrium with β(2.5 a/o Be) and $TiBe_2$. The solubility of a solute in solution phase which is in equilibrium with a compound phase can be calculated directly from Equations 91-94 of reference (6) when the solubility is very small as

$$RT\ell nx = (1-x_*)x_*^{-1}\Delta F_i^{\phi\to\theta} + \Delta F_j^{\phi\to\theta} + (1-x_*)(L-C) - \Phi \quad (1)$$

where x is the atom fraction of solute j in the solution phase ϕ (which consists mainly of i atoms) which is in equilibrium with the compound $i_{(1-x_*)}j_{x_*}$. The latter is characterized by a base phase θ and a compound parameter C. The interaction parameters for the liquid and ϕ solution phases in the i-j system are L and Φ, respectively. Table 7 summarizes the results obtained from Equation (1) based on the description of Table 2 for the solubility of Be in hcp Ti and Zr. Two solubility equations are shown for Be in hcp Ti. The first corresponds to E= 6000, the second to E= 10,000 in Table 2. The former equation yields a solubility of 0.002 atom fraction of Be in hcp Ti at the eutectoid in keeping with Hunter's finding (49) that the solubility is less than 0.28 a/o (0.0028 at.fr.) Be. Recent observations (50) indicate that the actual solubility may be as low as 0.03 a/o Be. This possibility is reflected by the second alternative. Present calculations for the Zr-Be system yield a maximum solubility of 0.024 atom fraction Be in bcc Zr at 1200°K. Elliot's compilation suggests a solubility less than 0.030 (46). Experimental evaluation of the solubility of Be in hcp Zr at 1100°K yields a value which is less than 0.001 at.fr. The equation in Table 7 yields 0.0005 at 1100°K. Limited thermochemical data are available at 1300°K to compare with the calculated free energy of formation shown in Table 4 for Zr-Be (48). This comparison yields reasonable results although a similar comparison of the associated heats of formation leads to substantial discrepancies. However, it should be noted that the experimental heats of formation (48) are inconsistent with the observation of a stable $ZrBe_2$ phase so that the fault may lie with the experimental data and not the current calculations. Reference to Table 2 shows that the interaction parameters for the liquid, bcc and hcp forms of titanium and Zirconium are similar.

The description of Ti-Si and Zr-Si summarized in Tables 5-7 and Figure 3 is in good agreement with the limited thermochemical (47, 48) and phase diagram data (44-46). Ten compound phases are considered including two sets (orthorhombic FeB type TiSi and ZrSi designated by ω and the hexagonal Mn_5Si_3 type Ti_5Si_3 and Zr_5Si_3 phases designated by ξ) which have the same structure. In addition, the face centered orthorhombic $TiSi_2$ phase is designated by ϕ and the five remaining Zr-Si phases include the tetragonal Ni_3P type Zr_3Si phase, τ; the bc tetragonal Al_2Cu type Zr_2Si phase, π; the Zr_3Si_2 σ phase which exhibits the tetragonal Si_2U_3 or CrFe σ structure; the orthorhombic CrB type Zr_6Si_5 phase, μ; and the orthorhombic $ZrSi_2$ phase, ρ. The solubility of silicon in hcp titanium derived from Table 7 (0.0064 vs. 0.0065 observed at 1123°K; 0.0054 vs. 0.0052 observed at 873°K (44-46))exhibits an unusually small temperature dependence leading to a temperature dependent interaction parameter for the hcp solid solution in Table 5. The Zr-Si system is characterized by very small solubilities of silicon in both bcc and hcp solutions. The maximum solubility at the bcc/Liquid/Zr_3Si eutectic is calculated to be 0.02 dropping to 0.00045 at 1155°K. At the latter temperature, Hansen (44) suggests a solubility which is less than 0.0065. At this temperature the observed solubility of silicon in hcp zirconium (44) is less than 0.0030, while Table 7 yields 0.00019.

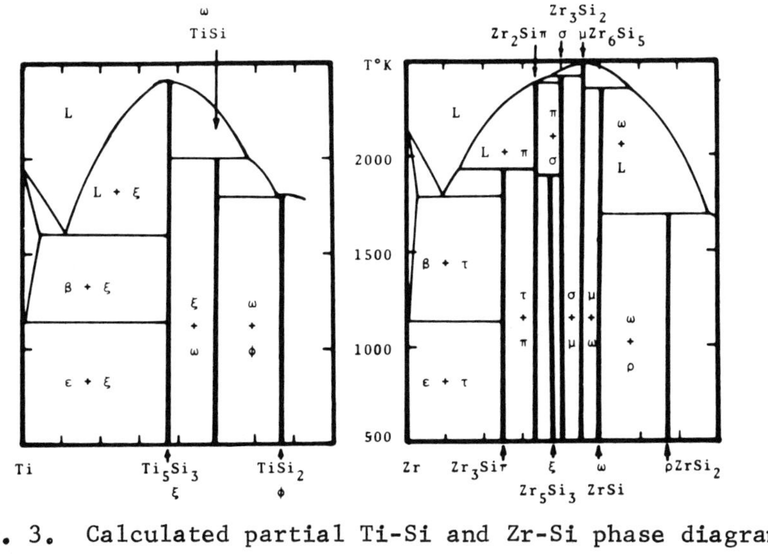

Fig. 3. Calculated partial Ti-Si and Zr-Si phase diagrams.

Analyses of the Ti-Zr, Ti-Al, Ti-Ga and Ti-Sn systems are summarized in Figures 4 and 5 and Tables 8-10. The Ti-Zr system considered earlier (26, 6), has been reconsidered with slight alteration of the B and L parameters to provide a better description. Similarly, the Ti-Al system has been reanalyzed and temperature dependent components have been included in the solution phase and compound phase parameters to improve the comparison between calculated and observed (47, 48) thermochemical properties as shown in Table 9. In addition, a compound phase is included at 25 a/o Al representing the hexagonal DO_{19} Ti_3Al structure, designated by ξ, which has counterparts in the ξ Ti_3Ga and ξ Ti_3Sn structures. Considerable controversy exists over the constitution of the Ti-Al system near Ti_3Al (44-46, 51-54). The present "line-compound" description precludes critical analysis of the controversy which

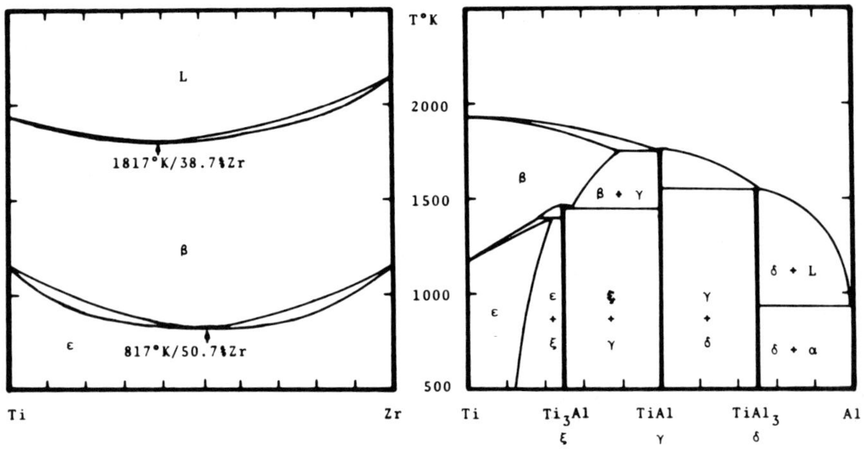

Fig. 4. Calculated Ti-Zr and Ti-Al phase diagrams.

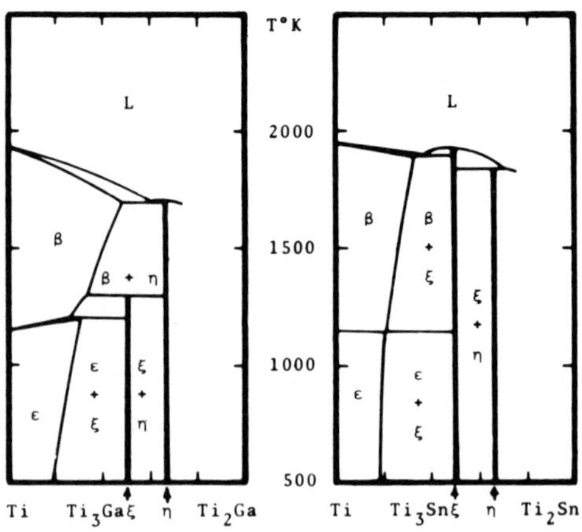

Fig. 5. Calculated partial Ti-Ga and Ti-Sn phase diagrams.

centers on ordering vs. miscibility gap decompositions. The current description of Ti_3Al leads to a peritectoid bcc/hcp/Ti_3Al at 1400°K in keeping with the findings of Clark et.al. (51). However, a small alteration of the Ti_3Al compound parameter to 7800 + 3.77T would reduce the stability of this phase sufficiently to eliminate the peritectoid leading to the phase diagram configuration suggested by Blackburn. Such an alteration would decrease the entropy of formation of Ti_3Al ξ shown in Table 9 by 0.043 cal/g.at.°K. Thus, the two alternatives are separated by a 4% entropy difference corresponding to 60 cal/g.at. or 0.0026 e.v./atom at 1400°K!

Partial Ti-Ga and Ti-Sn diagrams in Figure 5 are described in Tables 8 and 9, and include the Ti_2Ga and Ti_2Sn η phases which exhibit the hexagonal Ni_2In structure. Recent phase equilibria and vapor pressure measurements (50) suggest B =-13,000 for Ti-Ga and B= -21,000 for Ti-Sn, and that the composition of the β/β+L boundaries in Ti-Ga at 1770°K and 1700°K occur at 0.16 and 0.26 atom fractions of gallium respectively. The present solution model is too simple to reflect the minimum in the bcc/hcp phase boundaries reported for Ti-Sn (44-46). Before extending this description to ternary diagrams, it should be noted that the previous calculation of titanium phase diagrams (26, 6) characterized the Ti-Mo, Ti-V, Ti-Ta and Ti-Cb binary systems by positive interaction parameters shown in Table 10. These parameters lead to miscibility gap formation where the gap critical temperature is the interaction parameter divided by 2R. This simple description implies a symmetrical miscibility gap. These restrictions can be removed if more complex solution models are employed (6, p. 205). Nevertheless, the diagrams in Figure 6 include prediction of metastable miscibility gaps in the bcc phase. Thus, solute segregation in the bcc phase is predicted (26). This behavior has recently been observed in several titanium alloys (55, 56). Recent measurements of the activity of titanium in the bcc solution of the Ti-V and Ti-Mo systems (57, 58) yielded values of B =1800 and B =2900 respectively. Although these values support the predicted values (26) shown in Table 9 as far as sign and magnitude are concerned, the Ti-Mo value of B =2900 seems too large. Singhal and Worrell (59) have reported negative deviations from ideality in Ta-W, Cb-Mo and Ta-Mo corresponding to B = -6400, -8800 and -10,400 cal/g.at. respectively. These results are in keeping with B = -3178 suggested for Cr-V (1, p. 188). An independent analysis of the Cb-Mo-Cr ternary yields a value of B = -5500 for Cb-Mo which offers some support for the Singhal-Worrel result (59). These findings will be employed below to consider the stability of bcc solid solutions in ternary titanium alloys.

III. CALCULATION OF TERNARY SYSTEMS

Before turning to calculation of ternary systems it is instructive to consider the formation of isolated miscibility gaps (6, p.299). Such gaps can occur above the temperature where the gaps vanish in

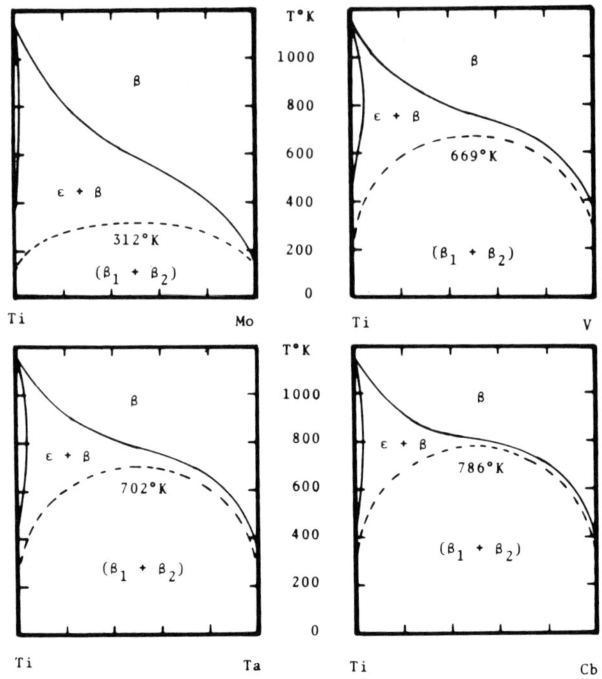

Fig. 6. Calculated BCC/HCP equilibria and metastable BCC miscibility gaps in binary titanium systems.

the binary components in systems where at least one of the interaction parameters is negative. Table 11 considers the Ti-Cb-Mo, Ti-Mo-V and Ti-Cb-V bcc phases employing the bcc interaction parameters for the titanium binary components listed in Table 10 and assuming identical values of B = -5500 for Cb-Mo, Mo-V and Cb-V. In all three cases isolated gaps are calculated with high summit temperatures. Such gaps have not been reported in the literature and will be discussed in detail below. In order to provide some confidence in the calculation, it should be noted that the isolated gap in the fcc phase of the Cu-Ni-Au system (60) occurs near 1273°K, at 20 a/o Cu, 60 a/o Ni, and 20 a/o Au. Current thermochemical data for the component binary systems (47-48) suggest interaction parameters of -5200, 2800 and 4400 cal/g.at. respectively for Cu-Au, Cu-Ni and Au-Ni. Thus the predicted summit conditions are 1264°K, at 17 a/o Cu, 50 a/o Ni and 33 a/o Au in reasonable agreement with the observed results.

Isothermal sections of the Ti-Ga-Al, Ti-Sn-Al, Zr-Si-Ti, and Ti-Be-Zr systems have been computed as shown in Figures 7-10 based on the forgoing description of the component binary systems employing

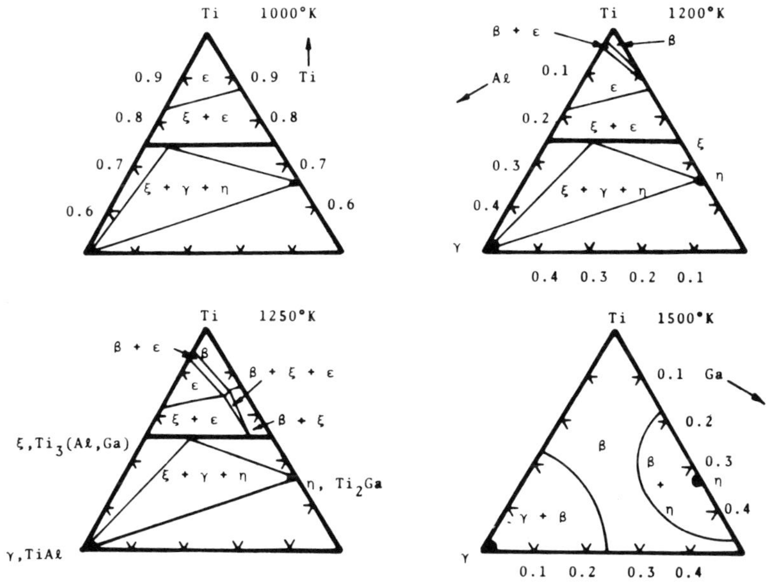

Fig. 7. Calculated partial isothermal sections in the Ti-Ga-Al system.

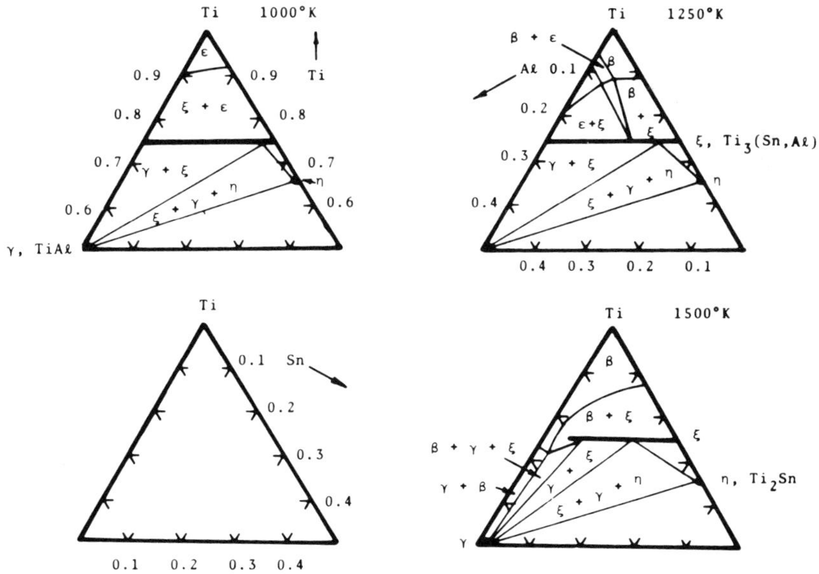

Fig. 8. Calculated partial isothermal sections in the Ti-Sn-Al system.

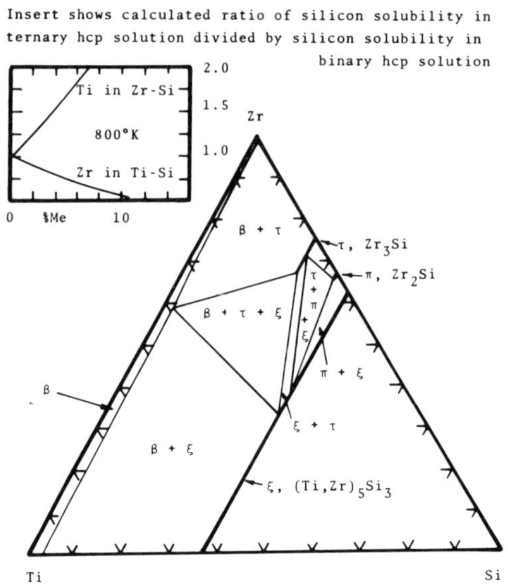

Fig.9. Calculated partial isothermal sections in the Zr-Si-Ti system at 1500°K.

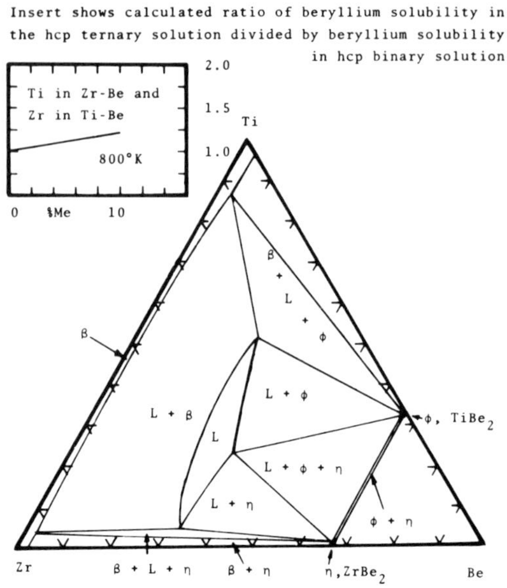

Fig.10. Calculated partial isothermal section in the Ti-Be-Zr system at 1180°K.

the techniques described earlier (6). In these calculations, the TiGa(γ) and TiSn(γ) counterphases of TiAl(γ) were based on the α structure with C = 0 while the Ti_2Al(η) counterphase of Ti_2Sn(η) and Ti_2Ga(η) was based on the ε structure with C = 0. Similarly, the Ti_3Si(τ) counterphase of Zr_3Si(τ) was based on the β structure with C=0, the $ZrBe_2$(ϕ) counterphase of $TiBe_2$(ϕ) was based on α with C=0 and the $TiBe_2$(η) counterphase of $ZrBe_2$(η) was based on the β structure with C = 0. The Al-Sn and Al-Ga interaction parameters in Table 10 were estimated from thermochemical data for these binary systems (47, 48)

Figure 10 shows the liquid phase present in the Ti-Be-Zr ternary at 1180°K. The inserted portions of Figures 10 and 11 show the effect of ternary additions on the solubility of silicon and beryllium in the hcp phase at 800°K. Addition of zirconium to titanium reduces the solubility. In the other cases, the reverse behavior is predicted. The reason for this behavior can be identified by deriving the ternary analogue of equation 1 by equilibrating the partial molar free energies of all three components, i, j and k, in the solution phase which contains 1-x-y atom fractions of i, and y atom fractions of k with those in the compound phase. The stoichiometry of the compound phase is $(i,k)_{1-x_*}j_{x_*}$. If consideration is restricted to cases where the solubility of j in the solution phase is small and solubility of k in the compound phase is small, then Equations (301) and (304) of reference (1) combine to yield the following simple results:

$$RT\ln \frac{x(1-y)^{\frac{1-x_*}{x_*}}}{x_o} = y[E_{ij}+E_{ik}-E_{jk}-yE_{ik}x_*^{-1}] \qquad (2)$$

Equation (2) describes the ratio of the solubility of element j (i.e. x/x_o) in the ternary to the corresponding ratio in the binary i-j system. Thus x_o is the solubility of j in i given by Table 7. Equation (2) is only valid for small values of x and x_o. This ratio depends upon the concentration of element j in the i-k solution which is given by y, and the stoichiometry $(i,k)_{1-x_*}j_{x_*}$ of the coexisting compound phases.

Ignoring the right side of Equation (2) for the moment we see that x/x_o will normally tend to increase with increasing y if the right side is zero. This is due to entropic effects and may be considered to reflect base line behavior. The ratio x/x_o can decrease or increase more rapidly than the ideal behavior as a function of the interaction parameters E_{ij}, E_{ik} and E_{jk} as illustrated in Equation (2). The ratio x/x_o will only decrease with increasing y if $E_{ij} + E_{ik} - E_{jk}$ is negative. If the solubility of k in the compound is not small (i.e. $(Zr,Ti)_5Si_3$ in the present case) and if the ratio of y in the compound divided by y in the solution phase equals $1-x_*$ (the solution-compound tie lines radiate from pure j), then the left side of Equation (2) reduces to $RT\ln x/x_o$. This eliminates the entropic advantage of the solution phase over the compound phase as regards solubility. Examination of Table 12, which tabulates the requisite parameters, reveals the reasons for the solubility behavior displayed in Figures 9 and 10. First the Ti-Si-Zr case (corresponding to additions of Zr to Ti-Si) is the only instance where negative values of $E_{ij} + E_{ik} - E_{jk}$ are obtained. In addition, this is the only case where extensive solubility of the metal addition in the compound is indicated (i.e. $(Ti,Zr)_5Si$). Both factors tend to reduce the solubility of silicon in titanium with increasing zirconium content.

At the beginning of this section, calculations of the isolated gap summit temperatures in Ti-Cb-Mo, Ti-Mo-V and Ti-Cb-V bcc phases were discussed (Table 11) on the assumption that Mo-V and Cb-V exhibited the same negative interaction parameter as Cb-Mo (i.e. B= -5500). The lower right corner of Figure 11 shows the calculated miscibility gaps for Ti-Cb-Mo derived on this basis. The remaining portions of Figure 12 illustrate the results which are obtained if the Mo-V, Mo-Cb and V-Cb bcc phases are treated as ideal solutions (i.e. B =0). Comparison of the calculated Ti-Cb-Mo gaps in the lower part of Figure 11 for the cases B(Mo-Cb) = 0 and B(Mo-Cb) = -5500 shows that the latter value leads to much more extensive gap stability.

Figure 12 shows the calculated bcc/bcc + hcp phase boundaries in the Ti-V-Mo and Ti-Cb-V systems computed on the basis of B(Mo-V) =E(Mo-V) = B(Cb-V) =E(Cb-V) = 0 as a function of temperature

Table 12. INPUT PARAMETERS FOR CALCULATION OF THE SOLUBILITY OF BERYLLIUM AND SILICON in HCP TITANIUM-ZIRCONIUM ALLOYS

(All units in cal/g.at.)

i	j	k	—	x_*	E_{ij}	E_{ik}	$-E_{jk}$	$-E_{ik}/x_*$
Ti	Si	Zr	Ti_5Si_3	0.375	-60,300+13.2T	+2091	+40,000	-5576
Ti	Be	Zr	$TiBe_2$	0.667	+10,000	+2091	-10,000	-3560
Zr	Si	Ti	Zr_3Si	0.250	-40,000	+2091	+60,300-13.2T	-8364
Zr	Be	Ti	$ZrBe_2$	0.667	+10,000	+2091	-10,000	-3560

System	$(E_{ij} + E_{ik} - E_{jk})$ (at 800°K)
Ti-Si-Zr	-8049
Ti-Be-Zr	+2091
Zr-Si-Ti	+12,231
Zr-Be-Ti	+2091

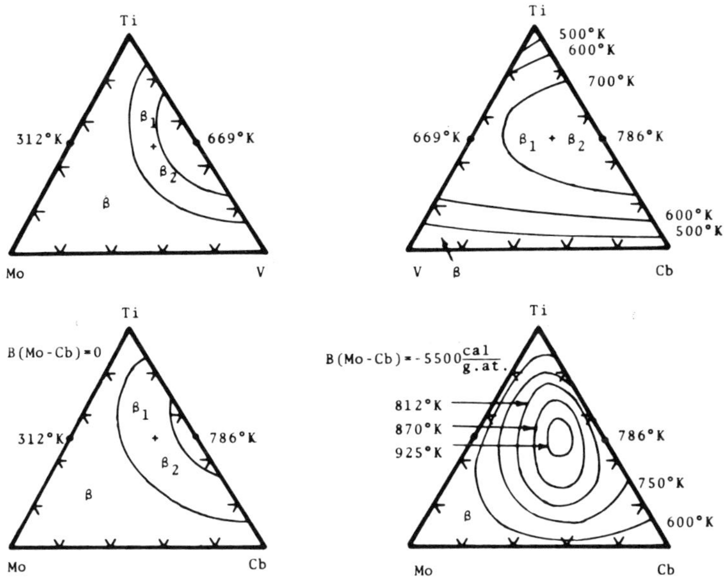

Fig. 11. Calculated miscibility gaps in the BCC phase of the Ti-V-Mo, Ti-Cb-V, and Ti-Cb-Mo systems.

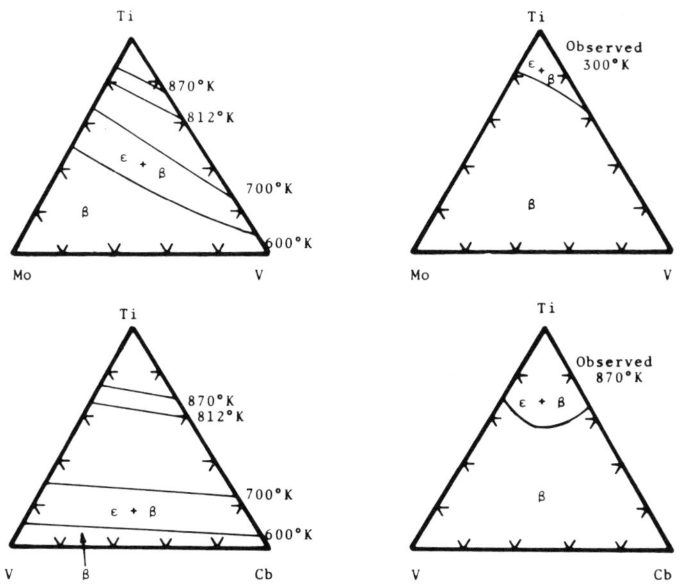

ig. 12. Calculated and observed BCC/(HCP + BCC) phase boundaries in the Ti-V-Mo and Ti-Cb-V systems.

compared with observed phase boundaries at 300°K in Ti-V-Mo (61) and 870°K in Ti-Cb-V (62). The marked divergence for Ti-Mo-V suggests that the observed 300°K section represents equilibrium near 800°K rather than equilibrium at 300°K. The bulging bcc/bcc + hcp boundary which characterizes the observed Ti-V-Cb diagram at 870°K shown in the lower right corner of Figure 12 is indicative of negative values for B(Cb-V).

Figure 13 shows the calculated bcc/bcc + hcp boundaries calculated for B(Mo-Cb) = E(Mo-Cb) = 0 (upper left corner) and B(Mo-Cb) = E(Mo-Cb) = -5500 cal/g.at. in the lower left and right corners. The upper right corner displays the observed bcc/bcc + hcp boundary (64), which is in better agreement with the ideal Mo-Cb description than the description of Mo-Cb based on negative interaction parameters shown in the lower part of Figure 13. Nevertheless, it should be noted that the Ti-Mo and Ti-Cb edge values for the bcc/bcc + hcp boundaries displayed in the "observed" diagram at 870°K differ substantially from the collated results to Ti-Mo and Ti-Cb (44-46) which are reflected in the calculated portions of Figure 13 and in Figure 6.

The complex configuration of the bcc/bcc + hcp boundary shown in the lower half of Figure 13 results from the interaction of the

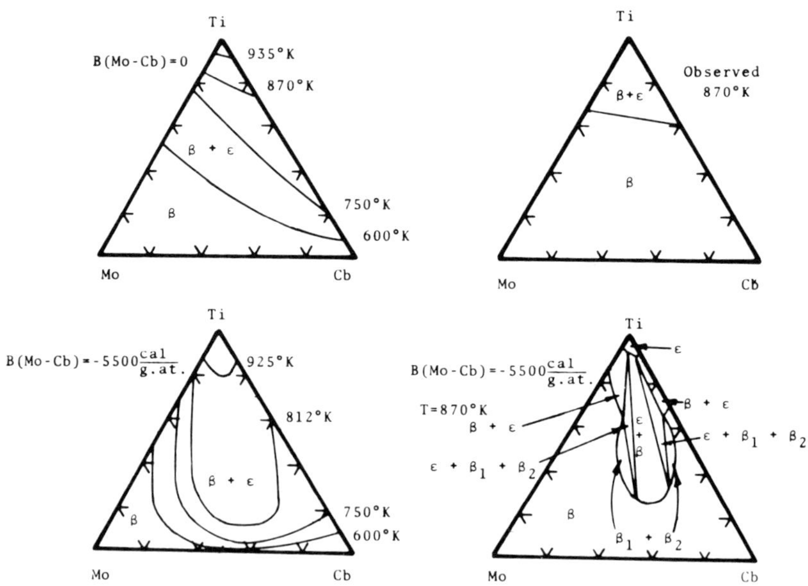

Fig. 13. Calculated BCC/(BCC + HCP) phase boundaries in the Ti-Cb-Mo system. Observed boundary at 870°K shown in the upper right corner.

hexagonal phase and the isolated bcc miscibility gap. These equilibria have not been observed. However, the fact that it has not been observed could well reflect the fact that no one has looked for it! In summary, the calculations presented here suggest that metastable miscibility gap formation in the bcc phase of titanium alloy systems can occur. Due to the low temperatures at which these reactions are expected to occur, their attainment may be impeded by kinetic factors or competitive reactions. However, further elucidation of the energetics of these reactions will undoubtedly lead to a better overall description of phase stability in titanium alloys.

ACKNOWLEDGEMENT

Significant technical contributions to this work have been made by Dr. L. R. Bidwell who is the contract monitor for the Air Force Materials Laboratory.

REFERENCES

1. V. Heine, Phase Stability in Metals and Alloys, P. S. Rudman, J. Stringer, and R. I. Jaffee Eds., McGraw-Hill (1967) p. 111.

2. L. Kaufman, Progress in Material Science (1969) 14, pp. 57, 58.

3. V. Heine and D. Weaire, Solid State Physics (1970) 24, pp. 433-439.

4. J. Inglesfield, International Symposium on Metallurgical Chemistry--Fundamentals and Applications, N.P.L.--London (1971) paper 2.1.

5. L. Kaufman, Ibid (Discussion to paper 2.1).

6. L. Kaufman and H. Bernstein, Computer Calculation of Phase Diagrams, Academic Press, New York (1970).

7. L. Kaufman and H. Bernstein, Phase Diagrams, Volume I, A. Alper, Ed., Academic Press (1970) p. 45.

8. L. Kaufman, Conference on In-Situ Composites, F. D. Lemkey and E. R. Thompson, Eds., (1972) to be published.

9. L. Kaufman, International Symposium on Thermodynamics of Alloys, Th. Heumann and B. Predel, Eds. (1972) Deutschen Gesselschaft für Metallkunde.

10. R. M. Pick, J. Phys. (France) (1967) 28, 539.

11. R. W. Shaw, J. Phys. C., (1969) 2 2335 (Proc. Phys. Soc.).

12. D. G. Pettifor, J. Phys. C.: Solid State Physics (1970) 3 367.

13. D. G. Pettifor, International Symposium on Metallurgical Chemistry--Fundamentals and Applications, N.P.L.--London (1971) paper 2.2.

14. L. Kaufman, Ibid paper 3.1.

15. E. W. Collings and J. C. Ho, The Science, Technology and Application of Titanium, R. I. Jaffee and N. Promisel Eds., Pergamon (1970) p. 331.

16. E. S. Fisher and D. Dever, Ibid p. 373.

17. I. I. Kornilov, Ibid p. 407.

18. A. D. McQuillan, Ibid (Discussion) p. 421.

19. R. I. Jaffee, Ibid (Discussion) p. 421.

20. N. V. Ageev and L. A. Petrova, Ibid p. 809.

21. W. Hume-Rothery, Progress in Material Science (1967) 13 229.

22. J. R. Mihalisin, C. G. Bieber and R. T. Grant, Tr. Met. Soc. A.I.M.E. (1968) 242 2414.

23. L. Kaufman, Acta Met. (1959) 7 575.

24. L. Kaufman, Phase Stability in Metals and Alloys, P. S. Rudman, J. Stringer and R. I. Jaffee Eds., McGraw-Hill (1967) p. 125.

25. H. Kaneko and Y. C. Huang, J. Japan. Inst. of Metals (1963) 27 387.

26. L. Kaufman and H. Bernstein, The Science, Technology and Application of Titanium, R. I. Jaffee and N. Promisel Eds., Pergamon (1970) p. 361.

27. K. S. Jepson, A. R. G. Brown and J. Gray, Ibid 677.

28. Y. C. Huang, S. Suzuki, H. Kaneko and T. Sato, Ibid.

29. L. Kaufman, Energetics in Metallurgical Phenomena, W. Mueller Ed., Gordon Breach Science Publishers, New York (1967) 3 53.

30. C. Hammond and P. M. Kelley, The Science, Technology and Application of Titanium, R. I. Jaffee and N. Promisel Eds., Pergamon (1970) p. 659.

31. M. J. Blackburn, Ibid p.633.

32. L. N. Guseva, L. A. Petrova and I. A. Ogloblina, Dokl.Akad. Nauk SSSR (1969) 185 799 (Tr. Sov. Phys. Dokl. (1969)) 14 367.

33. R. H. Ericksen, R. Taggart, and D. H. Polonis, Tr. Met. Soc. A.I.M.E. (1969) 245 359.

34. L. Kaufman, E. V. Clougherty and R. J. Weiss, Acta Met. (1963) 11 323.

35. A. P. Miodownik, International Symposium on Metallurgical Chemistry--Fundamentals and Applications, N.P.L.--London (1971) paper 2.6.

36. I. Ansara, Ibid paper 3.2.

37. H. Harvig, T. Nishizawa and B. Uhrenius, Ibid paper 3.5.

38. J. F. Counsell, E. B. Lees and P. J. Spencer, Ibid paper 3.6.

39. J. F. Counsell and O. Kubaschewski, Ibid paper 5.2.

40. C. H. P. Lupis and H. Gaye, Ibid paper 3.8.

41. K. A. Gschneider and R. B. Griffin, Met. Tr. (1971) 2 2525.

42. N. A. Vanderpuye and A. P. Miodownik, The Science, Technology and Application of Titanium, R. I. Jaffee and N. Promisel Eds., Pergamon (1970) p. 719.

43. J. F. Breedis and L. Kaufman (1971) 2 2359.

44. M. Hansen and K. Anderko, Constitution of Binary Alloys, McGraw-Hill, New York (1958).

45. R. P. Elliot, Ibid, First Supplement (1965).

46. F. A. Shrunk, Ibid, Second Supplement (1968).

47. R. Hultgren, R. L. Orr, P. D. Anderson and K.K. Kelley, Selected Values of Thermodynamic Properties of Metals and Alloys, John Wiley (1963).

48. R. Hultgren et. al., Supplement to Selected Values of Thermodynamic Properties of Metals and Alloys, University of California (Berkeley) (1966-1971).

9. D. B. Hunter, Tr. Met. Soc. A.I.M.E. (1966) 236 900.

50. L. Bidwell, AFML (Private Communication) May 1972. See also H. L. Gegel and M. Hoch, "Thermodynamics of Ti-X-Y System (this volume).

51. D. Clark, K. S. Jepson and C. J. Lewis, J. Inst. Metals (1962-1963) 91 197.

52. M. J. Blackburn, Tr. Met. Soc. A.I.M.E. (1967) 239 1200.

53. F. A. Crossley, Ibid (1968) 242 726.

54. M. J. Blackburn, Ibid (1968) 242 728.

55. G. H. Narayaman and T. F. Archibold, Tr. Met. Soc. A.I.M.E. (1970) 1 2281.

56. J. C. Williams, B. S. Hickman and D. H. Leslie, Met. Trans. (1971) 2, 477.

57. E. J. Rolinski, M. Hoch and C. Oblinger, Met. Trans. (1971) 2 2613.

58. M. Hoch and R. Viswanathan, Met. Trans. (1971) 2 2765.

59. S. C. Singhal and W. L. Worrell, International Symposium on Metallurgical Chemistry--Fundamentals and Applications, N.P.L.--London (1971) paper 1.5.

60. J. L. Meijering, Philips Research Reports (1950) 5 333.

61. I. I. Kornilov and R. S. Polyakova, Izvest. Akad. Nauk SSSR Otdel. Tekh. Nauk Met. I. Toplivo (1961) 4 76 (Report 183, DMIC, Battelle-Columbus (1963) p (162-1)-63).

62. Report 152, DMIC, Battelle-Columbus (1961) p. 113 (Office of Technical Services PB 171421).

63. I. I. Kornilov, Zhur. neorg. Khim. (1958) 3 879 (Report 152 DMIC Battelle-Columbus (1961) p. 105--Office of Technical Services PB 171421).

SOME RELATIONSHIPS BETWEEN THE ELECTRONIC AND MECHANICAL PROPERTIES OF Ti ALLOYS, DISCUSSED FROM THE STANDPOINT OF FUNDAMENTAL ALLOY THEORY*

E. W. Collings,† J. E. Enderby,†† H. L. Gegel,††† and J. C. Ho††††

The results of two general theories, which when taken together enable the structure-dependence of the cohesive energy of an alloy to be related to the electronic structures (atomic and band) of the pure constituents, are applied in a discussion of solid solution strengthening in α-stabilized and β-stabilized Ti-base alloys.

* Supported by the Air Force Materials Laboratory, Wright-Patterson Air Force Base, Ohio, under Contract AF33(615)69-C-1594, and the U. S. Air Force Office of Scientific Research (AFSC) under Grant No. 71-2084.

† Battelle, Columbus Laboratories, Columbus, Ohio, 43201.

†† Department of Physics, University of Leicester, Leicester, England, LEI 7RH.

††† Air Force Materials Laboratory, WPAFB, Ohio, 45433.

†††† Department of Physics, Wichita State University, Wichita, Kansas, 67208.

1. INTRODUCTION

The problem of finding some common ground between the metal physicist and the physical metallurgist has always been a difficult one, particularly when the electronic basis of mechanical strength in metals and alloys is under debate. Progress has been inhibited by, for example, the fact that some electronic properties of some materials are practically unaffected by mechanical deformation. Nevertheless, a proper description of a metal, as an assembly of positive ions in equilibrium with the conduction electrons, must eventually lead to an understanding of the principal interactions which control ionic motion during deformation. Accordingly we commence our discussion with an expression for the cohesive energy (E_{COH}) of a pure metal, expanded in terms of atomic potentials and structural arrangements. For an alloy, E_{COH} in this format may be regarded as being dependent on the atomic potential differences of the constituents, or some equivalent parameter. In applying this extension of the basic theory, particularly to alloys of transition metals with nontransition metals, we are guided by the results of Stern's[1,2,3] tight-binding theory of electronic structure.

Through the use of such concepts, *a postiori* qualitative explanations of various mechanical property trends in real metals (as distinct from perfect crystals) have begun to develop into useful alloy design criteria. The fundamental approach seems, at first inspection, to be applicable only to pure or single-phase materials, and as such, would be of only limited value. This difficulty can be overcome by applying the method separately to the individual components of a polyphase system. Since the particular mechanical behavior of interest in a given situation is usually dominated by the mechanical properties of only one of the component phases (e.g., brittle particles either precipitated uniformly or concentrated at grain boundaries), at least as much attention is paid to the intrinsic characteristics of such a phase as to those of the host alloy.

2. THE COHESIVE ENERGY OF METALS AND ALLOYS

2.1. Pure Metals

The cohesive, and hence the mechanical properties of metals and alloys, depend principally on the interaction of metallic ions with, and *via*, the conduction electrons. The cohesive energy is represented by E_{COH}, a component of the internal energy E, where:

$$E = E_{COH} + TS \ ; \tag{1}$$

with S and T representing entropy and absolute temperature respectively. If constant volume conditions prevail, E_{COH} may be identified as the Helmholtz free energy.

The key statement in this approach to an understanding of electronic-property/mechanical-property interrelationships, takes the form of a series expansion of E_{COH} into components representing the contributions by (a) the electron gas itself, and (b) n-body (n = 1,2 ...) ionic interactions; thus:

$$E_{COH} = \sum_{i=1}^{n} E_i \; . \tag{2}$$

Such an expression might well be used axiomatically[4], but has also been derived formally by Cohen[5] who pointed out how the free energy could in this way be resolved into structure-independent and structure-dependent components. The expansion for E_{COH} is dominated by E_1, which represents the contribution by the electron gas plus the self-energy of the ions imbedded in it. Being structure-independent it is this term which confers on metals and alloys their characteristic ductility. E_2 depends on pair-wise (central-force interactions, E_3 on triplet interactions, and so on. These are all structure-dependent, and therefore control the metal's resistance to deformation. The convergence of (2) depends on the strengths of the atomic potentials expressed in some suitable form.

For very weak potentials (such as pseudopotentials) the series may be terminated at n = 2[6]. Within the context of the topic under discussion such a truncated form of Equation (2) has been derived by Harrison[7] and by Blandin[8] using second-order perturbation theory. Under these conditions E_{COH} is only weakly structure-dependent. A second-order perturbation expression of E_{COH} is valid, for example, for the monovalent[9] pseudopotential metals Na and K, which are for this reason plastic at all temperatures.

If the atomic potentials are strong, perturbation theory may no longer be used and we return to the full expression for E_{COH}, entering the regime of interest to us here, in which third- and higher-order terms become significant. The resistance to deformation of a pure metal may thus be related to the strength of its atomic potential; one measure of which, as Cohen has suggested[5], being E_g/Δ, the ratio of a low-lying band gap to the width of the unperturbed energy band.

In this paper we employ these concepts, formerly applied in discussions of phase stability in metals and alloys, to gain a fundamental understanding of solid-solution strengthening in Ti-base alloys.

2.2. Alloys

As the above discussion indicates, strengthening in metals is accompanied by a breakdown of perturbation theory. This is the clue which leads to an understanding of solid-solution strengthening in alloys. As Stern[2] has pointed out, perturbation theory may be used in describing the electronic structures of alloys formed between two simple metals only if V_{12}/Δ is small, where V_{12} is the difference in the atomic potentials of the components. Thus V_{12} for an alloy plays the same role as E_g for a pure metal. By selecting solute elements whose potentials differ considerably from that of the host we can control the convergence of Equation (2) and consequently the degree of solid-solution strengthening.

When dealing with alloys, it may be more convenient to go directly to factors which govern the validity of perturbation theory, and hence which from this standpoint control the convergence of Equation (2) rather than to take the intermediate step of considering individual atomic potentials or their relative differences in the format V_{12}/Δ. This is because many alloys of interest are those formed between nontransition metals and transition metals. It is difficult to assign to a transition metal a meaningful "atomic potential", particularly one that would be valid in a mixed alloy. On the other hand it is more straightforward to consider directly, and perhaps in a more fundamental way, the strong perturbation of electron states that result when a nontransition metal is alloyed with a transition metal.

3. ELECTRONIC STRUCTURES OF ALLOYS

The atoms of a pure simple metal (metal 1) of valence Z each contribute Z electrons to the conduction band. Similarly for a pure simple metal (metal 2) of valence Z + 1. But this is no longer true when metal 2 is dissolved in metal 1; since screening or charge-neutrality considerations require the extra valence electron to be more-or-less localized in the vicinity of the metal-2 atoms. This example, discussed initially by Mott[10] and developed into theories of metallic virtual bound states by Friedel[11] and into the theory of "charging", through the work of Stern[1,2,3], shows clearly that the perturbation approach and the rigid-band model are inapplicable to alloys of simple metals differing in valence. The theory of charging implies more than just a cell-by-cell neutralization of net charge by screening. If charging is "large"[1] the theory requires in addition a drastic alteration in band shape as the electrons are redistributed in energy. Thus, for example, the screening electrons of atom 1 may be assigned mostly to low-lying states in the band, an conversely for atom 2. Such an arrangement would imply the existenc of a minimum in the n(E) (density-of-states *versus* energy) curve. I contrast to the motion of "neutral pseudo-atoms"[12] which carry their "own" screening electrons with them, the movements of atoms under conditions of large charging are resisted by the energy barriers associated with redistribution of the screening charge in space and energy with each change of local environment. This strong structure-dependence of electron states which accompanies large charging is expressive of solid-solution strengthening.

Three results from the above theory, particularly important in this context are summarized below.

3.1. Transition-Metal Binary Alloys (T_1-T_2 Alloys)

Charging is particularly small in alloys between adjacent transition elements near the middle of a long period(3,13). Excellent agreement with the rigid-band model has in fact been demonstrated by McMillan(14) for adjacent binary alloys from the sequence Hf-Ta-W-Re. We would not expect to find such good adherence to rigid-band principles for Ti-rich alloys with nearby transition elements, but nevertheless such alloys may still be classed as "low perturbation".

3.2. Transition-Metal/Nontransition-Metal Binary Alloys (T_1-B Alloys)

For isolated transition-metal ions dissolved in simple metals, the amplitudes of the d wavefunctions are large in the vicinity of the impurity. These are referred to as virtual bound states(11). Together with some extra broadening of the d-states this picture carries over to concentrated alloys, with the result that we again find maximal d-wave-function amplitudes at the transition-metal-ion sites. The tendency of the d wavefunctions to stay away from the nontransition-metal sites leads to an approximate representation of the alloy as a "diluted" or "expanded" transition metal lattice. This model is substantiated by several pieces of experimental evidence. For example, as a result of NMR studies of the V-Al system, Van Ostenburg, et al(15), were able to demonstrate that the d-band electrons tended to avoid the Al cells. It is useful to interpret this as strong scattering of the d wavefunctions by the Al ions. In addition, the above authors were able to conclude that the valence electrons of the Al appeared to contribute to a low-lying band; a suggestion which later appeared(1) as a part of Stern's tight-binding theory of disordered alloys. Although somewhat more refined, the conclusions reached by Lye(16) with regard to the electronic structures of intermetallic components of Ti with the nonmetallic "p-s" elements C, N, and O, based on the results of band-structure calculations for TiC, are also in general agreement with these ideas.

For a T_1-B alloy, therefore, the picture which emerges is one of strong noncentral interactions between the ions, resulting in a pronounced structure-dependence of the corresponding E_{COH}.

3.3. Influence of Order-Disorder on Electron States

Stern's theory(1) shows that in the presence of small charging an electron energy band splits into two when the alloy orders. On the other hand, when $V_{12}/\Delta >> 1$ the model yields a band gap in both the disordered and ordered states of the alloy. The real situation will presumably be less well defined than this; nevertheless, as some of us have pointed out elsewhere(4), it is clear that various degrees of electronic structure change can be expected to accompany order-disorder. It can be argued that in general (there is at least one exception) the disordered alloy will tend to be the more ductile.

4. APPLICATION TO ALLOYS OF TITANIUM

4.1. Alloys of Ti With a Nearby Transition Metal (Ti-T_2 Alloys)

As indicated in 3.1., Ti-base transition-metal-binary (Ti-T_2) alloys may be classified as "low-perturbation". If T_2 is close to Ti in the transition-metal block of the periodic table, many of the properties of Ti-T_2 alloys are determined by the average electron-to-atom ratio, $\mathfrak{z}$. For example, when $\mathfrak{z}$ exceeds a threshold value of about 4.4 the bcc (β) structure remains stable during quenching. We Therefore regard T_2 solutes as structural (β) stabilizers rather than solid-solution strengtheners, although some degree of strengthening may be contributed by their presence. Having obtained a suitable β-stabilized binary alloy, a second solute species may then be selected solely on the basis of its strengthening properties. This will generally be a nontransition element.

4.2. Alloys of Ti With Nontransition Metals -- Low-Concentration Ti-B Alloys

Significant solid-solution strengthening is achieved by adding nontransition metals (designated B-metals) to Ti. As indicated in 3.2., Ti-B alloys are characterized by a relatively high density of occupied d-states at the Fermi level. The d-wavefunctions are strongly scattered from the B-ions, which are themselves screened by electrons occupying states in the lower part of the band. During alloying, as the Ti-lattice becomes diluted or "expands", the influence on the Fermi density-of-states, $n(E_F)$, of a decreasing d-band width tends to be compensated for by a decrease in the average volume-density of d-states. The resultant relatively small concentration-dependence of $n(E_F)$ with alloying, in low-concentration Ti-Al alloys, has already been demonstrated(17).

It is possible to enumerate several important consequences of the rearrangement of electron states which accompanies the addition of B-metals to Ti.

4.2.1. The Cohesive Energy

Of primary concern to us here is the fact that the readjustment of the energy state of B which accompanies its solution in Ti can be designated as a "strong interaction" between B and its Ti neighbors. The increase in structure-dependence of E_{COH} through the necessity of including high-order terms in Equation (2) is an expression of solid-solution strengthening. Listed below are some of the various physical property measurement techniques which may be employed in studying the magnitude of this interaction strength.

4.2.1(a) Relative Vapor Pressure. The application of this technique in the determination of thermodynamic activities and pair-wise interaction parameters is discussed elsewhere in these proceedings[18].

4.2.1(b) The Electrical Resistivity. As pointed out above, it is possible to relate the solid-solution-strengthening capacity of B, when dissolved in Ti, to the scattering potential presented by B-ions to d-wavefunctions. Thus the relative magnitudes of the specific resistivities[19] of various metals when dissolved in Ti, can be used as indicators of the degrees of charging, and consequently of solid-solution strengthening. The composition-dependences of resistivity of several Ti-T_2 and Ti-B alloys are shown in Figure 1. It is seen immediately that the specific resistivity common to various T_2 ions in Ti is about an order of magnitude smaller than those for B-metals; a result consistent with Sections 4.1 and 4.2. Furthermore the resistivity data suggest that the interaction strengths between various B-metals and Ti, and hence the solid-solution-strengthening abilities of the B-metals in Ti-B alloys, increase in the sequence Al $\rightarrow$ Ga $\rightarrow$ Sn. This has indeed proved to be the case for Al and Ga additions to Ti, as indicated in Figure 2.

4.2.2. Phase Stability

In contrast to transition-metal solutes, additions of which increase the d-electron density and lead eventually to β-stabilization; the "p-s" elements (particularly the B-metals) by "diluting" the Ti lattice in the manner outlined in Section 3.2, yield solid solutions which remain "Ti-like" as alloying proceeds. This is the mechanism of α-stabilization.

In other papers[20] we have usually defined an electron-to-atom ratio ($\mathfrak{z}$) scale by considering the average number of valence electrons (i.e., the atomic d, s electrons for Ti; and the p, s electrons for Al). However, the dilution concept referred to above, together with the conclusions of other workers(21), suggest that a more physically meaningful $\mathfrak{z}$-type scale, valid for both Ti-T_2 and Ti-B alloys, might result if we restricted ourselves to a consideration of the average density of only the d-electrons.

4.2.3 Ternary Solid-Solution Strengthening

For a given total solute concentration we would generally expect to observe an improvement in solid-solution strengthening in going from a binary to a ternary (Ti-B_1-B_2) alloy, as the binary Ti-B_i-type interactions become augmented by interactions (B_1-B_2) between the solute atoms themselves. An example of this is given in Figure 2 which illustrates a "secondary" solid-solution strengthening of Ti-Ga accompanying the substitution of Al for half of the Ga formerly present. In that figure it is interesting to note that the Ti-xAl-xGa curve lies above that for Ti-Ga, rather than between those for Ti-Al and Ti-Ga as might otherwise have been expected. Numerical values of the pair-wise interaction parameters appropriate to the Ti-Al, Ti-Ga, and Al-Ga interactions respectively, are presented elsewhere in these proceedings[18].

Finally we note that these principles of solid-solution strengthening can be extended to β-stabilized (i.e., Ti-T_2) alloys, which may thus be strengthened through additions of the elements B_1, B_2, etc.

4.3. Alloys of Ti With Nontransition Metals -- High-Concentration Ti-B Alloys

The strong spacial correlations which characterize the structur[e] of Ti-B alloys lead eventually, as the alloys become more concentrat[ed] and hence as the atom-ratio n_{Ti}/n_B becomes smaller, to long-range-ordered sturcutres. A good example of this effect is to be found in the Ti-Al system. At, say, 700° C Ti-Al alloys with Al concentratio[ns] of 21 through 25 at.% assume the long-range-ordered α_2 structure[22] The influences of such ordering on the electronic transport and density-of-states properties of Ti-Al are described elsewhere[4,17,23], the experimental results being consistent with a considerable reduction of $n(E_F)$ with ordering, particularly in Ti_3Al.

4.3.1 Influence of Order-Disorder on Electronic Structure and Mechanical Characteristics

Referring to previous sections (the above, and Section 3.3) an[d] elsewhere[4] the results of both theory and experiment indicate tha[t] changes in electronic structure are expected to accompany order-disorder in most alloys except those characterized by small value o[f]

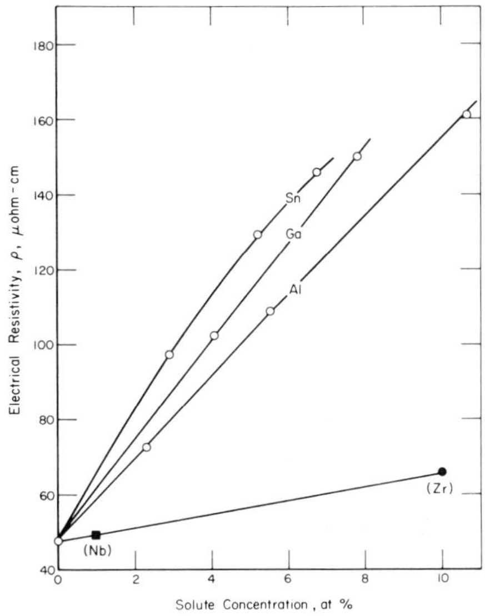

Fig. 1. Composition dependences of resistivity for Ti-T_2 and Ti-B alloys. The data for Ti-T_2 alloys are from S. L. Ames and A. D. McQuillan Acta. Met. 4 619 (1956). The results show that the interaction strengths of various solute elements with Ti proceed in the sequence T_2, Al, Ga, Sn.

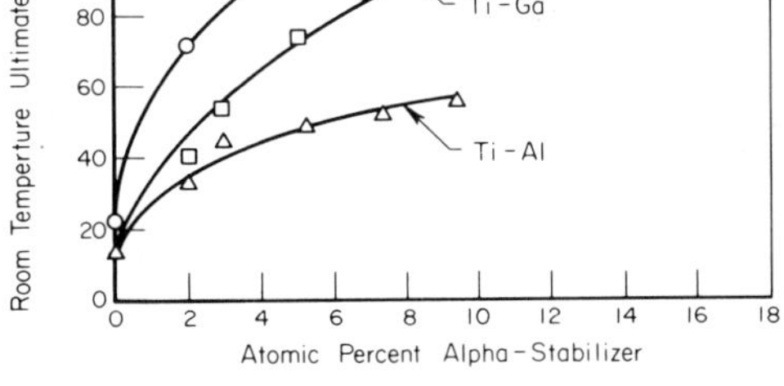

Fig. 2. The solid-solution strengthening capacities of B-metals dissolved in Ti increase in the sequence Al, Ga, in agreement with the data of Figure 1. In addition it can be seen that "secondary" solid-solution strengthening results in a ternary alloy (Ti-Al-Ga) which is stronger than Ti-Ga.

the parameter V_{12}/Δ. For example, in Ti_3Al the ordered α_2 structure supports, and is supported by, a significant degree of bond directionality(23,24).

It follows that deformation of a long-range-ordered structure results in a reduction in the order-parameter, and consequently a change in the electronic structure. This is not necessarily true for a disordered alloy, deformation of which merely results in another disordered structure. As a result, an ordered alloy tends to resist deformation to a greater extent than it otherwise would, if in the disordered state(25). A classical exception to this is Mg_3Cd(26) whose deformation properties require individual treatment.

Thus it is generally possible to alleviate the brittleness associated with a given intermetallic compound by destroying the long-range order. Two methods of disordering, or partially disordering, binary compounds are available: (a) isothermal disordering through the addition of a third atomic species; (b) thermal disordering by heating through the order-disorder critical temperature, T_c. Examples of these are as follows.

4.3.1.(a). A significant reduction in the brittleness associated with Ti_3Al is achieved by the substitution of Ga for some of the Al, within the "ordered" lattice. This procedure also results in a reduction in T_c from 850° C for Ti_3Al to 640° C for "$Ti_3Al_{1/2}Ga_{1/2}$".

4.3.1.(b). Disordering can be achieved thermally by heating through T_c. At the same time we would expect to find an improvement in ductility. That this does indeed take place in "$Ti_3Al_{1/2}Ga_{1/2}$" is evidenced by Figures 3(a) and (b) which show a correlation between the order-disorder and brittle-ductile transformation temperatures. Microscopically, we have interpreted this behavior in terms of a relatively high structure-dependence of E_{COH} in the ordered phase. Thermodynamically, in considering the effect of temperature, and ultimately disordering, on ductility we return to Equation (1). For $T > 0$, E_{COH} is augmented by the "structure-independent" term TS which normally increases smoothly with temperature; but which in addition undergoes a relatively large rapid increase as T approaches T_c.

5. SUMMARY

The resistance to deformation of a metal is gauged by the extent to which the cohesive energy, E_{COH}, is structure-dependent. E_{COH} may be expanded in the form:

$$E_{COH} = \sum_{i=1}^{n} E_i \tag{1}$$

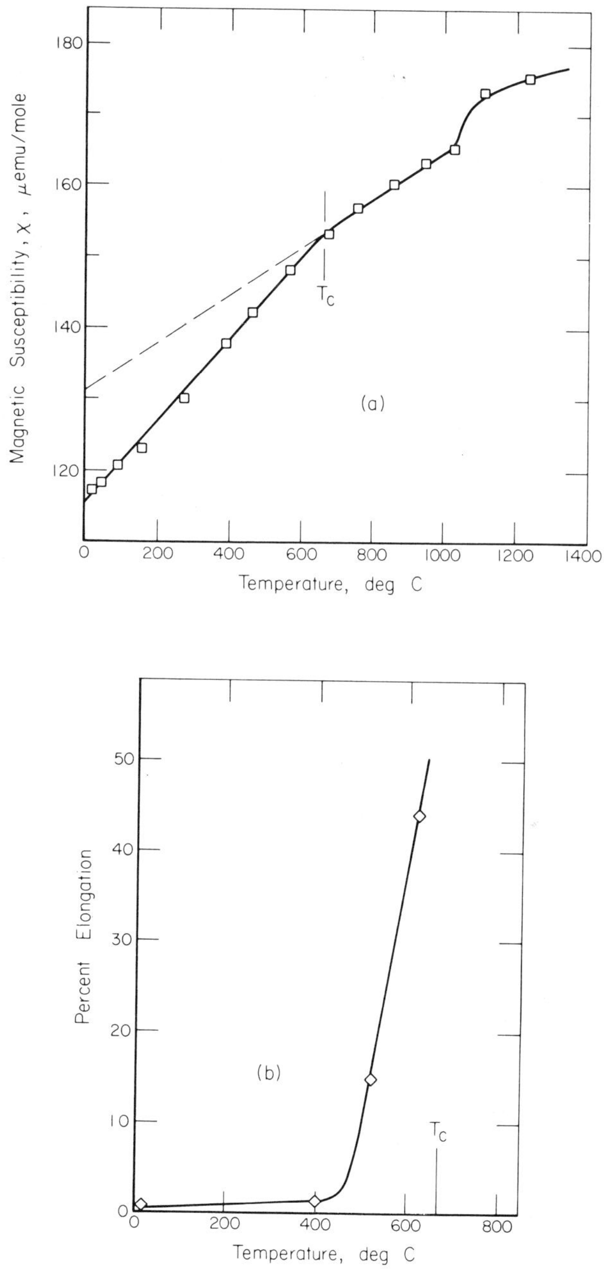

Fig. 3. (a) Magnetic determination of the order-disorder critical temperature, T_c, in "$Ti_3\ Al_{1/2}\ Ga_{1/2}$." (b) Increase in ductility in "$Ti_3\ Al_{1/2}\ Ga_{1/2}$" on approaching T_c.

in which E_n represents n-body interactions. The convergence of (1) is controlled by the strengths of the atomic potentials, in the case of pure metals; or the atomic potential differences between the constituents, in the case of alloys— in other words by the extent to which perturbation theory is valid. These principles have been employed to provide a fundamental understanding of phase stability and solid-solution strengthening in Ti-base alloys.

(a) When alloyed with B-metals the lattice remains "Ti-like" and consequently α-stable. The alloying of Ti with other transition elements β-stabilizes the lattice when the electron-to-atom ratio ($\mathfrak{z}$) becomes suitably large ($\mathfrak{z} \gtrsim 4.4$).

(b) Solid-solution strengthening may be achieved by adding B-metals either to Ti, resulting in strengthened α-alloys; or to $Ti-T_2$ ($\mathfrak{z} \gtrsim 4.4$) alloys, to produce strengthened β-alloys.

(c) "Secondary" solid-solution strengthening through B_1-B_2-type interactions may be achieved through the use of several B-metal additions to form $Ti-B_1-B_2$-etc. α-alloys, or $Ti-T_2-B_1-B_2$-etc. β-alloys.

(d) The embrittlement associated with "excessive" solid-solution strengthening of binary α-alloys to the point of ordered compound (e.g. α_2) formation may be alleviated by disordering, accomplished either through ternary alloying or thermally.

REFERENCES

1. E. A. Stern, Physics, 1, 255 (1965).

2. E. A. Stern, Phys. Rev., 144, 545 (1966).

3. E. A. Stern in "Energy Bands in Metals and Alloys", ed. by L. H. Bennett and J. T. Waber, Gordon and Breach (1968), p 151.

4. E. W. Collings, J. E. Enderby, and J. C. Ho in Proceedings of the 3rd Materials Research Symposium "Electronic Density of States", Nat. Bur. Stand. (U.S.) Spec. Publ. 323, Dec. 1971, p 483.

5. M. H. Cohen in "Alloying Behavior and Effects in Concentrated Solid Solutions", ed. by T. B. Massalski, Gordon and Breach (1965), p 1.

6. M. H. Cohen in Colloquium on Solid Metallic Solutions, Orsay, 1972, J. Phys. Radium, 23, 643 (1962).

7. W. A. Harrison, "Pseudopotentials in the Theory of Metals", Benjamin (1966).

8. A. Blandin in "Alloying Behavior and Effects in Concentrated Solid Solutions", ed. by T. B. Massalski, Gordon and Breach (1965), p 50; see also R. Pick and A. Blandin, Phys. kondens. Materie, 3, 1 (1964).

9. The situation becomes more complicated with polyvalent metals, because of the interrelationship between structure and band structure as discussed by V. Heine in "Phase Stability in Metals and Alloys", ed. by P. S. Rudman et al., McGraw-Hill (1967), p 103.

10. N. F. Mott and H. Jones, "The Theory of the Properties of Metals and Alloys" (1936), Dover Publications (1958), p 86.

11. J. Friedel, Advances in Physics, 3, 446 (1954) [e.g., p 477]; see also numerous subsequent papers, e.g., J. Friedel in "Proceedings of the International Conference on Electronic Transport in Metals and Solids (1956); Can. J. Phys., 34, 1190 (1956).

12. J. M. Ziman, Advances in Physics, 13, 89 (1964).

13. J. Friedel, discussion in "Phase Stability in Metals and Alloys", ed. by P. S. Rudman, et al., McGraw-Hill (1967), p 162.

14. W. L. McMillan, Phys. Rev., 167, 331 (1968).

15. D. O. Van Ostenburg, et al., Phys. Rev., 135, A455 (1964).

16. R. G. Lye in "Atomic and Electronic Structure of Metals", ed. by J. J. Gilman and W. A. Tiller, Chapman and Hall (1967), p 99.

17. E. W. Collings and J. C. Ho in "The Science, Technology, and Application of Titanium", Proceedings of an International Conference, 1968, ed. by R. I. Jaffee and N. E. Promisel, Pergamon Press Ltd. (1970), p 331.

18. H. L. Gegel and M. Hoch, "Thermodynamics of α-Stabilized Ti-X-Y Systems" -- paper 11 of this Session.

19. Specific resistivity is defined as the increment in resistivity per at.% solute.

20. See for example papers 2 and 3 of this Session.

21. E. S. Fisher and D. Dever, Acta Met., 18, 265 (1970).

22. M. J. Blackburn, Trans. AIME, 239, 1200 (1967).

23. J. C. Ho, P. C. Gehlen, and E. W. Collings, Solid State Comm., 7, 511 (1969).

24. P. C. Gehlen in "The Science, Technology, and Application of Titanium", Proceedings of an International Conference, 1968, ed. by R. I. Jaffee and N. E. Promisel, Pergamon Press Ltd. (1970), p 349.

25. N. S. Stoloff in "Fracture", ed. by H. Liebowitz, Academic Press (1969), pp 2-77 (in particular p 53).

26. N. S. Stoloff and R. G. Davies, Trans. ASM, 57, 247 (1964).

PHYSICS OF TITANIUM ALLOYS*-I: ALLOYING AND MICROSTRUCTURAL EFFECTS IN THE SUPERCONDUCTIVITY OF Ti-Mo

J. C. Ho† and E. W. Collings
BATTELLE
Columbus Laboratories
Columbus, Ohio 43201

In this, the first of a pair of papers reviewing some of the physical properties of Ti alloys, the conjoint influences of solute concentration and microstructural state on the superconducting transition temperature of Ti-Mo are discussed.

In a series of ice-brine-quenched Ti-Mo alloys whose structures are successively α' (0 to ~4-1/2 at.% Mo), β+ω (5 to ~15 at.% Mo), and β (> 15 at.% Mo) the superconducting transition temperature, T_c, first increases rather rapidly with increasing Mo concentration [Mo]; taking on <u>enhanced</u> values within the α' regime through the operation of a mechanism based on a high density of suitably "soft" vibrational states, a property of the α' structure. But immediately upon entering the (β+ω) field, T_c drops to a very low value due to the depressant influence of a high density of ω-phase precipitates. T_c then increases again with increasing [Mo] as the volume fraction of ω-phase precipitation decreases. At higher concentrations ($\gtrsim$ 20 at.%) T_c decreases with increasing [Mo], an effect which we deduce to be characteristic of <u>single-phase-bcc</u> transition-metal binary alloys within the electron-to-atom-ratio ($\mathfrak{z}$) range from 4 to 6.

*Supported by the Air Force Materials Laboratory, Wright-Patterson Air Force Base, Ohio, under Contract AF33(615)69-C-1594, and the U.S. Air Force Office of Scientific Research (AFSC) under Grant No. 71-2084.

†Present address: Department of Physics, Wichita State University, Wichita, Kansas, 67208.

1. INTRODUCTION

The results of some low-temperature specific heat and magnetic susceptibility measurements on a pair of representative Ti alloy systems: α-stabilized Ti-Al, and β-stabilized Ti-Mo, were presented and discussed at the First International Conference on Titanium (London, 1968)(1). In this and a companion paper (henceforth II), in which some of the subsequent developments of that work are reviewed, and some new results are outlined, the interdependences of various electronic and microstructural properties are emphasized.

The superconducting properties of a sequence of Ti-Mo alloys (designated TM-x, where x is the Mo concentration in at.%) were studied using low-temperature calorimetry as principal experimental technique. Interpretation of the specific-heat data was aided by magnetic-susceptibility investigations; and also by optical- and electron-microscopic studies of the as-quenched and aged microstructures. Figures 1 and 2 can be regarded as summarizing the joint compositional and microstructural dependences of the superconducting transition temperatures (T_c) of the quenched Ti-Mo alloys. As indicated by the nonequilibrium diagram of Figure 1 (b), the 30-40g specific-heat ingots of Ti-Mo after ice-brine quenching from 1300° C, exhibited structures which were thermal martensitic (α') up to 4-1/2 at.% Mo and basically bcc at higher compositions. However, according to the literature(2,3,4) and our own electron microscopic studies(1,5) quenched ingots of compositions through 10 at.% Mo, and possibly as high as 15 at.% support ω-phase precipitation [Figure 2(a)]. As is well known by now the ω-phase is submicroscopic in size(6) and hexagonal in crystal symmetry. The equilibrium and nonequilibrium structures indicated in Figures 1(a) and 1(b) represent extreme states of the Ti-Mo system between which many different structural states may exist depending on both mechanical and thermal treatments.

Figure 2(b), which shows that T_c depends strongly on the structural state as well as the alloy composition, serves to introduce the following discussion dealing with the influences on T_c of (i) thermal and deformation martensites (α' and α" respectively) and (ii) controlled ω-phase precipitation.

2. THERMAL AND DEFORMATION MARTENSITES

Figure 2(a) indicates that in quenched Ti-Mo the α' region terminates fairly abruptly within a narrow composition range near 4-1/2 at.% Mo while Figure 2(b) shows T_c to increase rapidly, at the rate of about 1 degree per at.% Mo, within that regime. A comparison of curves (b) and (d) in Figure 2 showed that the relative rate of increase of T_c with composition was significantly greater than the corresponding relative rate of increase to electronic-specific-heat coefficient, γ [a measure of the Fermi density-of-states, $n(E_F)$(7)],

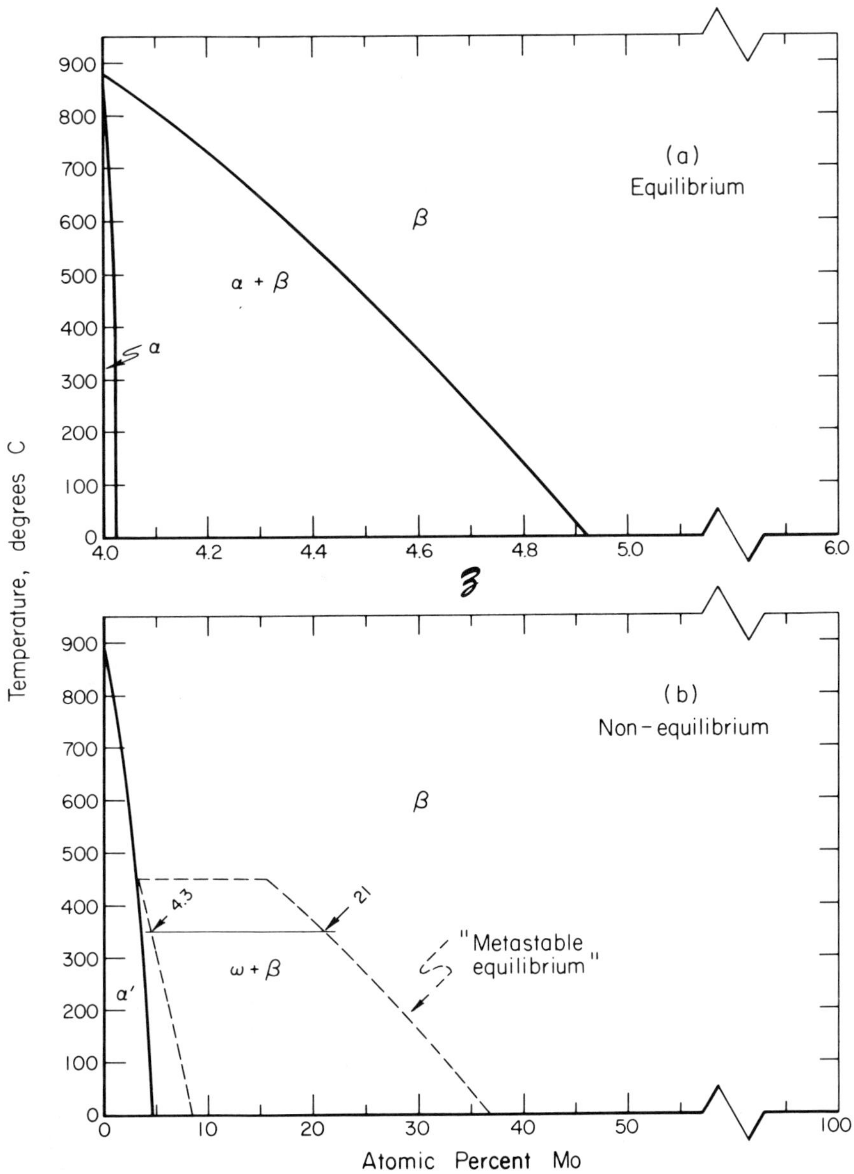

Figure 1. Equilibrium and nonequilibrium phase diagrams for Ti-Mo.

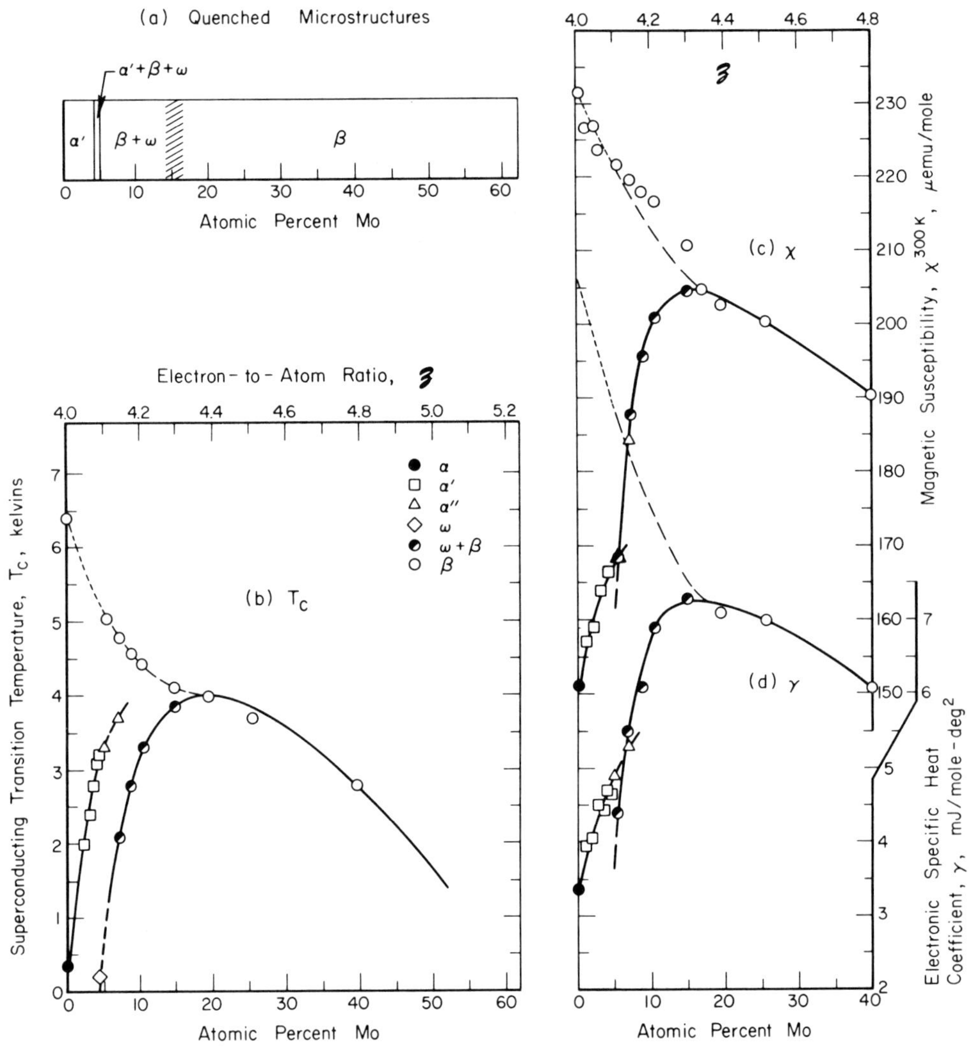

Figure 2. (a) Microstructural states of quenched (from $\lesssim 1000^{\circ}C$) Ti-Mo alloys. (b), (c), and (d) Superconducting transition temperature, T_c; magnetic susceptibility, χ; and electronic specific-heat coefficient, γ, for Ti-Mo in various microstructural states.

suggesting that some kind of T_c enhancement mechanism was operative.[8] Anomalies in the expected relationship between T_c and γ can best be expressed with the aid of a BCS-Morel-Morin-Maita[9] plot of log (T_c/Θ_D) versus $(0.212\ \gamma)^{-1}$, where Θ_D is the calorimetrically-measured Debye temperature. Presented in this way (Figure 3) the data clearly show that α' alloys are characterized by almost threefold enhancements of T_c in comparison with values expected for the same materials were they to possess nonmartensitic [i.e. β or (β+ω)] structures. An enhanced T_c is a quite general characteristic of α' Ti-T_2 and Zr-T_2 alloys (where T_2 represents another transition element)[10]. This is exemplified by Figure 4 in which the T_c-composition-dependence of Ti-Mo and Ti-Fe are compared. That figure also implies that it is no longer necessary to invoke a "magnetic" electronic interaction of the type previously suggested[11,12,13], in order to explain the enhanced transition temperatures, first thought to be properties only of dilute alloys of Ti with "magnetic" transition element solutes.

A detailed study was made of superconductivity in TM-4-1/2, whose composition placed it at the terminus of the quenched α' regime[14]. The results of that work, together with the known close relationship between thermal (α') and deformation (α") martensites suggested that it might be possible to induce enhanced values of T_c in low-concentration (β+ω)-Ti-Mo alloys by subjecting them to mechanical deformation of sufficient severity. Alloys of 5 and 7 at.% Mo were chosen for further experimentation[15]. The results of low temperature specific heat measurements on TM-5, for example, taken before and after compressive deformation are summarized in Figure 5[15,16]. In the quenched state (β+ω) the transition temperature, which was below the lower temperature limit of the apparatus, was estimated to be about 1 K based on Figure 3 and the measured γ and Θ_D values. After being deformed at room temperature, during which α" was induced, the same specimen yielded a threshold T_c of 3.3 K. Figure 3 indicates the transition temperatures of TM (5 and 7) as special points (Δ) falling near those for alloys possessing the α' structure. For α' and α" alloys, it is not possible to attribute these enhanced T_c's to electronic-density-of-states effects within the framework of the BCS-theory using the interaction parameter, derived from the slope of the line in Figure 3, which was common to all the nonmartensitic alloys. Instead we conclude that the enhanced transition temperatures occur as a result of the extreme lattice distortion or disorder, which is responsible for increased electron-phonon-electron coupling in the manner described by McMillan[17], and Garland *et al.*[18].

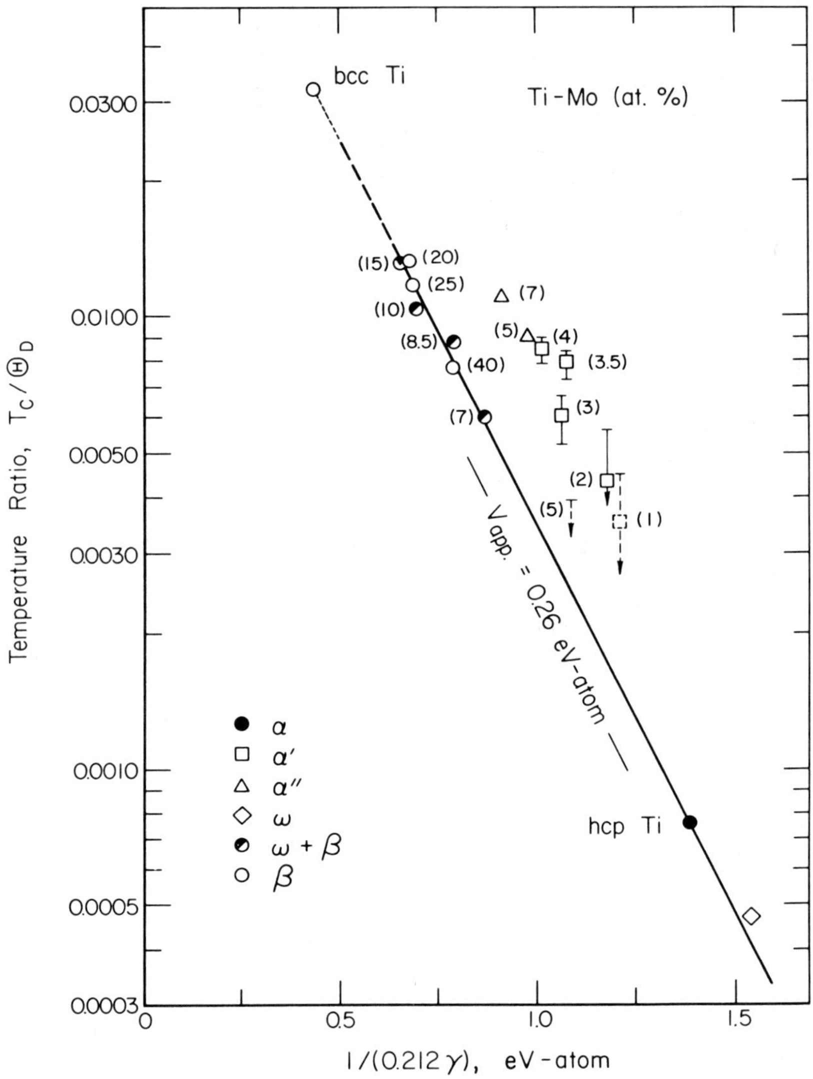

Figure 3. Plot of log (T_c/θ_D) *versus* $(0.212\ \gamma)^{-1}$ [i.e., a BCS-Morel-Morin-Maita(9) plot] for Ti-Mo in various structural states.

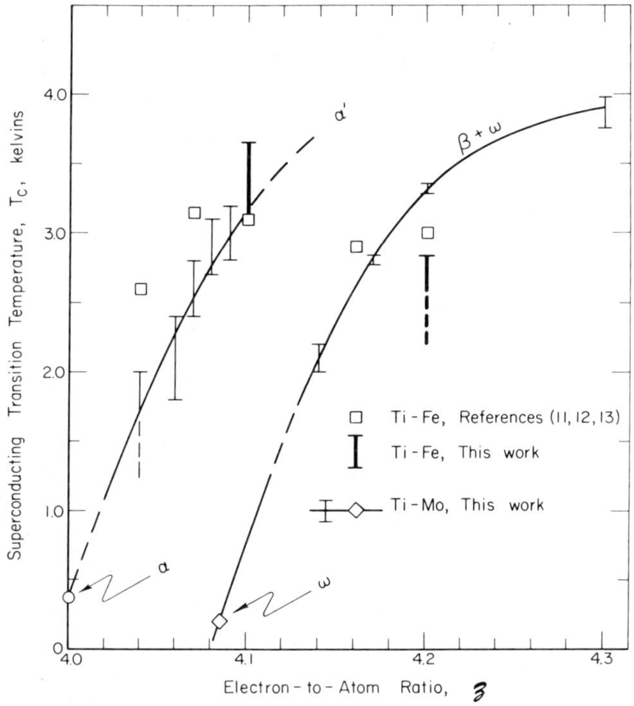

Figure 4. Superconducting transition temperatures of both Ti-Mo and Ti-Fe plotted as functions of the electron-to-atom ratio, @.

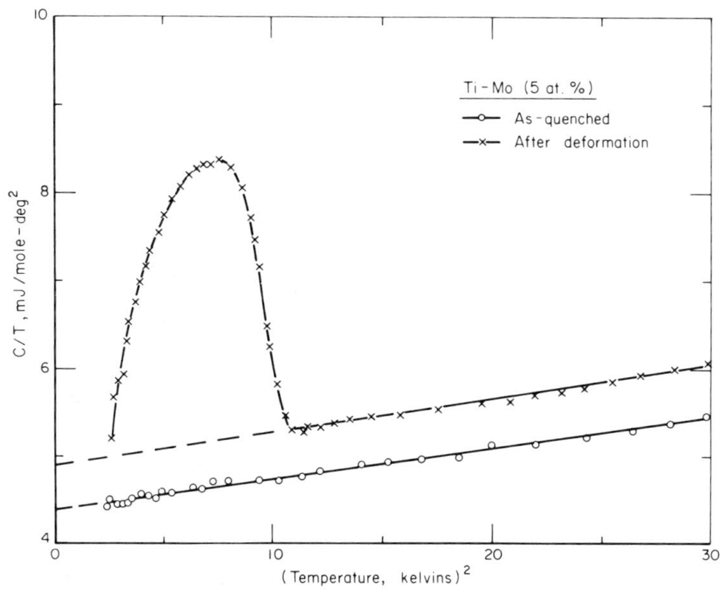

Figure 5. Influence of deformation on the superconducting transition in Ti-Mo (5 at.%).

3. ω-PHASE PRECIPITATES

3.1 Composition-Dependence of T_c in As-Quenched Ti-Mo Alloys

Returning to Figure 2(b) we see that after rising to a relatively high value at the terminus of the martensitic (α' and α") regime, T_c drops below the lower temperature limit (~1.5 K) of our calorimeter as the quenched microstructure changes to (β+ω). T_c then increases again with solute concentration, and passes through a maximum at an electron-to-atom ratio ($\mathfrak{z}$) near 4.4. A set of low-temperature specific heat data for the alloys TM-5 through 15 is included for reference as Figure 6. Particularly noticable are the small widths (~0.15 degrees) of the superconducting transitions. Moreover the relative heights of the specific-heat jumps at T_c, i.e. $[\Delta C/T_c)/\gamma]$, are sufficiently close to the BCS-prescribed value (viz. 1.43) as to be indicative of bulk transitions. The sharpness and homogeneity of the calorimetrically measured superconducting transitions for alloys in the range $4.1 \lesssim \mathfrak{z} \lesssim 4.3$ were misinterpreted by earlier workers, who believed that such alloys were single-phase bcc (β); and consequently that the maximum in T_c was the result of a maximum in $n(E_F)$ which was thus assumed to be characteristic of β-phase alloys at $\mathfrak{z} \cong 4.3$ or 4.4. However, the existence of ω-phase precipitation in representative alloys (e.g. TM-5 and 10) within $4.1 \lesssim \mathfrak{z} \lesssim 4.3$ was suspected[2,3,4] and eventually established by direct observation[1,5]. Accordingly, it was reasonable to postulate[19] that the maximum in T_c was <u>induced</u> in an otherwise monotonic function of T_c <u>versus</u> $\mathfrak{z}$ by a proximity-effect-induced depression of T_c brought about by the presence of the ω-phase. If this were true then it should be possible (a) to increase T_c by removal of the ω-phase; and (b) to bring about further depression in T_c by encouraging the ω-phase to develop. The results of this pair of tests of the initial postulate are outlined below.

3.2 Composition-Dependence of T_c in Hypothetical ("Unstable") Single-Phase bcc (β) Ti-Mo, $0 \leq \mathfrak{z} \leq 4.3$

To establish references, against which to compare the physical properties of ω-phase-containing alloys, and eventually ω-phase itself, we set out to make estimates of γ, χ, and T_c for hypothetical ["unstable"[20]] β-Ti-Mo in the temperature regime for which $\mathfrak{z} \lesssim 4.3$. These were based on the extrapolated results of magnetic-susceptibility-temperature-dependence, χ(T), studies carried out in the elevated-temperature equilibrium-β field, as indicated by the dashed line of Figure 2(c). After observing the manner in which the composition dependences of χ, γ,[5] (and T_c), for the as-quenched alloys tended to scale [c.f. the full lines of Figures 2(c), (d), (and (b)), respectively], we were reasonably confident in using the extrapolated magnetic data to deduce the composition-dependence of

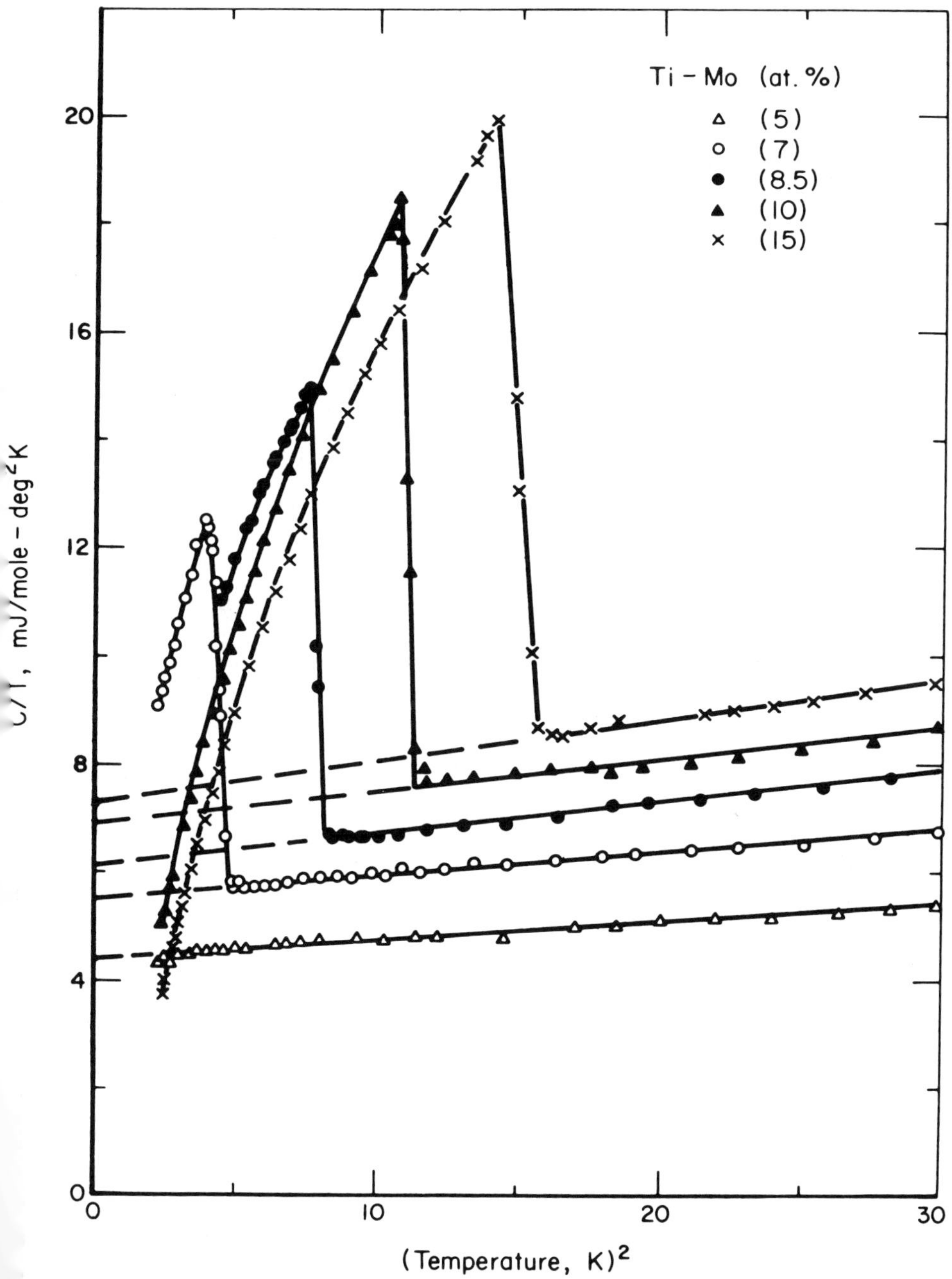

Figure 6. Low-temperature specific-heat data for several (β+ω)-phase Ti-Mo alloys. These alloys exhibit sharp, bulk, superconducting transitions.

γ(β-Ti-Mo) within 4.0 $\lesssim \mathfrak{z} \lesssim$ 4.3. The result is indicated by the dashed line of Figure 2(d). The γ's so obtained were then applied to a short extrapolation of the log (T_c/Θ_D) <u>versus</u> $1/(0.212\ \gamma)$ plot of Figure 3, in order to derive a set of (T_c/Θ_D)-values for unstable Ti-Mo. Calling once more on the χ, γ, T_c scaling relationship, separate sets of values of T_c and Θ_D were then extracted, using smoothness as one criterion. That is to say, following the behavior χ(β-Ti-Mo, 300 K), which was found to increase monotonically as $\mathfrak{z}$ decreased from 6.0 to 4.0 rather than to pass through a maximum at $\mathfrak{z} \approxeq 4.3$, T_c was extrapolated back into the unstable regime, increasing as $\mathfrak{z}$ decreased, as a smooth continuation of the directly measured stable-β locus [dashed line in Figure 2(b)]. At the same time the coupled quantity, Θ_D, dropped rather rapidly as $\mathfrak{z}$ decreased below 4.3. This is a physically realistic result, in that it expresses a continuous softening of the unstable β lattice as the condition for "absolute instability" [viz. $(c_{11}-c_{12}) \leq 0$ for $\mathfrak{z} \lesssim$ 4.1] is approached.

Table I. Superconducting-Transition, And Debye Temperatures [T_c, Θ_D, respectively], And Fermi Density-of-States [$n(E_F)$], in Ti-Mo Alloys

Alloy Designation		T_c (kelvins)		Θ_D (kelvins)		$n(E_F)$ States/eV-Atom	
		β	$\beta+\omega$	β	$\beta+\omega$	β	$\beta+\omega$
Pure Ti	4.00	[6.4]	--	[200]	--	[1.50]	--
TM-5	4.10	5.0_5	<1.5	207	380	1.35	0.75
TM-7	4.14	4.8_0	2.1	220	350	1.30	0.89
TM-8-1/2	4.17	4.5_9	2.8	232	320	1.25	0.97
TM-10	4.20	4.4_2	3.3	243	320	1.22	1.06
TM-15	4.30	4.1_0	3.8_5	281	295	1.13	1.10
TM-20	4.40	4.0	--	305	--	1.08	--
TM-25	4.50	3.7	--	320	--	1.07	--
TM-40	4.80	2.8	--	365	--	0.97	--
TM-70	5.40	<1.5	--	410	--	0.49_0	--
Pure Mo	6.00	<1.5	--	470	--	0.35_6	--

Recognizing that the conjoint χ and γ data contribute to a calculation of $n(E_F)$, the numerical final results of the above semi-empirical procedure may be summarized (Table I) as sets of values of T_c, Θ_D and $n(E_F)$. The latter pair of parameters will be referred to again in II, while T_c for β-Ti-Mo is plotted in Figure 2(b); and again in Figure 7 for comparison with the $\mathfrak{z}$ -dependence of T_c for transition-metal-binary (T_1-T_2) alloys of the second long period. The present work provides a rationale for the characteristic double-humped shape[21], which is also exhibited by the T_c- $\mathfrak{z}$ curves for the first-period and fifth-period transition-metal binaries. For Ti-Mo we have shown the maximum in T_c to be <u>induced</u> by ω-phase precipitation taking place below $\mathfrak{z} \eqsim 4.3$; a regime, moreover, in which all T_1-T_2 alloys exhibit such precipitation. We, therefore, suggest that the low-$\mathfrak{z}$ maxima present in the T_c curves for 3d-3d, 4d-4d, and 5d-5d T_1-T_2 alloys are strongly, if not entirely, contributed to by such precipitation. Secondly, it is highly significant that the other set of T_c maxima occur near $\mathfrak{z} \eqsim 6.7$, the upper limit[22] of bcc stability. We conclude that superconductivity in T_1-T_2 alloys is favored by a bcc lattice, especially if accompanied by a high $n(E_F)$, and that the conditions for superconductivity became even more favorable as $\mathfrak{z}$ approaches either limit of bcc stability; furthermore, that were it not for the intervention of structural change, T_c <u>versus</u> $\mathfrak{z}$ would tend to be U-shaped.

3.3 Depression of the Superconducting Transition Temperature in Ti-Mo Through Thermal Aging of Precipitated ω-Phase

If ω-phase precipitation were indeed responsible for the depression of T_c in quenched Ti-Mo alloys for which $\mathfrak{z} < 4.3$, we postulated that even greater decreases would be incurred as the precipitate developed to saturation in response to suitable moderate-temperature heat treatment. Accordingly, an alloy was selected whose composition (10 at.% Mo) placed it near the center of the $\mathfrak{z}$ -range of interest; and magnetic susceptibility and low-temperature specific heat parameters were measured at various times during aging at 350° C. Figure 8 shows how T_c decreased continuously as the structural state of TM-10 proceeded from β towards "saturated" (or "metastable equilibrium"[23]) (β+ω). At this point it is useful to note that by following such changes in physical parameters during aging, the onset of decomposition into (α+β), if it does eventually take place, can be readily distinguished.

Focussing attention now on fully-aged (β+ω) TM-10, the conservation equation

$$\langle \Phi \rangle_{av} = f_\omega \, \Phi_\omega + (1-f_\omega) \, \Phi_\beta \ , \qquad (1)$$

(in which Φ symbolizes a property of a component phase) was first of all employed (in the format Φ ≡ [Mo], the "lever rule") to determine f_ω, the mole-fraction of ω-phase present after 880 hours at 350° C. Following Hickman[4], $[Mo]_\omega$ was set at 4.3 at.%; and guided by the

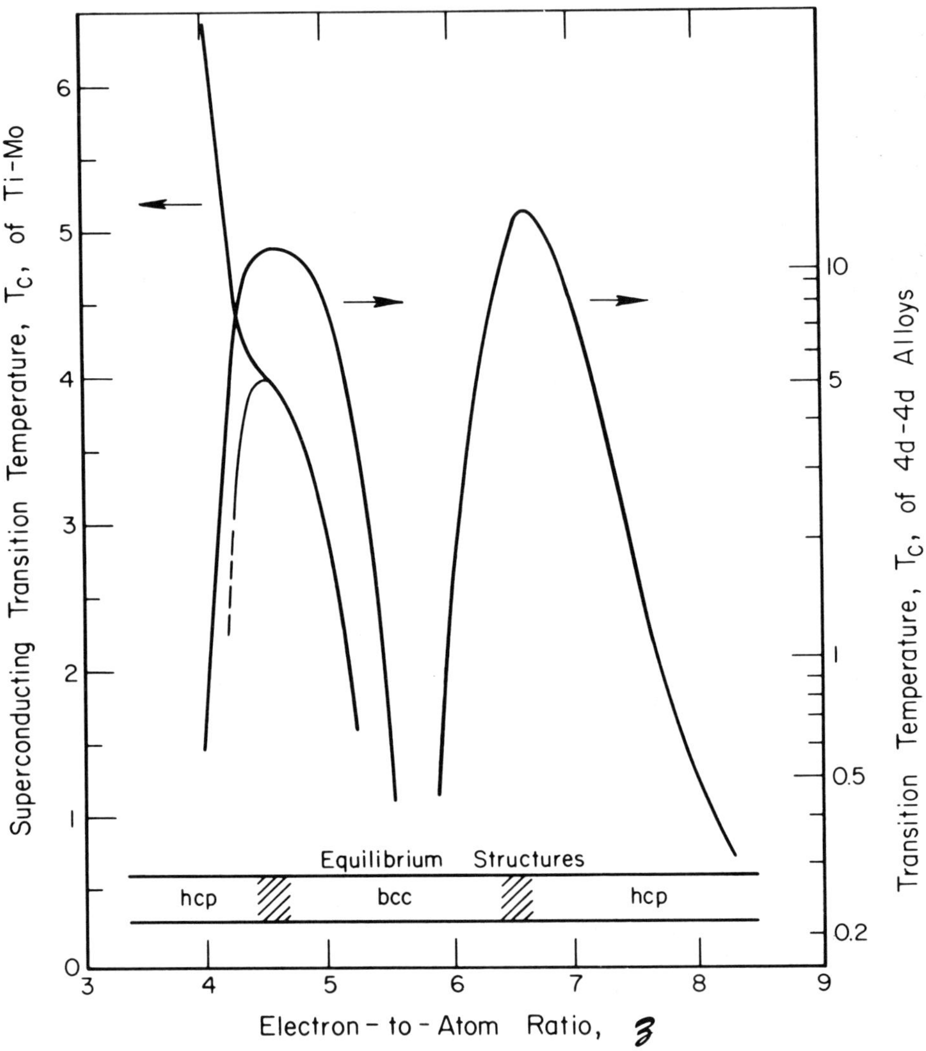

Figure 7. @ -dependence of the superconducting transition temperatures of Ti-Mo alloys, compared with that of transition metal-binary alloys of the second long period.

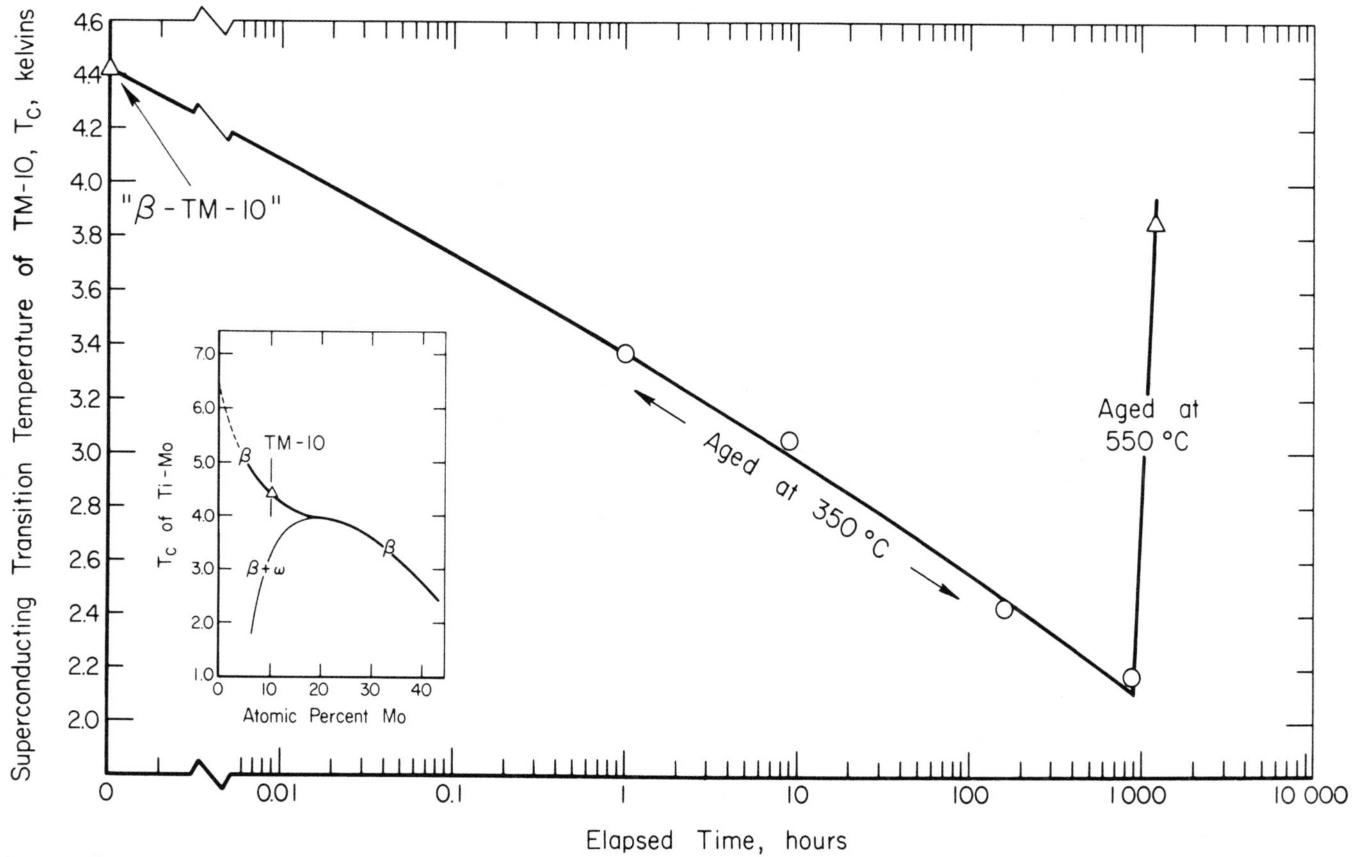

Figure 8. Influence of microstructural change through aging on the superconducting transition temperature of a Ti-Mo (10 at.%) alloy. The inset, based on Figure 2(b), indicates that for <u>single-phase</u> β-Ti-Mo (10 at.%) $T_c \approx 4.4$ K. The main figure shows this value to be consistent with those observed in the (β+ω)-phase alloys as a function of aging.

work of de Fontaine et al[23] [reproduced as the metastable-equilibrium $(\beta+\omega)$ diagram included in Figure 1(b)] the value of $[Mo]_\beta$ was taken as 21 at.%. Equation (1) then yielded $f_\omega = 0.64$. A re-application of the equation with $\Phi \equiv \gamma$, then enabled γ_ω and consequently $n(E_F)_\omega$ to be determined.

In order to obtain $T_c(\omega\text{-TM-4.3})$ in which we are primarily interested here, a proximity-effect type of averaging equation of the form[25]

$$\langle T_c \rangle_{av} = \Sigma_i v_i \, n(E_F)_i \, T_{ci} / \Sigma_i v_i \, n(E_F)_i \qquad (2)$$

was employed, where v_i represents the superconductive volume fraction of the ith component. Since even for the maximally aged alloy, the magnitude of $\Delta C/\gamma T_c$ was indicative of a bulk transition at T_c we assumed the ω-phase precipitate to be completely penetrated by the electron-pair wave-functions. It followed that v_ω was equal to the actual volume-fraction of ω-phase present which in turn is practically identical[4] with the mole-fraction, f_ω. From the measured $\langle T_c \rangle_{av}$ and the known values of the other quantities in Equation (2), T_c $(\omega\text{-TM-4.3})$ was calculated and found to be 0.2_0 K. The relationship of this value to those for the quenched (β and β+ω) Ti-Mo alloys, is indicated in Figure 2(b), and again in Figure 3, by special points which are seen to fit nicely on smooth continuations of the experimentally-derived curves.

Figure 2(b) also shows that the previously described extrapolation procedure applied to single-phase bcc TM-4.3 yields $T_c = 5.2_5$ K. This demonstration with respect to TM-4.3, that a hypothetical structural transformation from β to ω would result in a 26-fold decrease in T_c, verified our initial argument that the depression in T_c exhibited by Ti-Mo alloys for $\mathfrak{z} < 4.3$ was a result of ω-phase precipitation.

4. SUMMARY

Conditions particularly favorable for superconductivity in T_1-T_2 alloys exist near both the upper and lower limits of the regime of bcc stability. Thus, for example, as $\mathfrak{z}$ decreases below 6.0 (a value which corresponds to maximal bcc stability) T_c first increases towards relatively high values, but then drops as the structure changes to hexagonal through ω-phase precipitation and finally martensitic transformation.

Although lower than if the structure were single-phase bcc, T_c in the α' regime is much larger than it would be if the structure were undistorted hcp, and in that sense it can be regarded as "enhanced". This enhancement, due to the operation of a "soft-phonon" mechanism which accompanies the structural disorder characteristic of α', can also be induced in formerly (β+ω)-phase alloys by mechanical deformation to α".

Finally, that the T_c maximum occurring in Ti-Mo alloys near $\mathfrak{z}$ = 4.4 is, in fact, induced by ω-phase precipitation, has been demonstrated by two essentially independent experiments--one involving the hypothetical removal of ω-phase, and another based on a study of the decrease in T_c which accompanies ω-phase development during moderate-temperature aging. The fact that the linear dimensions of even the aged precipitate are commensurate with a coherence distance, is responsible for the bulk nature of the superconducting transition in the two-phase (β+ω) alloys.

REFERENCES

1. E. W. Collings and J. C. Ho, "The Science Technology and Application of Titanium", Proceedings of an International Conference, London, May, 1968; ed. R. I. Jaffee and N. E. Promisel, Pergamon Press Ltd. (1970) p. 331.

2. E. K. Molchanova, "Phase Diagrams of Titanium Alloys", ed. S. G. Glazunov (Daniel Davey & Company, Inc., N. Y., (1965) p. 29.

3. M. J. Blackburn and J. C. Williams, Trans, Met. Soc. AIME, 242 2461 (1968).

4. B. S. Hickman, Trans Met. Soc. AIME, 245 1329 (1969).

5. E. W. Collings and J. C. Ho, Proceedings of the 3rd Materials Research Symposium, November, 1969, "Electronic Density of States", Ed. L. M. Bennett, Nat. Bur. Stand. (U.S.) Special Publication 323, December, 1971, p. 587.

6. For example, according to Reference (3) the aging of Ti-Mo (11.6 wt.%) for 960 hours at 350° C resulted in the formation of ellipsoidal precipitates of geometries defined by a:b = 2.2 with b = 270 ± 100 Å.

7. In the appropriate units [Reference (5)] $n(E_F) = 0.212\ \gamma/(1 + 0.212\ \gamma\ V_{app})$; where $-1/V_{app}$ is the slope of the line in Figure 3.

8. E. W. Collings and J. C. Ho, Physics Letters, 29A 306 (1969).

9. J. Bardeen, L. N. Cooper, and J. R. Schrieffer (BCS), Phys. Rev. 108 1175 (1957); P. Morel, J. Phys. Chem. Solids, 10 277 (1959); F. M. Morin and J. P. Maita, Phys. Rev. 129 1115 (1963); see also Reference (5).

10. See Table I of Reference (14).

11. B. T. Matthias, V. B. Compton, H. Suhl, and E. Corenzwit, Phys. Rev. 115 1597 (1959).

12. B. T. Matthias, in "Superconductors", Ed. M. Tanenbaum and W. V. Wright, Interscience Publishers, 1962, p. 1.

13. B. T. Matthias, Proceedings of the 8th International Conference on Low Temperature Physics, London, 1962, Ed. R. O. Davies, Butterworths, 1963, p. 135.

14. E. W. Collings and J. C. Ho, Phys. Rev. 1 4289 (1970).

15. J. C. Ho and E. W. Collings, J. App. Phys. 42 5144 (1971).

16. E. W. Collings, J. D. Boyd, and J. C. Ho, Proceedings of the 12th International Conference on Low Temperature Physics, Kyoto, Japan, September, 1970, p. 316.

17. W. L. McMillan, Phys. Rev., 167 331 (1968).

18. J. W. Garland, K. M. Bennemann, and F. M. Mueller, Phys. Rev. Letters, 21 1315 (1968).

19. J. C. Ho and E. W. Collings, Physics Letters, 29A 206 (1969).

20. We refer to single-phase-bcc alloys, at room temperature, in the composition range 0-15 at.% Mo as "unstable"; and divide them into two classes: those in the range 5 < at.% Mo < 15, which although normally unstable do not yield unphysical elastic moduli, we designate "virtual" β-Ti-Mo; while those below the M_s composition (4-1/2 at.% Mo) we refer to as "absolutely unstable", since for these the extrapolated elastic constants yield $c_{11} < c_{12}$. Data obtained by extrapolation into this latter region must be interpreted with considerable caution.

21. B. T. Matthias, American Scientist, 58 80 (1970).

22. F. Heiniger, E. Bucher, and J. Muller, Phys. kondens. Materie, 5 243 (1966)--see in particular Figures 2 and 3.

23. D. de Fontaine, N. E. Paton, and J. C. William, Acta Met., 19 1153 (1971).

24. E. W. Collings and J. C. Ho, "Experimental Studies of Proximity-Effect-Induced Depression of the Superconducting Transition Temperatures of Two-Phase Ti-Mo Alloys"--in preparation.

25. See, for example, E. A. Lynton, "Superconductivity", Methuen and Company, 2nd Ed., 1964, pp. 150-152.

PHYSICS OF TITANIUM ALLOYS*-II: FERMI DENSITY-OF-STATES PROPERTIES, AND PHASE STABILITY OF Ti-Al AND Ti-Mo

E. W. Collings, J. C. Ho†, and R. I. Jaffee
BATTELLE
Columbus Laboratories
Columbus, Ohio 43201

The results of magnetic and calorimetric measurements leading to evaluations of the Fermi densities-of-states, $n(E_F)$ and Debye temperatures, Θ_D, of Ti-Al and Ti-Mo alloys in various structural states are presented. Phase stability in these systems is correlated with the observed variations of $n(E_F)$ and Θ_D with electron-to-atom ratio, $\mathfrak{z}$; and the paper concludes with some general comments on the relationships between physical properties and phase stability in transition-metal binary alloys, and Ti-base alloys in particular.

*Supported by the Air Force Materials Laboratory, WPAFB, Ohio, under Contract AF33(615)69-C-1594; and the U. S. Air Force Office of Scientific Research (AFSC) under Grant No. 71-2084.

†Present address: Department of Physics, Wichita State University, Wichita, Kansas, 67208.

1. INTRODUCTION

1.1 Specimen Materials - Experimental Techniques

In an investigation of the electronic bases of phase stability and solid-solution strengthening in Ti alloys, physical property measurements were made on specimens of Ti-Al and Ti-Mo alloys as typical examples of "α-stabilized" and "β-stabilized" systems, respectively. Although a great deal of useful information was derived from measurements of a pair of electrical transport properties viz. resistivity(1), and Hall coefficient(2), the principal experimental techniques used in the study were low-temperature calorimetry augmented by magnetic susceptibility.

The low-temperature specific heat, $C = \gamma T + \beta T^3$, when plotted in the format C/T vs T^2 yields as intercept, the electronic specific heat coefficient γ [leading to $n(E_F)$, the Fermi density-of-states] and as slope ($\beta \propto \Theta_D^{-3}$) the Debye temperature Θ_D. Also, as discussed in the companion paper (henceforth I) if the specimens become superconducting within the temperature range of the calorimeter, the specific heat measurements yield a wealth of information regarding the superconducting transition. Magnetic susceptibility, χ, was used as adjunct to the low-temperature specific heat investigations. Since χ contains a separable component, χ_{spin}, which also depends on $n(E_F)$ the magnetic susceptibility technique was used to confirm and extend the results of the calorimetric measurements. Magnetic susceptibility is a simple, rapid, and economical measurement, requiring only relatively small specimens (100 mg compared with ~30g for specific heat). It can be employed advantageously when a detailed study is required of the response of $\langle n(E_F)\rangle_{av}$ to a wide range of alloy solute concentrations and heat treatments (such as quench temperatures and aging conditions). In addition the susceptibility technique can be used to investigate the $n(E_F)$ properties of alloy phases stable only at elevated temperatures and, therefore, inaccessible to low-temperature calorimetry.

1.2 Experimental Program

(a) The results of magnetic susceptibility and low-temperature specific heat measurements on "β-quenched" Ti-Mo alloys; and "α-quenched", and otherwise heat-treated, Ti-Al alloys were presented in the previous International Titanium Conference.(3) (b) In order to determine the $n(E_F)$ behavior of single-phase-bcc (i.e., β) Ti-Mo alloys in the electron-to-atom-ratio ($\mathfrak{z}$) range below about 4.3, in which they are not stable at ordinary temperatures, use was made of elevated-temperature magnetic susceptibility data in the manner indicated in the companion paper (I). A comparison of the results with those for the as-quenched alloys showed that $\langle n(E_F)\rangle_{av}$ was relatively low in the latter due to the presence of ω-phase precipitates.

(c) As confirmation of this conclusion, and in order to determine some of the physical properties of ω-phase itself, magnetic and calorimetric measurements were made as a function of ω-phase development to "metastable equilibrium"(4) through moderate-temperature (350° C) aging.

1.3 Data Analysis

An aim of the analytical procedure was to determine $n(E_F)$, a fairly fundamental quantity, and one which may be compared with the results of band structure calculations when such information is available. In suitable units(5) $n(E_F) \cong 0.212\ \gamma/(1 + 0.212\ \gamma\ V_{app})$, where the denominator is the electronic-specific-heat electron-phonon enhancement factor. When dealing with a family of transition-metal alloy superconductors the calorimetric data itself may contain sufficient information which, if expressible in the form of a linear $\log (T_c/\Theta_D)$ vs $(0.212\ \gamma)^{-1}$ plot of slope $-1/V_{app}$, enables the electron-phonon correction to be readily applied.(6) Such was, in fact, the case for Ti-Mo alloys, including α-Ti, as has already been shown(7). Values of $n(E_F)$ [$\cong 0.212\ \gamma/(1 + 0.212\ V_{app}\ \gamma)$], calculated by applying the electron-phonon correction with V_{app} = 0.26 eV-atom,

Table I. Debye Temperature, Θ_D, And Fermi Density-of-States, $n(E_F)$, For Alloys in Electron-to-Atom-Ratio ($\mathfrak{z}$) Ranges $3.63 \leq$ " $\mathfrak{z}$ " ≤ 4.08 And $6.05 \leq \mathfrak{z} \leq 6.76$

Alloy*	"$\mathfrak{z}$"	Phase	Θ_D	$n(E_F)$	Alloy	$\mathfrak{z}$	Phase	Θ_D	$n(E_F)$
HP-37	3.63	[α_2]	495	0.50_1	Pure Ti	4.00	α	420	0.60
HP-33.3	3.66	[α_2]	506	0.48_2	TM-4.3	4.08	ω	420	0.54_6
HP-30	3.70	[α_2]	485	0.45_6					
HP-28	3.72	[α_2]	518	0.41_6	MR-5	6.05	β	450	0.38_4
HP-25	3.75	α_2	495	0.36_8	MR-10	6.10	β	440	0.45_0
HP-23	3.77	α_2	485	0.30_3	MR-20	6.20	β	420	0.59_6
HP-20	3.80	α_2	485	0.35_4	MR-25	6.25	β	405	0.62_6
HP-15	3.85	α	442	0.64_4	MR-30	6.30	β	395	0.63_4
HP-13	3.87	α	445	0.66_0	MR-40	6.40	β	340	0.67_1
HP-10	3.90	α	413	0.64_0	MR-50	6.50	β	320	0.66_4
HP-5	3.95	α	427	1.62_8	MR-58	6.58	σ	351	0.54_0
HP-3	3.97	α	413	0.61_8					
HP-2	3.98	α	403	0.61_5	MO-38	6.76	σ	371	0.54_2

*HP, TM, MR, and MO-x represent Ti-Al, Ti-Mo, Mo-Re, and Mo-Os (x at.%) respectively.

to the measured γ's for β-phase and $(\beta+\omega)$-Ti-Mo alloys have been presented in Table I of I. Also tabulated there are the directly-measured values of Θ_D, together with a list of extrapolated Θ_D's for "unstable"[8] β-Ti-Mo ($4.0 < \mathfrak{z} < 4.3$), derived from a set of T_c/Θ_D values[9]. The α-Ti-Al alloys on the other hand were "nonsuperconducting" (i.e., $T_c < 1.5$ K). But since pure α-Ti was represented by a point on the above-mentioned semilog plot, $V_{app} = 0.26$ eV-atom was also used in correcting γ_{Ti-Al} for electron-phonon effects. The values of $n(E_F)$ so obtained are listed (together with the directly-measured Θ_D's) in Table I of this paper.

2. RESULTS AND DISCUSSION

2.1 Physical Properties of α-, and α_2-Ti-Al and ω-, $(\beta+\omega)$-, and β-Ti-Mo Alloys

The experimental results, expressed in terms of the physical properties γ, χ, and Θ_D for α, α_2, ω, $(\beta+\omega)$, and β-phase alloys, are summarized in Figure 1.

2.1.1 Θ_D in Ti-Mo

Since as described in I, the Debye temperatures were separated out from a set of empirically-derived (T_c/Θ_D)-values, some confirmation of the validity of the form of the resulting composition-dependence of Θ_D was warranted.

Accordingly a comparison was made with Θ_D(Ti-Cr), calculated by applying Anderson's method[10], to Fisher and Dever's[11] measured elastic constants. The results of this procedure are summarized in Figure 2. Taking into consideration the fact that, in the fairly massive ingots measured, ω-phase seemed to make its appearance in Ti-Mo at a higher $\mathfrak{z}$ than was the case for Ti-Cr[12] Figure 2(c) shows that the deduced Θ_D-composition-dependence for "virtual"[8] β-Ti-Mo was in good accord with that calculated for virtual β-Ti-Cr. Θ_D is sensitive to the shear modulus $C' = (c_{11}-c_{12})/2$ which as pointed out by Fisher and Dever[11] decreases relatively rapidly with decreasing $\mathfrak{z}$ upon entering the virtual-β regime. This behavior is reflected in Figure 2(c). For Ti-Cr, martensitic transformation which has been observed within $4.10 < \mathfrak{z} < 4.12$[13] seems to be related to the vanishing of C' which Figure 2(a) shows as occurring at Ti-Cr(5.03 at.%) or $\mathfrak{z} = 4.1$. At the same time, Anderson's formula yields $\Theta_D = 208$ K; a value which tends to confirm the independently deduced behavior of Θ_D (Ti-Mo) which drops to 207 K at 5 at.% Mo, just prior to martensitic transformation at 4-1/2 at.% Mo[14]. Whereas the martensitic transformation is undoubtedly associated with $C' = 0$[4,11] (i.e., "softening"[15]), de Fontaine et al[4] have shown that the $\beta \rightarrow \omega$ transformation can be achieved formally by altering

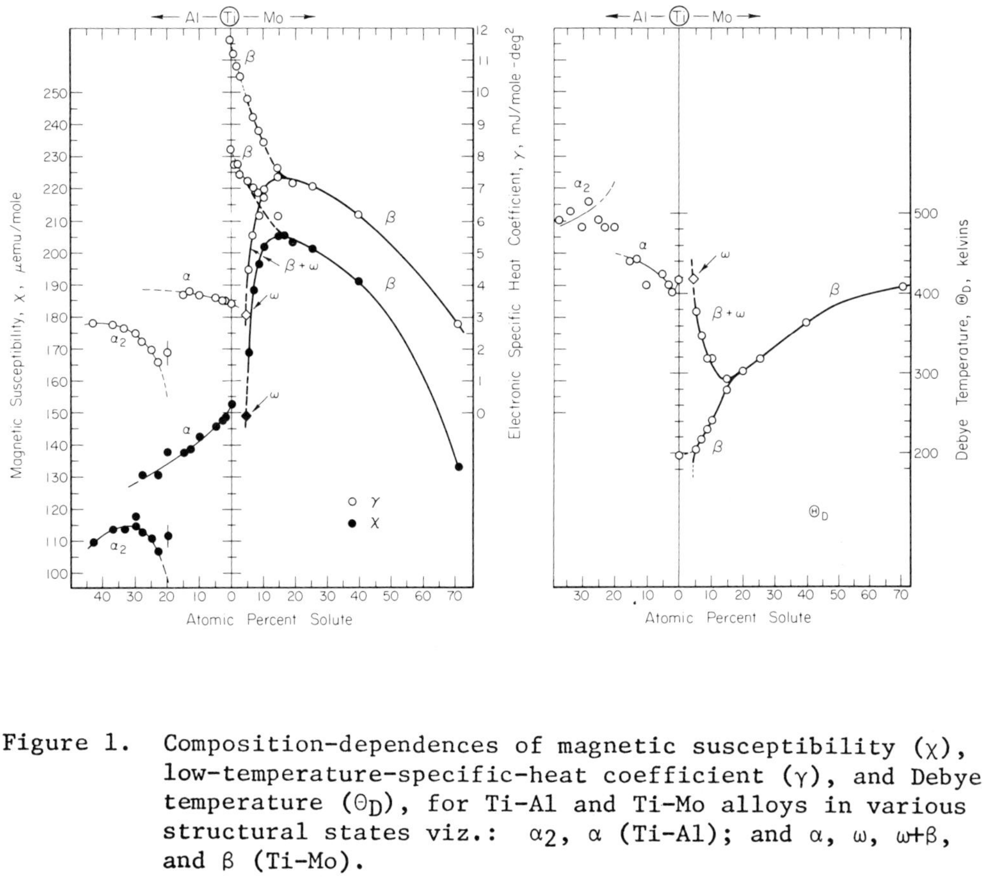

Figure 1. Composition-dependences of magnetic susceptibility (χ), low-temperature-specific-heat coefficient (γ), and Debye temperature (Θ_D), for Ti-Al and Ti-Mo alloys in various structural states viz.: α_2, α (Ti-Al); and α, ω, ω+β, and β (Ti-Mo).

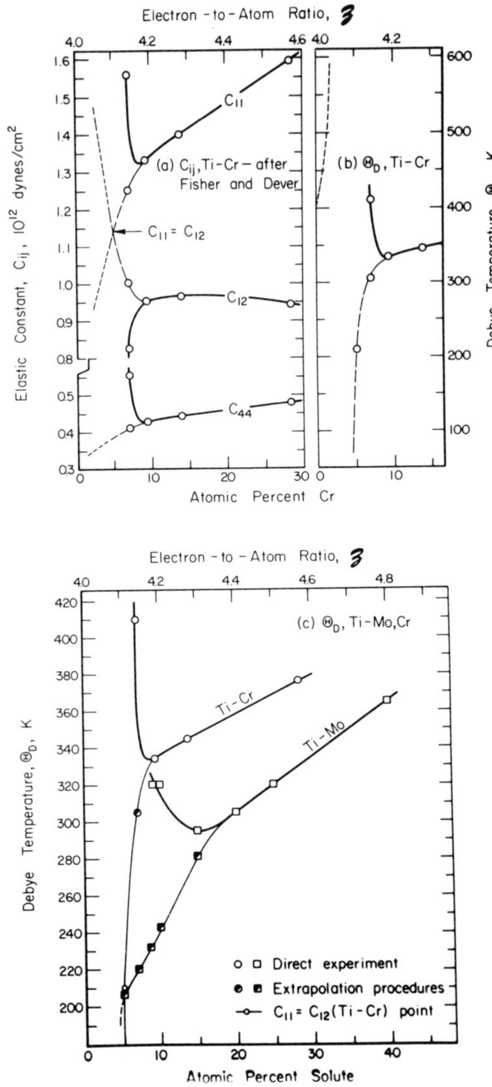

Figure 2. (a) Elastic constants c_{11}, c_{12} and c_{44} of Ti-Cr - after Fisher and Dever(11,16). (b) Θ_D for Ti-Cr calculated from the data of the previous figure, using Anderson's(10) method. (c) Comparison of the temperature-dependence of Θ_D for Ti-Cr and Ti-Mo. At the observed room-temperature threshold of martensitic transformation for Ti-Mo (viz. 4-1/2 at.%), and when $c_{11} = c_{12}$ for Ti-Cr, we find $\Theta_D]_{Ti-Mo} \approxeq \Theta_D]_{Ti-Cr} = 208$ K.

the atomic force constants at constant c_{ij}. In practice, however, it seems that it is in fact a "weakening"(16) of C' (as the softening-point, C' = 0, is approached) which leads to displacement fluctuations(4) of sufficient amplitude to excite the $\beta \to \omega$ diffusionless transformation.

2.2 Physical Properties of ω-Phase in Ti-Mo

In an aging study, Ti-Mo (10 at.%) (TM-10) was annealed at 350°C for 880 hours, by which time it was assumed that a condition approximating quasi-equilibrium had been attained as represented by the metastable phase diagram of Reference (4). The evaluation of f_ω, the mole-fraction of ω-phase present in the aged TM-10 has been discussed in I. If Φ is then taken to represent the physical properties χ, γ, and Θ_D in turn, the conservation equation:

$$\langle\Phi\rangle_{av} = f_\omega \Phi_\omega + (1-f_\omega)\,\Phi_\beta\ , \qquad (1)$$

together with the previously estimated values of Φ_β, may be used to determine χ_ω, γ_ω, and $\Theta_{D,\omega}$ for ω-TM-4.3. The values so obtained, plotted as special points in Figure 1, are seen to be in excellent agreement with the directly measured quantities for (β+ω)-Ti-Mo. These results serve to confirm that the decreases in χ and γ with decreasing $\mathfrak{z}$ ($\lesssim 4.3$) as well as the corresponding stiffening of Θ_D are, in fact, due to the continuously increasing influence of ω-phase precipitation. The $n(E_F)$ for ω-TM-4.3, derived from γ_ω using V_{app} = 0.26 eV-atom as before, is listed in Table 1.

2.3 $n(E_F)$ and Θ_D in Ti-Al and Ti-Mo

The variations of $n(E_F)$ and Θ_D with $\mathfrak{z}$, derived as a result of the magnetic and calorimetric experiments on the α and α_2 phases of Ti-Al, and the α, ω, β+ω, and β phases of Ti-Mo are shown in Figure 3. In this Figure, $\mathfrak{z}$ is not intended to have a real physical meaning but is to be interpreted as an arbitrary horizontal scale, based on the average number of atomic s, p and d electrons in the case of Ti-Al, and s+d electrons in the case of Ti-Mo. Particularly noticeable are (a): that the more stable (at say, room temperature) of a pair of allotropes, possesses the lower $n(E_F)$ together with the higher Θ_D--tightly-bound α_2-Ti-Al compared to α-Ti-Al is a very good example of this--and (b): that discontinuities or turning points in the $\mathfrak{z}$-dependence of $n(E_F)$ [or Θ_D, which scales inversely with it] occur at the limits of the various regimes of structural stability.

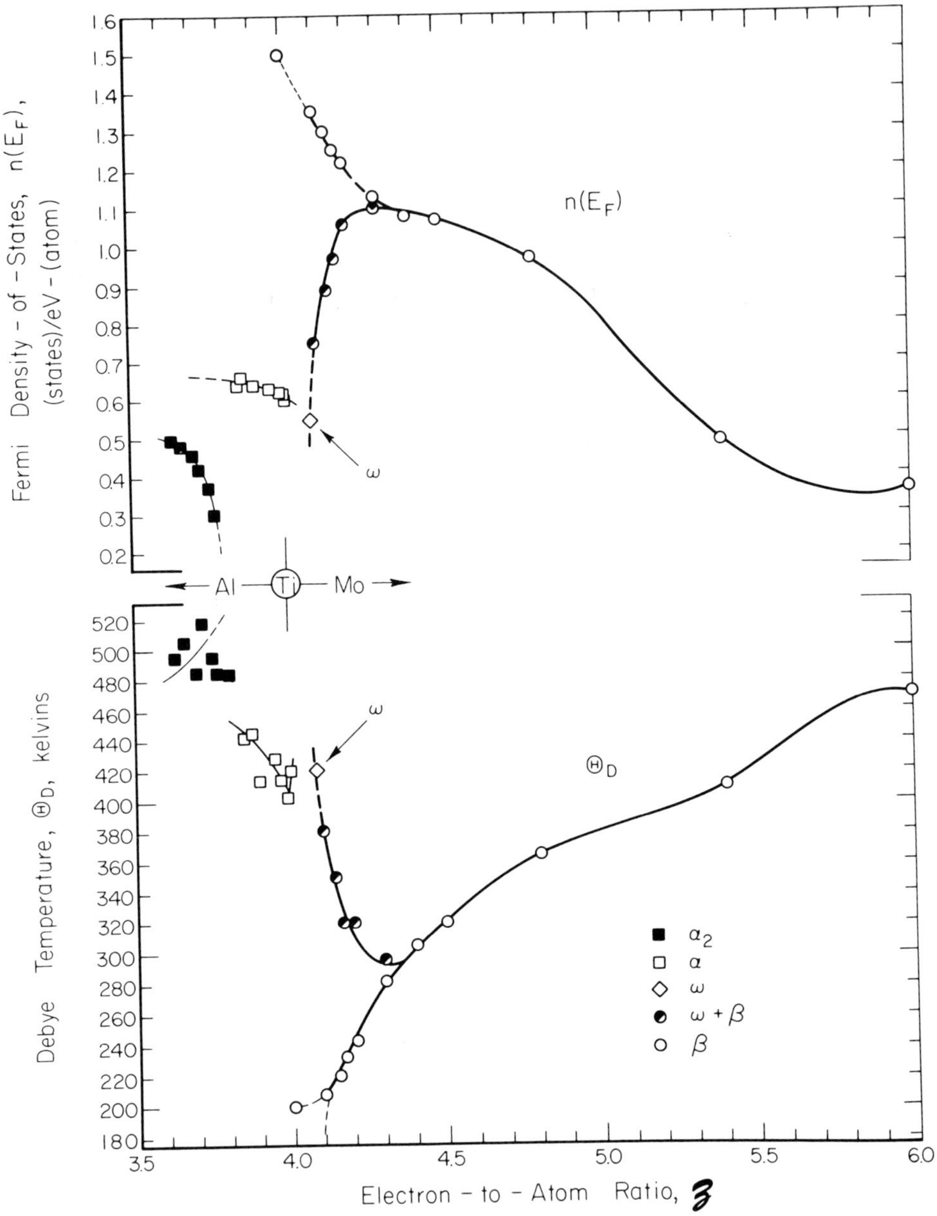

Figure 3. Empirically-derived Fermi density-of-states, $n(E_F)$, and Debye temperature, Θ_D, as functions of electron-to-atom ratio, $\mathfrak{z}$, in Ti-base alloys.

2.4 Phase Stability in Titanium Alloys

Figure 4 presents an experimentally-determined $n(E_F)$ curve for some binary transition-metal alloys within $4 \leq \mathfrak{z} \leq 7$. The $4 \leq \mathfrak{z} \leq 6$ regime was covered by Ti-Mo, as described above. In order to extend the range to $\mathfrak{z} = 7$, use was made of the calorimetrically determined data for Mo-Re, and Mo-Os from the work of Morin and Maita[17] and Bucher *et al.*[18]. In applying an electron-phonon correction in this range Morin and Maita's[17] $V_{app} = 0.42_6$ was employed. The experimentally-derived curve may be compared with a "calculated" $n(E_F)$-curve based on computed[19] values of $n(E_F)$ for bcc Ti, V, Cr, Mn, and Fe, and for the alloy system Ti-Mo[20]. Also included for comparison as an inset, are reproductions of Waber's[20] calculated $n(E)$ profiles for the bcc 3d and 4d elements Cr and Mo. Such curves for all bcc 3d transition elements as well as some bcc 4d transition elements are characterized by a pair of dominant peaks[21]. At this stage it is appropriate to point out that, even within the context of rigid-band-type behavior [as discussed in (19)], the twin main peaks of the N(E) curves of pure bcc metals (see inset) are not to be confused with those apparent in the calorimetrically-derived $n(E_F)$ loci. The latter occur "closer in" and are induced by structural transformations of the β-phase, which take place both for $\mathfrak{z} \lesssim 4.3$ and $\mathfrak{z} \gtrsim 6.7$. Thus, the experimentally-determined maxima (e.g. that for $\mathfrak{z} \sim 4.3$) vanish when $n(E_F)$ is extrapolated semi-empirically into the unstable β regime.

De Fontaine[22] in studying the lattice dynamics of the bcc structure has computed with remarkable success the instability conditions leading to spontaneous ("athermal") ω-phase formation. Presumably the upper-$\mathfrak{z}$ limit of bcc stability could also be treated in terms of atomic force constants. The ions of course interest *via* the electrons[23], but so far a rigorous electronic interpretation of transition-metal-alloy phase stability, using modern theories of metals, has not yet appeared. Figures 3 and 4, however, contain some empirical results which could perhaps serve as a starting point viz.:

(a) For a given $\mathfrak{z}$ the more stable ($\lesssim$ room temperature) of a pair of allotropes is associated with the lower $n(E_F)$. Thus, for three specific single-phase situations (Figure 3):

$$n(E_F)_{\alpha_2\text{-}Ti_3Al} < n(E_F)_{\alpha\text{-}Ti_3Al} \quad (T \lesssim 830°\ C)$$

$$n(E_F)_{\alpha\text{-}Ti} < n(E_F)_{\beta\text{-}Ti} \quad (T \lesssim 883°\ C)$$

$$n(E_F)_{\omega\text{-}TM\text{-}4.3} < n(E_F)_{\beta\text{-}TM\text{-}4.3} \quad (T \lesssim 500°\ C)$$

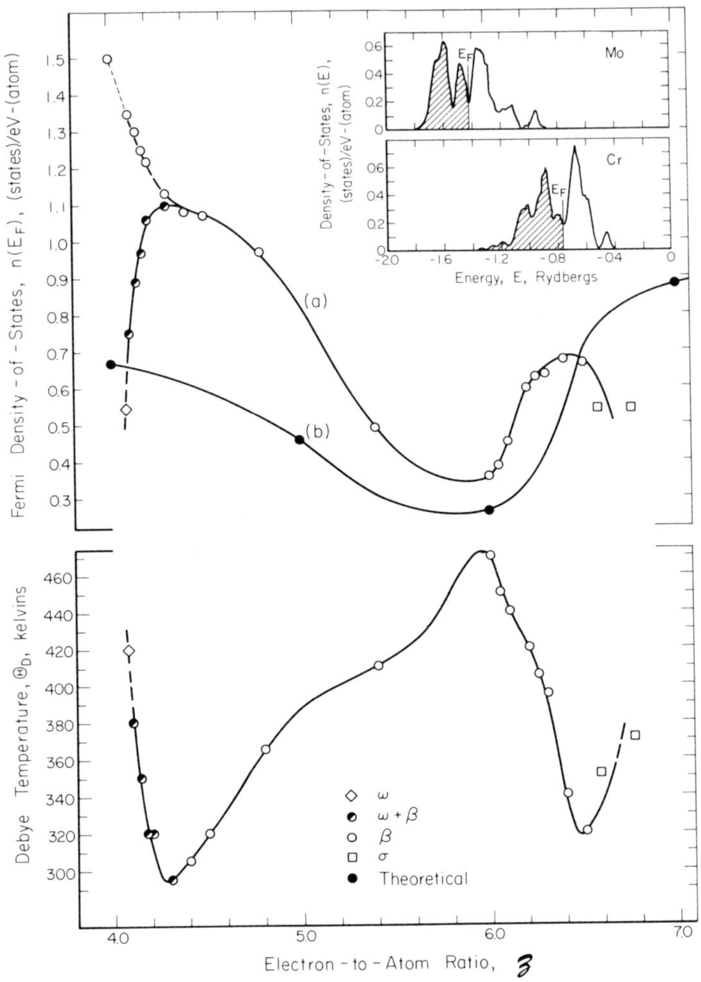

Figure 4. 𝔷-dependence of $n(E_F)$ for transition-metal-binary (i.e. T_1-T_2) alloys: (a) Empirically-derived curve for Ti-Mo-Re, based on the present data and those of References (17) and (18); (b) "calculated" $n(E_F)$ for bcc T_1-T_2 alloys based on published data points[19] for bcc Ti, V, Cr, Mn and Fe. For comparison, the 𝔷-dependence of Θ_D for Ti-Mo-Re is also shown. The remarkable inverse scaling exhibited by the $n(E_F)$ and Θ_D curves owes its origin to phase stability-considerations.

Inset are calculated n(E) curves for bcc Ti and Mo from the work of Waber and colleagues[20]. $n(E_F)$ is at a local minimum in n(E) for both bcc Ti and Mo; however, the latter shows an extra peak, which seems to be reflected in the calculated $n(E_F)$ for Ti-Mo[20] but not in the experimentally derived curve presented here.

(b) As $\mathfrak{z}$ is varied, maximal bcc stability is achieved with a half-filled d-band. The decrease in stability as $\mathfrak{z}$ either increases or decreases from this position is reflected in a decreasing Θ_D. This, in turn, is the result of a weakening in $C' = (c_{11}-c_{12})/2$ which has been shown to occur in each case[24,11, respectively], and which precedes structural transformation to σ-phase on one hand, and ω-phase on the other.

For Ti-base alloys, therefore, since the only way to increase the occupancy of the d-band is to add conduction electrons, transition elements tend to be β-stabilizers. On the other hand non-transition metals such as Al, Ga, and Sn reduce the occupancy of the Ti d-band, not only because they possess fewer conduction electrons but through the formation of bound states. These are, of course, α-stabilizers.

References

1. J. C. Ho and E. W. Collings, "Anomalous Electrical Resistivity in Titanium Molybdenum Alloys", Acta. Met.,---submitted for publication.

2. J. C. Ho, P. C. Gehlen, and E. W. Collings, Solid State Comm. 7 511 (1969).

3. E. W. Collings and J. C. Ho, "The Science Technology and Application of Titanium", Proceedings of an International Conference, London, May, 1968; ed. R. I. Jaffee and N. E. Promisel, Pergamon Press Ltd. (1970) p. 331.

4. D. de Fontaine, N. E. Paton, and J. C. William, Acta Met., 19 1153 (1971).

5. The units of $n(E_F)$ are taken to be states/eV-atom considering a single spin direction (i.e., half the total density-of-states); and those of γ are mJ/mole-deg^2 K.

6. E. W. Collings and J. C. Ho, phys. stat. sol. (b) 43 K123 (1971).

7. See Figure 3 of Paper I.

8. Referring to single-phase-bcc (β) Ti-Mo we have designated several electron-to-atom-ratio ($\mathfrak{z}$) regimes as follows:
$4.0 \le \mathfrak{z} \le 4.3$ - "unstable"; a regime which includes
(i) $4.0 \le \mathfrak{z} \le 4.05$ - "absolutely unstable", and
(ii) $4.05 \le \mathfrak{z} \le 4.3$ - "virtual".

9. E. W. Collings, J. C. Ho, and R. I. Jaffee "Superconducting Transition Temperature, Lattice Instability, and Electron-to-

Atom Ratio in Transition Metal Binary Solid Solutions", Phys. Rev.--to be published.

10. O. S. Anderson, J. Phys. Chem, Solids, 24 909 (1963).

11. E. S. Fisher and D. Dever, Acta Met., 18 265 (1970).

12. The microstructures of quenched Ti-Cr and Ti-Mo are reviewed in Appendix 3 of Reference (9).

13. P. Duwez and J. L. Taylor, Trans. ASM, 44 495 (1952).

14. E. W. Collings and J. C. Ho, Phys. Rev. B, 1 4289 (1970).

15. Whereas a significant decrease in C' might reasonably be referred to as "softening", Fisher and Dever(16) would refer to this as "weakening"; while de Fontaine et al.(4) reserve the term "softening" for C' → 0.

16. E. S. Fisher and D. Dever, "The Science Technology and Application of Titanium", Proceedings of an International Conference, London, May, 1968; ed. R. I. Jaffee and N. E. Promisel, Pergamon Press Ltd. (1970) p. 373.

17. F. J. Morin and J. P. Maita, Phys. Rev., 129 1115 (1963).

18. E. Bucher, F. Heiniger, and J. Muller, Phys. kondens. Materie, 2 210 (1964).

19. E. C. Snow and J. T. Waber, Acta Met., 17 623 (1969).

20. J. T. Waber, in discussion to G. Brodén et al., Proceedings of the 3rd Materials Research Symposium, "Electronic Density of States", Nat. Bur. Stand. (U.S.), Spec. Pub. 323, December, 1971, p. 217 (pp. 221-223).

21. The n(E) profiles for Nb and Mo possess less-pronounced centrally-situated peaks; a property which seems to show up in the calculated $n(E_F)$ curve for Ti-Mo alloys, but which is not shared by the experimentally observed $n(E_F)$ for Ti-Mo.

22. D. de Fontaine, Acta Met., 18 275 (1970).

23. M. H. Cohen in "Alloying Behavior and Effects in Concentrated Solid Solutions", T. B. Massalski (ed.) Gordon and Breach (1965) pp. 1-23; A. Blandin. loc. cit. pp. 50-84; W. A. Harrison, "Pseudopotentials in the Theory of Metals", W. A. Benjamin, Inc. (1966).

24. D. L. Davidson and F. R. Brotzen, J. App. Phys., 39 5768 (1968)

RELATIONSHIPS BETWEEN MICROSTRUCTURE, SUPERCONDUCTIVITY AND MECHANICAL PROPERTIES OF Ti-6Al-4V

Ernest G. Wolff, Richard Lepper and George J. Mills

Northrop Corporation
Hawthorne, California

Introduction

In recent years, there has been a growing awareness that superconducting properties are greatly influenced by microstructure. The interrelationships are exceedingly complex, especially as a result of the trend towards ternary systems of which at least 200 have been investigated to date in the superconducting range. In many low temperature applications, such as magnets, computer switches, levitation devices, rotating machinery and power transmission, the workability of the superconductor is also an important consideration. The objective of the present program therefore was to establish the existence of correlations between superconducting material properties and mechanical properties of interest to metallurgical processing and structural applications, such as tensile and yield strength and hardness.

A ternary alloy (Ti-6Al-4V) was chosen for this study for several reasons. A superconducting phase was detected at $4^{o}K$ by resistivity measurements (1). It has adequate ductility and impact toughness for many structural applications to liquid hydrogen temperatures (about $20^{o}K$). It is a leading materials candidate for high performance aircraft and is commonly used for pressure vessels below $80^{o}K$. Although its components are prominent in many useful superconducting alloys, and many binary titanium alloys have been studied, the Ti-Al-V system, to our knowledge, has not been investigated in the superconducting range. The forming difficulties, and the influence of aging treatment on stress corrosion susceptibility are well known. Cryogenic testing offers a possible NDT indicator of overaging, for which a hardness test would be an ambiguous guide.

Theory

The disappearance of electrical resistance below a critical temperature originally defined the superconducting state, which is the region lying between the origin and a critical surface in a space defined by three orthogonal axes; temperature (T), applied field intensity (H) and impressed transport current density (J). The magnetic corollary to the disappearance of electrical resistance is perfect diamagnetism, where the magnetic permeability (μ) vanishes, so that the internal field $B = \mu H$ is also zero. Most materials have low N-S boundary energies (Type II superconductors) which allows partial flux penetration and results in a mixed N + S state (called the Shubnikov phase). The lower and upper boundaries to this state are H_{c1}-H_{c2}, T_{c1}-T_{c2} and J_{c1}-J_{c2}, the technologically most important superconducting properties. Magnetic flux interactions with microstructural features are complex and affect these superconducting properties in different ways. For example, high densities of flux pinning defects produce the highest critical currents. Critical temperatures (and critical fields) are strongly dependent on atomic ordering and stoichiometry in single phases, and on the morphology as well as composition of multi-phase microstructures (2). Other properties, such as the A.C. susceptibility and hysteresis are also highly sensitive to microstructure.

In dilute binary titanium alloys it has been found that elements which do not form localized magnetic states, but which increase the density of states at the Fermi level, tend to increase T_c, both for α-hcp and β-bcc primary phases (3,4). It has often been postulated that minority phases can cause complete diamagnetism (2). Metallographic evidence for this has rarely been presented, but an exception is work on Ti-Mn alloys (3) where interconnected filaments of the β phase in the α matrix accounted for the bulk T_c corresponding to that of the β phase. Similar considerations are used to explain minimum resistance phenomena in the normal state for dilute binary Ti solid solutions (1).

Ti-6Al-4V is an alpha-beta alloy which is normally solution treated in the upper half of the $\alpha + \beta$ region (about 1350°F-1830°F), quenched to induce β to transform to α' martensite, and aged in the range 900°F-1100°F. Strength and hardness are developed as a result of α' decomposition to $\alpha + \beta$ and to ordering of the primary α phase to form α_2, or Ti_3Al. Overaging, weakening and stress corrosion susceptibility occur as a result of β precipitate agglomeration (5).

Experimental Procedures

Ti-6Al-4V test specimens were cut from two nearly identical billets. The 2000 series of samples contained 6.13% Al, 3.93% V, 0.15% Fe, 0.19% O, 0.019% N, 0.028% C and 0.0010% H. The 3000 series contained 6.10% Al, 4.20% V, 0.23% Fe, 0.11% O, 0.020% N, 0.030% C and 0.0065% H. Bars and pull tabs (6) were subjected to various solution temperatures, quench delay times, and aging temperatures and times (Table I). The bars were machined to ellipsoids of revolution 1" long by 1/8" diameter for low temperature testing. Figures 1-4 illustrate typical microstructures.

The techniques used for the measurement of T_{c1}-T_{c2}, H_{c1}-H_{c2} and the complex susceptibility $\chi^* = \chi' + j\chi''$ were outlined at a previous conference (7). The A.C. permeability technique requires no leads, the specimen shape requirements are flexible and no moving coils or samples are involved. Figure 5 illustrates the basic circuitry for balancing, amplifying, and filtering the voltage signals. Two coil systems are employed, a sample and a standard, and a difference measurement determines the condition of the sample. The peak voltage signal is

$$V'_{pk} = \frac{N'}{10^8}\left(\frac{d\phi}{dt}\right)_{pk} \cdot \mu^*_{sample} \sim 10^{-4}\ \mu^*_{sample}$$

N' is the number of turns of the sense coil, $\mu^* = 1 + 4\pi\chi^*$, and χ^* approaches $-1/4\pi$ in the superconducting state. Since Al, V and Ti are all known to have small, positive susceptibilities at ambient temperatures and below, one may assume that the Ti-6Al-4V alloy is paramagnetic in the normal state, with its permeability slightly greater than one. The ellipsoidal sample is installed in one of the coil assemblies, and after thermal and magnetic stabilization at 4.2°K, the electronic system is balanced in phase and amplitude to give a zero output signal. The temperature, controlled by the He^4 pressure, is then decreased in 0.01°K increments. The recorded in-phase (P) and quadrature (Q) output signals change below T_{c2} until T_{c1} is reached. μ^*, χ' and χ'' are proportional to $(P^2 + Q^2)^{1/2}$, Q and P, respectively. A D.C. magnetic field can be superimposed on the system until the sample is returned to the normal state, thus defining H_{c2} for a given T. The temperature could be varied over the range T_{c1}-T_{c2} at a fixed H (down to T = 1°K), or H could be varied at a constant rate for a fixed T to study hysteresis effects, if any.

Mechanical Properties

Table I shows excellent correlation between yield and tensile strengths and R_c hardness, which achieve a maximum after a 4-hour aging treatment at 900°F. The 2000 series of samples are softer than the 3000 series, probably because of the lower interstitial content. The gradual decrease in R_c with time of aging at 1000°F

Table I. Sample Heat Treatments, Mechanical Properties and Critical Temperatures

Sample No.	Solution Temp. (^{o}F)	Solution Time (hrs)	Quench Delay (sec)	Aging Temp. (^{o}F)	Aging Time (hrs)	Yield Strength (ksi)	Ultimate Strength (ksi)	Hardness R_c	T_{c2} ^{o}K at 0 Field
2004	1600	2.0	F.C.	None	-	116.00	129.0	26.4	<1.1
2005	1730	0.5	5	1000	4	149.20	161.3	35.5	3.41
2008	1730	0.5	40	1000	4	119.70	136.2	29.7	4.14
2009	1600	0.5	5	1000	4	119.20	138.7	31.0	4.04
2013	1800	0.5	5	1000	4	152.10	165.8	33.0	1.70
2014	1725	0.5	5	1000	1	158.50	173.2	37.8	2.33
2020	1725	0.5	5	1000	6	154.40	169.1	34.7	2.73
2022	1725	0.5	5	900	4	155.90	173.2	38.0	2.48
2025	1725	0.5	5	1300	4	130.40	142.4	31.6	3.27
2026	1725	0.5	5	1400	4	113.00	129.9	28.6	3.23
3002	1725	1.0	5	None	-	152.05	170.7	34.0	1.96
3003	1725	1.0	5	800	4	165.04	181.1	39.0	1.71
3004	1725	1.0	5	900	4	170.15	185.1	41.0	2.52
3005	1725	1.0	5	1000	4	171.15	181.2	40.0	2.63
3006	1725	1.0	5	1100	4	165.55	173.4	36.5	2.53
3007	1725	1.0	5	1210	4	147.15	156.6	34.5	2.93
3008	1350	1.0	F.C.	None	-	144.55	150.7	30.0	1.73

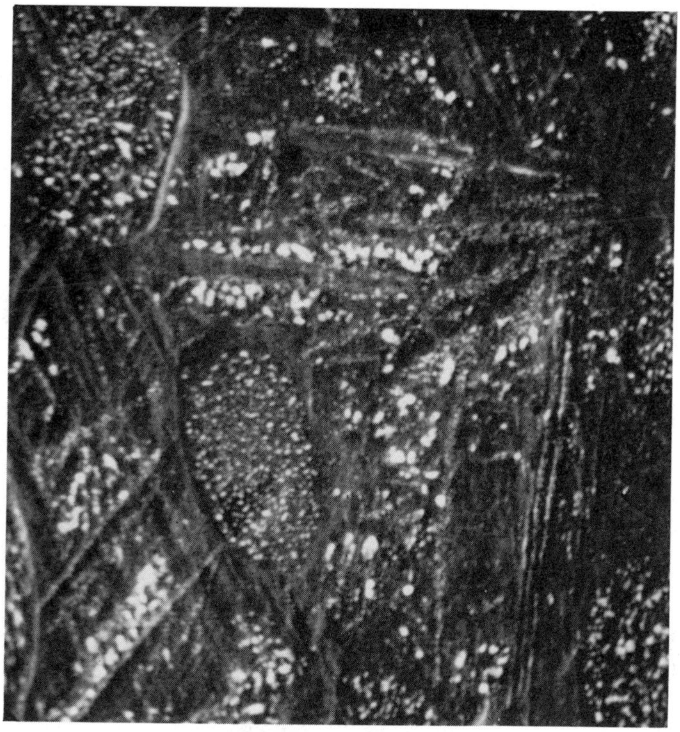

Fig. 1. Typical as-quenched (A.Q.) structure from intermediate $\alpha + \beta$ region. The intergranular β transforms to acicular martensite α' and the retained β appears as fine white dispersion in primary α grains. Sample 3002. 5000X. Austin Etch.

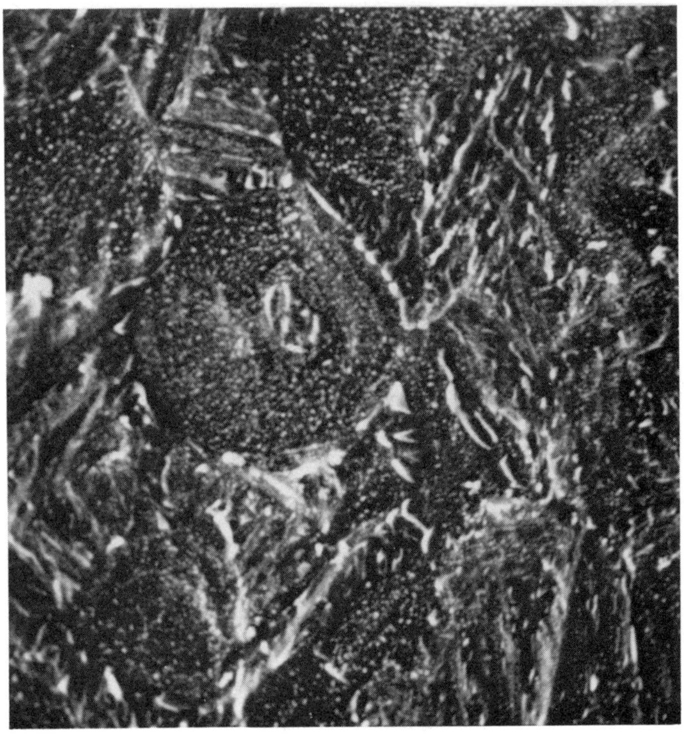

Fig. 2. Sample of Figure 1 slightly over-aged (1200°F, 4 hours). Martensite α has decomposed at α-α' interfaces to $\beta + \alpha$. Some β retains acicular structure but also shows signs of agglomeration. Sample 3007. 5000X.

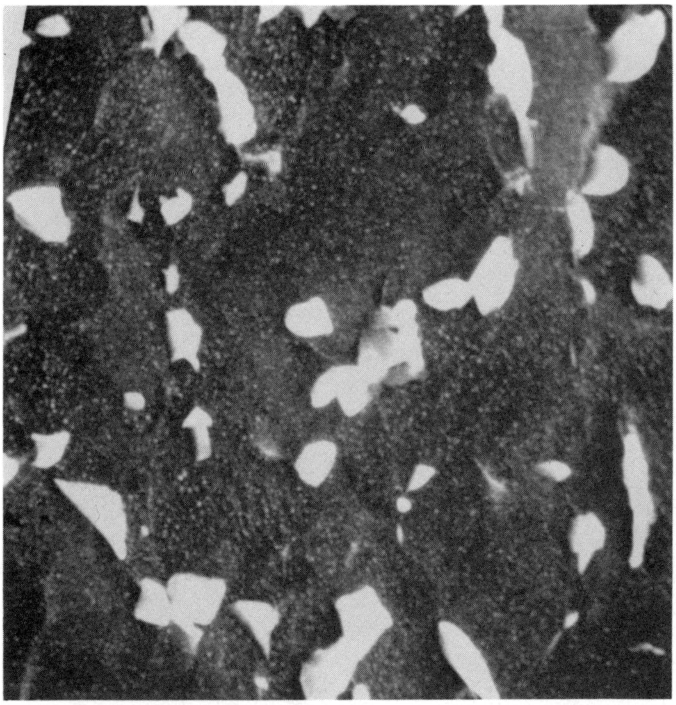

Fig. 3. Furnace cooled from 1350°F. A higher solution temperature would increase fraction of intergranular β, with some increase in primary α and β grain sizes. Sample 3008. 500X.

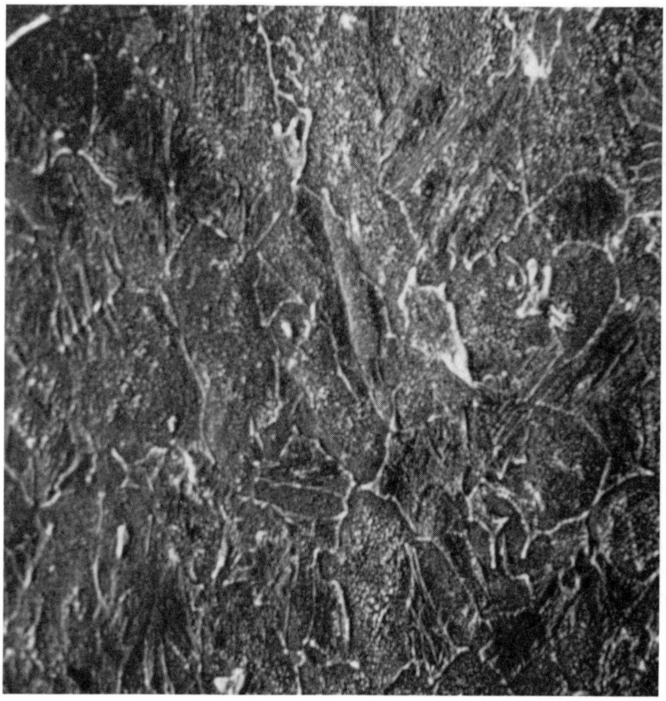

Fig. 4. Effect of a 40 second quench delay on a 1000°F-4hr aging treatment. β has formed from α' decomposition during quench. Fairly continuous β around primary α grains. Sample 2008, 2200X.

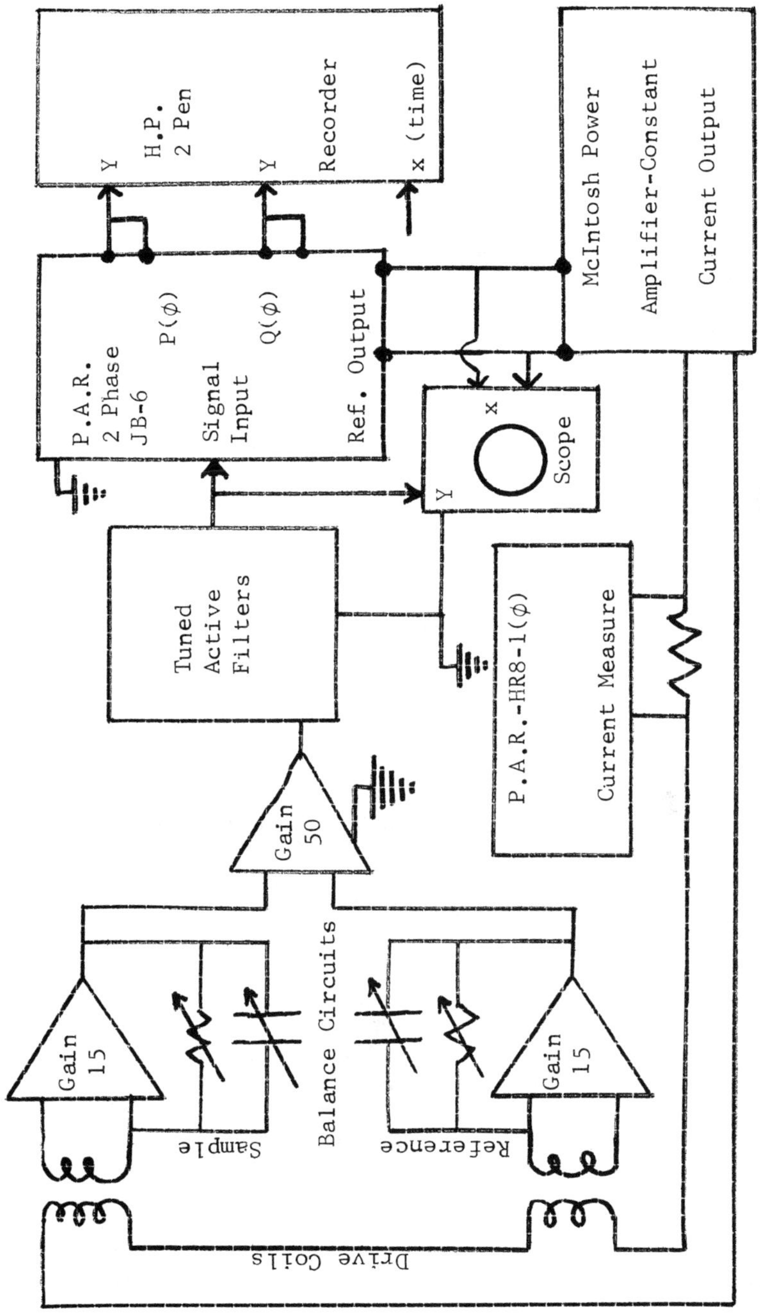

Fig. 5. Circuit diagram of A.C. permeability test system.

(Samples 2014, 2005 and 2020) is expected (5). The hardness increases with increasing solution temperature after a 4-hour age at $1000^{\circ}F$ (Samples 2009, 2005 and 2013). A high solution temperature means a high fraction of β before quenching, and α' martensite after, so that there is more decomposed α' on aging. This is the initial hardening mechanism in this alloy. The large amount of martensite means that only fine β is formed on further aging and the high initial nucleation rate of β reduces its subsequent growth or agglomeration rate (5).

Critical Temperature at Zero Field

A simple relationship of T_c with aging temperature, useful for nondestructive testing, does not exist. For example, Samples 3004 and 3006 represent a strengthened alloy ($R_c \sim 41$) and one that is on the border line of being overaged ($R_c \sim 36.5$). Yet the difference in T_c between these samples is an essentially negligible $0.01^{\circ}K$ (2.52-$2.53^{\circ}K$). Again, Samples 3003 and 3008 represent an underaged, partially hardened condition ($R_c = 37$) and a soft annealed state ($R_c = 30$), respectively, but the T_c values are virtually identical ($1.71^{\circ}K$ and $1.73^{\circ}K$, respectively. A more significant observation, is that the various heat treatments have varied T_c over a relatively wide range (from about $1^{\circ}K$ to $4.14^{\circ}K$) and, as no special effort was made to maximize this range, it is feasible that it could be increased even further. Extensive control and variation of the superconducting properties of alloys require adequate explanations of the present data.

Since the β phase is bcc and has the highest V content it is expected to have the highest T_c (2,4). The α' phase has a higher V content than primary α, mainly because it transforms from the β by a diffusionless process. It might therefore be expected to have an intermediate T_c between the β and α phases. The phase with the highest T_c that is interconnected determines the T_c of the alloy. Consequently, T_c should increase once the β-phase has formed and becomes interconnected, e.g., at overaging. There is a general trend of increasing T_c with aging temperature at 4 hours (Samples 3003, 3004, 3005, 2002, 2005, 3007, 2025, and 2026). The α' is fairly interconnected on quenching (see Figure 1) and when β becomes interconnected (Figure 2) an increase in T_c is obtained. The loss of α' interconnectedness during the early decomposition stages ($\alpha' \rightarrow \alpha + \beta$) at $800^{\circ}F$ may explain the decrease in T_c from Samples 3002 to 3003. The fine dispersed β in primary α (Figure 1) is not considered to affect T_c because (a) it may be smaller than the coherence length, or thickness of an N-S boundary, and (b) its dimensions increase only slightly during aging even at a $1400^{\circ}F$ aging temperature. A depletion of α' can both lower T_c (α formation) or raise it (β formation) but if the main result is a loss of α' interconnectedness, a temporary lowering of T_c could occur, as with samples 3005 and 3006. The latter has a higher aging temperature (1100°F) but a lower T_c. The assumption that T_c (α') is intermediate leads to the prediction that a higher solution temperature causes a lower T_c after aging due to slower rate of β interconnectedness or agglomeration. This is illustrated in Figure 6, where the % β at the solution temperature varies from about 90% at 1800°F (Sample 2013) to 28% at 1600° F (Sample 2009).

The two furnace cooled (F.C.) samples have very low T_c values and these are compared to other low T_c samples.

Sample	2004	3008	3002	3003
Solution T°F	1600	1350	1725	1725
% α	70	95	40	40
% β	30	5	60	60
% V in β	7	>15	5	5
Age T - t	F.C.	F.C.	A.Q.	800-4
R_c	26.4	30	34	39
T_c °K	1.1	1.73	1.96	1.71

The microstructure of Sample 2004 consists of massive β which is not interconnected (similar to Figure 3 but more β and more elongated). There is also evidence of the reversion of β to primary α on cooling. The β-phase has a lower vanadium content than the β in Sample 2008; consequently a slightly lower T_c may be expected. The high V content of the β in 3008 may explain the T_c of 1.73 compared to 1.71°K for Sample 3003, which has more α' but in its underaged condition very little β as formed from α' decomposition. Sample 3002 has less primary α than 3008; a higher T_c is therefore expected. Sample 3002, with 40% α' (from the β) should be higher in T_c than 3003 because in the latter some reversion of α' to α + β has occurred with β being as yet not interconnected. The hardness values are consistent with these explanations.

The occurence of α ordering, or α_2 (Ti_3Al) formation, which would increase T_c, would be significant if the α phase had a high Al content, but this is achieved only by a high solution temperature where the total amount of α is small compared to β, the major influence on T_c. Effects attributable to α_2 formation are small (5) and in the present work can be expected only to enhance the observed T_c trends (e.g. Figure 6). Vanadium in general retards α_2 precipitation, especially at α/β boundaries (8), while aluminum suppresses ω or β′ phase formation as found, for example in Ti + ≥ 10% V alloys.

A longer quench delay allows α' decomposition before aging, so that β development is further advanced and a lower hardness and higher T_c are observed. (Samples 2008, 2005, 3005.) Similarly,

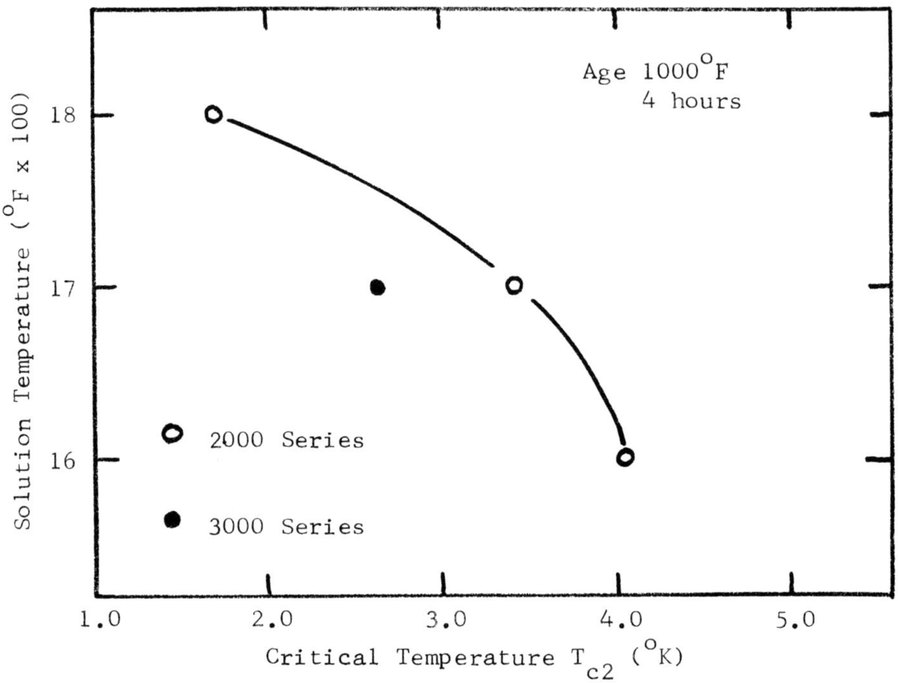

Fig. 6. Variation of critical temperature T_{c2} with solution temperature after 5 sec quench and 1000°F, 4 hr aging.

data for Samples 2020, 2014 and 2005 suggest that increased time at aging temperature lowers R_c, but a maximum in T_c may occur. The competing effects of loss of α' interconnectedness and β formation may be responsible but more data are required.

Complex Susceptibility

The complex relationships between T_c and hardness after aging prompted the search for another superconducting indicator. The real and imaginary components (χ', χ'') of the complex susceptibility can be determined from the output data (7) and their physical interpretation has been outlined (9). The imaginary component χ'' normally changes from a finite value in the normal state to zero in the superconducting state. It is dependent on the skin depth of the eddy currents produced by an external alternating magnetic field, which in turn depends on the frequency and sample conductivity. Normally, a maximum in χ'' (and inflection in χ') occur when the sample (cylinder) radius to skin depth ratio is about 1.8. If this maximum has not occurred in the normal state, there is a high probability that it will occur at the normal-superconducting transition, because there is such a large increase in the sample conductivity at this time. It can be shown that a maximum in χ'' should occur during the N-S transition for Ti-6Al-4V, especially since the T_{c1}-T_{c2} range is relatively large. A large χ'' peak occurs if superconducting inclusions or regions increase the average conductivity without appreciably reducing the fraction of normal metal present, e.g. by being randomly dispersed, or in a thin flake or filamentary configuration. Large spheroidal inclusions, especially if interconnected, will not give a large χ'' maximum on cooling because the decrease in resistance for the eddy currents could not increase the average dissipation due to the fact that the effective fraction of normal metal would be reduced proportionately. The latter case resembles structurally the overaged heat treated microstructure. Figure 7 shows the variation of the relative peak height of the in-phase voltage signal (proportional to χ'') during a N-S transition in an A.C. applied field of about 0.5 gauss, with various aging temperatures. The microstructure responding to initial β formation is filamentary in nature; hence small regions of transformed α' (to β) give local dissipation of eddy currents without affecting the bulk conductivity. As overaging progresses, the β regions become bulky and interconnected, resulting in a decrease in the average χ''. The 6 hour aging and 40 second quench delay results are consistent with this general interpretation.

Conclusions

Correlations between the superconducting critical temperature, mechanical properties, microstructures and various heat treatments were determined for the ternary alloy Ti-6Al-4V. The importance of interconnected minority phases with high values of T_c has been

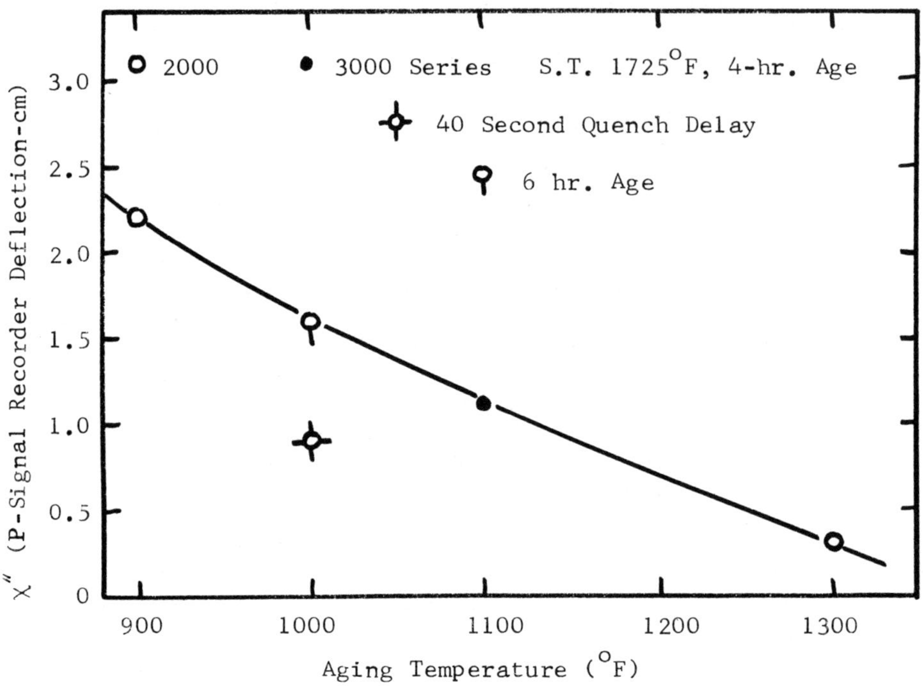

Fig. 7. Variation of χ'' with 4-hr aging temperature after a 1725°F solution treatment.

substantiated and the concepts involved were extended to account for the T_c values of a wide variety of microstructures. The imaginary component of the complex susceptibility has been found to be a more accurate indicator of the aging process than the critical temperature. It was also shown that a maximum T_c can be attained with a minimum value of χ''. This suggests possibilities for minimizing A.C. losses in superconducting magnets or machinery.

Possible NDT applications include the cryogenic monitoring of Ti alloy foils (10), thin films, fibers, wires, or powders. Ti-6Al-4V foils are used for honeycomb structures for which the impurity content is critical for the required ductility. Both T_c and χ^* are sensitive to composition as well as heat treatment. Particle or fiber thicknesses greater than 1000 Å (1 μ) are not likely to show significant size effects on superconducting properties.

The principles outlined in relating heat treatments and microstructures to high values of T_c ($\geq 4.14^oK$ for Ti-6Al-4V) could be applied to other ternary systems, e.g. those based on β-W compounds. The alloy could be heat treated first to a soft condition for workability and then an interconnected filamentary high T_c phase could be precipitated for optimum T_c and H_c.

References

1. Clark, A. F., Childs, G. E., and Wallace, G. H., "Electrical Resistivity of Some Engineering Alloys at Low Temperatures," Cryogenics, Vol. 10, August 1970, pp. 295-305.

2. Livingstone, J. D., and Schadler, H. W., "The Effect of Metallurgical Variables on Superconducting Properties," Progress in Materials Science, Vol. 12., No. 3, 1964, pp. 183-287.

3. Cape, J. A., "Superconductivity and Localized Magnetic States in Ti-Mn Alloys," Physical Review, Vol. 132, No. 4, November 1963, pp. 1486-1492.

4. Raub, C. J., and Zwicker, V., "Superconductivity of α-Titanium Solid Solutions with Vanadium, Niobium and Tantalum," Physical Review, Vol. 137, No. 1A, January 1965, pp. A142-A143.

5. Fopiano, P. J., Bever, M. B., and Averbach, B. L., "Phase Transformations and Strengthening Mechanisms in the Alloy Ti-6Al-4V," ASM Transactions Quarterly, Vol. 62, No. 2, June 1969, pp. 324-332.

6. "Heat Treatment of Titanium and Titanium Alloys," Boeing Process Specifications, BAC 5613, 5753, 5756 (1959) Northrop Revisions HT-10, 31 March 1967.

7. Lepper, R., Wolff, E. G., and Mills, G. J., "A.C. Permeability Studies of Ternary Alloys at Cryogenic Temperatures," 1972 Applied Superconductivity Conference (American Physical Society), Annapolis, Maryland, 1-3 May 1972.

8. Crossly, F. A., "Kinetics of Ti_3Al Grain Boundary Precipitation in Ti-Al Binary and Ti-Al-X Ternary Alloys and Correlation with Mechanical Properties," Metallurgical Transactions, Vol. 1, July 1970, pp. 1921-1929.

9. Maxwell, E. and Strongin, M., "Filamentary Structure in Superconductors," Physical Review Letters, Vol. 10, No. 6, 15 March 1963, pp. 212-215.

10. Smith, H. R., Kennedy, K., and Boericke, F. S., "Metallurgical Characteristics of Titanium-Alloy Foil Prepared by Electron-Beam Evaporation," Journal of Vacuum Science and Technology, Vol. 7, No. 6, Nov/Dec 1970, pp. 548-551.

INFLUENCE OF PINNING FORCES OF TITANIUM-NIOBIUM BASE ALLOYS ON THE CRITICAL CURRENT UP TO 10 TESLA+

R. Löhberg, W. Heller and U. Zwicker
Institut für Werkstoffwissenschaften II
(Werkstoffkunde - Technologie der Metalle)
Universität Erlangen-Nürnberg
8520 Erlangen, Egerlandstraße 5, W-Germany

To determine the amount of specific pinning forces of different pinning centers in Ti-Nb-base alloys the effect of vacancies, dislocation network, precipitation of intermetallic compounds, of ω- and α-phase on the critical current density J_c as a function of the external transverse magnetic field up to 10 Tesla was investigated. Results of critical current measurement are as follows:

Pinning centers	J_c (A/cm^2) 5 Tesla	J_c (A/cm^2) 10 Tesla	Pinning centers	J_c (A/cm^2) 5 Tesla	J_c (A/cm^2) 10 Tesla
Vacancies random distribution TiNb50; quenched	2,5x 10^1	1x 10^2	ω-precipitation TiNb50; homogen. 400°C annealed	4x 10^4	7x 10^2
Agglom. of vacancies TiNb50; quenched + 700°C ann.	1x 10^3	6x 10^2	ω-precipitation TiNb49Ge2,3; hom. 400°C annealed	2x 10^4	4x 10^3
Disloc. network TiNb50; 97% cold worked	1,5x 10^2	3x 10^3	ω- + α-precip. TiNb50; 97% cold worked 400°C ann.	4x 10^4	4x 10^3
Precipitation of Ti_2Cu; TiNb49Cu3 700°C annealed	3,5x 10^4	3,5x 10^3	ω- + α-precip. TiNb49Ge2,3; 97% c.w. 400°C ann.	4x 10^4	1,3x 10^4
Precipitation of Ti_5Ge_3; TiNb49Ge2,3 700°C annealed	5x 10^4	2x 10^3	ω- + α-precip. TiNb49Cu3; hom. or 97% c.w. 400°C ann.	4x 10^4	1,3x 10^4

+Most part of the work was performed under the auspices of the Deutsche Forschungsgemeinschaft

Introduction

Although superconducting TiNb-alloys are inferior in their superconducting properties (H_c at 4,2 ^{o}K ~12 Tesla, T_c ~10 ^{o}K) compared with intermetallic compounds such as Nb_3Sn or $Nb_3(Al, Ge)$ (H_c at 4,2 ^{o}K above 20 Tesla, T_c ~ 20 ^{o}K), preference is given to them as a material for the construction of magnets with fields up to 10 Tesla because of their good formability and workability. The most often used commercial alloy containing about 50 % Nb can be easily cold worked by swaging and drawing with more than 95 % area reduction (1). This produces a spirallike lamellar structure (2),which consists of numerous quasi-directed substructures (3,4). These inhomogeneities and the dislocations originated by the cold working process increase the critical current.

The current carrying capacity of the TiNb-superconductors is furthermore increased by the presence and the properties of other inhomogeneities, which can originate pinning centers in the matrix of the superconducting material. The efficiency of a certain inhomogeneity depends on its specific pinning force and its distribution in the matrix. It is thought,that the precipitation of the ω- and α-phase during annealing (5) are most responsible for the high critical current of TiNb-base-alloys (6,7). Other pinning centers can be originated by small particles of oxides (8) or by intermetallic compounds (9,10) and the pinning force can be improved by segregations (11). This paper describes the attempt to produce separately the following inhomogeneities: 1. vacancies, 2. dislocation networks, 3. ω- and α-precipitation and 4. intermetallic compounds, and to investigate their specific influence on the pinning forces by measuring the critical current density J_c.

Experimental

The alloys (20 g buttons) were melted in an arc-furnace in a high purity argon atmosphere.The oxygen content of the alloys was about 500 ppm. From these buttons, rods 6 mm in diameter were cast in a water-cooled copper-mould as described earlier (12). Microprobe analysis of the cast rods showed segregations with a variation of the Nb content of 5 % over a distance of about 5 - 10 μm. The rods were swaged into wires 1 mm in diameter (97% degree of area reduction). Some of the rods were homogenized at 1200°C prior to cold working. Homogenisation and heat treatment of the wires above 400°C up to 1000°C were carried out in evacuated quartz tubes, at 400°C and below in evacuated glass tubes. The specimens were water quenched after each heat treatment. J_c(H)-measurements were done in a superconducting magnet with a maximum field of 10 Tesla in the Siemens Research Laboratory, Erlangen. Because of the residual magnetization of the superconducting coil the lowest field in the bore was 1 Tesla after a previous excitation. Wire specimens of 300 mm in length were bent into U-shapes indium tinned by ultrasonic soldering and fixed with screws and clamps to an sample holder. The transition into the normal conducting state was measured by the first voltage drop over the part of the wire perpendicular to the field. A 1000 A rectifier produced the specimen current, which was switched off by a voltage drop of 100 µV over the specimen. The accuracy of J_c is $\pm$ 5 %.

Results

Vacancies

It was suggested that a random distribution of vacancies is achieved by water quenching TiNb50 (1 mm diam wire) after homogenization of 3 h at 1000°C. By additional annealing at 700°C in the temperature range of the β-phase field and water quenching, the vacancies are probably able to agglomerate. Fig. 1 shows J_c(H)-curves of specimens, which were homogenized at 1000°C and heat treated at 700°C. A heat treatment of 1,25 h at 700°C can increase the critical current density at 5 Tesla from $2{,}5 \cdot 10^1$ to 10^3 A/cm^2 and at 10 Tesla from 10^2 to about $6 \cdot 10^2$ A/cm^2.

Dislocation networks

Rods of the composition TiNb50 and TiNb49Ge2,3 were homogenized at 1200°C and subsequently swaged into wires (area reduction 97 %), to obtain samples which contained mainly dislocation networks as inhomogeneities. The J_c(H)-curves of these wires and those of wires at 1000°C annealed and water-quenched (H_2O) to dissolve most of the dislocation networks are presented in Fig. 2. The cold worked wires show an increase in the critical current carrying capacity compared with the annealed wires. The minima at medium fields in the J_c(H)-curves are typical for cold worked and homogenized material. The addition of germanium increases the critical current in the homogenized and in the cold worked condition.

Precipitation of intermetallic compounds

Precipitation of intermetallic compounds was produced in the ternary alloys TiNb49Cu3 and TiNb49Ge2,3 by

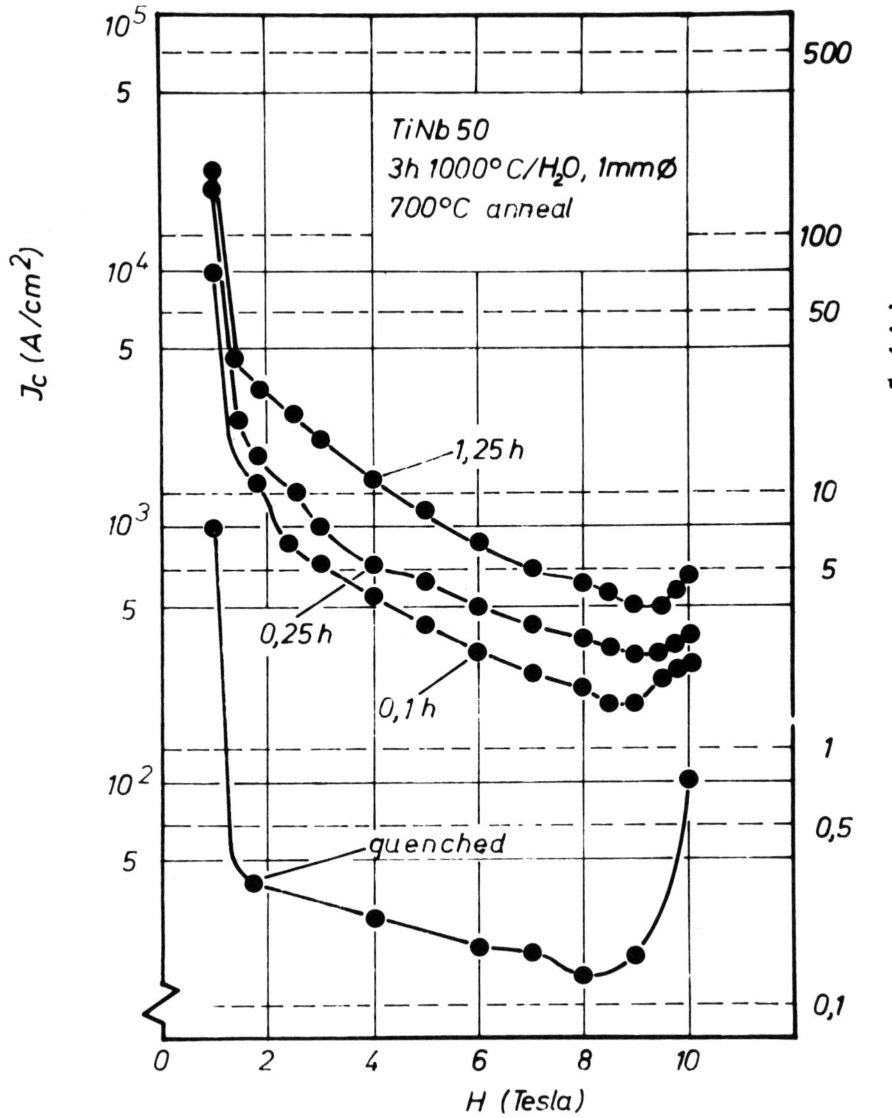

Fig. 1. J_c(H)-curves of 1 mm dia wire samples (TiNb50) homogenized 3 h 1000°C/H_2O and annealed 0.01 - 1.25 h at 700°C.

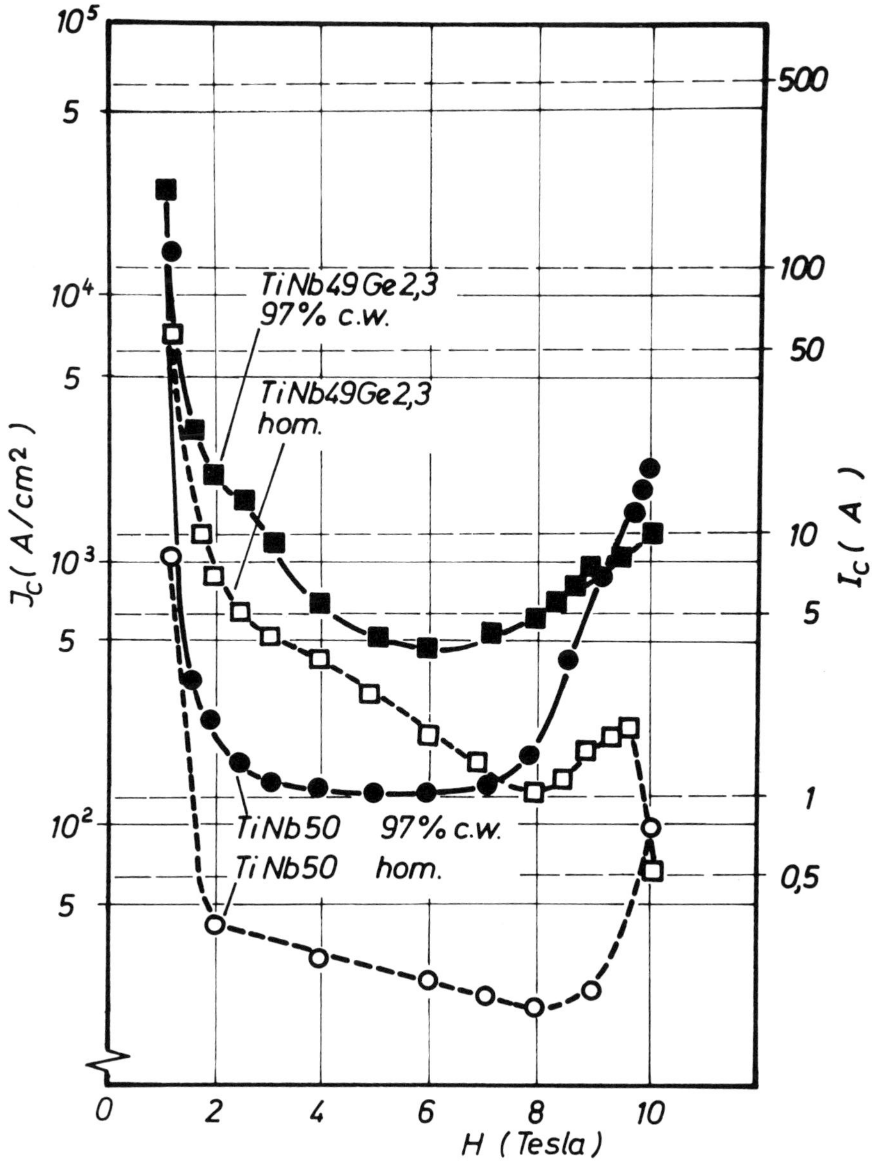

Fig. 2. Influence of cold working (97% area reduction) on the J_c(H)-curves of TiNb50 and TiNb49Ge2,3, homogenized at 1200°C before cold working.

homogenization of wire samples in the temperature range of the β-phase field and subsequent annealing at 700°C. This temperature is above the β/(α+β) transformation temperature of the alloys, so that only Ti_2Cu or Ti_5Ge_3 can be precipitated as small particles. Fig. 3 shows the J_c(H)-curves achieved with these specimens after different annealing times. At the annealing temperature of 700°C ≅ 1 % Cu and ≅ 0,2 % Ge are soluble in the TiNb-matrix. Although the solid solution of Cu and Ge in the TiNb-matrix decreases the critical field of TiNb50 to about 10 Tesla, the critical current is increased from 1,5 - $5\cdot10^1$ A/cm^2 up to more than 10^3 A/cm^2 at 10 Tesla(13) by the precipitation of the intermetallic compounds, compared with an increase from $2,5\cdot10^1$ up to 10^2 by vacancies agglomeration in Fig. 1.

ω-precipitation

ω-precipitation is obtained by a 400°C heat treatment of specimens previously homogenized. The 5- and 10-Tesla curves of the alloys TiNb50, TiNb49Cu3 and TiNb49Ge2,3 with different annealing time are illustrated in Fig. 4. In medium fields the ω-precipitation rises in all alloys the current carrying capacity to more than 10^4 A/cm^2, whereas in high fields at 10 Tesla the ternary alloys - especially the alloy containing copper which probably also precipitates the α-phase - are superior to TiNb50.

Dislocation networks and ω- + α-precipitation

α-precipitation in addition to ω-precipitation and dislocation network is achieved in all alloys after sufficient cold deformation preceding the 400°C heat treatment. After 97,2 % cold swaging and 400°C ageing (about 20 h) TiNb50 shows a critical current density in Fig. 5 at 10 Tesla of about $4\cdot10^3$ A/cm^2, TiNb49Cu3 of $1,3\cdot10^4$ A/cm^2 (about the same value as without cold deformation) and TiNb49Ge2,3 of $1,3\cdot10^4$ A/cm^2.

Other investigations

Measurements of transition temperature, of hardness, of specific electrical resistivity, of modulus of elasticity and electron microscopic and X-ray investigations have shown, that the kinetic of the precipitation of ω and α from the TiNb solid solution can be increased by copper addition and can be diminished by the addition of germanium (9,10). The homogenized copper containing alloy precipitates therefore during 400°C annealing already the α-phase together with the ω-phase (Fig. 4). In this case

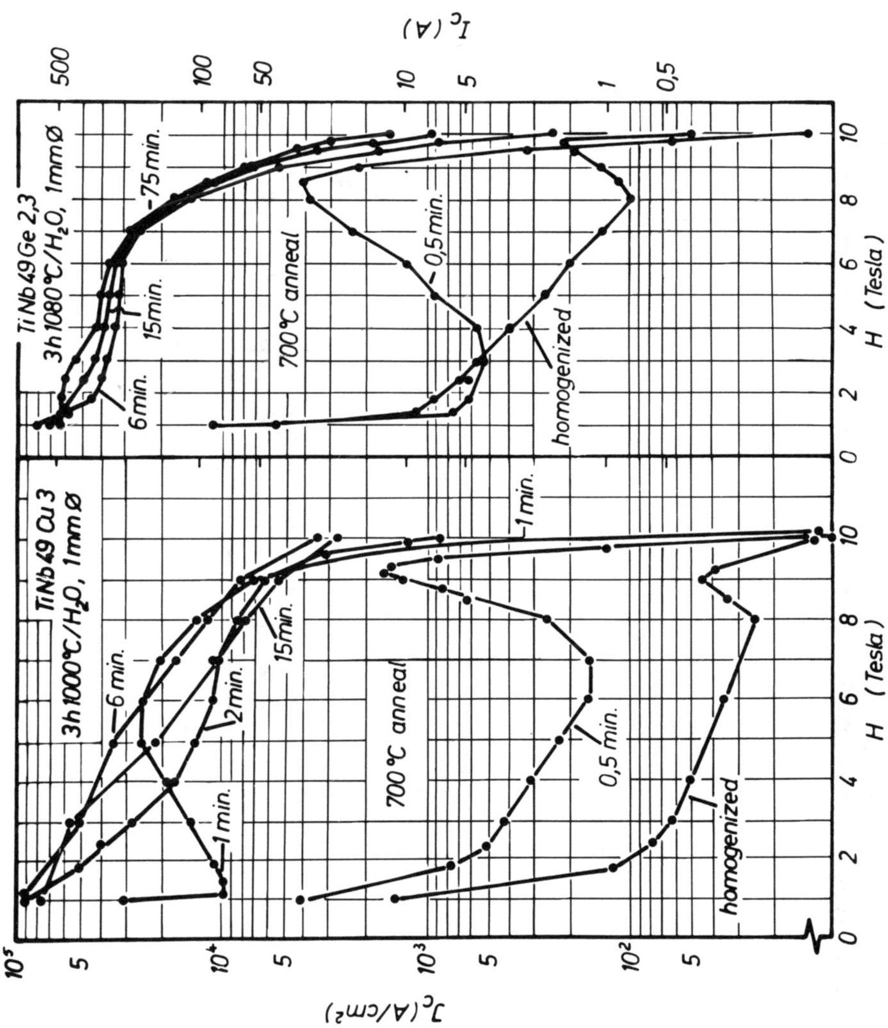

Fig. 3. Influence of precipitation of Ti_2Cu or Ti_5Ge_3 at 700°C on the J_c(H)-curves of TiNb49Cu3 and TiNb49Ge2,3.

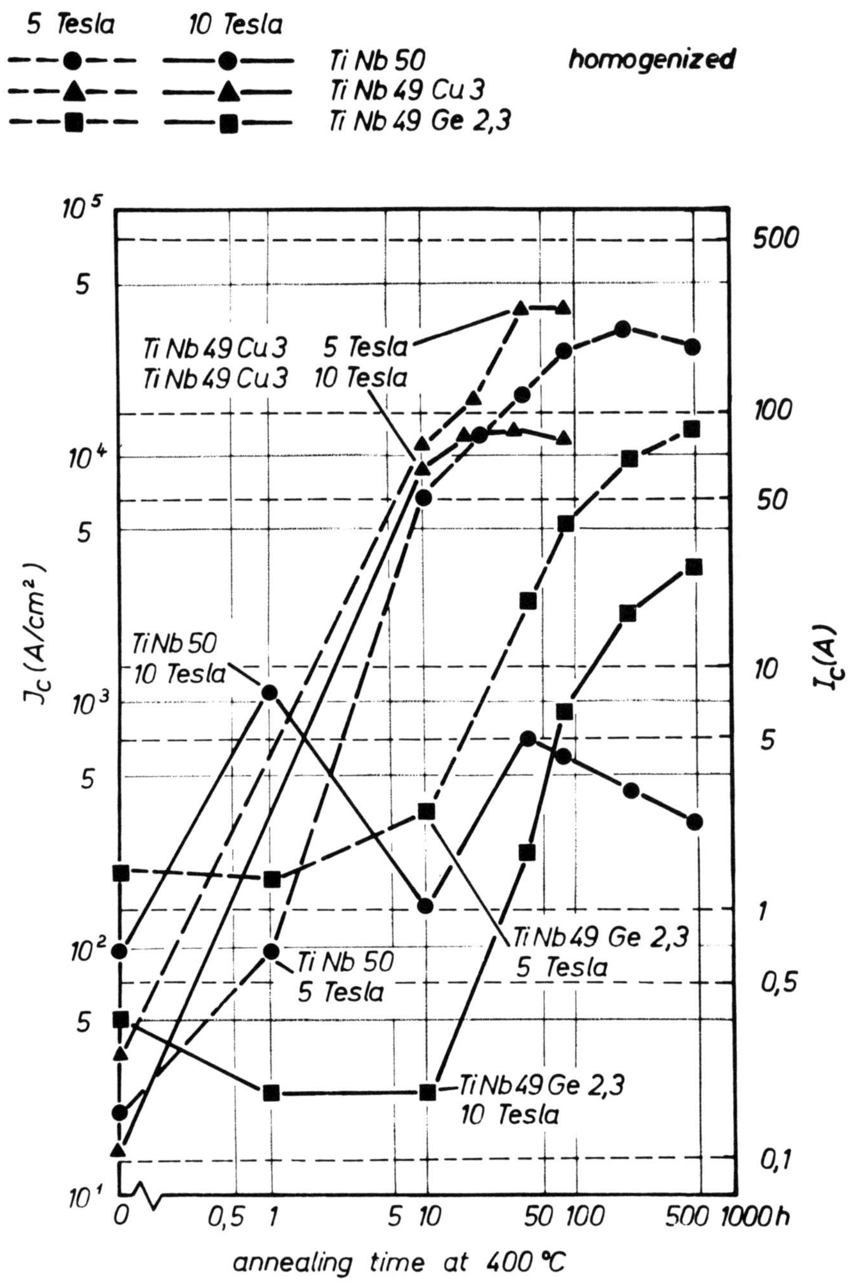

Fig. 4. Effect of annealing time at 400°C on the critical current of homogenized 3 h 1000°C - 1080°C/H_2O 1 mm dia wire samples of TiNb50, TiNb49Cu3, and TiNb49Ge2,3.

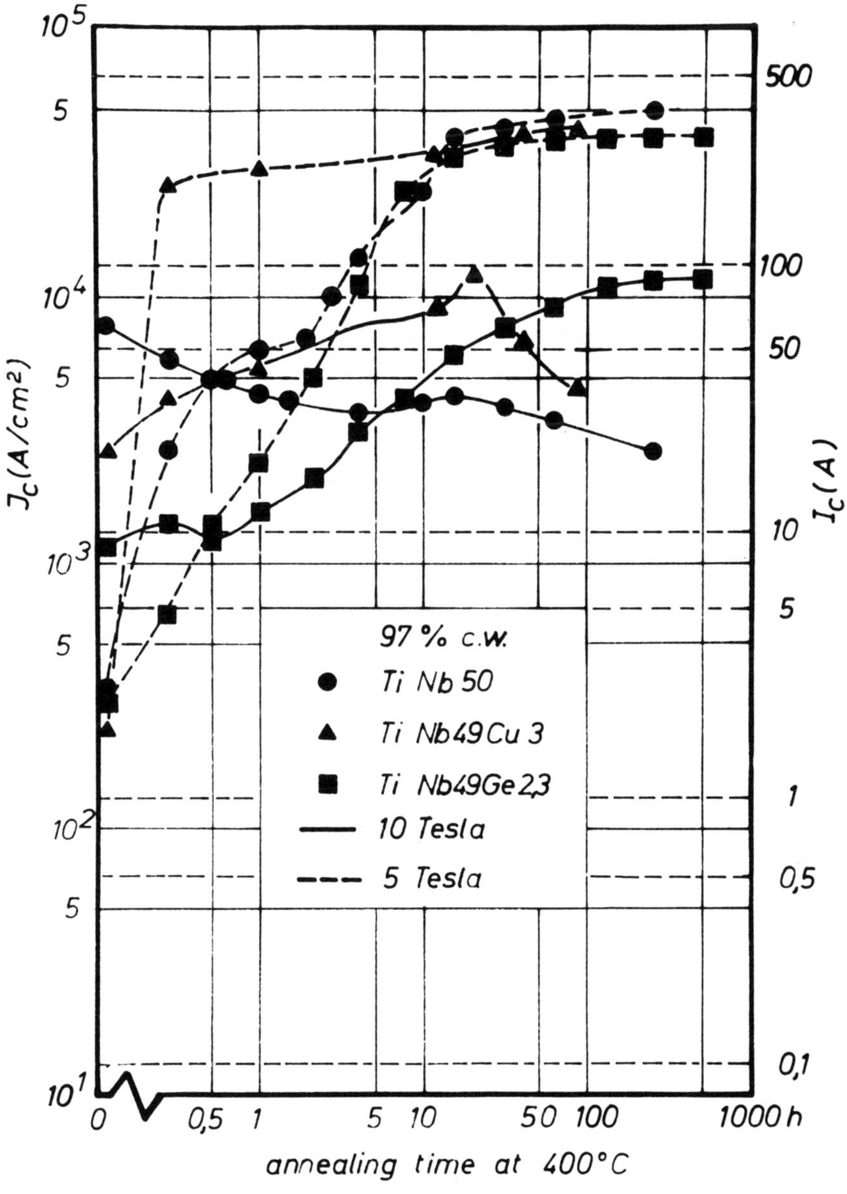

Fig. 5. Effect of annealing time at 400°C on the critical current of as cast cold worked (97% area reduction) 1mm dia wire samples of TiNb50, TiNb49Cu3, and TiNb49Ge2,3.

the critical current cannot be increased by additional cold working (Fig. 5) as is the case in alloys containing germanium addition.

Conclusions

The relation between the Lorentz-force J·H which acts under a current flow on the fluxons in a high current superconductor and the pinning forces of the different inhomogeneities can be approximated in the critical state by $J_c \cdot H \alpha \mathbf{F}_p (N_p)^{1/3}$, where F_p is the flux pinning parameter and N_p the number density of inhomogeneities. In a certain fixed field H the critical current J_c is a measure of the pinning forces of the precipitations, and with a known N_p the flux pinning parameter F_p can be determined as investigations of TiNb quaternary alloys have shown (8). It was impossible to make electron microscopic investigations on thinned samples of alloys with a high content of copper or germanium. As an electron microscopic investigation of a sample (TiNb49Ge1,2, as cast, cold worked, 97 % area reduction and 32 h 400°C annealing, $J_c \sim 7 \cdot 10^3$ A/cm² at 10 Tesla) in Fig. 6 shows, it is very difficult to estimate the number density of inhomogeneities, because dislocation networks, α- and ω-precipitation are intermixed. It is thought, that the beneficial effect

Fig. 6. Electron microscopic investigations of thin foils of TiNb49Ge1,2 (as cast, 97% area reduction, 32 h 400°C 1mm dia wire), left: parallel, right: perpendicular to wire axis.

of germanium and copper additions to the critical current is originated by dissolving of copper or germanium in the β-matrix and in the ω- and α-precipitation, hence increasing the pinning forces of these two sorts of inhomogeneities and of the dislocation network.

References

1. Dietrich, I., Weyl, R. and Zwicker, U., "Untersuchungen an Drähten aus supraleitenden Legierungen der Systeme Niob-Titan und Niob-Zirkonium", Z. Metallkunde, Vol. 53, 1962, pp. 721-728

2. Zwicker, U., "Supraleitung von Titan und seinen Legierungen", Z. Metallkunde, Vol. 54, 1963, pp. 477-483

3. Pfeiffer, I., Hillmann, H., "Der Einfluß der Struktur auf die Supraleitungseigenschaften von NbTi50 und NbTi65", Acta Met., Vol. 16, 1968, pp. 1429-1439

4. Gidley, J.A., Richards, P.N. and Barber, A.C., "Preliminary observations on the microstructure of the Nb-44%-Ti superconductor", Nature, Vol. 219, 1968, pp. 479-480

5. Reuter, F.W., Ralls, K.M.and Wulff, J., "Microstructure and superconductivity of a 44,7 at % Nb-54,3 at % Ti alloy containing Oxygen", Trans. Met. Soc. AIME, Vol. 236, 1966, pp. 1143 - 1151

6. Kramer, D., Rhodes, C. G., "Omega phase precipitation and superconducting critical transport currents in Ti-22 at % Nb", Trans. AIME, Vol. 239 1967 , pp. 1612-1615

7. Vetrano, J.B., Boom, R.W., " High critical current superconducting Ti-Nb-alloys", J. Appl. Phys., Vol. 36, 1965, pp.1179-1180.

8. Picketts, R.L., Courtney, T.H., Shepard, L.A., Wulff, J., "Correlation of structure and superconducting properties in Nb(Cb)-Ti quaternary alloys", Metal. Trans. AIME, Vol. 1, 1970, pp. 1537-1544

9. Zwicker, U., Löhberg, R., Heller, W., "Metallkundliche Probleme und Supraleitung bei Legierungen auf Basis Titan-Niob, die als Werkstoffe für die Herstellung von supraleitenden Magneten dienen können", Z. Metallkunde, Vol. 61, 1970, pp. 836-847

10. Löhberg, R., Heller, W., Zwicker, U., "Investigations on superconducting Ti-Nb alloys with additions of third elements", Proceedings Third Int. Cryogen. Eng. Conference, Berlin, 20.-25.5.1970

11. Baranov, I.A., Boikowa, K.I., Guts, Z.A., Kurakina, L.A., Kogan, A.V., Fogel, A.A., Tsypkin, S.I., Shmulevich, R.S., "Investigations of superconducting characteristics in Nb-Ti-alloys", Fizika Tverd. Tela, Vol. 10, Nr. 2, 1968, pp. 404-410

12. Zwicker, U., "Einfluß von Platinmetallen auf technologische Eigenschaften und Leitfähigkeit von Titan-Niob-Legierungen", Z. Metallkunde, Vol. 56, 1965, pp. 222-228

13. Heller, W., Löhberg, R., Babiskin, J., "Einfluß von Zusätzen dritter Elemente auf das kritische Magnetfeld von Ti-Nb-Legierungen, J. Less Common Metals, Vol. 24, 1971, pp. 265-276

PECULIARITIES OF CHANGES IN ELASTIC PROPERTIES OF TITANIUM MARTENSITE

S. G. Fedotov

A. A. Baikov Institute of Metallurgy

USSR Academy of Sciences, Moscow

As is well known, alloys of titanium with many transition elements, such as vanadium, niobium, tantalum, chromium, molybdenum, tungsten, iron, manganese, nickel, etc., when quenched from the β -region, form several metastable phases. The phases may be as follows: martensite α' , ω -and supercooled β -solid solution.

Owing to the considerable effect of these phases on the properties of alloys, the distinguishing features of their crystalline structure and the mechanism of their formation, the problem of the metastable state and phase transformations in titanium alloys continues to this day to attract the attention of many researchers. In this connection the results of a systematic study of the module of elasticity of metastable titanium alloys with β-isomorphous and eutectoid-forming elements may apparently evoke a certain interest. This is so much the more that elastic properties proved to be sensitive to changes in the composition and phase structure of quenched alloys.

Materials and methods of investigation

Alloys weighing 30 g were melted in an arc furnace with a non-consumable tungsten electrode in an atmosphere of purified argon. Each of the alloys was remelted at least three times. A titanium sponge with a 99.85% degree of purity and titanium iodide were used as the base of the alloys. Impurities of the sponge, w/o: H - 0.002; N - 0.04; Si - 0.09.

The charge material, electrolytic chromium and manganese, were loaded into little titanium glasses specially made for the purpose, covered during arc smelting with titanium lids.

The chemical composition of the alloying elements was as follows (w/o): Mo - 99.9; V - 99.5 with impurities: Al - 0.005; Fe - 0.04; Si - 0.02; C - 0.1; O - 0.3; N - 0.01; Nb - 99.4 with impurities: Ta - 0.2; Si - 0.06; Fe - 0.05; Nickel - 99.99; Iron - 99.99; chromium and manganese (98.83) - electrolytic; cobalt - 99.5.

The alloy ingots were forged into rods, 9 mm in dia., and heated in the air up to 900-1000°C. Cylindrical specimens 6 mm in dia. were turned on lathes and these were burnished. The composition of the alloys, shown on charts according to calculation, was controlled by weight and density determined on the specimens by the hydrostatic method; a selective chemical analysis was also made. The specimens were annealed and heated for quenching in evacuated quartz ampoules.

The specimens were quenched, breaking the ampoules in water of room temperature, after 24 hrs of ageing time at 900°C. The specimens were annealed in accordance with the following conditions: 900°C - 24 hrs; 800°C - 100 hrs; 700°C - 200 hrs, and at 600°C - 500 hrs, with subsequent cooling down to room temperature with the specimens remaining in the furnace.

The module of normal elasticity E and the module of transverse elasticity G were calculated by the density of the alloys and by the frequencies of the longitudinal and torsional oscillations of the specimens, determined by the resonance method on an "Elastomat" device.

The accuracy of determining the modules of elasticity is within 1%.

Elastic properties of alloys of β-isomorphous systems

The results of studying the modules of elasticity in alloys of the Ti-Mo, Ti-V and Ti-Nb systems in an annealed state and after quenching from a single-phase β-region from 900°C are given in Fig. 1. The properties change in a similar way in each of the above states, which is due to the like interaction of titanium and molybdenum, vanadium and niobium. We also obtained similar relationships for another β-isomorphous system Ti-Ta.

The above data, which are in agreement with the results of paper [1], make it possible to single out three regions in each system. In the first one the modules of elasticity of the alloys intensively and continuously diminish with an increase in the content of the alloying elements. All the alloys of the first region have a needle-shaped microstructure typical of martensite. The lines of the α-phase on the X-ray diffraction patterns then become blurred, and some of them even bifurcate as supersaturation increases [2]. That is why the drop in the elastic characteristics of the martensite region alloys is brought about by supersaturation of α-Ti with alloying elements.

The second region is distinguished by its relatively high indices of modules of elasticity. It occupies an intermediate position between the alloys which undergo a martensite transformation upon being quenched, on the one hand, and those which did not undergo any transformation, on the other. The alloys of the intermediate region are of a more intricate phase structure. The alloys corresponding by composition to the ascending branch of curves exhibit a needle-shaped α'-phase along with non-transformed crystals of a high-temperature β-solid solution. As the content of the alloying elements rises, the relative share of the α'-phase in them diminishes, and that of the β-phase increases. The structure of the alloys corresponding by composition to a descending branch, as a rule, reveals microscopically only a β-solid solution. X-ray examination additionally discloses in them an ω-phase which precisely induces an increase in the modules of elasticity. The amount of the ω-phase can be appraised only qualitatively, for example, by the level of the modules of elasticity of the alloys and by the intensity of the lines on the X-ray diffraction patterns.

The third region is made up of alloys with considerably greater alloyage, with a structure of supercooled high-temperature β-solid solution. The alloying elements increase the modules of elasticity of alloys in the β-region, and the greater their own

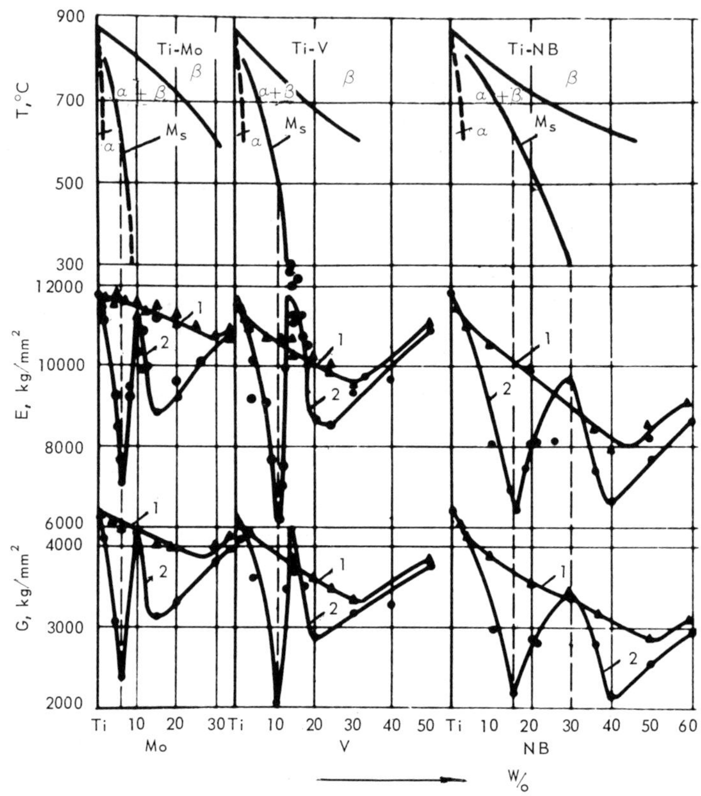

Fig.1. Change in Young's module E and in the module of transverse elasticity G of alloys of the Ti-Mo, Ti--V and Ti-Nb systems, depending on the composition and phase structure:

1 - annealed state; 2 - after quenching from the β-region.

elastic constants, the more intensively. In less alloyed β -solid solutions, the decomposition processes are underway even at room temperature [3].

It is noteworthy that in the first martensite region the modules of elasticity in alloys of all systems decrease to an almost similar minimal level. For a module of normal elasticity this level is close to 6000 kg/mm^2.

The wide occurrence of the first martensite region is determined by the solubility of the components in α-Ti; the less the component dissolves in α-Ti, the less its composition. The established regularities of changes in modules of elasticity allow to confine this region to approximately the following concentrations, w/o (at.%): 6 (3.1) Mo; 12 (11.3) V; 17 (9.6) Nb; 28 (10) Ta. The maximum solubility of these components in α-Ti does not exceed, as is well known, w/o(at.%): 1 (0.5) Mo, 2 (1.9) V; 3.5 (1.6) Nb. The data on the solubility of Ta in α-Ti are very contradictory [4-6].

A definite quantitative interconnection is revealed between the boundary concentrations of C α' and the maximum solubility of C α components by comparing them: the ratio C α': C α for all the systems amounts to the same magnitude, namely to about six.

The results illustrating the continuous decrease in the modules of elasticity in connection with supersaturation of α-Ti to a similar minimal level in all the systems, when the alloying elements limiting this region are directly linked with their maximum solubility in α-Ti, warrant the assumption that these supersaturated α' -solid solutions and concentrations are maximal. Taking into consideration the above correlation, it may be supposed that the solubility of Ta in α-Ti should not exceed 5 w/o.

Supersaturated α' -solid solutions of a predetermined composition are formed as a result of martensite transformation of the alloy at a martensite temperature appropriate for this composition. For this reason it is more correct to relate the compositions C α' and C α to be compared to one and the same temperature. For example, at 600°C, the following is dissolved in α-Ti (w/o): 2 V, 1 Mo and 3.5 Nb, while at temperature Ms = 600°C, the following compositions are dissolved: 9 w/o V, 4.5 w/o Mo and 15 w/o Nb. The ratio of these concentrations changes to some extent, but it is also of a similar magnitude for all the systems.

It is worth noting that in the Fe-Ni system, similar by the type of the diagram of the state to the β-isomorphous systems of titanium, the elastic properties of the martensite region alloys likewise diminish intensively with an increase in supersaturation, while the intermediate region is characterized by relatively high indices modules of elasticity (Fig. 2). Typical of iron-nickel martensite is two-fold supersaturation.

The appearance of needle-shaped martensite α'-phase in titanium alloys quenched from the β-region is not confined, however, to the above limits. It is also revealed with a considerably higher content of the alloying elements. In the systems Ti-Nb and Ti-Ta, the α'-phase can be observed, for instance, in the case of 30 w/o Nb or of 50 w/o Ta and even of higher niobium and tantalum content.

Attempts to determine the chemical composition of the α'--phase by the X-ray method on a MS-46 microassayer did not lead to an unambiguous conclusion because of the lack of a local solution. However, the change in the modules of elasticity in the course of continuous heating of metastable titanium alloys at the rate of about five deg/min indicates that the temperature of the onset of martensite decomposition, as appraised by the intensive growth of E and G, is determined by its supersaturation.

In all β-isomorphous systems, the temperature of the onset of the first region martensite decomposition regularly drops with an increase in supersaturation, and at maximum concentrations of the alloying elements in alloys it reaches the following level: with 12 w/o V $\sim$ 280°C, with 6 w/o Mo $\sim$ 400°C, with 17 w/o Nb $\sim$ 530°C, and with 28 w/o Ta $\sim$ 620°C.

At higher contents of the alloying elements in the alloys, the temperature at the onset of decomposition of the martensite α'-component practically does not change and remains at the one indicated level for the given system. This fact may originate from the closeness of the α'-phase composition to the maximum concentration. Such a conclusion is supported by the increment in the modules of elasticity upon martensite decomposition. In the first region it increases with a rise in the content of the alloying elements the alloys up to maximum concentration. With additional alloying, the increment in E and G, quite reversely, diminishes. In this case there is a greater change in the modules of elasticity,

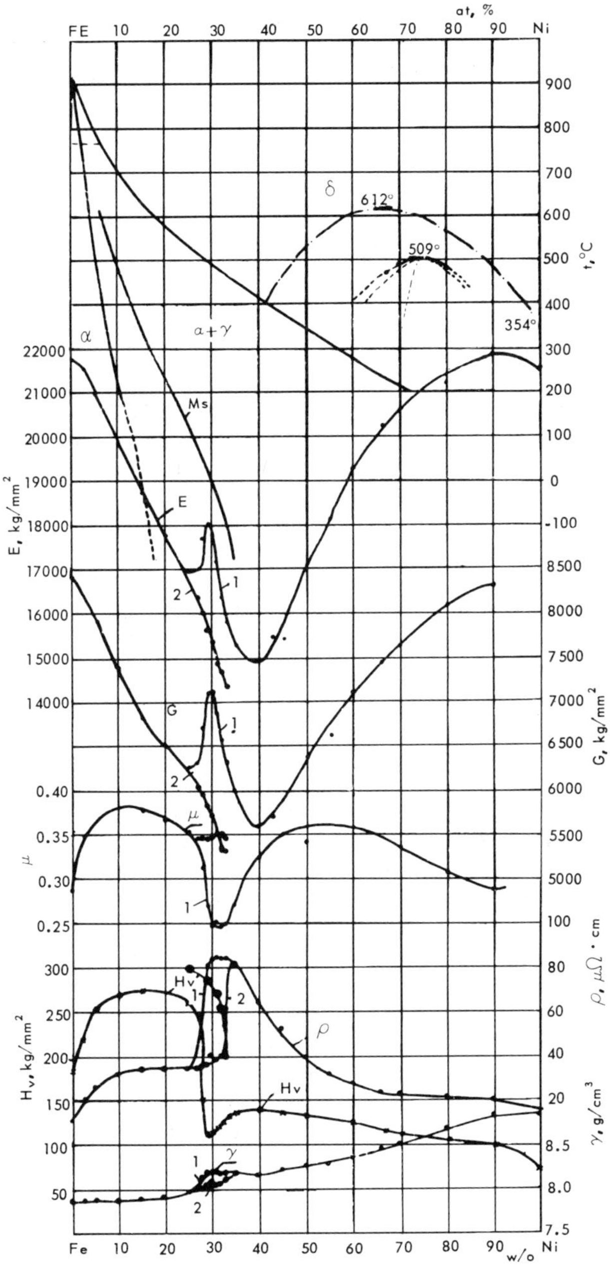

Fig.2. Change in Young's module E, the module of transverse elasticity G, Poisson's ration, the specific electrical resistance ρ , hardness Hv and density γ of alloys of the Fe-Ni system, quenched in water from 1000°C -1, after their additional cooling in liquid nitrogen -2.

brought about by decomposition of the unstable β-solid-solution, since its relative amount in the structure of hardened alloys of the intermediate region increases, while that of the α'-phase diminishes.

Elastic properties of quenched titanium alloys of eutectoid systems

In the case of eutectoid systems, the modules of elasticity in quenched alloys of the first martensite region decrease upon alloying to a different level in each of the systems, and to a lesser degree as compared with alloys with β-isomorphous components (Fig. 3). For example, in the Ti-CO and Ti-Ni systems, the elastic properties of martensite remain at the level of unalloyed titanium and do not depend on the composition of the alloy. We account for the smaller decrease in the modules of elasticity of eutectoid system martensite by the fact that in martensite transformation of alloys during quenching from the β-region, it is not possible to produce in these systems supersaturated α'-solid solutions which would correspond to the initial composition of the alloys. Such a conclusion follows from examining the diagrams of the state and solubility of the components in α-Ti.

The diagrams of the state of eutectoid systems with regard to solubility consist, as it were, of two parts. The upper part up to eutectoid temperature is similar to the diagrams of the β-isomorphous systems. It points to an increase in the solubility of the components in α-Ti upon a drop in temperature. Below the eutectoid temperature of the elements the solubility quite the contrary, goes down. The composition of martensite is related to the solubility of the components in α-Ti, as demonstrated in the case of β-isomorphous systems.

We therefore assume that upon martensite transformation of eutectoid system alloys, first α'-solid solutions supersaturated to the maximum are likewise formed and then, upon a change in solubility, they decompose, as the temperature drops below the eutectoid one. The higher the temperature of eutectoid transformation, the greater the degree of martensite decomposition. Thus supersaturation of martensite is eliminated to a different extent for the different systems, as well as the cause which induced such an appreciable decrease in the modules of elasticity.

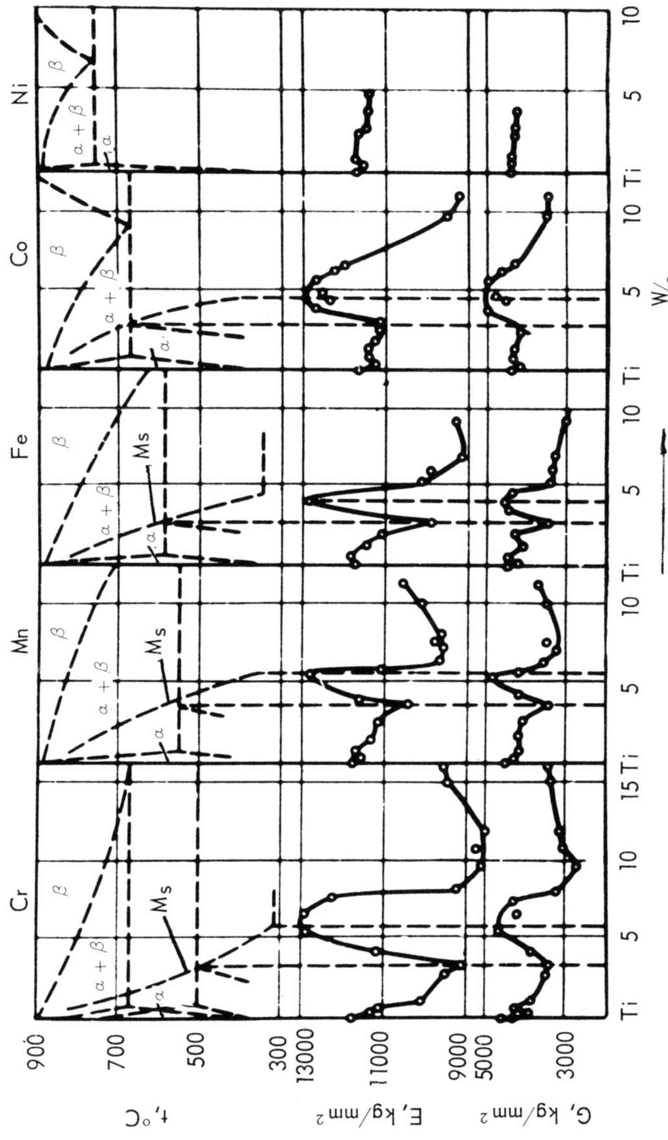

Fig.3. Change in Young's module E and in the module of transverse elasticity G of alloys of the Ti-Cr, Ti-Mn, Ti-Fe, Ti-Co and Ti-Ni systems after quenching from the β -region from 1000°C.

For this reason no anomalies pointing to a decomposition of supersaturated α' -solid solutions during the heating of alloys are exhibited on the temperature dependence of the modules of elasticity and the specific electrical resistance of martensite of most eutectoid systems.

The greatest supersaturation of martensite is achieved in the Ti-Cr system; and it is only the Ti-Cr system that demonstrates martensite decomposition upon heating accompanied by an increase in the modules of elasticity., It follows from our data that the Ti-Cr system requires a lower temperature of eutectoid transformation as also referred to earlier paper [8].

Conclusions

A systematic study was made of the elastic characteristics of titanium alloys in an annealed state and after quenching from a β -region and of establishing the regularities of their change, depending on the composition and phase structure of the alloys. As a result, an interconnection has been shown between the metastable and equilibrium state of alloys of the β -isomorphous and eutectoid systems. Our findings also point to the inevitability of decomposition of eutectoid system martensite in the process of martensite transformation of alloys, unlike martensite with β -isomorphous components.

References

1. Knorr, W. and Scholl, H. "Untersuchungen zum Umwandlungsverhalten von Titan-Molybdän und Titan-Vanadium - Legirungen," Zts. Metallkunde, 51, 10, 1960, S. 605.

2. Bagaryatsky, Yu.A., Tagunova, T.V. and Nosova, G.I., "Metastabilnyye fazy v splavakh titana s perekhodnymy elementami," sb. "Problemy metallovedeniya i fiziki metallov," vyp.5, Metallurgizdat Publishing House, 1968, p. 210-234.

3. Fedotov, S.G. and Konstantinov, K.M., "Raspad nestabilnovo β -tverdogo rastvora titana s 18 ves.% vanadiya," Doklady AN SSSR, Vol. 192, No. 3, 1970, p. 555-558.

4. Summers-Smith, D., "The Constitution of Tantalum-Titanium Alloys, J. Institute of Metals," Vol. 81, 1952, pp. 73-76.

5. Summers-Smith, D., "The Constitution of Tantalum-Titanium Alloys, J. Institute of Metals, Vol. 81, 1953, p. 426.

6. Maykuth, D.J., Ogden, H.R. and Jaffee, R.I., "Titanium-Tungsten and Titanium-Tantalum Systems," J. Metals, Vol. 5, No. 2, 1953, pp. 231-237.

7. Fedotov, S.G., Konstantinov, K.M. and Sinodova, E.P., "Raspad titanovo-vanadiyevo-martensita pri nepreryvnom nagreve," Zhurnal "Fizika metallov i metallovedeniye," Vol. 25, vyp. 25, 1968, pp. 860-866.

8. McQillan, M.K., "A Redetermination and Interpretation of the Titanium-Rich Region of the Titanium-Chromium System," J. Institute of Metals, Vol. 82, 1954, pp. 433-439.

INTERNAL FRICTION STUDIES IN TITANIUM

AND TITANIUM BASE ALPHA AND BETA ALLOYS

S. Mishra and M.K. Asundi
Metallurgy Division
Bhabha Atomic Research Centre
Trombay, Bombay-85
India

Internal friction studies have been carried out on pure titanium and titanium base alpha, alpha plus beta and beta alloys. It has been shown that interstitial solutes in pure titanium cannot cause internal friction peaks either individually or in association with one another. A substitutional solute must be present along with the interstitial solute to cause internal friction. Interaction peaks associated with substitutional-interstitial complexes are observed and the activation energy for the oxygen atom jump is found to depend upon the nature of the substitutional atom. By selecting solutes like Si, Ge and Sn belonging to the same group, the elastic contributions to internal frictions have been measured. Larger the insolution size difference between the substitutional solute atom and the titanium atom, lower are the peak temperature and the activation energy. Snoek peaks are observed for the bcc beta phase in alloys of Ti-V, Ti-Mo and Ti-Al-V, containing oxygen, nitrogen and carbon. Internal friction and modulii ratio changes with temperature have been used in studying the location of the phase boundaries and also the nature of the phase transitions occurring in these alloys. Preliminary investigations have been made to study the omega transformations occurring in some of these alloys with different cooling rates. Martensitic transformation in Ti-6Al-4V has also been investigated.

Introduction

Internal friction studies in titanium have attracted the attention of several investigators (1-4). Miller and Brown (4) have suggested that interstitial impurities like oxygen and carbon can cause internal friction peaks in titanium. On the contrary, others like Bratina (3), Gupta & Weinig (2) and Pratt et al (1), have observed that interstitial solutes by themselves cannot give rise to internal friction peaks without the presence of substitutional solutes in the matrix. Thus this work was taken up to clarify this situation in very pure titanium and in alloys with controlled interstitial and substitutional solutes. Internal friction and modulii ratio changes with temperature have also been made to study the metastable and equilibrium phase transformations in some of the binary and ternary titanium alloys.

Experimental

Iodide pure titanium, containing about 0.01(wt%) oxygen, and iron 0.024 (wt%) as the largest single impurity was used. Other twelve metallic impurities totalled to less than 0.0350(wt%). The same iodide material was used for making titanium-boron, and titanium-boron-germanium alloys. For the rest of the alloys containing higher amounts of alloying elements sodium reduced titanium crystals were used. The chemical analysis of the titanium used in the investigations is given in Table I.

Table I. Chemical Analysis of Titanium (Sodium Reduced & E.B. Melted)

Chemical Composition (%)

Al, Cu, Mg, Cr & Fe each	Co	Mn, Pb, Sb & Sn each	Si	V	Ni	O	N	C
0.005	0.001	0.001	0.012	0.013	0.002	0.065	0.001	0.011

A Kê type torsional pendulum enclosed in a vacuum chamber was used for measurements at and above room temperatures while an inverted type equipment, similar to the one developed by Koiwa (5), was used for low temperatures. The range of frequencies used varied from 0.5Hz to 10Hz. Visual observations were made with the Ke type apparatus while automatic recording of the decay of oscillations was made with the Koiwa type pendulum.

Results and Discussion

Internal friction spectrum (Fig. 1) for iodide pure titanium shows no peak other than the grain boundary plateau when the oxygen content is low. As the oxygen content is raised to 1.6 wt%, a peak appears at 470°C (1.3Hz) which is due to the Fe-O complex, as already reported by Gupta and Weinig (2) and by Bratina (3). The peak at 420°C (0.8Hz) reported by Miller and Brown (4) for iodide pure titanium (0.02 wt% Mn) has not been observed. An alloy prepared from iodide titanium and .07 wt% manganese shows (Fig. 2) the Mn-O peak at 430°C (at 1Hz). Thus it is very likely that the peak obtained by Miller and Brown (4) is due to Mn-O complex rather than due to oxygen pairing mechanism as suggested by them. Fig. 2 also shows peaks due to Si-O, Sn-O and Sn-N complexes. Fig. 3 shows the peaks due to Ge-B and Ge-O interactions. Increased solubility of boron by increasing the solutionising temperature is reflected in the increased peak heights of the Ge-B complex. If the oxygen content is high, then the boron peak gets suppressed. There is no B-O complex in the temperature range studied and hence boron is assumed to occupy interstitial positions. Boron alone also does not give any peak in titanium. In the titanium-0.5 wt% Mo alloy the internal friction peaks due to Mo-C, Mo-O and Mo-N interactions are shown in Fig. 4.

The Snoek relaxation peaks for the molybdenum, and vanadium stabilised beta-titanium are shown in Figs. 5 and 6. In addition to these peaks, the Ti-43 wt% Mo alloy shows a maxima around 270°C, which is due to the beta-omega transformation and occurs during normal heating and cooling cycles but is found to give higher peak heights during faster cooling rates. The Ti-56 wt% V alloy also shows a maxima at 300°C and is due to the beta-alpha transformation, or beta to omega transformation which is unlike the earlier one is diffusion controlled (observing the effect of cooling rates on the peak heights associated with the transformation). The change in the shear modulii ratio is also shown in Figs. 5 & 6. Table II gives the pertinent data regarding activation energy, frequency factor and diffusion constant.

Omega transformations were studied in a Ti-17 wt% Mo alloy and Fig. 7 shows a slow cooled sample from the beta phase. Peaks occurring at 530°C, 470°C and 235°C are due to beta-alpha, beta-omega and omega-alpha transformations respectively. Modulii ratio has also been indicated in the diagram. When the same alloy is cooled at a faster rate from various soaking temperatures (in effect, changing the composition of the beta phase), the height of the beta-omega transformation peak, changes as indicated in Fig. 8. Peaks appear at 440°C for samples cooled from 620°C and is found

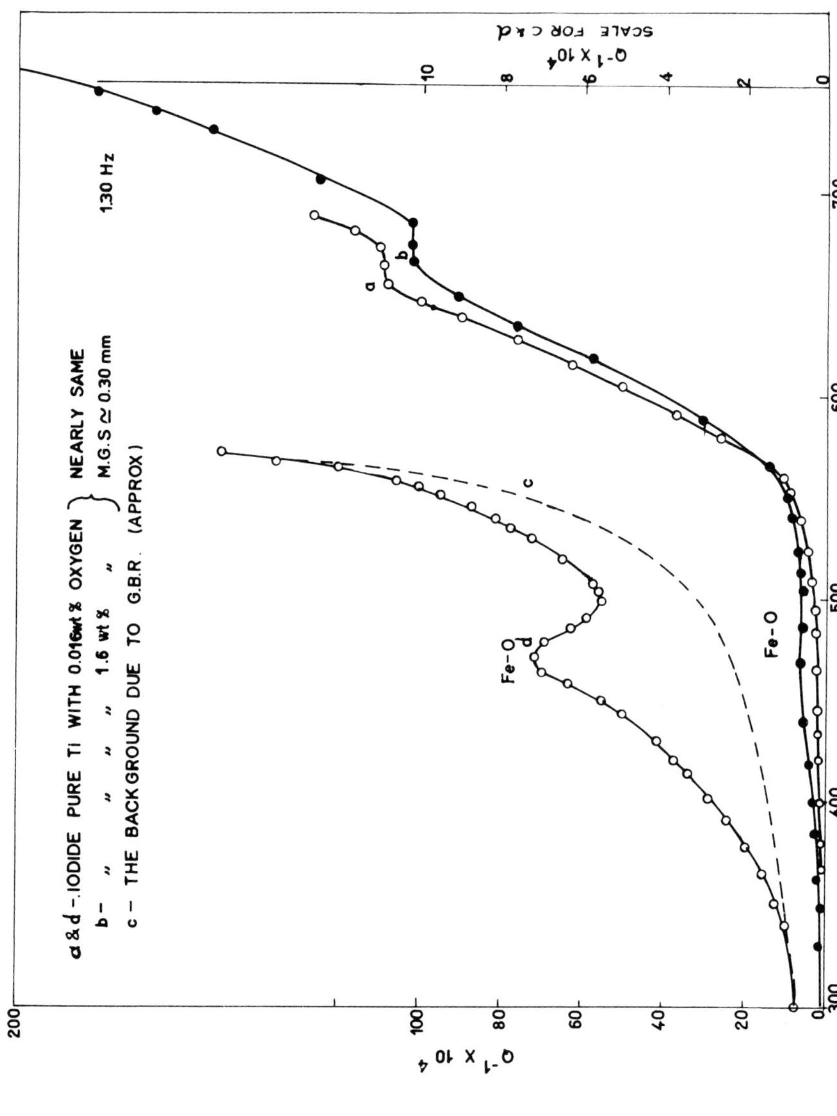

Fig. 1. Internal friction curve for pure titanium containing small amount of iron as impurity. Fe-O complex gives the peak.

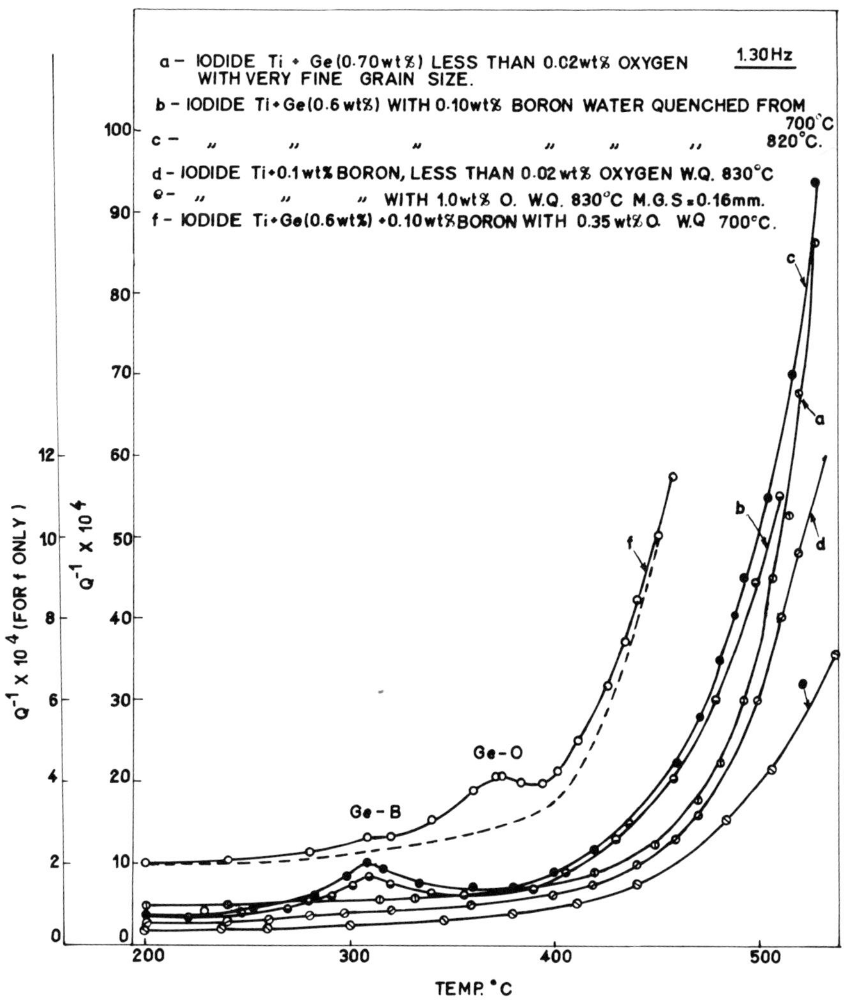

Fig. 2. Peaks due to Ge-O and Ge-B complexes. Effect of oxygen on boron is shown.

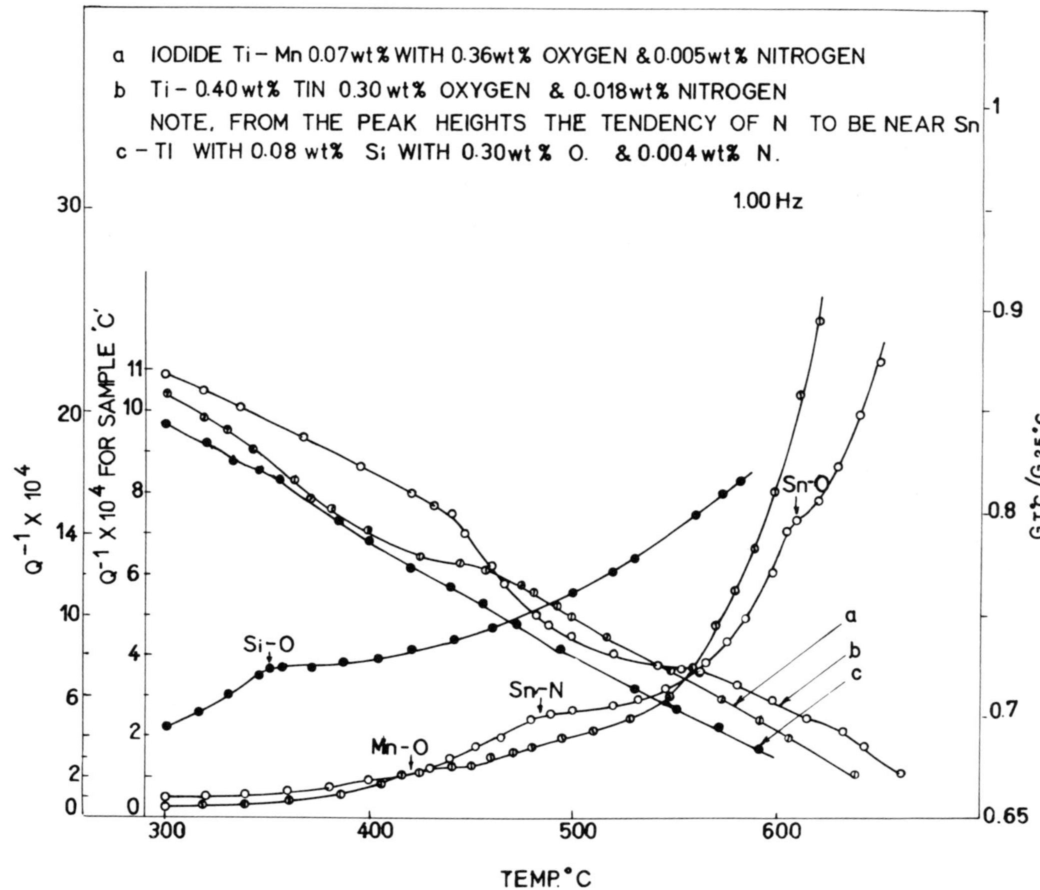

Fig. 3. Peaks due to Mn-O, Sn-O, Si-O, and Sn-N. Modulii ratio also indicated.

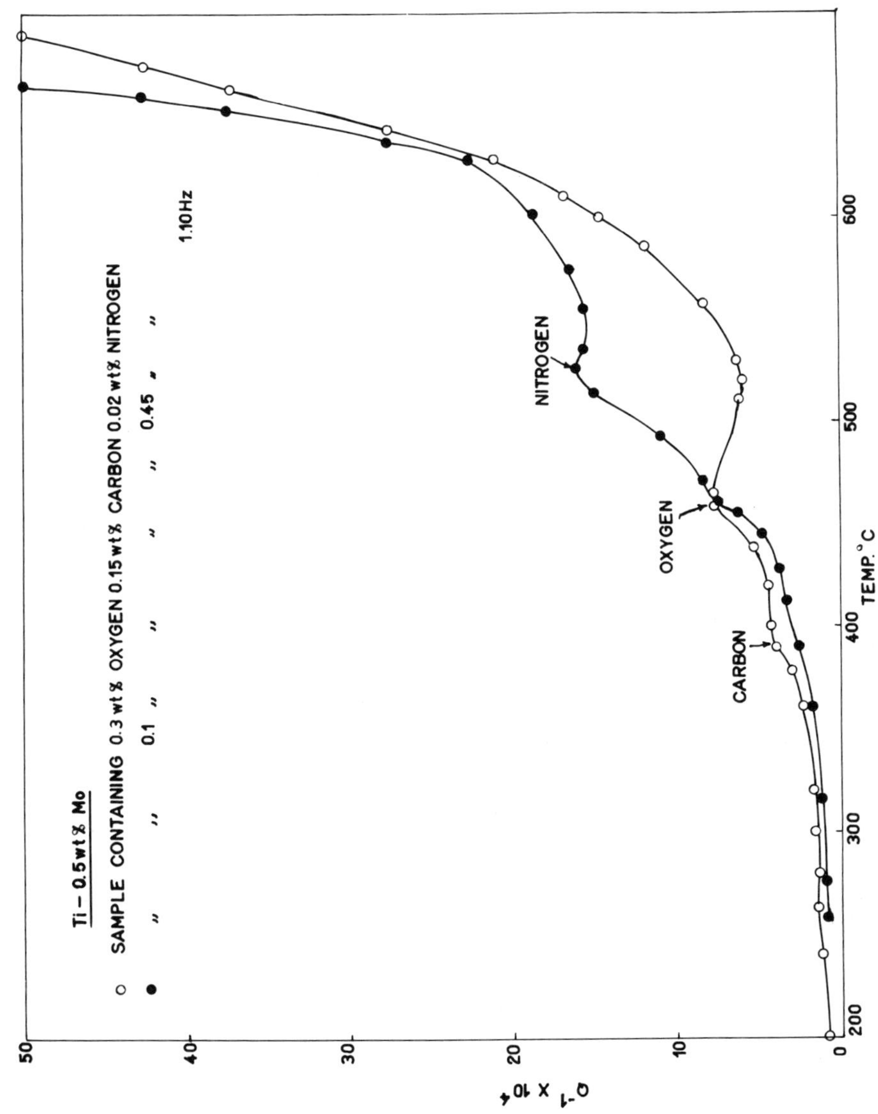

Fig. 4. Peaks due to Mo-N, Mo-O, and Mo-C.

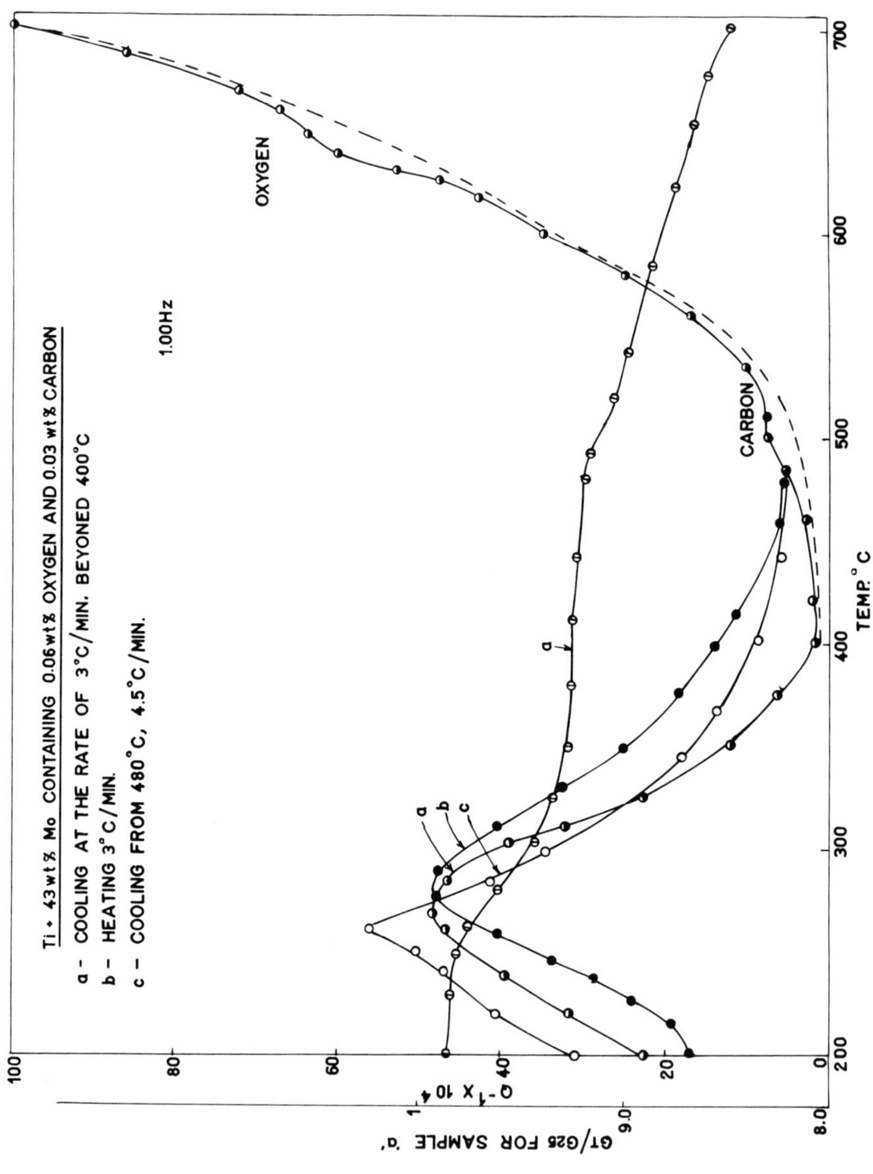

Fig. 5. Snoek peak for oxygen and carbon in beta, and omega transformation peaks for different cooling rates. Modulii ratio also shown with temperature.

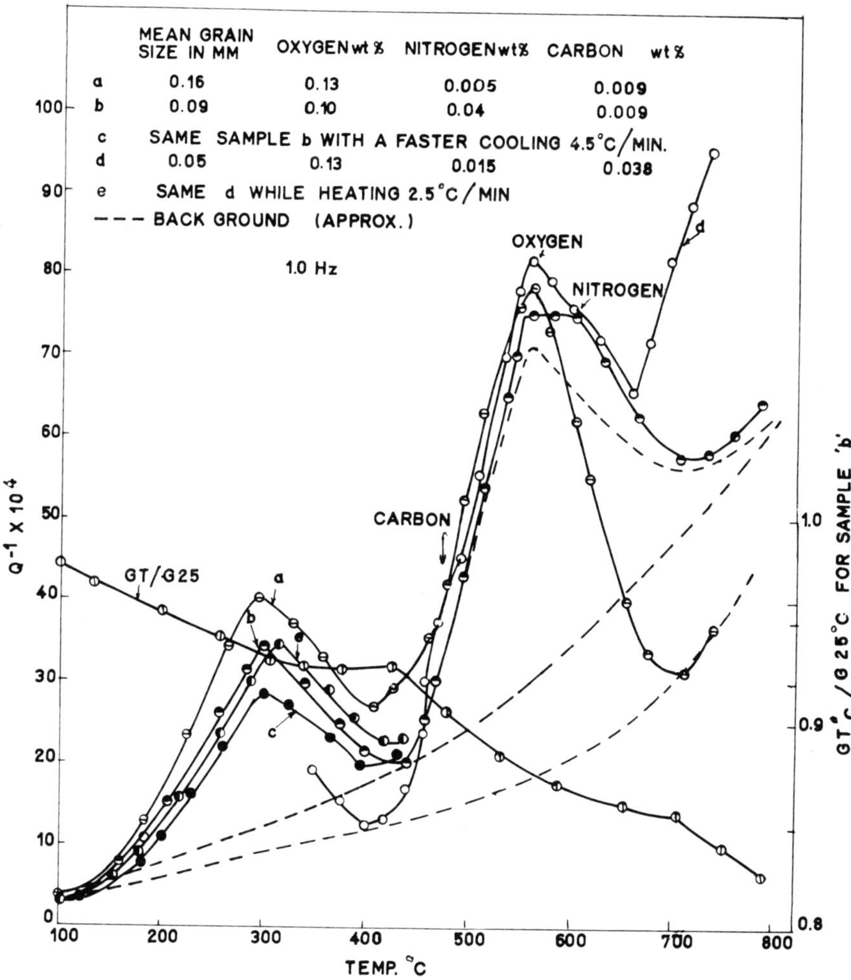

Fig. 6. Ti-56 wt % V alloy showing Snoek peaks due to carbon, oxygen, and nitrogen. Transformation peaks appear at 300°C (heating and cooling). Modulii ratio also shown.

Table II. Activation Energy, Relaxation Time, and Diffusion Constant.

Alloy System	ΔH Kcals/mol	τ_o Frequency factor sec.	Do cm^2/sec Ref.(21)	Remarks
1. Iodide Ti (Fe-0.024%, Oxy-1.60%)	52.5	$5x10^{-17}$	-	-
2. Ti+Ge(0.6%) B-0.1% Oxy 0.02%	39.0	$4x10^{-16}$	-	-
3. Same as 2 Oxy-0.35%	42.5	-	-	-
4. Ti-Si(0.08%) Oxy-0.35%	40.5	-	-	-
Ti-Sn(0.4%) Oxy-0.35%	65.0	-	-	ΔH for the oxygen jump process is larger for smaller the size difference in case of Si, Ge and Sn, "Insolution radii" are 1.41, 1.44 & 1.48 Å resptvly.
5. Ti-Mo(0.5%)				The diffusion parameter is not representative of the long range diffusion. Similar results are found for Zr.(20)
(a) O-0.3%	46.0	$8x10^{-17}$		
(b) N-0.45%	53.0	$1.1x10^{-17}$		
(c) C-0.15%	43.0	$2x10^{-16}$		
6. Ti-43%Mo				The ΔH value is close to that given by William et al (18) and the Do value lies in between that given by(17) &(18)
O-0.06%	Oxy-68±2	$5x10^{-21}$	$5.8x10^{3*}$	
C-0.03%	C-50	$6x10^{-18}$	4.85	
7. Ti-56V(%)				
(a) O-0.13%	Oxy - 56	$3x10^{-20}$	-	-
(b) N-0.04%	N - 60	-	-	-
(c) C-0.04%	C - 49	-	-	-
8. Ti-6Al-4V beta quenched	Plateau at 675°C-70	-	-	The plateau may be due to needles of martensites sliding with each other, the ΔH close to the ΔH for G.B.R(1) in Ti.

* The lattice parameter "a" taken from Hansen et al (19) at room temperature for the alloy Ti-43 Mo

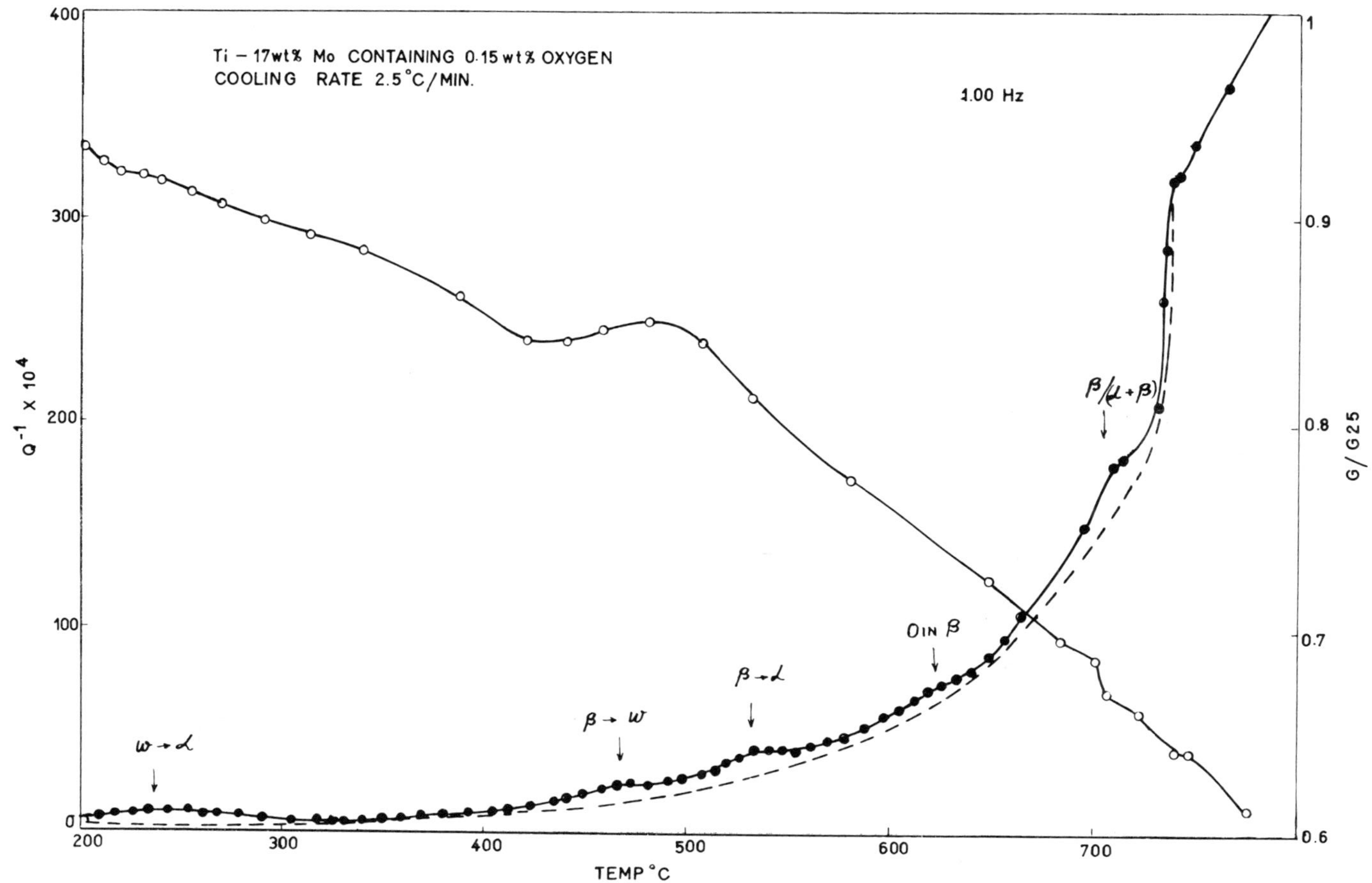

Fig. 7. Transformation peaks in Ti-17 wt % Mo on slow cooling. Modulii ratio is also indicated.

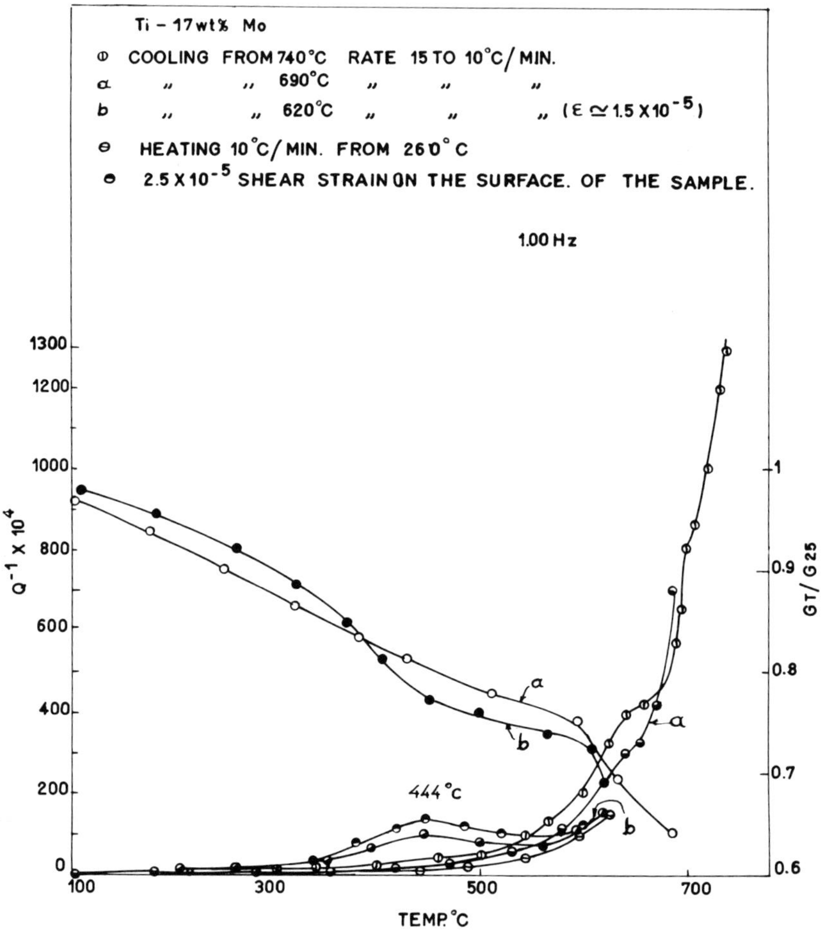

Fig. 8. Same as for Fig. 7, but faster cooling rate. Beta-omega transformation peak at 444°C. Seen only for temperature of soaking below 620°C. Disappear on heating.

to be strain amplitude dependent. However, when the soaking temperature is lowered from 540°C to 430°C, the beta-omega transformation temperature is observed to shift from lower to higher temperatures, for the same cooling rates (Fig. 9). The magnitude of the peak is dependent upon the quantity of beta at the soaking temperature. For the same soaking temperature, faster cooling rates shift the peak position to low temperature sides but the peak heights increase. On the other hand, the beta-alpha transformation, being diffusion controlled, shows reverse trends in internal friction maxima. Same is the case for omega to alpha transformation for slow cooling.

Figs. 10, 11 and 12 show the internal friction and modulii ratio vs temperature curves for ternary Ti-Al-V alloys. Sharp change in modulii ratio occurs at the alpha-plus beta to alpha phase boundary at 580°C as seen in Fig. 10. It also shows the internal friction peak at this temperature. Fig. 10 shows peaks due to Al-O, and V-O complexes in alpha titanium as well as Snoek peak due to oxygen in the beta phase. The shift in vanadium peak is ascribed to a change in the vanadium content of the alpha phase during different cooling rates. The internal friction peaks obtained for the martensitic transformations in Ti-6Al-4V are shown in Fig. 12. There are two peaks P_1 and P_2 occurring around -75°C and -25°C respectively. Peak P_1 is not affected by deformation or by change in frequency whereas peak P_2 is very much amplified. Hence it is suggested that this is due to stress-induced twin boundary sliding. An activation energy of 13 Kcals/mole for the twin boundary peak (P_2) has been obtained by plotting frequency vs reciprocal of absolute temperature. The frequency factor was found to be 2 x 10^{-14} sec.

Relaxation Processes

Substitutional-interstitial interactions are responsible for the internal friction peaks in titanium as in zirconium (6). The appropriate values of the sizes of the substitutional solute atoms have been calculated on the basis of Friedel's (7) equation. It is observed that for solutes in the same group, as the size difference between titanium and the solute atoms increases, the activation energy for the oxygen jump linearly decreases (8). Thus the magnitude of the elastic strain plays an important role in the oxygen jump process. Due to the high oxygen content of these alloys only the low temperature peak (9) (i.e. single oxygen jump mechanism) is observed in titanium.

It is observed that eventhough the nitrogen content is low in Ti-Sn alloy the interaction peak due to Sn-N complex is quite prominent and occurs at a lower temperature than for oxygen. Tin, having a larger atomic size that titanium, this effect is analogous

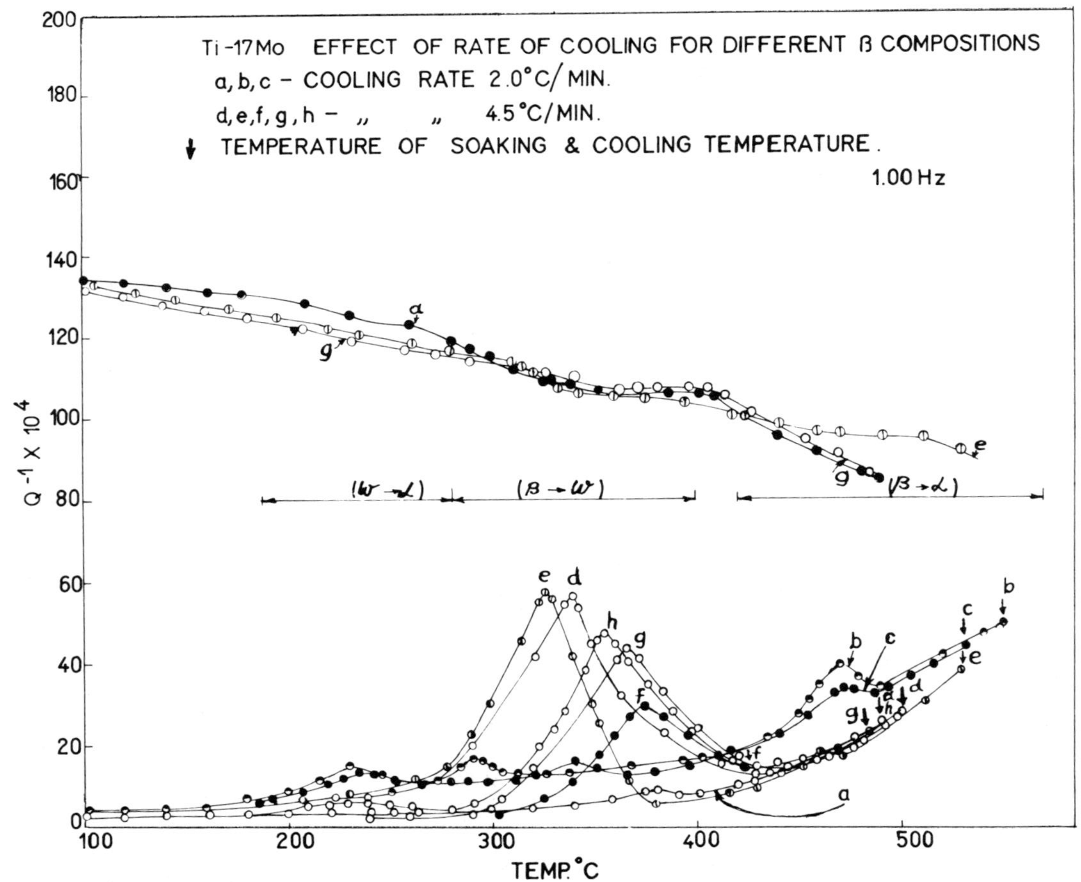

Fig. 9. Effect of cooling rate and composition on the transformation shifts of omega. Beta to alpha and omega to alpha appear on slow cooling.

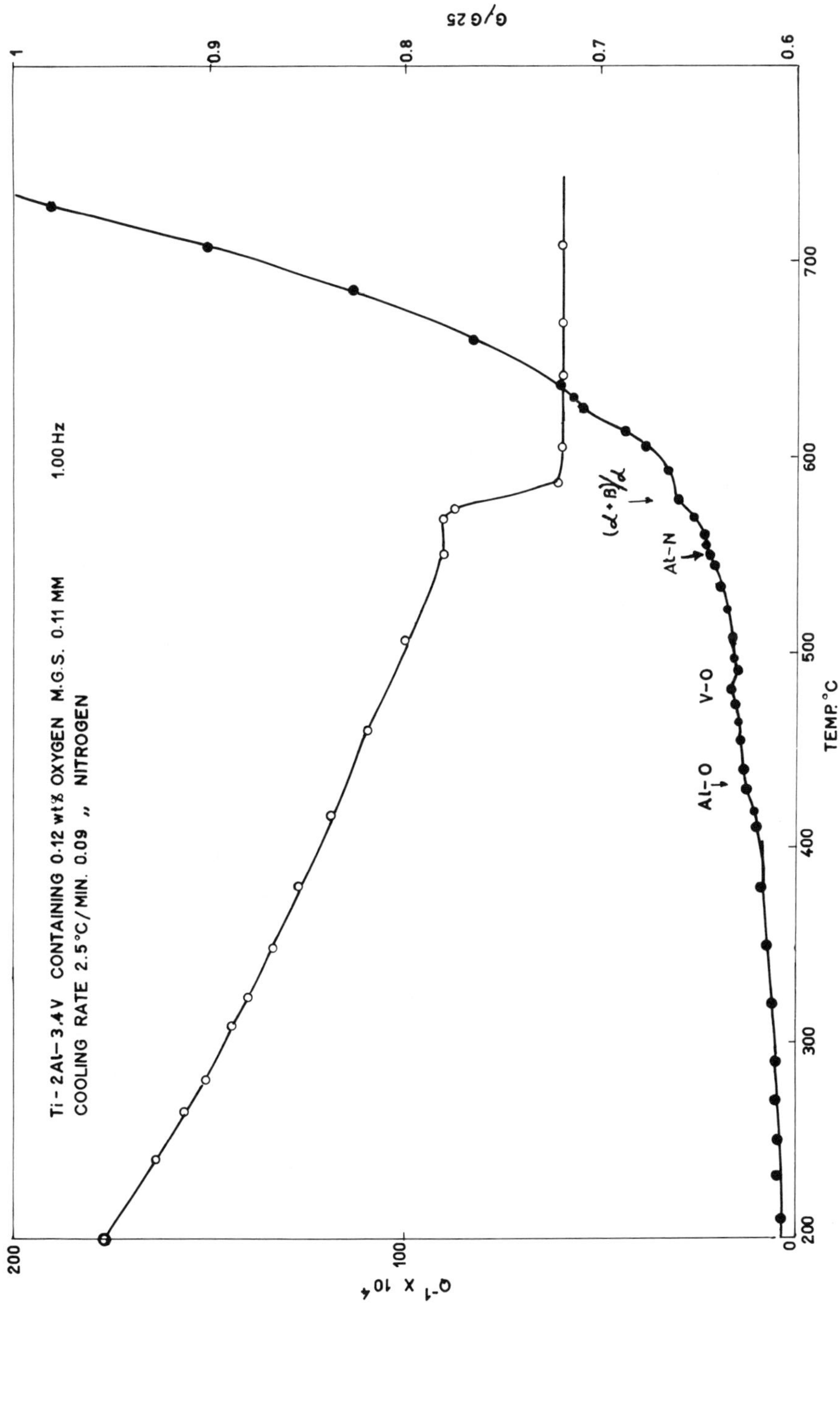

Fig. 10. Ti-2Al-3.4 V alloy: Interaction peaks due to Al-O, V-O, Al-N and the phase boundary transformation peak shown. Note the moduli change.

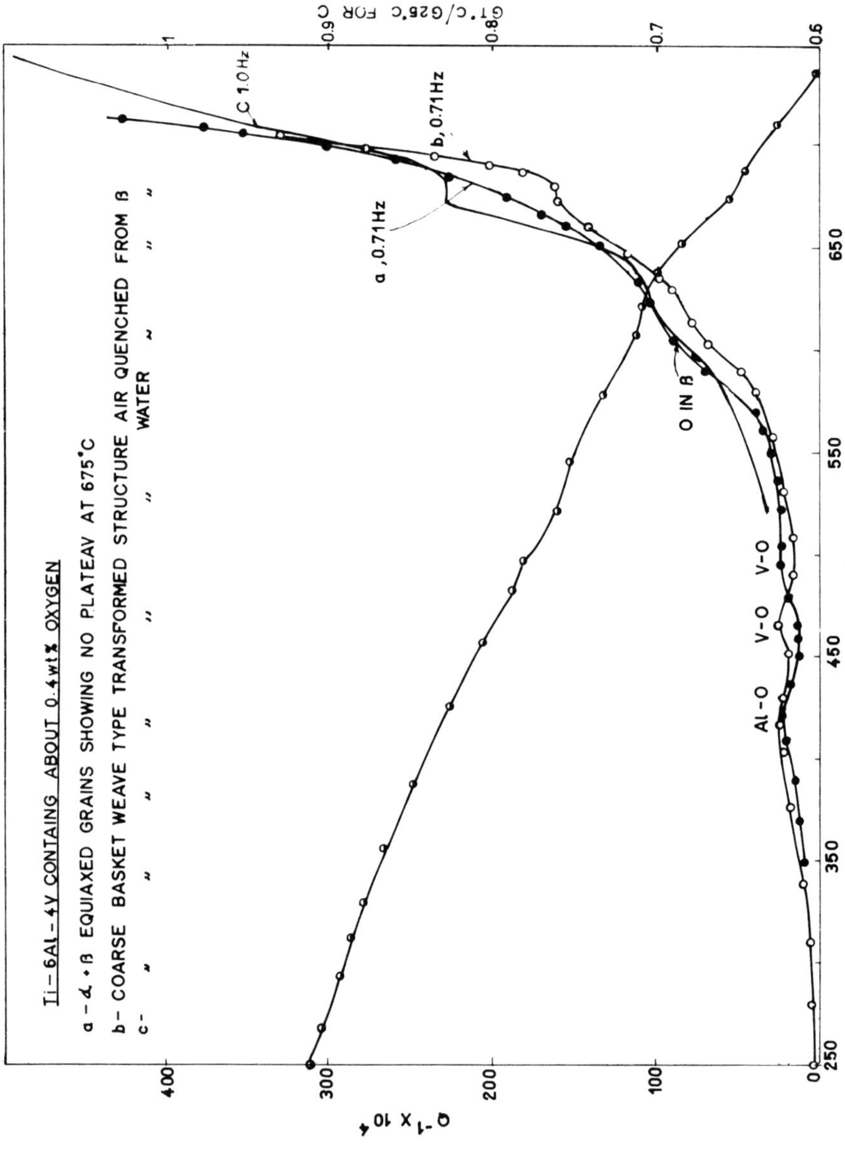

Fig. 11. Ti-6Al-4 V alloy: Showing interaction peaks and plateau of unknown origin in transformed structure.

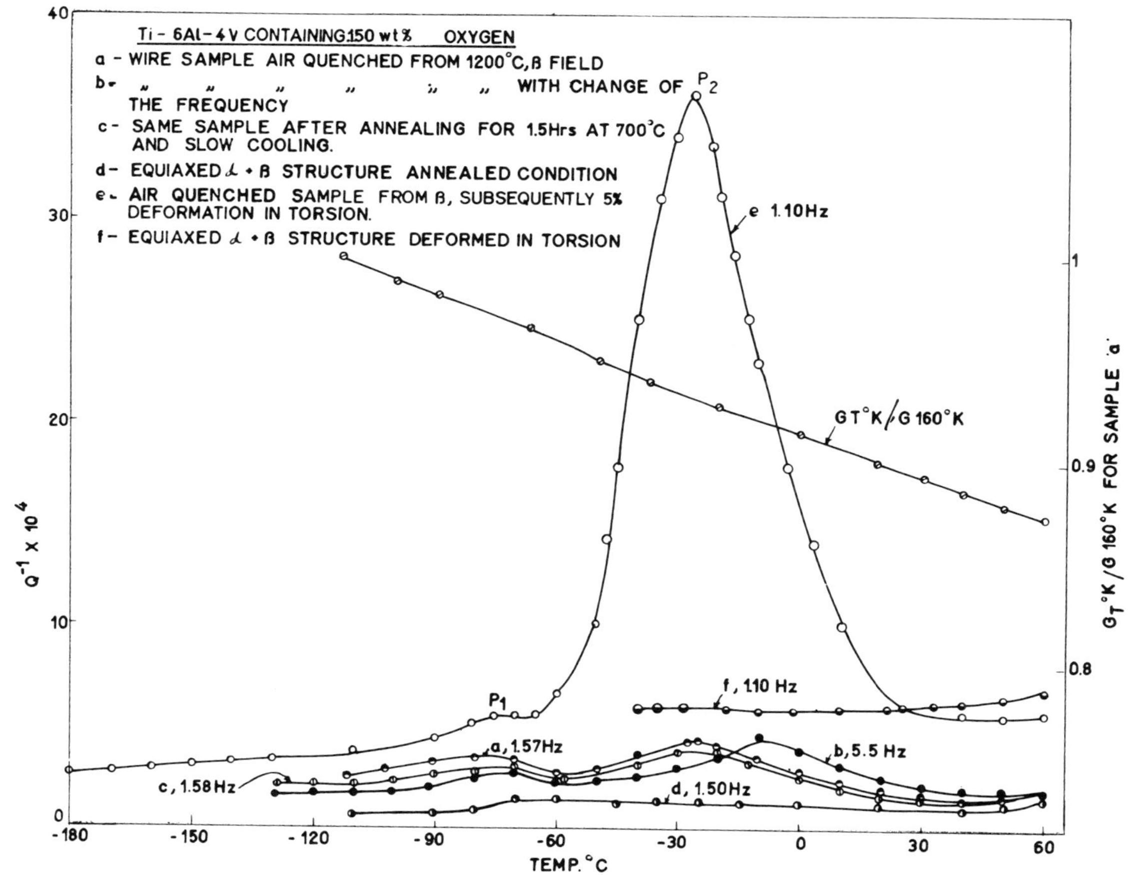

Fig. 12. Ti-6Al-4 V alloy: Twin boundary peak P_2 seen after transformation-cum-deformation. Oxygen content 0.15 wt %. Modulii change is shown.

to the results of zirconium alloys containing tin, yittrium and gadolinium (9).

Snoek relaxation peaks for oxygen, nitrogen and carbon occur in the beta phase of titanium from which diffusion data have been compiled. Published data for oxygen diffusion in beta titanium agree well with the present data. No data are available for carbon and nitrogen diffusion. Dilute bcc alloys give rise to two Snoek relaxation peaks, one due to the jump around the matrix atoms and the second around the substitutional solute atoms. In the present alloys since the solute concentration was high only one peak was observed. The diffusion rate refers to that in the alloy rather than to the pure titanium beta phase. Activation energies for diffusion of oxygen in Ti-Mo and Ti-V reflect the influence of molybdenum and vanadium on the melting point of the alloy.

Internal friction peaks due to twin boundary movement (of $10\bar{1}1$ type) have been observed in martensite of Ti-6Al-4V alloy. Such peaks have been reported earlier in other systems also (10,11). The effect is very marked in specimens quenched and deformed at room temperature due to the increase in the twin density in the structure. This observation appears to be different from that of Ayoga and Sumino (10). The origin of the peak P_1, which is independent of cold work, is unknown and it could be due to the transformation of retained beta (12). If the original structure is an annealed one, then cold work simply raises the background internal friction.

Omega Transformation

It has been demonstrated (13) that phase transformations in pure metals can be studied using internal friction and modulii ratio changes with temperature. Same technique has been followed in observing transformation at phase boundaries and for transformations involving diffusion as well as shear mechanisms. Phase boundaries are located by observing, changes in slopes of the modulii ratio graphs and the internal friction peak (Figs.6,7 & 9). This, in turn, is a function of the number of atoms taking part in the phase separations. Omega transformations in titanium base alloys have been observed both during cooling and during ageing of quenched beta structure (14). It can form by diffusionless as well as diffusion controlled processes. Internal friction techniques have been used to study the diffusionless transformations in Fe-Mn (15) and Co-Ni (16) alloys. In the present investigations the omega transformation is found to be a function of the beta composition and the rate of cooling. Fig. 8 shows peaks due to omega transformation at 440°C in specimens cooled from 620°C while those cooled from above this temperature fail to reveal the transformation. It appears that omega that is formed from high molybdenum beta is assisted by the elastic strains during oscillations. Internal friction maxima, due to the elastic stress associated omega, shifts to higher temperatures with increasing molybdenum content of the beta for the same cooling rate. It is also noticed that the maxima is dependent upon shear strain on the sample and shows a 'time effect' similar to the diffusionless transformation in Co-Ni (16) alloy. Omega transformation maxima around 280°C in Ti-43 wt% Mo alloy are not sensitive to strain amplitude due to the inherent hardening of the lattice.

Acknowledgements

The authors would like to thank Prof. T.R. Anantharaman, Head, Metallurgy Department, Banaras Hindu University for useful discussions and Dr V.K. Moorthy, Head, Metallurgy Division, Bhabha Atomic Research Centre for his keen interest in the work. They are also grateful to Mr. C.V. Sundaram, Head, Extractive Metallurgy Section, B.A.R.C. for the supply & analysis of sodium reduced titanium crystal and to Dr M. Koiwa for helping us to build the apparatus.

References

1. Pratt, J.N., Bratina, W.J. and Chalmers, B., "Internal friction in Ti and Titanium Oxygen Alloys", Acta Metallurgica, Vol. 2, Mar. 1954, pp.203-208.

2. Gupta, D., and Weinig, S., "Interaction between Interstitial and Substitutional Solutes in an H.C.P. Lattice", Acta Metallurgica, Vol. 10, April 1962, pp. 292-298.

3. Bratina, W.J., "Contribution to the Relaxation Maximum due to Oxygen in Alpha-titanium", Acta Metallurgica, Vol. 10, April 1962, pp. 332-333.

4. Miller, D.R., and Browne, K.M., "Stress-induced Diffusion of Carbon and Oxygen in Titanium", The Science Technology and Application of Titanium, Editors - Dr R. Jaffee and N. Promisel, Pergamon Press - Oxford and New York - 1970, pp. 401-405.

5. Koiwa, M. and Hirabayashi, M., "Recording Equipment for Measurements of Internal Friction and Elastic Modulus of Metals at low Frequency", The Science Reports of the Research Institutes, Tohoku University, A-Vol. 18, Supplement Dec. 1966, pp. 344-355.

6. Mishra, S. and Asundi, M.K., "Internal Friction Due to Oxygen-Substitutional Impurity Atom Complexes in Zirconium", Trans. Indian Institute of Metals 23 1970 pp.73-79.

7. Swalin, R.A., "A Model for Solute Diffusion in Metals Based on Elasticity Concepts", Acta Metallurgica, Vol. 5, August 1957, pp. 443-448.

8. Mishra,S., and Asundi, M.K., "Elastic and Electronic Effects of Substitutional Solute Atoms on the Interstitial Oxygen Migration in Alpha Zirconium", unpublished results.

9. Mishra, S, and Asundi, M.K., "Internal Friction Studies in Dilute Zirconium Alloys", International Symposium on Zirconium Alloys, Montreal, August 1971, To appear in Canadian Metallurgical Quarterly.

10. Aoyagi, T., and Sumino, K., "Mechanical Behaviour of Crystals with Twinned Structure", Physica Status Solidi, 33, 1969, pp. 317-326.

11. Povolo, F., & Bisogni, E.A., "Internal Friction in Zirconium-Hydrogen Alloys at Low Temperatures", Journal of Nuclear Materials, 29, 1969, pp. 82-102.

12. Williams, J.C., and Blackburn, M.J., "A Comparison of Phase Transformation in Three Commercial Titanium Alloys", Transactions of the A.S.M., Vol. 60, 1967, pp.373

13. Selle, J.E. and Focke, A.E., "An Internal Friction Study of the Allotropic Transformations of Plutonium, Uranium and Cobalt", Journal of Nuclear Materials, 33, 1969, pp. 149-158.

14. Hickman, B.S., "The Formation of Omega Phase in Titanium and Zirconium Alloys", A Review Journal of Materials Science, 4, 1969, pp. 554-563.

15. Wang Yeh-Ning and Chu Chiang-Chung, "The Behaviour of Internal Friction Associated with the Process of Martensite-type Transformation", Scientia Sinica, Vol. 9, No.2, 1960, pp.197-212.

16. Belko, V.N., Darinskiy, B.M., Postnikov, V.S., and Sharshakov, I.M., "Internal Friction during Diffusionless Phase Transformations in Co-Ni Alloys", Fiz. Metal Metalloved, 27, No.1, 1969, pp.141-147.

17. Rosa, J.C., "Oxygen Diffusion in Alpha and Beta Titanium in the Temperature Range of 932° to 1142°C", Transactions of the Metallurgical Society of AIME, Vol. 9, September 1970, pp. 2517-2522.

18. William, P.R., Howard, R., and William, R.O., "Diffusion of Oxygen in Alpha and Beta Titanium", Transactions of the A.S.M. Vol.52, 1960, pp.191-200.

19. Hansen, M., Kamen, E.L., Kessler, H.D., and McPherson, D.J. "Systems Titanium-Molybdenum and Titanium-Columbium", Transactions AIME, 191 October, 1951, pp.881-888.

20. Mishra, S. and Asundi, M.K., "Comments on Stress Induced Diffusion of Oxygen in Alpha Zirconium", Scripta Metallurgica, 5(11), 1971, pp.973-976.

21. Powers, R.W. and Doyle, V.M., "Diffusion of Interstitial Solutes in the Group V Transition Metals", Journal of Applied Physics, Vol.30, No.4, April 1959, pp.514-524.

INVESTIGATION OF LATTICE DEFECTS, ELECTRONIC STRUCTURE, AND DIFFUSION MOBILITY IN TITANIUM

L. G. Korneluk, L. M. Mirsky, and B. S. Bokshtein

Institute of Light Alloys
Moscow, USSR

1. Titanium, a metal of great practical importance, is characterized by various diffusive anomalies. Among these first of all should be noted the high diffusive penetrability and the failure to observe the Arrhenius law under diffusion in high-temperature β-modification in combination with low values of the activation energy and preexponential factor.

As all anomalous bcc metals, titanium undergoes a polymorphic α (hcp) — β(bcc) transformation proceeding by a martensite mechanism. This transformation upon both heating and cooling leads to the formation of strong structural and chemical heterogeneity. A strong segregation of β-stabilizers (e.g., nickel, iron) along laminae of α-phase was observed by Bokshtein et al.(1) The laminated structure occurring after the $\beta \rightarrow \alpha$ transformation, is characterized by considerable thermal stability. The high density of defects (dislocations) that form on heating as a result of the $\alpha \rightarrow \beta$ transformation considerably increases and remains up to pre-melting temperatures.(2)

The formation of an extensive defect structure and chemical nonuniformity may result in diffusive anomalies. It is also natural to assume that the occurrence of excess vacancies, which may be connected with impurity atoms and the transfer of which defines the mobility, results from structural changes, and concentration nonuniformity has a considerable effect on diffusion if it proceeds by the vacancy mechanism. Kidson(3) has carried out experiments in an effort to study the influence of dislocation on the self-diffusion of zirconium, which is analogous to titanium. However, no influence of repeated phase transformations on zirconium self-diffusion was discovered in this work. The absence of this effect is probably associated with the extreme difficulty of the experimental procedure, and with some omissions in the methodical order prescribed by Le Clair(4); the preliminary annealing was not too long or it proceeded at not very high temperature.

2. The connection between the fine structure arising in titanium during the polymorphic transformation and the diffusion mobility was carefully studied by Bokshtein *et al.*(4) by the methods of tracer atoms, X-ray, and optical and electron microscopy. Further work along these lines by the same authors(5) showed that structure imperfections occurring during the $\alpha \rightleftarrows \beta$ transformation were the main reason for the accelerated diffusion in titanium. Structure defects of the old, original phase pass over to the new phase formed in the $\alpha \rightarrow \beta$ transformation. The structure imperfections observed during the polymorphic transformation are very stable in the region near the transformation temperature ($\sim$ 900-1000°C). When the preannealing temperature increases - from 1200°C to the premelting temperature - the defects gradually vanish and the diffusion mobility decreases (Table 1). The experimental scheme is shown in Fig. 1.

We now observe some peculiarities of the experiment. The subject of the investigation was titanium of the following composition: H_2 0.0044; O_2 0.188; Ni 0.0026; balance titanium. No grain size effect on diffusion was possible - specimens were subjected to preannealing at 1000°C for 3 hr. To decrease the defect density, the specimens were subjected to a secondary annealing at 800°C for 115 hr. Diffusion of the isotope Sn^{113}, which refers to a group of neutral elements, and, furthermore, effectively increases the strength of titanium at elevated temperatures, was studied.

Special equipment was built for the experiment. The equipment provided (a) a vacuum $3\text{-}4.10^{-5}$ torr and liquid nitrogen supply to the chamber during the entire experiment; (b) radioisotope spraying on the specimen by an evaporation method in a wide temperature range, in both the α- and β-fields; (c) putting the isotope into a special

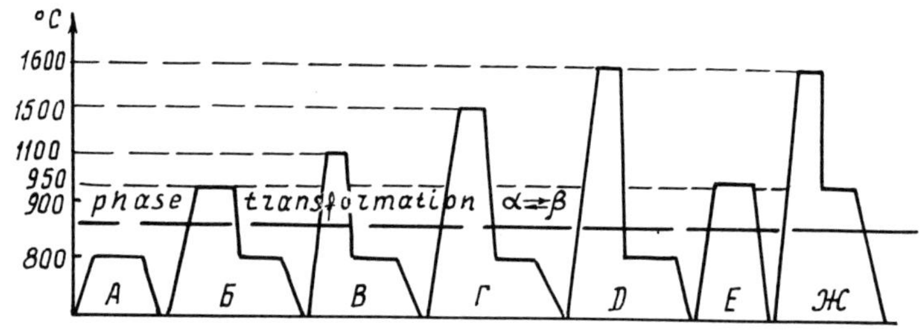

Fig. 1. Scheme of the experimental procedure.

TABLE 1

Group Studied	Scheme of treatment	Temperature, °C			Coefficient of diffusion $D \times 10^{-10}$ $cm^2 \cdot sec^{-1}$
		Preliminary annealing	Excess over polymorphic transformation temperature	Diffusion annealing	
A	α		-	800	2.60
B	$\alpha \to \beta \to \alpha$	1950	68	800	31.0
C	$\alpha \to \beta \to \alpha$	1100	218	800	17.0
D	$\alpha \to \beta \to \alpha$	1500	618	800	5.8
E	$\alpha \to \beta \to \alpha$	1550	668	800	4.1
F	$\alpha \to \beta$	-	-	950	9.1
G	$\alpha \to \beta \to \alpha$	1550	668	950	0.24

tantalum boat with a cap which opens automatically only while the isotope is sprayed on the specimen but closes during the diffusive annealing to maintain the boundary conditions of the experiment. The temperature of the specimen was controlled by a platinum-rhodium thermocouple accurate to $\pm$ 5°C. A thin uniformly distributed layer of radioactive tin of high specific activity ($\sim$ 8000-10,000 pulses/min) was obtained.

3. The diffusion coefficient of α-titanium at 800°C was determined by the method of skimming and measuring the integrated intensity of the remainder ($D = 2.6 \times 10^{-10}$ cm^2/sec). The specimens were then subjected to intermediate annealing in the β-field for 3 hours at different temperatures, followed by measuring the diffusion coefficient at 800°C, again in the α-field, the specimens having undergone two polymorphic transformations ($\alpha \to \beta \to \alpha$). Intermediate annealing at 950°C led to an increase in the diffusion coefficient of more than an order (31.0×10^{-10} cm^2/sec). Annealing at higher temperatures gave lower D values as the temperature increased: 17.0 after annealing at 1100°C; 5.8 for 1500°C, and only 1.5 times more than the initial value, 4.1×10^{-10} cm^2/sec, after annealing at 1600°C.

Similarly, the diffusion coefficient measured at 950°C (after an $\alpha \to \beta$ transformation) proved to be 9.1×10^{-10} cm^2/sec, and intermediate annealing at 1600°C for 25 hours decreased it to 0.23×10^{-10} cm^2/sec. Thus, D at 950°C proved to be an order less than at 800°C and 40 times less if there was no preannealing at high temperatures. These results are conclusive evidence of the strong influence on diffusion mobility of defects occurring during polymorphic transformations in titanium, and of the considerable temperature stability of these defects.

4. Comparison studies of the α- and β-titanium diffusive permeability and studies of iron mobility concentration dependence in titanium-iron alloys (2.8, 4.8, and 9.5 wt% Fe) were performed. Measurements were made in the temperature ranges 700-850° (α-titanium) and 1000-1400° (β-titanium). Radioactive iron-59 was sprayed at the measurement temperature, i.e., in studies of diffusion in β-phase above the $\alpha \to \beta$ transformation temperature, directly in a diffusive cell, on a specimen that was first subjected to prolonged (60 hours) high-temperature annealing at 1650°C, so that the lattice defects occurring as a result of the transformation were practically eliminated. Diffussion coefficients were determined by the method of residual activities, for β-phase according to the data obtained in the temperature range 1000-1300°C, since at 1400°C the values deviated strongly upward from the linear dependence (from T^{-1}). The results are given in Table 2.

The iron mobility in the α-phase at the $\alpha \to \beta$ transformation point is considerably (approximately five times) greater than in the β-phase. The activation energy of iron diffusion increases under the transformation from α- to β-titanium and with increased iron concentration in the β-phase.

5. With the aim of studying the relationship between the observed effect and the electronic structure the values of chemical shifts of Mössbauer absorption lines of iron-57 nuclei in α- and

TABLE 2

Alloy composition	Temperature range, °C	Do, cm^2/sec	Q, kcal/g-at
Ti	700-850 (α)	1.2×10^{-4}	26.4
Ti	1000-1200 (β)	5.6×10^{-3}	31.3
Ti + 2.8 wt% Fe	1000-1300 (β)	0.15	41.2
Ti + 4.8 wt% Fe	1000-1300 (β)	1.7	48.0
Ti + 9.5 wt% Fe	1000-1300 (β)	3.6	51.6

β-titanium were measured. The source of γ-quanta was Co^{57} in stainless steel. The chemical shift value in β-titanium is considerably less than in α-titanium. Since the chemical shift (δ) is proportional to the difference between the electron densities on an iron nucleus ($\psi^2(0)$) in the absorber (α- or β-titanium) and in the source (steel); $\delta = A\{\psi^2_{absorb}(0) - \psi^2_{source}(0)\}$ and for iron $A < 0$. The increase in s-electron density on the iron nucleus in β-titanium compared to α-titanium corresponds to the chemical shift decrease. This result may be treated as an increase of the covalent contribution to the interatomic bond in β-titanium, and that is what leads to the decrease in diffusive mobility.

6. The concentration dependence of the chemical shift in β-titanium on the iron content was also measured.(6) The Mössbauer absorption spectra were taken at 900°C in a specially made high-temperature furnace. High-temperature measurements were necessary because in the studied alloys the β-phase was not stable at room temperature and specimens had complex phase composition (β, ω, etc.). The furnace design permitted measurements in an inert atmosphere (Ar) and at constant temperature. Absorber specimens were made of Ti alloys with 4.8, 9.8, and 15 wt % Fe; they were sprayed as a powder onto 0.3 mm thick beryllium foil; the density of Fe was about 4 mg/cm^2. The specimens were subjected to prolonged (more than 10 hours) exposure at 900°C prior to the measurements. The absorption spectrum of iron-57 nuclei in -Ti comprised a doublet of the quadrupole splitting in all alloys. Spectral parameters are given in Table 3.

The s-electron density falls as the content of Fe in β-Ti rises; it will be clear from Table 2 that the quadrupole splitting which characterizes the degree of lattice distortion also decreases as the Fe content increases.

TABLE 3

No.	Fe, wt %	Chemical shift (δ, mm/sec) of source at 20°C	900°C	Quadrupole splitting, , mm/sec
1	4.8	- 0.62	- 0.08	0.28
2	9.8	- 0.56	- 0.02	0.22
3	15.0	- 0.48	+ 0.06	0.18

A correlation between diffusion measurements and chemical shift shows that the diffusion activation energy of Fe in β-Ti increases as the density of electrons on a nucleus decreases.

It has been shown(7) that if iron diffuses by a common vacancy mechanism, the inverse relationship takes place; the diffusion activation energy of iron in different solvents increases as the density of s-electrons on an iron nucelus increases. The relationship obtained for β-Ti leads to the assumption that the anomaly occurring in β-Ti in a dilute solution is connected with a partial change of the diffusion mechanism. It should be emphasized that as the Fe content increases, the diffusion becomes more "normal." At 15 wt % Fe both the activation energy and the chemical shift value stay within a correlative range found by Bokshtein and Juchovitskii.(7)

The sum total of the data is consistent with the supposition that iron in dilute solutions in β-Ti partially diffuses through the interstitial mechanism. According to Anthony and Turnbull,(8) the interstitial diffusion mechanism in substitutional solutions in favored by the similar electronic structure of the impurity and the solvent, with electrons efficiently participating in the bond considered. β-Ti and metallic iron have four such electrons each; they have a d^3s^1 configuration. The difference between this configuration and that usually assigned to metallic Fe (d^7s^1) lies in the fact that four paired d-electrons don't participate in the bonding. The indicated configuration favors occurrence of tetrahedral hybridization, giving maximum overlapping of electron shells.

The interstitial mechanism is also favored by geometrical factors. The β-Ti lattice period of 3.32 Å is the largest of all the elements of the fourth period of the periodic table (V, 3.03 Å; Cr, 2.88 Å) and, consequently, it has the largest tetrapore that lightens diffusion by the interstitial mechanism. However, the placement of Fe atoms in tetrapores without causing a strong lattice deformation is possible only with a small concentration of Fe; an increase in Fe content presents difficulties for the Fe atom output in the tetrapore and for the diffusion by the interstitial mechanism. As Fe content increases, the diffusion activation energy also increases, approaching values corresponding to the interstitial diffusion mechanism.

The assumption of the occurrence of hybrid type bonds permits us to explain the dependence between the diffusion activation energy and the electron density on a nucleus. In fact, d-electrons of Fe atoms in tetrapores forming d^3s hybridized orbitals take an active part in the bond and slightly screen their own ^{3}s-electrons, which leads to a density increase of s-electrons

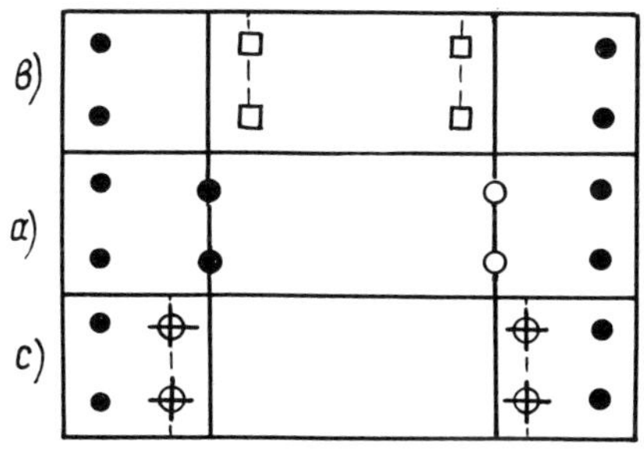

Fig. 2. Scheme of tungsten mark location: (a) before diffusive annealing, (b) after annealing at 1400°C in the system Ti — (Ti+2 wt % Ni) —Ti; (c) after annealing at 1400°C in the system Ti — (Ti+2 wt % Ni+0.6 wt % O_2) - Ti

on the nucleus. As Fe atoms fill β-Ti crystal lattice points, d-electron participation in bonding decreases since the interatomic distance increases. The density of d-electrons inside the s-sphere increases, the screening created by them increases, and the density of s-electrons decreases. Simultaneously the quadrupole splitting also decreases; since d-orbitals become more elongated, more symmetric, the tensor of the electric field gradient decreases.

7. The results of other experiments, on the other hand, speak in favor of the impure vacancies hypothesis. The Kirkendall displacement of tungsten marks located in the plane of the boundary between A and AB in trilaminated samples of the A-AB-A "sandwich" type was measured. The extreme A laminates were enhanced-purity titanium; the intermediate laminate AB in one case was a titanium alloy with 2 wt % Ni and in the other Ti-2 wt % Ni-0.6 wt % O_2. Diffusive annealing was performed at 1400°C, and mark displacement was measured with a microscope and a cathetometer. The mark displacements has different signs: marks are brought closer together in systems without oxygen but diverge in the presence of oxygen. According to a common interpretation of the Kirkendall effect, this means that the excess of vacancies occurs in the middle zone, in the absence of oxygen and in the outer zones in the presence of oxygen.

The results of the present work along with the deceleration of nickel diffusion(7) in the subsurface zone of titanium (where oxygen is dissolved) compared with deeper layers (where there is no oxygen) must become the subject of a serious analysis.

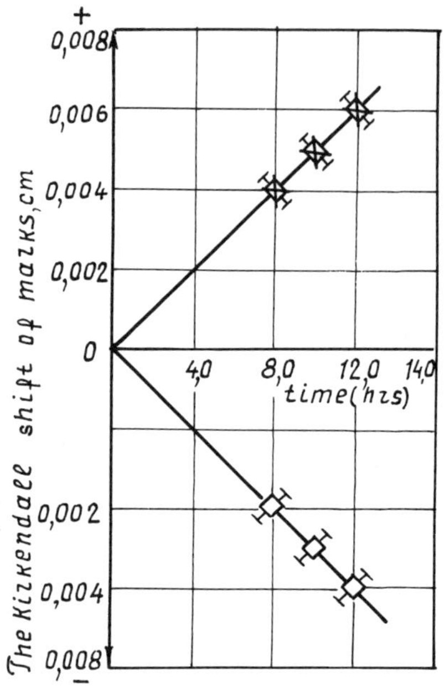

Fig. 3. Kirkendall shift of marks for
Ti - (Ti+2 wt % Ni) - Ti
Ti - (Ti-2 wt % Ni-0.6 wt % O_2) - Ti

In this connection the data obtained in our work when studying the effect of the polymorphic transformation on a concentration of vacancies in titanium is also of interest. Using the pulse method for the determination of vacancy properties, the authors found that the $\alpha \rightarrow \beta$ transformation in titanium resulted in an excess (compared to equilibrium thermal) concentration of vacancies. The authors attribute this elevated concentration of vacancies to their interaction with oxygen absorbed by titanium. In any case, the nonequilibrium flow of vacancies occurring during the phase transformation in titanium may explain the abnormally high diffusion coefficient in β-titanium at temperatures near the polymorphic transformation point.

Thus, the analysis of the present empirical data shows that the anomalies of the diffusion mobility in titanium may be associated as with the development of the defect structure and the excess vacancies resulted from the $\alpha \rightarrow \beta$ transformation, and with the features of the titanium electronic structure.

Conclusions

1. It is shown that during the polymorphic $\alpha \rightarrow \beta$ transformation the lattice defects of the old, original phase are carried over to the newly formed phase, and this phenomenon may be one of the reasons for the accelerated diffusion in β-phase titanium. These imperfections are very stable near the $\alpha \rightarrow \beta$ transformation temperature (900 - 1100°C). When the material is preannealed above this temperature range (particularly near the premelting temperature) the imperfections are gradually annihilated. The diffusion mobility decreases to approximately one-fortieth of its original value (i.e., at 750°C, $D_\beta = 9.1 \times 10^{-10}$ cm²/sec, and after the preliminary thermal treatment at 1500°C for 25 hours, $D_\beta = 0.24 \times 10^{-10}$ cm²/sec).

2. It is shown that the mobility of iron in the α-phase at the $\alpha \rightarrow \beta$ transformation point is considerably (about five times) higher than in the β-phase. The activation energy of iron diffusion increases when passing from α- to β-titanium and when the iron content in the β-phase increases.

3. The density of electrons on the iron nucleus in β-titanium (evaluated by the Mössbauer effect) is considerably higher than in α-titanium. This result is treated as equivalent to an increase of the covalent contribution to the interatomic bond in β-titanium, which results in a decrease of the diffusion mobility.

4. Mössbauer studies have shown that the density of s-electrons decreases when the iron content in β-titanium increases (the quadrupole splitting of the doublet which characterizes the degree of lattice distortion also decreases). Since with the usual vacancy mechanism the diffusion activation energy increases when the density of s-electrons increases (rather than decreases), it has been proposed that iron in dilute solutions in β-titanium partially diffuses by the interstitial mechanism.

5. The measurement of the Kirkendall effect in the Ti-Ni-O system points to the important role of oxygen. Similar conclusions may be drawn from the data obtained when studying the effect of the polymorphic transformation on the characteristics of vacancies in titanium.

References

1. S. Z. Bokshtein, T. A. Yemeljanova, S. T. Kishkin, L. M. Mirsky, Izv. Akad. Nauk USSR, Metally, 1968, No. 5, p. 144.

2. W. C. Hagel, Trans. AIME, 1963, No. 227, p. 267.

3. G. Kidson and J. McGurn, Can. J. Phys. Vol. 39, 1964, p. 1146.

4. S. Z. Bokshtein, T. A. Yemeljanova, S. T. Kishkin, and L. M. Mirsky, The Science, Technology, and Application of Titanium, 1970, London.

5. S. Z. Bokshtein, S. T. Kishkin, and L. M. Mirsky, Izv. Akad. Nauk USSR, Metally, 1971, No. 5.

6. A. N. Aljoshin, B. S. Bokshtein, and G. S. Nickolskii, Izv. Akad. Nauk USSR, Metally, in press.

7. B. S. Bokshtein and A. A. Juchovitskii, Zh. Eksp. Teor. Fiz., in press.

8. T. R. Anthony and D. Turnbull, Appl. Phys. Letters, 1966, 8, p. 120.

INTERACTION OF OXYGEN AND HYDROGEN IN TITANIUM

A. D. McQuillan

Professor of Physical Metallurgy
University of Birmingham
England

Although it has long been known that titanium forms extensive interstitial solutions with the elements hydrogen, oxygen, nitrogen and carbon, no theory has yet emerged which explains why the early transition metals exhibit this atypical property. Recently the author put forward a theoretical interpretation of hydrogen solutions in β-titanium (1). It was not, however, immediately obvious how this could be extended to other interstitial elements without recourse to further experiments. The work described here is an attempt to gain some insight into the electronic state of oxygen in the metal by observing its interaction with hydrogen.

The addition of oxygen to titanium reduces the solubility of hydrogen in the metal (2), but this does not appear to have been investigated in a systematic way. In the present work a series of titanium-oxygen alloys has been used to obtain hydrogen pressure-composition-temperature relationships in β-phase (b.c.c.) ternary Ti-H-O solid solutions. The resulting data have been used to suggest a possible mechanism for the effectiveness of oxygen in reducing hydrogen solubility in the metal.

EXPERIMENTAL

Titanium-oxygen alloys were prepared by arc-melting high purity titanium with the required additions of zone-refined titanium dioxide. After homogenisation at 1000°C for 48 hours in high vacuum, the alloy specimens were analysed by mass-spectrometry and shown to contain less than 1 p.p.m. of impurities.

Hydrogen pressure-composition-temperature relationships were obtained for each of the six specimens (1, 2, 3, 4 and 6 at.% oxygen) in apparatus which is now quite standard (3). In order to be sure that no contamination of the specimens occurred during the measurements, the hydrogen was removed and the experiments repeated with a further charge of hydrogen. In every case quoted in the paper the two sets of results so obtained lay within the experimental errors of ± 0.1% in the variables, pressure, temperature and hydrogen content.

The constitutional diagrams of the titanium-oxygen (4) and titanium-hydrogen (3) systems indicate that whereas oxygen additions have the effect of elevating the temperature of the α-β transformation, hydrogen brings about a depression. Since our detailed knowledge of the behaviour of hydrogen solution in titanium was confined to the β-phase, it was important that our results for the ternary alloys also related only to this field. In such alloys, the onset of the (α+β) phase region at lower temperatures cannot readily be determined from the hydrogen pressure results, and high temperature magnetic susceptibility measurements in a hydrogen atmosphere were used instead, since the appearance of α-phase in the alloys is clearly marked in this method. Apparatus previously described (5) was used in this work and the positions of the β/(α+β) boundaries for the series of oxygen alloys as a function of hydrogen content and temperature are as shown in Fig. 1. Using this data, it was possible to select those regions of the hydrogen-pressure-temperature-composition results which related to the β-phase. Within the limitations of the apparatus, which could not be used above 1000°C without danger of contaminating the specimen, only the alloy containing 6 at.% oxygen could not be obtained in the β-phase condition during hydrogen measurements.

RESULTS AND DISCUSSION

As shown in Fig. 2 the presence of oxygen dissolved in β-titanium brought about an increase in hydrogen equilibrium pressure compared with the pressure over the pure metal at equal temperatures and hydrogen concentrations. The logarithmic pressure increase appears to be proportional to the oxygen content in single phase β-regions of the ternary system. Thus it would appear that dissolved oxygen and hydrogen atoms repel each other in titanium. There are six tetrahedral interstices per metal atom in the b.c.c. structure, hence if there are N metal atoms and n dissolved hydrogen atoms in tetrahedral sites, the site occupancy by hydrogen will be n/6N. If an oxygen atom in an octahedral site effectively prevents hydrogen occupying z adjacent tetrahedral sites, and there are m dissolved oxygen atoms, then their presence will reduce the site occupancy by hydrogen to n/(6N - zm). Assuming that the presence of small quantities of oxygen has no appreciable effect on the heat of solution of

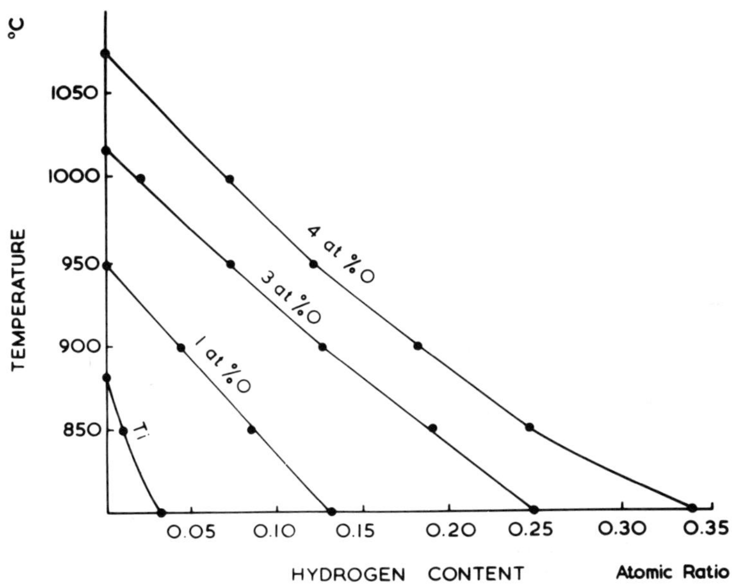

Fig. 1. The $\beta/(\alpha+\beta)$ phase boundaries as a function of temperature and hydrogen content for a series of titanium-oxygen alloys. The β-phase region for each alloy lies above the appropriate boundary.

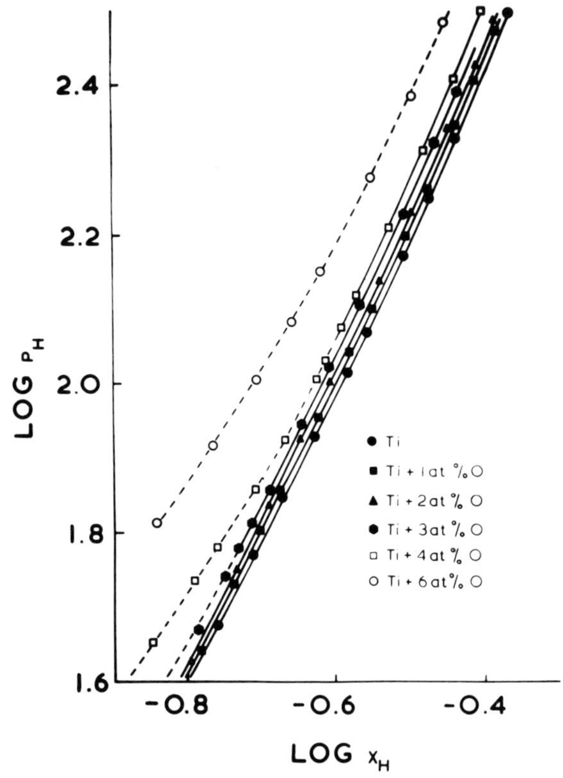

Fig. 2. Isothermal curves at 850°C for the series of β-phase oxygen alloys in which the logarithms of hydrogen equilibrium pressure p in torr and hydrogen concentration x_H in atomic ratio (n'/N) are plotted against each other. The broken lines denote regions in which the alloys are in the (α + β) condition.

hydrogen in the metal, then at a given temperature, we may equate equal site occupancy by hydrogen at constant hydrogen equilibrium pressure in the metal and in oxygen alloys. Hence for oxygen-free titanium and an oxygen alloy

$$\frac{n}{6N} = \frac{n'}{(6N-zm)} ,$$

where n' is the number of dissolved hydrogen atoms in N metal atoms containing m oxygen atoms which gives the same hydrogen equilibrium pressure at the same temperature as n hydrogen atoms in N metal atoms of the pure metal. Expressing this in terms of hydrogen and oxygen concentrations in atomic ratios,

(i.e. $\frac{n}{N} = x_H$, $\frac{n'}{N} = x_H'$ and $\frac{m}{N} = x_o$) gives the relationship

$$x' = x_H(1 - \frac{z}{6} x_o).$$

If $\frac{z}{6} x_o$ is small, this can be approximated to

$$\log x_{H/x_H'} \approx \frac{z}{6} x_o$$

which explains why the series of experimental curves in Fig. 2 lie parallel to each other with about equal spacing, except where the alloys are no longer in the single β-phase.

The experimental results enable us to obtain a large number of values for x_H' for each oxygen content x_o over a wide range of temperatures, and hydrogen concentrations. These mean values of $\log x_H'/x_H$ as a function of x_o are plotted in Fig. 3, and give rise to a linear relationship for which z = 11.8 ± 0.5.

We must now consider the distribution of tetrahedral interstices around an octahedral interstice in the body-centred cubic metal lattice. As shown in Fig. 4 there are four tetrahedral sites at a distance of only a/4 from an octahedral site, where a is the lattice parameter of the metal lattice. These are followed by eight further tetrahedral sites at a distance of $\sqrt{5}$ a/4, and then sixteen at $\sqrt{7}$ a/4. Depending on the degree of repulsion between dissolved oxygen and hydrogen atoms, the number of tetrahedral sites z not available for hydrogen occupation around an oxygen atom should be 4, 12 or 28. The experimental results shown in Fig. 3 clearly indicate that z = 12 is the appropriate number of excluded hydrogen sites.

Examination of the spatial relationship between an octahedral

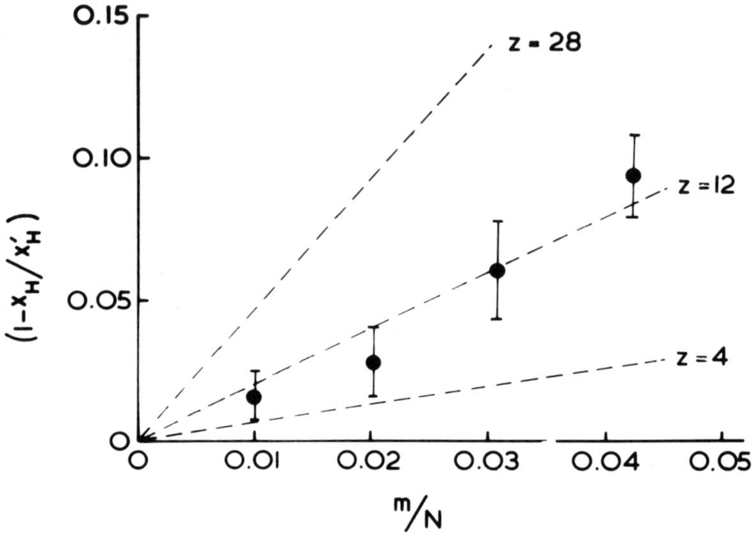

Fig. 3. Average results obtained from the experimental data of the quantity $(1 - x_H'/x_H)$ against m/N for the series of titanium-oxygen alloys. The three broken straight lines correspond to z = 4, 12, and 28. The most probable straight line has $z = 11.8 \pm 0.5$.

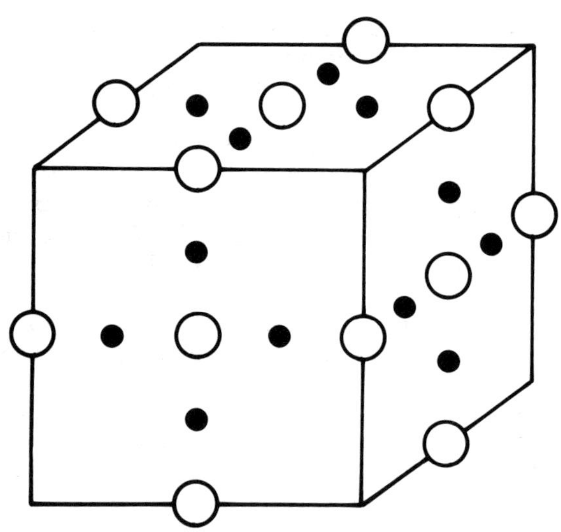

Fig. 4. Spatial arrangement of the tetrahedral • and octahedral ○ interstices in the b.c.c. metal lattice. All interstitial sites are on the surfaces of the unit cell.

site and tetrahedral sites at distances a/4 and $\sqrt{5}$ a/4 from it, shows that no metal atom (assumed to have a diameter of $\sqrt{3}$ a/2) lies between the line joining the octahedral and tetrahedral sites, whereas a metal atom does lie between an octahedral site and the tetrahedral sites $\sqrt{7}$ a/4 away from it. This is the same rule as that which is found (1) for the repulsive interactions between hydrogen atoms in the pure metal, and it would seem, therefore, that the interaction between oxygen and hydrogen atoms in the metal may well also be due to the scattering of conduction electrons by the interstitial atoms. If this were the case, then oxygen atoms would, like hydrogen, have to donate electrons to the conduction band. This would give rise to a O^+ ion instead of the more chemically usual O^{2-} ion. A O^{2-} ion is spherically symmetrical and large, whereas the O^+ ion is not only smaller but has the form of an ellipsoid of revolution. The octahedral interstice in a body-centred cubic structure is surrounded by an asymmetrical distribution of metal atoms, four lying at a distance of a/$\sqrt{2}$ and two at a/2. Thus an O^+ ion would more readily fit into an octahedral site. Furthermore the ordering of dissolved oxygen atoms in α-titanium described by Kornilov et al (6) could also be attributed to a long-range repulsive interaction between oxygen atoms by means of an overlap of the cloud of scattered screening electrons surrounding adjacent O^+ ions.

REFERENCES

1. McQuillan, A.D., "Interpretation of Hydrogen Solutions in b.c.c. Early Transition Metals", Journal of Chemical Physics, Vol. 53, No. 1, July 1970, pp. 156-164.

2. McQuillan, A.D., "The Titanium-Hydrogen System for Magnesium-Reduced Titanium", Journal of the Institute of Metals, Vol. 79, Part 5, 1951, pp. 371-378.

3. McQuillan, A.D., "An Experimental and Thermodynamic Investigation of the Hydrogen-Titanium System", Proceedings of the Royal Society, A. Vol. 204, 1950, pp. 309-323.

4. Jaffee, R.I., Ogden, H.R. and Maykuth, D.J., "Alloys of Titanium with Carbon, Oxygen and Nitrogen", Transactions of the Metallurgical Society of AIME, Vol. 188, 1950, pp. 1261-1266.

5. Jones, D.W., "High Temperature Magnetic Susceptibility Measurements on the Titanium-Hydrogen and Niobium-Hydrogen Alloy Systems", Journal of Less Common Metals, Vol. 6, 1964, pp. 100-107, and private communication.

6. Kornilov, I.I., Vavilova, V.V., Fykin, L.E., Ozerov, R.P.,

Solowiev, S.P. and Smirnov, V.P., "Neutron Diffraction Investigation of Ordered Structures in the Titanium-Oxygen System", Metallurgical Transactions AIME-ASM, Vol. 1, No. 9, 1970, pp. 2569-2571.

THERMODYNAMICS OF α-STABILIZED Ti-X-Y SYSTEMS*

H. L. Gegel
Air Force Materials Laboratory
Wright-Patterson Air Force Base, Ohio

M. Hoch
Department of Materials Science
and Metallurgical Engineering
University of Cincinnati
Cincinnati, Ohio 45221

ABSTRACT

The thermodynamic properties of the titanium-aluminum, titanium-gallium, titanium-tin, and titanium-aluminum-gallium systems were investigated. Thermodynamic activity measurements and binary interaction parameters were obtained using a high temperature Knudsen cell coupled with a time-of-flight mass spectrometer. Samples were placed in the Knudsen cell and the ratios of the signals from the various elements were determined as a function of composition.

The data were analyzed with the aid of a solid solution model and yielded for body centered cubic titanium:

*This work has been supported in part by the Air Force Materials Laboratory, Wright-Patterson Air Force Base, Ohio 45433, under Air Force Contract No. AF33615-69-C-1529.

$$\Omega^*_{Ti-Al} = -51.8 \pm 3.4 \text{ kJ/mol}$$

$$\Omega^*_{Ti-Ga} = -51.9 \pm 2.9 \text{ kJ/mol}$$

$$\Omega^*_{Ti-Sn} = -92.5 \pm 3.3 \text{ kJ/mol}$$

$$\Omega^*_{Al-Ga} = 7.1 \pm 7.9 \text{ kJ/mol}$$

INTRODUCTION

The present paper summarizes work on thermodynamic activity measurements for binary and ternary titanium-base systems with alloying elements aluminum, gallium, and tin which are α stabilizers. From these activity measurements interaction parameters, Ω^*, were calculated for the binary systems of titanium-aluminum, titanium-gallium and titanium-tin, and for aluminum-gallium in the ternary titanium-aluminum-gallium system.

If an element exists in two allotropic phases, α and β, and the β phase is the high temperature phase, then an alloying element which stabilizes the α phase will have a negative pairwise interaction energy whereas one which stabilizes the β phase will have a positive pairwise interaction energy(1). In the present case the interaction parameters were measured in the β phase, but their values will be equal to those for the α phase, within experimental error.

The thermodynamic activity and interaction parameter measurements, for binary titanium-base systems with β stabilizing elements were reported(2,3,4).

THERMODYNAMIC EQUATIONS

The thermodynamic and mass spectrometer equations were described earlier(3,5,6,7) and need only to be summarized here.

The free energy difference between an element, i, in solution and in its pure form is given by(6)

$$\overline{G_i^\circ} - G_i^\circ \equiv RT\ln a_i = \Delta G_i^{tr} + RT\ln N_i + \overline{G_i^{xs}} \tag{1}$$

In addition, it should be noted that(5)

$$\overline{G_i^{xs}} = N_j\Omega^*_{ij}\ (1-N_i) + N_k\Omega^*_{ik}\ (1-N_i) - \Omega^*_{jk}N_jN_k$$
$$+ \Omega^*_{ijk}N_jN_k - 2N_iN_jN_k\Omega^*_{ijk} \tag{2}$$

Relating the ion currents in a mass spectrometer to the partial pressures of the components, and making use of the definition of the activity coefficient, equation 2 for the i-j set becomes(7)

$$RT\ln[\frac{I_iN_j}{I_jN_i}] + RT\ln[\frac{Q_jR_j(E-A)_j(IA)_jp_j^\circ}{Q_iR_i(E-A)_i(IA)_ip_i^\circ}] + \Delta G_j^{tr} - \Delta G_i^{tr}$$

$$= (\Omega^*_{ij} + N_k\Omega^*_{ijk})(N_j - N_i) + N_k(\Omega^*_{ik} - \Omega^*_{jk}) \tag{3}$$

For convenience we define

$$Y_{ij}^\circ = RT\ln[\frac{Q_jR_j(E-A)_j(IA)_jp_j^\circ}{Q_iR_i(E-A)_i(IA)_ip_i^\circ}], \tag{4}$$

and

$$Y_{ij} = RT\ln[\frac{I_iN_j}{I_jN_i}] \tag{5}$$

In all equations

ΔG_i^{tr} = free energy difference between i in its standard state and that of the pure element when it has the structure and lattice parameter of the solution

I_i = ion current

Q_i = electron impact ionization cross-section

R_i = electron multiplier efficiency $\propto M_i^{-1/2}$

$(E-A)_i$ = energy of the ionizing electrons minus appearance potential

$(IA)_i$ = isotope abundance

$a_i = p_i/p_i^\circ$ = activity of element i

p_i° = pressure in the standard state

p_i = partial pressure of i over the alloy

N_i = atom fraction of i

Ω_{ij}^* = pairwise interaction energy

M_i = mass of isotope

If in equation 3 N_k is now held constant, Y_{ij} can be plotted against (N_j-N_i), with the slope of the resulting straight line equalling $(\Omega_{ij}^*+N_k\Omega_{ijk}^*)$. Since $N_k\Omega_{ijk}^*$ is a higher order term, this slope is a good approximation of Ω_{ij}^*.

EXPERIMENTAL METHOD

The experimental method consisted of coupling the Knudsen effusion technique with a time-of-flight mass spectrometer measurements of ion intensity ratios as a function of composition. A detailed description of this method has been given(3,5,6).

The particular data reported here were obtained on two separate, but similar, experimental setups. The titanium-aluminum data and the titanium-aluminum-gallium data were both obtained using system 1, but by different persons. The titanium-tin and titanium-gallium data were obtained by a third experimenter using system 2. The consistency of the data was checked through evaluating the ionization cross-section ratio between aluminum and gallium twice, once directly from the titanium-aluminum-gallium measurements, and then a second time using the ratios of the data obtained in the titanium-aluminum and titanium-gallium measurements.

MATERIALS

Samples in the form of 15-50g buttons, were prepared by arc melting (with nonconsumable electrodes) under an argon atmosphere. Each ingot was inverted and remelted six times to ensure homogeneity. Afterwards the sample compositions were determined by chemical analysis. Knudsen cells made of tungsten were used.

EXPERIMENTAL RESULTS

The experimental data for the titanium-aluminum and titanium-gallium systems are plotted in Figures 1 and 2 using equation 3. (Similar plots were obtained for the other systems.) The data for all the systems are summarized in Table I. ΔG^{tr} for aluminum, the energy required to transfer pure liquid aluminum at 1780°K to a body centered cubic structure with titanium's lattice parameter was determined(6) to be 6.7±2.9 kJ/mol. This same value was used for gallium and tin.

Table I contains the values of Ω^* and Y°_{ij} obtained at various temperatures as well as the ratio $\frac{Q_j}{Q_i}$ calculated using the vapor pressure data of Hultgren(8). A "best" value was arrived at, both for Ω^* and $\frac{Q_j}{Q_i}$, by determining a value which would be within one standard deviation of all the results. As a consequence the experimental data with small standard deviations had greater weight.

The standard errors in Ω^*_{ij} and Y°_{ij} and thus $\frac{Q_j}{Q_i}$ are influenced by the composition range (a wide composition range giving small error in Ω^*_{ij}) and the length of the extrapolation (a short extrapolation giving small errors in Y°_{ij}). Table I also contains the theoretical values of $\frac{Q_j}{Q_i}$ calculated by the methods of Otvos and Stevenson(9), Mann(10) and Gryzinski(11).

The value of 7.1 ± 7.9 kJ/mol for Ω^*_{Al-Ga} in the ternary titanium-aluminum-gallium system agrees quite well with the value of 2.9 kJ/mol calculated from the data summarized by Hultgren(8), if we keep in mind that

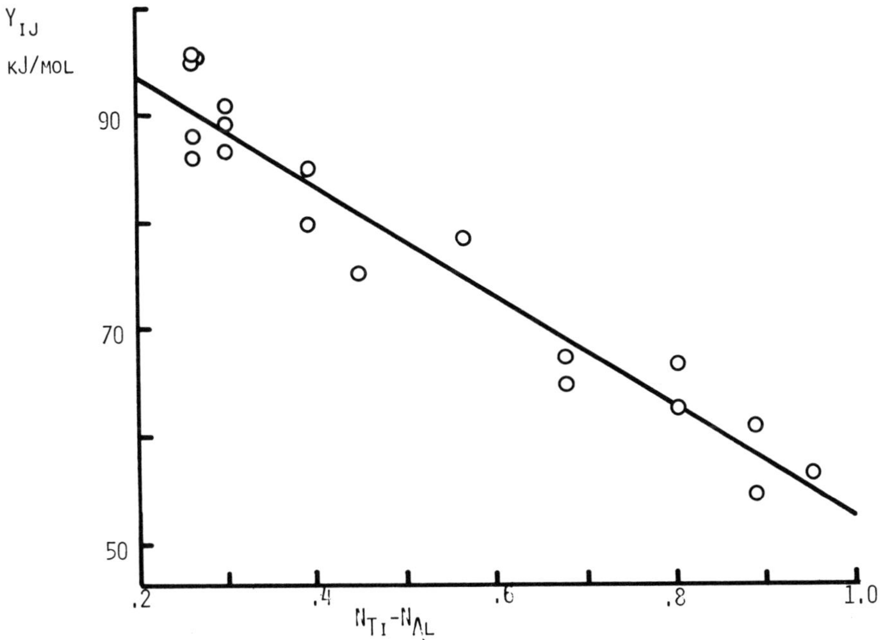

FIGURE 1. INTERACTION PARAMETER IN THE SYSTEM Ti-Al.

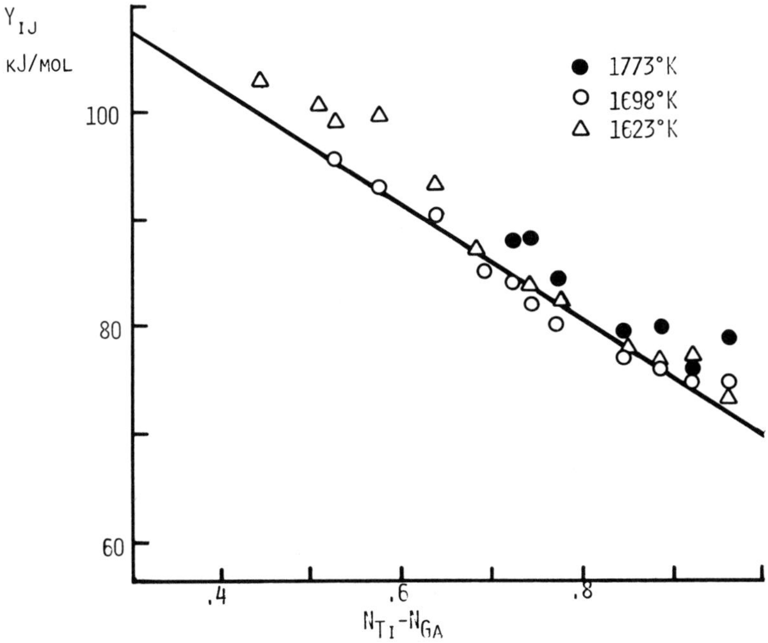

FIGURE 2. INTERACTION PARAMETER IN THE SYSTEM Ti-Ga.

TABLE I

INTERACTION PARAMETERS Ω^*_{ij} AND IONIZATION CROSS-SECTION RATIOS $\frac{Q_j}{Q_i}$ FOR Ti, Al, Ga and Sn

System	Temp.	Ω^*	Y°_{ij}	$\frac{Q_j}{Q_i}$	"best value" Ω^*	"best value" $\frac{Q_j}{Q_i}$	Theoretical $\frac{Q_j}{Q_i}$ Otvos-Stevenson	Theoretical $\frac{Q_j}{Q_i}$ Mann	Theoretical $\frac{Q_j}{Q_i}$ Gryzinski
j i	°K	kJ/mol	kJ/mol		kJ/mol				
Ti-Sn	1573	-94.3± 4.6	-105.4± 7.5	0.59					
	1623	-93.0± 6.4	-106.4± 6.5	0.58					
	1673	-91.6± 3.8	-110.2± 3.3	0.56	-92.5±3.3	0.56±0.12	1.43	0.85	1.16
	1723	-87.5± 7.6	-111.2± 5.4	0.64					
	1773	-86.8± 8.8	-113.0± 3.8	0.60					
Ti-Ga	1623	-62.2±15.4	-136.4±11.1	4.2					
	1698	-51.9± 3.1	-129.0± 2.4	5.7	-51.9±2.9	5.7 ±1.3	2.1	1.17	1.55
	1773	-47.4± 8.7	-128.5± 7.3	4.6					
Ti-Al	1780	-51.8± 3.4	-110.8± 2.0	3.6	-51.8±3.4	3.6 ±0.8	2.4	1.10	1.54
Ga-Al	1518	7.1± 7.9	16.6± 0.4	0.56	7.1±7.9	0.56±0.01	1.6	0.94	0.99
Ga-Al $[\frac{\text{Ti-Al}}{\text{Ti-Ga}}]$						0.63±0.23			

A_i, appearance potentials: Ti 6.82, Ga 6.00, Sn 7.34, Al 5.98 eV.
35 eV electron energy for Ti-Al, Ti-Sn, Ti-Al-Ga; 30 eV for Ti-Ga.

Ω^*_{Al-Ga} in the titanium-aluminum-gallium solid system can be expected to be more positive than in liquid aluminum-gallium. Deviation from ideality is normally greater (more positive or more negative) in a solid than in a similar liquid.

In contrast, the agreement between the theoretical calculations and the experimental data is not very good, for $\frac{Q_j}{Q_i}$, the relative ionization cross-sections except in the case of aluminum-gallium. This implies that the theoretical calculations for titanium are in error.

In the last line of Table I, an ionization cross-section ratio for aluminum-gallium is shown as calculated from the ratio of the values for titanium-gallium and titanium-aluminum. The value obtained of 0.63 ± 0.24 is in very good agreement with the directly measured value of 0.56 ± 0.01. This indicates that our thermodynamic model describes the systems satisfactorily, and that variations between the two experimental systems and the three experimenters are negligible, or, less probably, cancelling.

ACKNOWLEDGEMENT

These measurements were carried out by R. J. Usell, M. Yamawaki and C. J. Oblinger, with important help from L. R. Bidwell.

REFERENCES

1. Hoch, M., Hackworth, J.V., Usell, R.J., and Gegel, H.L., "Thermodynamic Properties of the Body-Centered-Cubic Beta-Phase in the Titanium-Copper and the Titanium-Aluminum Systems", The Science Technology and Application of Titanium, Dr. R. Jaffee and N. Promisel, Editors, Pergamon Press, New York, 1970 pp. 359-360.

2. Hackworth, J.V., Hoch, M., and Gegel, H.L., "Thermodynamics of Titanium Alloys: I. Development of a Triple Knudsen Cell and Its Use to Study the Activity of Copper in the Ti-Cu System", Met. Trans. 2, 1971 pp. 1799-1805.

3. Rolinski, E.J., Hoch, M., and Oblinger, C.J., "Determination of Thermodynamic Interaction Parameters in Solid V-Ti Alloys Using the Mass Spectrometer", Met. Trans., 2, 1971 pp. 2613-2618.

4. Hoch, M., and Viswanathan, R., "Thermodynamics of Titanium Alloys: III. The Ti-Mo System", Met. Trans. 2, 1971 pp. 2765-2767.

5. Rolinski, E.J., Hoch, M., and Oblinger, C.J., "Determination of Thermodynamic Interaction Parameters in Solid Vanadium-Titanium-Chromium Alloys Using the Mass Spectrometer", Met. Trans. To be published.

6. Hoch, M., and Usell, Jr., R.J., "Thermodynamics of Titanium Alloys II. Titanium and Aluminum Activities in the BCC β Phase of the Ti-Al System", Met. Trans. 2, 1971 pp. 2627-2632.

7. Rolinski, E.J., Oblinger, C.J., and Hoch, M., "Advances in Mass Spectrometry", Vol. 5, The Institute of Petroleum, "Relative Ionization Cross-sections of Transition Metals Using Time-of-Flight Mass Spectrometry", Printed in Great Britain by Adlard & Son Ltd, Dorking, 1971.

8. Hultgren, R., Orr, R.L., Anderson, P.D., and Kelley, K.K., "Selected Values of Thermodynamic Properties of Metals and Alloys". John Wiley & Sons, New York, 1963. Supplements 1966-1971.

9. Otvos, J.W., and Stevenson, D.P., "Cross-Section of Molecules for Ionization by Electrons", J.Am.Chem. Soc., 78, 1956, pp. 546-551.

10. Mann, J.B., "Ionization Cross Sections of the Elements Calculated from Mean-Square Radii of Atomic Orbitals", J.Chem.Phys., 46, 1967 pp. 1646-1651.

11. Gryzinski, M., "Classical Theory of Atomic Collisions. I. Theory of Inelastic Collisions", Phys. Rev., 138, 1965 A. 336-347.

THE TITANIUM-ALUMINUM-GALLIUM SYSTEM*

M. Hoch, T. Sakai, J. J. Krupowicz, and M. Delahanty
Department of Materials Science
and Metallurgical Engineering
University of Cincinnati
Cincinnati, Ohio 45221

ABSTRACT

The titanium-aluminum-gallium system was investigated using x-ray diffraction, electron diffraction, optical and electron microscopy, and microhardness. Every alloy composition contained equal atomic percents of aluminum and gallium. The titanium-aluminum-gallium system with equal atomic amounts of aluminum and gallium seems to behave qualitatively in the same way as the equivalent titanium-aluminum system. The big difference between titanium-aluminum alloys and titanium-aluminum-gallium alloys lies in the kinetics of precipitation.

The x-ray diffraction patterns for all samples

*This work has been supported in part by the Air Force Materials Laboratory, Wright-Patterson Air Force Base, Ohio 45433, under Air Force Contract No. AF33615-69-C-1529.

showed only hexagonal close-packed α phase, whereas the electron diffraction of $Ti_{.84}Al_{.08}Ga_{.08}$ and $Ti_{.80}Al_{.10}Ga_{.10}$ alloys showed α mixed with α_2 (an ordered phase). The same two alloys, under all annealing conditions, showed (both by x-ray diffraction and electron diffraction) a very strained lattice. This highly strained lattice structure was taken to be the explanation for the observed decrease in ductility with increasing temperature since stress induced precipitation seems to be needed to reach "equilibrium" conditions.

Precipitation is also suggested as the reason that in all titanium-aluminum-gallium alloys the hardness remained unchanged after annealing, in contrast to equivalent titanium-aluminum alloys where the hardness first increased and then decreased with additional annealing.

INTRODUCTION

In earlier papers(1,2,3), the development of a "better" hexagonal close-packed, 100% α phase titanium alloy was proposed to provide a material with superior high temperature properties. Modification of this "super" α-Ti alloy through the addition of small amounts of modifying elements was expected to make it commercially useful. The alloy showing the most promise was a titanium-aluminum-gallium combination. The present paper summarizes the physical properties of various alloys within the titanium-aluminum-gallium system. Details concerning the experimental procedures and results can be found elsewhere(4,5).

The alloys were prepared by melting the components in a vacuum arc melting furnace. They were then worked in three ways: forging, rolling, and swaging. Their properties, however, did not vary greatly with the working process. Annealing temperatures ranged from 600 to 1270°C, and the annealing time from 3 to 1000 hours.

EXPERIMENTAL RESULTS

1. X-ray Diffraction

The results of the x-ray diffraction analysis are given in Table I. No super-lattice lines were observed in any of the ternary alloys. In $Ti_{.84}Al_{.08}Ga_{.08}$ and $Ti_{.80}Al_{.10}Ga_{.10}$, however, a broadening of the back reflection lines was observed. In neither of these compositions, though, regardless of preparation and heat treatment, was it possible to resolve the K_{α_1} and K_{α_2} back reflection lines - with one exception. Resolution was obtained in $Ti_{.84}Al_{.08}Ga_{.08}$, when it was annealed at 1270°C (β region) for 3 hours. Apparently then, the precipitation of α_2 during annealing in the α region occurs very slowly for high alloy compositions.

Line broadening indicates nonuniform strain.

TABLE I

LATTICE PARAMETER OF TITANIUM ALLOYS ANNEALED AT 1270°C FOR 3 HOURS

Alloy	a (Å)	c (Å)	c/a	Vol.
$Ti_{.98}Al_{.01}Ga_{.01}$	2.947 ± 0.0014	4.685 ± 0.0048	1.589	35.248
$Ti_{.96}Al_{.02}Ga_{.02}$	2.941 ± 0.0010	4.685 ± 0.0036	1.593	35.091
$Ti_{.92}Al_{.04}Ga_{.04}$	2.932 ± 0.0014	4.679 ± 0.0043	1.596	34.844
$Ti_{.88}Al_{.06}Ga_{.06}$	2.923 ± 0.0014	4.673 ± 0.0046	1.599	34.575
$Ti_{.84}Al_{.08}Ga_{.08}$	2.912 ± 0.0039	4.662 ± 0.0049	1.601	34.240
$Ti_{.75}Al_{.125}Ga_{.125}$	2.883 ± 0.0025	4.640 ± 0.0031	1.610	33.389
$Ti_{.75}Al_{.25}$ (a/2)	2.896 ± 0.0010	4.654 ± 0.0028	1.608	33.799
$Ti_{.75}Ga_{.25}$ (a/2)	2.875 ± 0.0048	4.629 ± 0.0052	1.610	33.141

Unfortunately, quantitative evaluation was not possible. However, visual inspection did reveal that the line broadening was greater in some samples than in others. The alloy, $Ti_{.84}Al_{.08}Ga_{.08}$, for example, was tensile tested at room temperature, 518 and 620°C, and showed elongations of less than 1%, 9.2% and 2.6%. The corresponding line broadening, measured on material very close to the fracture surface, was the same as that in annealed samples for the room temperature and 518°C sample, but was much larger in the 620°C sample, indicating additional strain in the latter due to precipitation during testing.

2. Hardness Measurement

Microhardness data are shown in Figures 1 and 2.

Figure 1 shows that the hardness increases rapidly with increasing amounts of the alloying elements, up to a peak for $Ti_{.80}Al_{.10}Ga_{.10}$ annealed at 1000°C for 25 hours and at 1270°C for three hours. Subsequently, a hardness drop was observed in $Ti_{.75}Al_{.125}Ga_{.125}$.

The unexpectedly high hardness of $Ti_{.84}Al_{.08}Ga_{.08}$ and $Ti_{.80}Al_{.10}Ga_{.10}$ is due to the strained conditions of the samples.

Figure 2 shows hardness versus annealing time for the (strain free) $Ti_{.88}Al_{.06}Ga_{.06}$ alloy, and the (strained) $Ti_{.84}Al_{.08}Ga_{.08}$ alloy as well as for $Ti_{.84}Al_{.16}$ with various types of preparation. In the titanium-aluminum-gallium alloys, there is no sharp

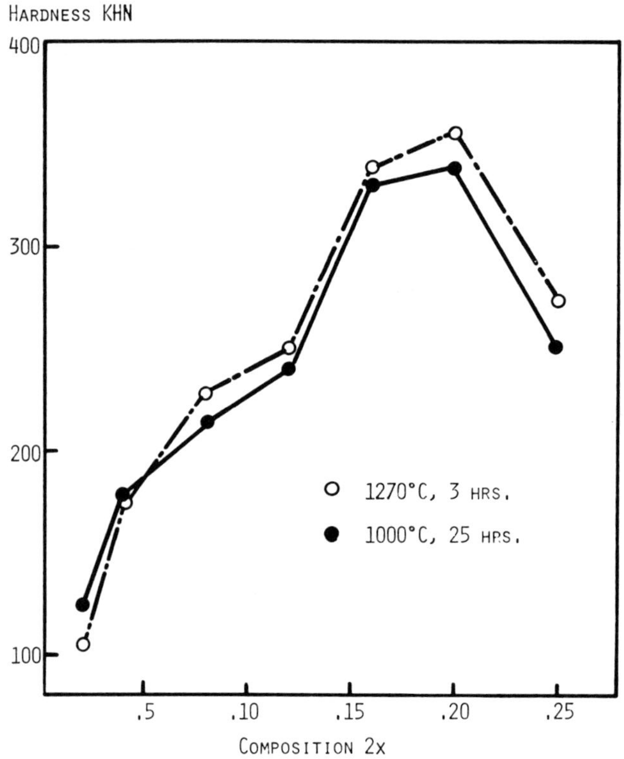

Fig. 1. Hardness versus composition of $Ti_{1-2x}Al_xGa_x$ alloys annealed at 1000°C and 1270°C.

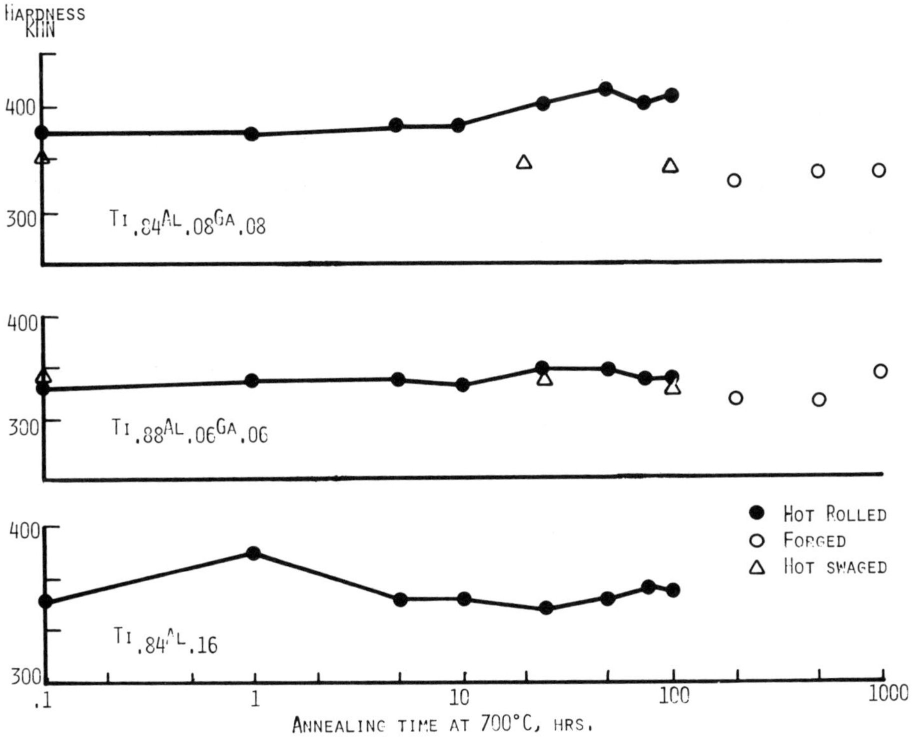

Fig. 2. Microhardness versus annealing time at 700°C, for $Ti_{.84}Al_{.08}Ga_{.08}$, $Ti_{.88}Al_{.06}Ga_{.08}$ and $Ti_{.84}Al_{.16}$, prepared by hot rolling, forging, and hot swaging.

hardness peak comparable to that observed in titanium-aluminum alloys(6) containing equal amounts of α stabilizer(6). The latter's hardness is normally ascribed to the precipitation of α_2.

3. Microscopic Examination

Figure 3 illustrates the microstructure of the six titanium-aluminum-gallium alloys after quenching from 1000°C. Martensitic or serrated structures were observed in those alloys because they were heated above the α-transus temperature.

$Ti_{.88}Al_{.06}Ga_{.06}$ annealed at 100°C for 25 hours was found to be a single phase α (Figure 3d), while $Ti_{.84}Al_{.08}Ga_{.08}$ showed weak striated lines in some grains after the same annealing process (Figure 3e). $Ti_{.80}Al_{.10}Ga_{.10}$ exhibited an even more complex micro-structure, which was described as a "textured" or "striated" structure in certain titanium-aluminum binary alloys previously investigated(7-13).

4. Fracture Surface

The fracture surfaces of the three $Ti_{.84}Al_{.08}Ga_{.08}$ alloys were inspected after testing at room temperature, 518 and 620°C. The surface of the sample tested at room temperature appeared uniformly brittle, while the surface of the sample tested at 518°C had both brittle and ductile areas and the sample tested at 620°C had a few ductile areas, but was mostly brittle.

5. Electron Diffraction

The $Ti_{.92}Al_{.04}Ga_{.04}$ and $Ti_{.88}Al_{.06}Ga_{.06}$ alloys did not show ordering by electron diffraction, but all $Ti_{.84}Al_{.08}Ga_{.08}$ and $Ti_{.80}Al_{.10}Ga_{.10}$ samples showed the diffraction spots of ordered alloys. Many more patterns were necessary, however, to find the ordered regions in the hot swaged alloys than in the forged and rod rolled samples.

In addition to the ordered patterns of α_2 another, as yet unindexed, pattern showed up in the specimen that was tensile tested at 620°C. Since this pattern was only found next to the fracture face of this sample (not in the grip section), and this sample showed considerable embrittlement when tested at 620°C, it is reasonable to assume that the strain nucleated another phase, perhaps Ti_2Ga.

6. Transmission Electron Microscopy

Figures 2a and 2b show transmission electron micrographs of alloys annealed at 700°C. Figure 4a shows a $Ti_{.88}Al_{.06}Ga_{.06}$ sample, with a limited dislocation network, and very little stress. Figure 4b shows an $Ti_{.84}Al_{.08}Ga_{.08}$ alloy with a maze of dislocations and a highly strained structure.

DISCUSSION

The titanium-aluminum-gallium system with equal atomic amounts of aluminum and gallium seems to behave

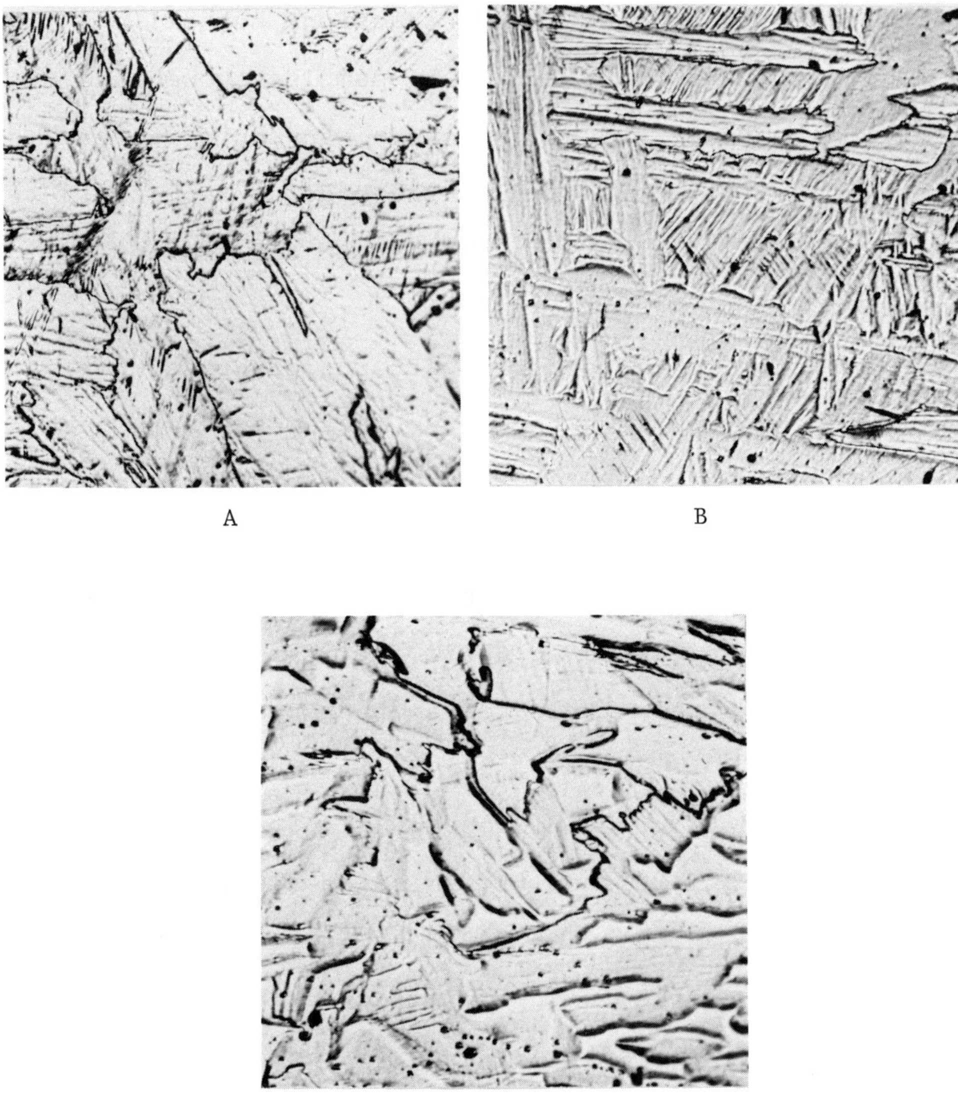

A B

C

Fig. 3. Microstructure of $Ti_{1-2x}Al_xGa_x$ alloys annealed at $1000^{\circ}C$ for 25 hours x200.

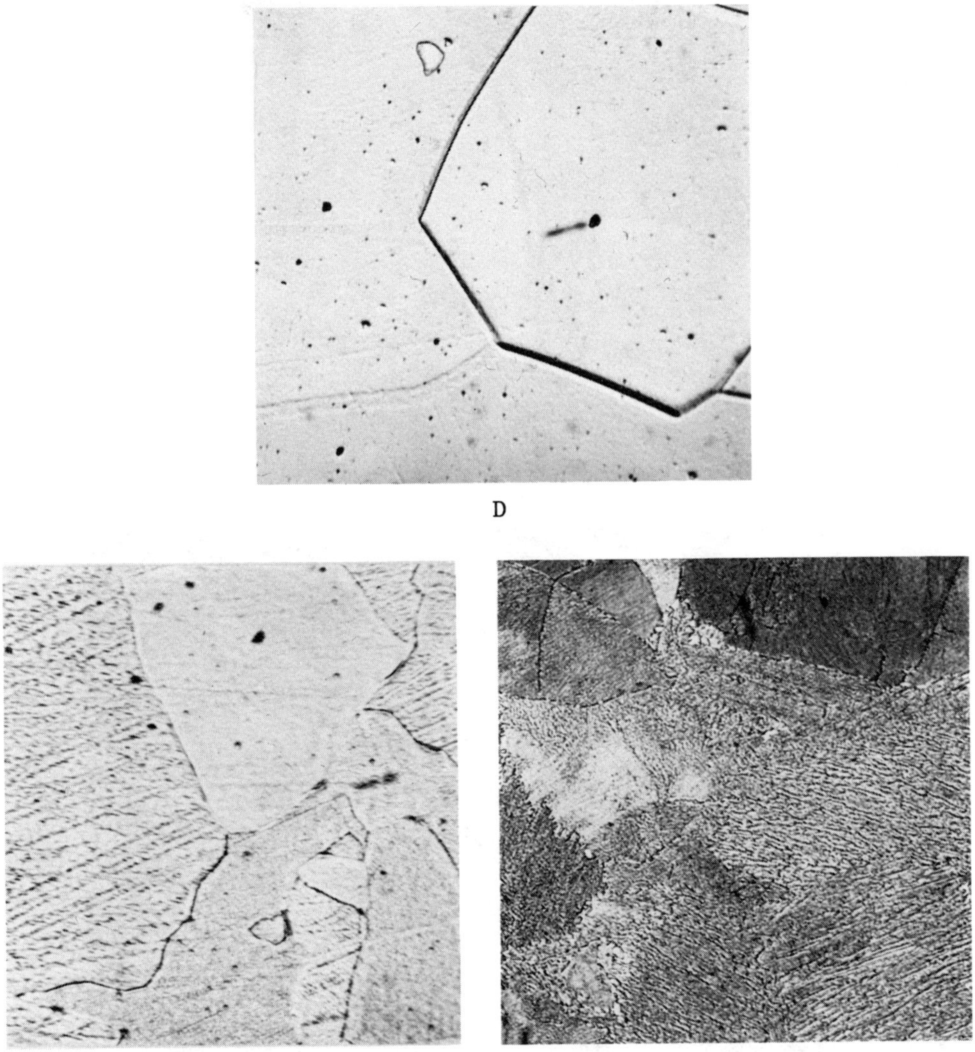

D

E

F

Fig. 3. (cont'd)

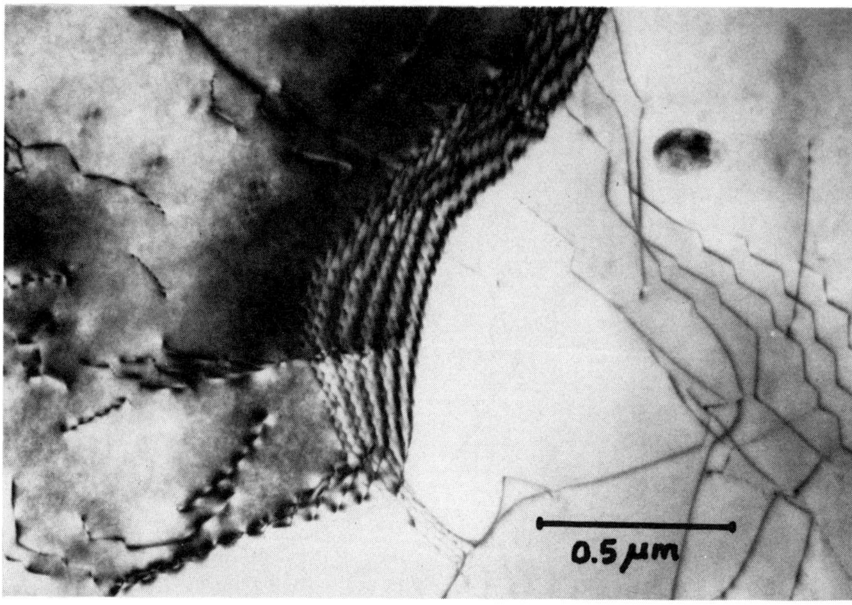

A

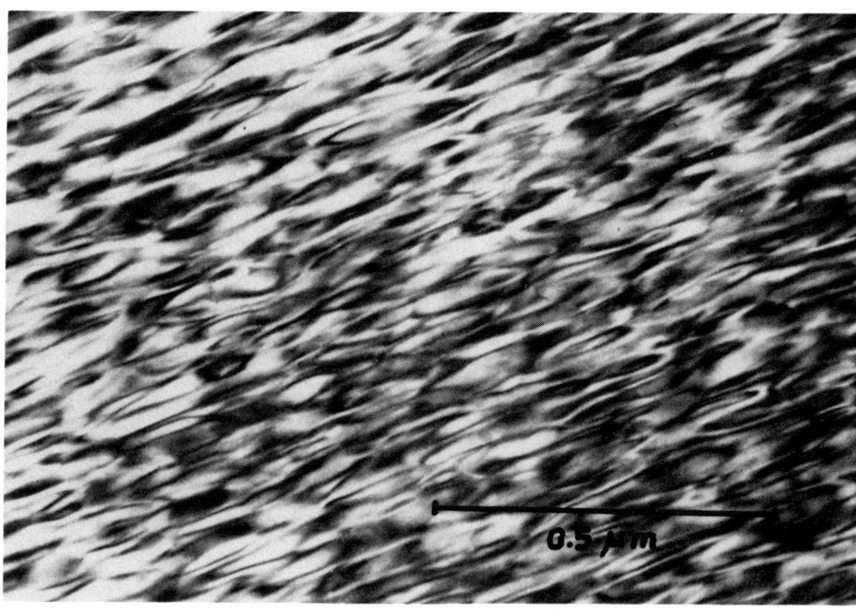

B

Fig. 4. Transmission electron micrograph, (a) $Ti_{.88}Al_{.06}Ga_{.06}$ low stress; (b) $Ti_{.84}Al_{.08}Ga_{.08}$, maze like structure, high stress.

qualitatively in the same way as the equivalent titanium-aluminum system. (Equivalency in this case, means the presence of the same atomic amounts of α stabilizers. Thus, the $Ti_{.88}Al_{.06}Ga_{.06}$ alloy corresponds to a 7.4 wt. % aluminum alloy, the $Ti_{.84}Al_{.08}Ga_{.08}$ to a 9.8 wt. % alloy and the $Ti_{.80}Al_{.10}Ga_{.10}$ to a 12.3 wt. % alloy.)

As with the titanium-aluminum system, the second phase of the titanium-aluminum-gallium precipitates when the composition gets into the region between $Ti_{.88}Al_{.06}Ga_{.06}$ and $Ti_{.84}Al_{.08}Ga_{.08}$. However, the quantitative aspects of the precipitation are quite different. Tests of the mechanical properties indicate that even after long annealing, strain induced precipitation takes place. In the $Ti_{.80}Al_{.10}Ga_{.10}$ sample the precipitation could not even be completed since the sample under all annealing conditions remained strained. In the case of the $Ti_{.84}Al_{.08}Ga_{.08}$ alloy only annealing at 1270°C (the β region) even lessened the strain enough so that back reflection lines in the x-ray diffraction pattern could be resolved. When the annealing took place at 700°C the alloys maintain constant hardness, not even showing the precipitation peak which occurs in titanium-aluminum alloys.

The big difference therefore between titanium-aluminum alloys and titanium-aluminum-gallium alloys lies in the kinetics of precipitation. The reason may be either that there is better coherence between the precipitate and the ternary alloy, or that the composition of the α_2 is quite different from the composition of the matrix in aluminum-gallium ratio. It may also be that the first material to precipitate in the ternary system is $Ti_2(Ga,Al)$, which then transforms into $\alpha_2[Ti_3(Al,Ga)]$.

Further work is needed to determine what controls the precipitation of α_2 so that some treatment can be found which permits complete α_2 precipitation in $Ti_{.80}Al_{.10}Ga_{.10}$. The $Ti_{.88}Al_{.06}Ga_{.06}$ alloys, with their smaller additions of alloying elements, have undergone additional testing of their mechanical properties and look promising for higher temperature applications. However, more α stabilizing elements must be added to improve the properties further before modifiers, such as silicon are added. This though, requires more knowledge of the precipitation mechanism.

REFERENCES

1. Hoch, M., Hackworth, J.V., Usell, R.J., and Gegel, H.L., "Thermodynamic Properties of the Body-Centered-Cubic Beta-Phase in the Titanium-Copper and the Titanium-Aluminum Systems", The Science Technology and Application of Titanium, Dr. R. Jaffee and N. Promisel, Editors, Pergamon Press, New York 1970, pp. 359-360.

2. Gegel, H.L., and Hoch, M., "Thermodynamics of α-Stabilized Ti-X-Y Systems" to be given at the 2nd International Conference on Titanium, Cambridge, Mass., May 2-5, 1972.

3. Gegel, H.L., and Hoch, M., "High Temperature Titanium Alloy Development" to be given at the III Inter-American Conference on Materials Technology, Rio de Janeiro, Brazil, August 14-17, 1972.

4. Sakai, Takeo, "Study on the Titanium-Rich Region of The Titanium-Aluminum-Gallium Ternary System", M.S. Thesis, University of Cincinnati, August, 1969.

5. Krupowicz, Jr., John J., "Study of the Titanium-Rich Regions of the Titanium-Aluminum-Gallium and Titanium-Aluminum-Gallium-Tin Alloy Systems", M.S. Thesis, University of Cincinnati, 1971.

6. Cavallaro, J.L., and Wilcox, R.C., "Embrittlement of Ti-7Al Binary Alloy in Sea Water", Corrosion 27, 1971 pp. 157-163.

7. Sagel, K., Schulz, E., and Zwicker, J., "Investigation of the Ti-Al System", Z. Metallkde 47, 1956, pp. 529-534.

8. Ogden, H.R., Maykuth, D.J., Finlay, W.L., and Jaffee, R.I., "Constitution of Ti-Al Alloys", Trans. AIME 191, 1951, pp. 1150-1155.

9. Sato, T., Huang, Y., and Kondo, Y., "A Metallographic Study on Ti-Al-Mn Alloy. Part 1: Equilibrium of the Ti-Al System", 1960, pp. 44 Sumitomo Light Metal Technical Reports.

10. Ence, E. and Margolin, H., "Phase Relations in the Ti-Al System", Trans. AIME 221, 1961, pp. 151-157.

11. Clark, D., Jepson, K.S., and Lewis, G.I., "Study of Ti-Al System up to 40 at. % Al", Insti. of Metals Journal 91, 1962-1963, pp. 197-203.

12. Crossley, F.A., "Titanium-Rich End of the Ti-Al Equilibrium Diagram", Trans. AIME 236, 1966, pp. 1174-1184.

13. Tsujimoto, T., and Adachi, M., "Reinvestigation of the Ti Rich Region of the Ti-Al Equilibrium Diagram", Inst. of Metal Journal 94, 1966, pp. 358-363.

THE INVESTIGATION OF PHASE EQUILIBRIUM AND SOME PROPERTIES OF Ti-Al-Zr-Mo-Fe ALLOYS

I. I. Kornilov, M. A. Volkova, and T. T. Nartova

A. A. Baikov Institute of Metallurgy

USSR Academy of Sciences, Moscow

Introduction

Aluminium is an important alloying element for titanium and is contained in most of titanium alloys. The studies of Ti-Al binary system alloys containing 8-IO w/o Al and the effect of some other elements on these alloys is of theoretical and practical interest because these alloys undergo the phase transition β bcc $\rightarrow$ α hcp the solid state transformation during cooling. Besides, in these alloys, the precipitation from α-solid solution takes place according to the solubility curve of α_2 phase based on Ti_3Al intermetallic compound (I).

Earlier we studied the effect of zirconium (2), molibdenum (3), vanadium (4) and the joint effect of Mo, Zr (5,6) on the phase composition and properties of titanium alloys with 9-IO w/o Al; thus, the multithermal cross-sections of corresponding ternary and quaternary systems were constructed.

The present work was carried out as a continuation of these studies. It was worthwhile to study the effect of iron, forming eutectoid with titanium, on the phase composition and properties of alloys of Ti-Al-Mo-Zr system (molibdenum and zirconium form continuous solid solutions with β-titanium; and zirconium continuous and molibdenum-limited solutions with α-titanium).

The alloys of the section of five-component system Ti-Al-Mo-Zr-Fe have been studied at different iron content from 0 to I0 w/o. The initial alloy composition was chosen on the section of Ti-Al-Zr-Mo system passing over the system Ti-Al, and the systems Ti-Zr--Mo and Zr-Mo-Al at the ratio of Zr:Mo = 5:I and containing 9 w/o Al, 5 w/o Zr, I w/o Mo and 85 w/o Ti.

The results from our studies on the state diagrams of the following systems: Ti-Al (I); Ti-Al-Fe (6) and Ti-Al-Zr (2) were used as the initial data for constructing the multithermal section of the five-component system. The initial alloy crystallizes as a β solid solution and during cooling passes through the β , $\alpha + \beta$ and $\alpha + \alpha_2 + \beta$ phase regions. Its phase composition is shown on the ordinate of multithermal section (Fig.I).

Experimental Procedure

For the preparation of alloys the following starting materials were used: iodide titanium, electrolytic iron, aluminium type AB-000, molibdenum of 99.9 w/o purity, iodide zirconium. The alloys containing 0, 0.3, 0.5, I, 2, 3, 5, 7 and I0 w/o Fe were studied.

The thermal, microstructural, X-ray phase analysis, specific electrical resistance, hardness methods were used in the work.

The alloys were prepared by the arc and by levitation meltings in argon. The difference in the specimen weights before and after melting was not more than 0.5%. The thermal analysis of the alloys was carried out with Kurnakov pyrometer by differential method. For heat treatment the alloys with titanium cuttings were sealed up into evacuated quartz ampules and annealed according to the following conditions: II00°C - 6 hrs, I000°C - 48 hrs, 800°C - 200 hrs, 700°C - - 300 hrs, 600°C - 400 hrs, 500°C - 500 hrs. After heating the specimens were quenched at every temperature in water. The polished specimens were etched with diluted hydrofluoric acid. Electrical resistance was measured by a potentiometer; the hardness on a diamond tipped tester at I0 kg load.

Results

Thermal analysis showed that the alloys of the investigated section of the five-component system undergo solid state transformations associated with ($\alpha \rightarrow \beta$) transformation in titanium and ($\beta + \alpha_2 \rightarrow \alpha$) in Ti-Al system and ($\beta \rightarrow \alpha$ + TiFe) in Ti-Fe system. The transformation temperatures obtained are shown on the multithermal section (Fig.I).

The alloy microstructure was studied after various quenching temperatures. Fig.I shows the phase composition of the alloys, depending on the chemical composition and temperature. After quenching at II00, I000, 800, 700, 600 and 500°C the alloys exhibited the following phase structures β , $\alpha + \beta$, $\alpha_2 + \beta$, $\alpha + \alpha_2 + \beta$, $\alpha + \alpha_2 + \beta$ + TiFe. In three- and four-phase regions $\alpha + \alpha_2 + \beta$ and $\alpha + \alpha_2 + \beta$ + TiFe the phase α_2(Ti_3Al) was observed not always distinctly. In most alloys quenched at 600 and 500°C shown in microphotographs TiFe compound is observed (in the finely dispersed mixture between grains of α -solid solution).

After quenching at 800, 700 and 600°C the alloys were examined under X-rays. The additional time of alloy annealing for X-ray analysis was as follows: 800°C - I00 hrs, 700°C - 200 hrs, 600°C - 340 hrs. The X-ray patterns were taken under copper radiation. The table shows the compositions of the analysed alloys and the phases observed.

From the tabulated data, it is obvious that α , β and TiFe phases are reflected in the X-ray patterns of the studied alloys. However, the state diagrams of the systems Ti-Al, Ti-Al-Zr, Ti-Al-Fe and Ti-Al-Mo-Zr, which are the components of the five-component system, as well as the thermal and microstructural analysis suggests that the phase α_2 based on Ti_3Al compound must exist in all alloys. The phase α_2 can be distinguished from the solid solution based on α -titanium only by the superstructural line I0I because the intense lines of the phases α and α_2 coincide. But the line I0I is very weak, moreover, its intensity decreases in the presence of other phases. It was discovered by us on the X-ray patterns of the titanium alloy with 9 w/o Al, while it was not observed on the X-ray patterns of other analysed alloys.

The solid solution α_2 based on the Ti_3Al compound which we think must exist in alloys was not corroborated by X-ray phase analysis (presence of line IOI in the Table is designated as (α_2).

The TiFe compound was detected by X-ray analysis in the alloys containing 5-IO w/o Fe (see Table).

In the alloys containing 0.5-2 w/o Fe the presence of TiFe compound from temperature 700°C and below is confirmed only by the microstructural analysis.

According to the thermal, microstructural and, partially, X-ray analysis, a multithermal section of the system Ti-Al-Mo-Zr-Fe was constructed (Fig.I). In the cross-section there exist β, $\alpha + \beta$, $\alpha_2 + \beta$, $\alpha_2 + \beta + \alpha$, α_2 + TiFe and $\alpha + \alpha_2 + \beta$ + TiFe phase regions. From Fig.I it is clear that on increasing the iron content the temperature of transition $\beta \rightarrow \alpha + \beta$ (or $\beta \rightarrow \alpha_2 + \beta$) decreases from 1030°C for an alloy not containing iron down to 790°C for an alloy with IO w/o Fe. The alloys containing 0-7 w/o Fe undergo solid state transformations with the precipitation of α or α_2 phase, associated with the transformation $\alpha_2 + \beta \rightarrow \alpha$ in Ti-Al binary system.

Table. X-ray Phase Analysis Data

Chemical composition w/o					Phase composition of the alloys		
Al	Mo	Zr	Ti	Fe	800°C	700°C	600°C
9	0	0	9I	0	$\alpha + \alpha_2$	–	–
9	0	5	86	0	$\alpha + (\alpha_2)$	–	–
9	I	5	85	0	–	$\alpha + (\alpha_2)$	$\alpha + (\alpha_2)$
9	I	5	83	2	$\alpha + (\alpha_2)$	–	$\alpha + (\alpha_2) + \beta$
9	I	5	80	5	–	–	$\alpha + (\alpha_2) + \beta$ + TiFe
9	I	5	78	7	$(\alpha_2) + \beta$	–	$\alpha + (\alpha_2) + \beta$ + TiFe
9	I	5	75	IO	–	–	$\alpha + (\alpha_2)$ + TiFe

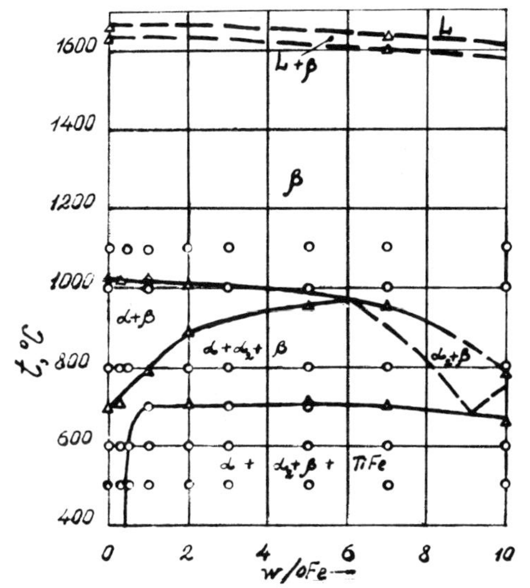

Fig. 1. Multithermal section of Ti-Al-Zr-Mo-Fe system with 15 w/o ΣAl, Mo, Zr, and variable iron content. — 0 — microstructural and X-ray analysis, — Δ — thermal analysis data.

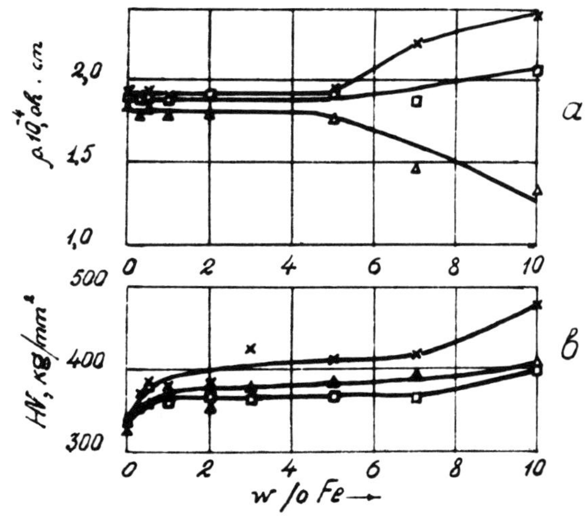

Fig. 2. Specific electrical resistance (a) and hardness (b) vs. iron content for the alloys of Ti-Al-Mo-Zr-Fe system over a section with 15 w/o ΣAl, Mo, Zr. x - quenched at 1000°C, □ - quenched at 800°C, and Δ - quenched at 500°C.

The alloys containing I-IO w/o Fe at 7IO-660°C undergo the transformation $\beta \rightarrow \alpha$ + TiFe with precipitation of TiFe compound as in the alloys of Ti-Fe binary. At 500°C TiFe compound precipitates at concentration ~ 0.4 w/o Fe.

The electrical resistance of the alloys were studied on the specimens, prepared by the levitation melting, after quenching at I000, 800 and 500°C (Fig.2a). On increasing the iron content from 0 to 5% the values of "ρ" for the alloys quenched at I000, 800 and 500°C remain constant. On further increasing the iron content up to 7-I0% the specific electrical resistance of the alloys increases up to $(2.2 \div 2.7) \cdot 10^{-4}$ ohm.cm. For the alloys quenched at 500°C containing 7-IO w/o Fe the specific electrical resistance decreases up to $(1.5 \div 1.3) \cdot 10^{-4}$ ohm.cm, respectively. The decrease of "ρ" values at quenching temperature 500°C is evidently connected with a more stable equilibrium state of the alloys after long time heating at this temperature.

The hardness of alloys quenched at I000, 800 and 500°C is shown in Fig.2b. The hardness of alloys quenched at I000°C, with increased iron content from 0 to IO w/o, increases from 350 to 470 kg/mm^2. For the alloys quenched at 800 and 500°C the hardness at the same iron content increases slightly and equal to 330--370 kg/mm^2. The hardness curves of the alloys quenched at I000°C are located higher than those of the alloys quenched at 800 and 500°C which is obviously associated with the increased hardness of the alloys quenched at the region β solid solution and with a more stable equilibrium state of the alloys annealed at 800 and 500°C.

Conclusions

I. The alloys of five-component Ti-Al-Zr-Mo-Fe system have been studied over the section with constant I5 w/o of Al+Zr+Mo and variable iron content from 0 to IO w/o by microstructural, thermal, X-ray phase analysis, measurements of electrical resistance and hardness. The multithermal section was constructed from the experimental results.

2. All the alloys of this section undergo solid state transformation associated with transformation of β-solid solution and precipitation of phases α, α_2 (Ti_3Al) and TiFe compound which begins to precipi-

tate at concentrations ~0.4 w/o Fe.

3. The specific electrical resistance and hardness of alloys quenched at 1000°C are higher than those of alloys quenched at 800 and 500°C because of the non-equilibrium state of the alloys under quenching conditions.

References

1. Kornilov, I.I., Pylayeva, E.N., Volkova, M.A., Krypyakevich, P.I., Markiv, V.Ya., "Phase Structure of the alloys of Ti-Al Binary System Containing from 0 to 30 w/o Al". Doklady AN SSSR, 161, 4, 1965, 843.

2. Pylayeva, E.N., Volkova, M.A., "Studies of Alloys Ti-Al-Zr Ternary System". Zb. Metallovedenie Titana, Izd.Nauka, 1964, 26.

3. Ge-Chzhi-Min, Pylayeva, E.N., "Studies of Phase Transformations in Ti-Al-Mo-System". Zb. Titan i ego Splavi. Vyp.10, Izd. AN SSSR, 1963, 22.

4. Kornilov, I.I., Volkova, M.A., "Phase Equilibrium in Ti-Al-V System". Izv. AN SSSR. Metally, No.1, 1968, 176.

5. Volkova, M.A., Kornilov, I.I. "Phase Equilibrium in the Alloys of Ti-Al-Mo-Zr System". Izv. AN SSSR, Metally, No.6, 1968, 195.

6. Volkova, M.A., Kornilov, I.I., "Studies of Phase Transformations in the Alloys of Ti-Al-Fe System in the Part Rich in Titanium". Izv. AN SSSR, Metally, No.4, 1969, 236.

DISCUSSION

SESSION VI. PHYSICAL PROPERTIES, ELECTRONIC STRUCTURE, PHASE STABILITY, PHASE EQUILIBRIA

Co-Chairmen

Professor A.D. McQuillan
University of Birmingham
Birmingham, England

Professor H. Margolin
New York University
Bronx, New York

Rapporteur

Dr. E.W. Collings
Battelle-Columbus Laboratories
Columbus, Ohio

The papers contributed to Session VI were grouped by the Rapporteur into three sections entitled, respectively, Superconductivity, Complex Elasticity, and An Electronic Basis for Mechanical Properties (Table 1). They were, therefore, discussed in a slightly different order from that in which they appeared in the Extended Abstract booklet, and in which they appear in these Proceedings. The Rapporteur emphasized that the goal of physical property measurements which, in the context of the Titanium Conference, was the development of alloys with improved mechanical properties. Accordingly, he pointed out that, although the results of physical property studies were of interest in their own right, it would be appropriate to indicate how they could lead eventually to the design of improved technical materials, the central theme of the conference. In his discussion, the Rapporteur attempted to introduce some coherency by emphasizing wherever possible the ways in which the various papers related to each other and to the central theme.

For example in the first section, the superconducting transition temperatures (T_c) of alloys of Ti with other transition elements were shown to be responsive electron/atom-ratio-induced microstructure. It was also pointed out that superconductivity could not only be regarded as an electronic (i.e., density-of-states-related) property but was also dependent on the electron-phonon interaction; for which reasons a kind of parallelism existed between the composition-dependence of T_c, the Debye temperature (θ_D), hardness, and an elastic modulus (e.g., Young's modulus).

TABLE 1. THE PAPERS OF SESSION VI

Physical Properties, Electronic Structure, Phase Stability, and Phase Equilibria

Section	Topic	Authors
Superconductivity	Primary superconducting properties, T_c	• J.C. Ho and E.W. Collings • E.G. Wolff and R.Lepper
	Tertiary superconducting properties, J_c	• R. Löhberg, W. Heller, and U. Zwicker
Complex elasticity	Elastic modulus	• S.G. Fedetov • S. Mishra and M.K. Asundi
	Internal friction	• S. Mishra and M.K. Asundi
	Diffusion	• L.G. Korneluk, L.M. Mirsky, and B.S. Bokshtein • S. Mishra and M.K. Asundi
	Basic theory	• E.W. Collings, J.E. Enderby, H.L. Gegel, and J.C. Ho
	Atomic interactions	• A.D. McQuillan • H.L. Gegel and M. Hoch
	Phase stability	• E.W. Collings, J.C. Ho, and R.I. Jaffee
Electronic basis for mechanical properties	Phenomenology of solid-solution strengthening and phase stability	• E.W. Collings, J.E. Enderby, H.L. Gegel, and J.C. Ho • E.W. Collings, J.C. Ho, and R.I. Jaffee
	Alloy phases	• M. Hoch, T. Sakai, J.J. Krupowicz, and M. Delahanty • I.I. Kornilov, M.A. Volkova, and T.T. Nartova

In a similar vein the complicated curves representing the composition-dependences of the elastic moduli of sequences of β-isomorphous and β-eutectoid Ti-base alloys could be simply explained in terms of microstructural state and electron/atom-ratio-controlled lattice electron/atom stability.

A fundamental treatment of solid-solution strengthening requires a knowledge of atomic-interaction strengths in alloys. In that context is was possible to couple papers dealing on one hand with thermodynamically measured interaction parameters for substitutional elements in Ti; and on the other hand, with interactions between interstitial elements in ternary interstitial alloys.

Finally the rapporteur delineated a phenomenology of phase stability and solid-solution strengthening in Ti alloys and indicated how the principles outlined might be applied to practical systems, offering Ti-Al-Ga and Ti-Al-Mo-Zr-Fe as examples.

The ensuing discussion centered about the thermodynamic and electronic bases of alloy theory; and the mechanical and structural properties of those alloys which might be regarded as having provided a proving-ground for some of the basic concepts. Thus, questions were raised concerning (a) applicability of electron/atom ratio in phase stability (b) the application of thermodynamics in phase diagram construction and in the determination of atomic interactions (c) electronic interactions between interstitial atoms themselves and between interstitials and the solvent, and (d) the mechanical and microstructural properties of the Ti-Al-Ga system.

In the context of phase stability, Kaufman questioned the validity of electron/atom ratio as a descriptor. He pointed out that a variety of approaches were currently being employed in defining electron/atom ratio in addition to those discussed in Collings and Ho's contribution. For example, the Engle-Brewer and PHACOMP methods respectively employ different bases. Moreover, recent application of PHACOMP techniques to sigma-phase formation in Ni-Cr-Mo required changing the group number of Mo from 4.66 to 9.66 (retaining a value of 4.66 for Cr) in order to conform with the phase boundary between the sigma and fcc phases.[1] Consequently the use of electron/atom ratios to define critical phase stability regions is questionable. This is further complicated by the fact that electron/atom based criteria do not define the relative stability of competing phases in quantitative terms. Kaufman asked Collings and Ho if their $n(E_F)$ versus electron/atom considerations could be employed to define the relative energies of bcc and hcp titanium alloys quantitatively.

On being invited by the chairman to respond to these comments, Collings discussed the determination and significance of electron/atom ratio as it was used in the third paper of this session.

In the simplest possible sense an electron/atom ratio (e/a or ζ) or Average Group Number in the case of transition-metal-binary alloys, may be employed (as indicated in the third paper at this session) as an arbitrary horizontal scale along which various physical properties can be displayed. Obviously such a parameter is not intended to provide information about the relative energies of the two principle allotropic forms of Ti. The Group Number of a transition element is equal to the number of d+s electrons in its outer atomic shells. This leads to a convenient way of assigning an electron/atom ratio to a transition metal binary alloy; and is the scheme employed in the first three papers of this session. Alternatively we could adopt an approach due to Matthias, who in one of his papers[(2)] used to some advantage an "electron density" (i.e., electrons/cc) as a descriptor of the superconductive and magnetic properties of transition-metal binary alloys.

Returning to electron/atom ratio (ζ), since the d-electrons make the largest contribution to the density-of-states at the Fermi level, it could be argued that it would be preferable to define ζ as the average number per atom of only the d-electrons. A scheme of this type was, in fact, adopted recently by Fisher and Dever.[(3)] An electron/atom ratio approach has considerable merit near the middle of a transition metal series where the rigid-band model has to be shown[(4)] to have some validity. Thus, when applied to transition-metal binary alloys in this regime, ζ can be regarded as defining a "new metal".

The problem of assigning a meaningful value to ζ becomes more difficult when dealing with alloys of a transition metal with a nontransitional element. In such strong-perturbation cases, as was pointed out in the first paper of this session, the electrons which screen the nontransition-ion cores go into states deep down in the conduction bands and the wavefunctions at the Fermi level derive essentially from the Ti d-electrons. Thus, the lattice remains "Ti-like". It is clear then that in considering the effective ζ, we must regard the B metal as diluting the Ti lattic; in which case ζ should perhaps be taken as the average number of (Ti) d-electrons per atomic (i.e., Ti plus B) site.

A lengthy discussion of this subject is, of course, not warranted, since a bare ζ, however, it is assigned, can never be very much more than a label, especially when applied to transition-metal/nontransition-metal alloys; and in considering electronic bases for physical properties we must look at the details of the electronic structures in each case.

During the discussion period, the subject of phase stability in bcc metals was taken up. In discussing the "degree" of stability of a given phase two possible variables could be considered: (a) On one hand phase stability could be treated as a function of structure at fixed composition (or ζ). An example might be a consideration of the relative stabilities of a pair of allotropic states of a pure metal; (b) on the other hand, we might want to consider the variation in degree of stability of a given structure as a function of alloy composition or $\mathfrak{z}$. In dealing with transition-metal binary alloys we are frequently interested in this latter aspect of lattice stability. In such a context, relative lattice stability could be defined as representing the degree to which a given lattice resists deformation at constant volume. The deformation involved in a conceptual test of stability could vary considerably in extent; from elastic deformations, through interplanar slip, to crystallographic transformation itself; which is alternative (a) above.

In Figure 1 the relative magnitude of the shear modulus, $C' = (c_{11}-c_{12})/2$, is used as a measure of the stability of the bcc structure. The figure shows the bcc phase to be most stable near an electron/atom ratio of 6, corresponding to a half-filling of the spectroscopic free-atom d-states (viz. $3d^5$, $4d^5$ and $5d^5$ for Cr, Mo, and W respectively). There is an irrefutable correlation between structural stability in the three long periods of transition-metal-binary alloys and $\mathfrak{z}$, however it is defined. We still await an interpretation of this effect in terms of modern electron theory.

Kaufman replied to several questions of a thermodynamic nature. In response to a question by Conrad as to whether thermochemical data could be predicted from phase diagram data, Kaufman replied affirmatively, citing several references from his paper[5,6,7], as examples. Conrad also wanted to know what relevant information could be obtained from high-pressure studies of quenched-in phases. In reply Kaufman pointed out that high-pressure studies of the ω-phase in Ti have been employed to characterize its lattice stability[5,8]. Similar considerations have been applied in the case of iron[5]. The question of metastable phase retention is a complex one which has not received adequate attention. Several interesting studies have been reported by Soviet workers in connection with superconductivity[9] and transitions in compounds.[10]

In response to a question by Nutting, Kaufman went on to discuss the fcc phase in Ti alloys. The calculated lattice stability of fcc titanium indicates that it is approximately 800 cal/g.at. less stable than the hcp form[5] Moreover, in addition to widespread observations of this structure in thin foils of titanium alloys[11-14], the fcc structure can be produced by vapor depositing thin films of Zr, Hf, Nb, Ta, Mo, and W.[5] Consequently, it is not surprising to find the fcc structure in titanium alloys appearing under proper conditions. A member of the audience suggested that a calculated phase diagram could be regarded as stating what "cannot exist" rather than what "can exist", since phase equilibrium conditions were calculated without detailing the kinetics or metastable phases. Kaufman replied that the

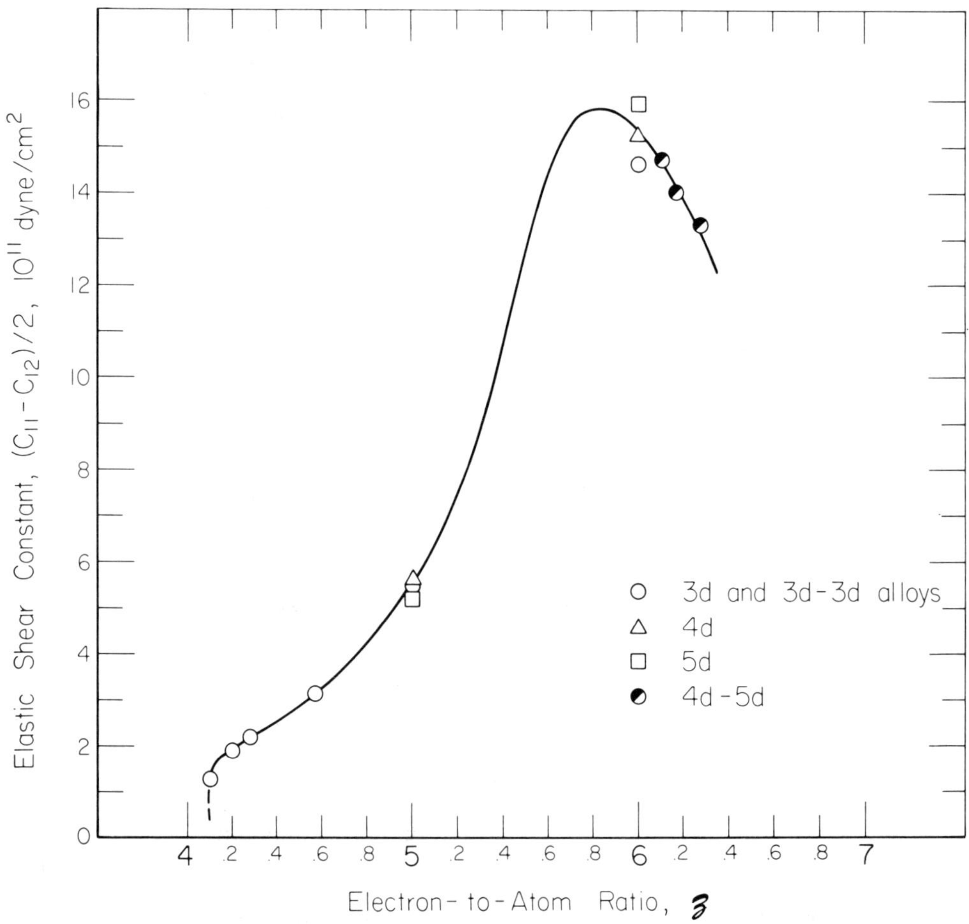

Fig. 1. Relative stabilities of transition-metal binary alloys in the electron/atom ($\mathfrak{z}$) range $4 < \mathfrak{z} < 6.4$ Symbols -○- refer to bcc Ti-Cr[3], V[15], and Cr[16]; the -△- refer to Nb[15] and Mo[17]; the -□- refer to Ta[18]; and the -◐- refer to three Mo-Re alloys[17].

calculations did not provide kinetic data but did provide a quantitative description of metastable and unstable phases in addition to the most stable phase relations. The resulting information could be used in kinetic considerations to define driving forces.

Considerable interest was shown in the mechanical, thermodynamic, and microstructural properties of the Ti-Al-Ga system, questions concerning which were addressed to Hoch. Petrova asked for the composition of the best alloy and its temperature range of applicability. Hoch replied that once some problems relating to the $Ti_{.80}Al_{.10}Ga_{.10}$ and $Ti_{.84}Al_{.08}Ga_{.08}$ alloys were resolved, they would probably be the best alloys of this type, with a useful temperature range of up to 1800°F. The presently developed alloys which are based on the $Ti_{.88}Al_{.06}Ga_{.06}$ alloys (see Gegel's work) will be for somewhat lower temperatures.

An important point was raised by Higbie of the Bureau of Mines, who questioned the availability of Ga and asked whether one should try to store it. Hoch in reply pointed out that Ga is quite an abundant element, although at present there is only a small market for it. The present method of extraction is electrolytic and involves large amounts of mercury. The method was developed over 30 years ago by a Ph.D. student in Marsille, or Lyon, France, whose father worked for Alusuisse and who needed a thesis topic for his Ph.D. degree. If the need arises, new methods will of course be developed for extracting gallium from Bauxite.

In reply to a question from the floor concerning the relationship between the interaction parameter, Ω, and dislocation density and dislocation movement, Hoch commented that he was not aware of any calculations connecting these quantities. He went on to add that elastic properties can, however, be calculated using the interaction parameter; citing his own work on ternary carbide systems as an example.

References

(1) J.R. Mihalisin, C.G. Bieber and R.T. Grant, Trans. Met. Soc. AIME, 242 2414 (1968).

(2) M.A. Jensen, B.T. Matthias, and K. Andres, Science, 150 1448 (1965).

(3) E.S. Fisher and D. Dever, Acta Met., 18 265 (1970).

(4) W.L. McMillan, Phys. Rev., 167 331 (1968).

(5) L. Kaufman and H. Bernstein, Computer Calculation of Phase Diagrams, Academic Press, New York (1970).

(6) L. Kaufman, International Symposium on Thermodynamics of Alloys, Th. Heumann and B. Predel, Eds. (1972) Deutschen Gesselschaft für Metallkunde.

(7) L. Kaufman, International Symposium on Metallurgical Chemistry--Fundamentals and Applications, N.P.L.--London (1971 paper 3.1.

(8) N.A. Vanderpuye and A.P. Miodownik, The Science, Technology and Application of Titanium, R.I. Jaffee and N. Promisel Eds., Pergamon (1970) p. 719.

(9) N.B. Brandt and N.I. Ginsburg, Scientific American, 224 83 (1971).

(10) E.I. Estrin, "Phase Changes under High Pressure", Study of Metals for High Temperature and Metallworkings, Moscow (1971).

(11) C. Hammond and P.M. Kelley, The Science, Technology and Applicatic of Titanium, R.I. Jaffee and N. Promisel Eds., Pergamon, (1970) p. 659.

(12) M.J. Blackburn, The Science, Technology and Application of Titani R.I. Jaffee and N. Promisel Eds., Pergamon (1970) p. 633.

(13) L.N. Guseva, L.A. Petrova, and I.A. Ogloblina, Dokl. Akad. Nauk SSSR, 185 799 (1969) (Tr. Sov. Phys. Dokl., 14 367 (1969).

(14) R.H. Ericksen, R. Taggart, and D.H. Polonis, Trans. Met. Soc. AIM 245 359 (1969).

(15) (Figure 1) D.I. Bolef, J. App. Phys., 32 100 (1961).

(16) (Figure 1) D.I. Bolef and J. de Klerk, Phys. Rev., 129 1063 (1963

(17) (Figure 1) D.L. Davidson and F.R. Brotzen, J. App. Phys., 39 5768 (1968).

(18) (Figure 1) F.H. Featherston and J.R. Neighbours, Phys. Rev., 130 1324 (1963).

SECTION VII

DEFORMATION AND FRACTURE

CRITICAL REVIEW

DEFORMATION AND FRACTURE

H. Conrad, M. Doner and B. de Meester

Metallurgical Engineering and Materials Science Department
University of Kentucky
Lexington, Kentucky 40506
U.S.A.

Abstract

The plastic flow and fracture of titanium are reviewed with special attention to the mechanisms which are controlling. The low temperature ($T<0.3T_m$) plastic deformation is described by

$$\tau = \tau^* (T, \dot{\gamma}, C_i) + 0.5\mu b\rho^{1/2} + \tau_\mu (C_s)$$

where τ is the resolved shear stress, T the temperature, $\dot{\gamma}$ the strain rate, C_i the interstitial solute content, μ the modulus, b the Burgers vector, ρ the dislocation density and C_s the substitutional solute content. For α-stabilizing substitutional solutes

$$\tau_\mu(C_s) = 34.1 \quad I\ C_\alpha^{1/2} \qquad (\mathrm{Kg/mm^2})$$

where I is the interaction parameter derived from individual atomic interaction coefficients. The rate controlling dislocation mechanism for both unalloyed titanium and titanium alloys at low temperatures is the thermally activated overcoming of interstitial solute obstacles. Fracture at low temperatures is most generally of the shear type and appears to be controlled by the same mechanism as plastic flow.

Static and dynamic strain aging occurs in the intermediate temperature range of 0.3 to 0.45 T_m. At higher temperatures the deformation can be described very well by Weertman's dislocation climb model. Fracture at high temperatures appears to be controlled by the same mechanism.

The ratio of fatigue limit to yield stress is approximately

0.65, relatively independent of grain size, interstitial content and temperature, suggesting that fatigue is controlled by the same mechanism as plastic flow.

Introduction

This paper is a review of the present state of our understanding of the plastic flow and fracture of titanium. Special attention is given to identifying the dislocation mechanisms which are operative and controlling. It is felt that an understanding of the mechanical behavior of titanium in terms of recent advances in dislocation theory and materials science will provide a base for the more effective use of available alloys and the design of new and improved alloys.

Of importance to the mechanical behavior of titanium are certain physical properties. Melting occurs at 1941°K and a phase change from BCC to CPH crystal structure takes place at 1156°K (1). The Debye temperature is reported to be 402° to 428°K (2, 3, 4) yielding a Debye frequency of approximately 9×10^{12} sec^{-1}. Some of the other physical properties of importance are the effect of temperature on the elastic moduli (Fig. 1) and diffusion coefficients, Tables I and II†. From Fig. 1 it is seen that the moduli of titanium exhibit a relatively strong temperature dependence. In general, both interstitial and α-stabilizing substitutional solutes tend to increase the elastic moduli (3, 4, 7, 9); their effect on the temperature dependence is not completely clear, but they do not appear to alter it significantly. The addition of β-stabilizing solutes generally leads to a decrease in modulus (4).

Table I. Diffusion of Interstitial Solutes in α-Titanium (10)

Interstitial Solute	D_o cm^2/sec	Q Kcal/mole
O	0.78 - 16	45 - 52.5
N	0.2 - 6.8	57 - 68
C	5.1	43.5 - 48.5
H	0.3 - 0.9	12.4 - 14.7

An excellent review of the crystallography and deformation modes in titanium and other CPH metals is provided by Partridge (11). The effects of interstitial solutes on the lattice parameters

† The diffusion coefficients listed in Tables I and II are those considered to be the most realistic (10).

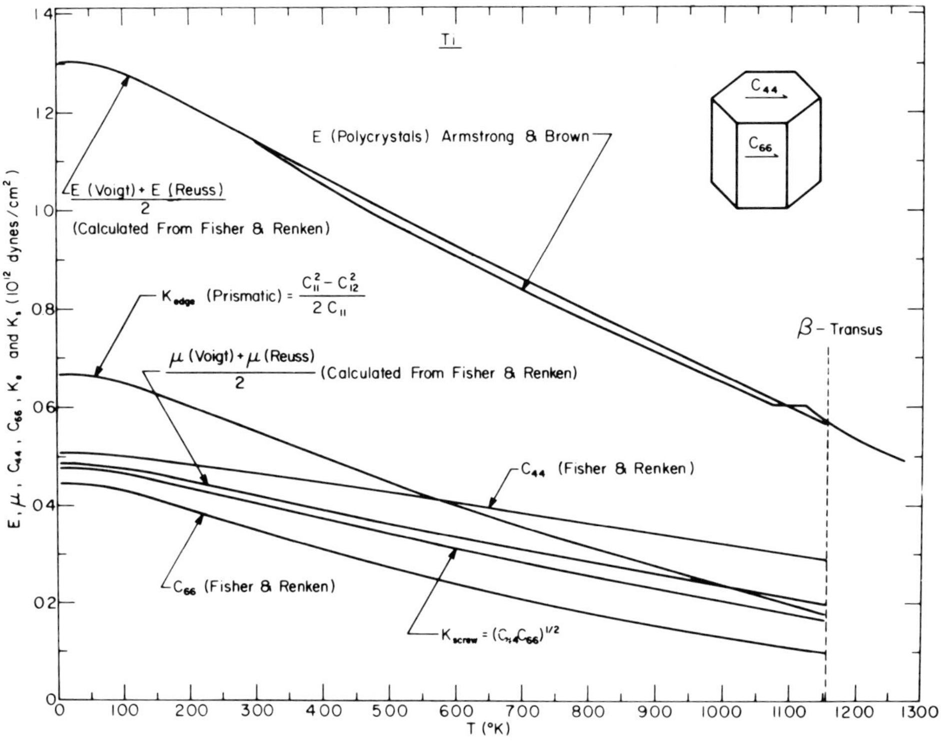

Fig. 1. Effect of temperature on the various elastic constants in titanium. Data from Refs. 5 and 6.

Table II. Self-Diffusion and Substitutional Solute Diffusion in Titanium (10)

Atom Species	D_{O_1} cm^2/sec	Q_1 Kcal/mole	D_{O_2} cm^2/sec	Q_2 Kcal/mole
		α-Ti		
Self	-	58†	-	-
	-	71.3††	-	-
		β-Ti		
Self	1.1	60	$3.6x10^{-4}$	31.2
Substitutional Solutes	0.2-20	51-73	$3x10^{-4}$-$80x10^{-4}$	29-43

Note: Two values of D_O and Q are given because of curvature in the Arrhenius plots. The subscript 1 refers to the high temperature results, the subscript 2 to low temperature data.

† Energy for the formation of vacancies plus that for migration determined from the recovery of electrical resistivity after plastic deformation: private communication K. Okazaki.

†† Twice the energy to form vacancies determined from heat capacity measurements; V. O. Shestopol, Soviet Phys. Solid State 7 232 (1966).

of CPH titanium are reviewed by Tyson (12); the elastic properties of dislocations in titanium are described by Teutonico (13).

Plastic Flow

1. Single Crystals

a. Deformation Modes (11, 14): Numerous investigators have established that the most common slip modes in CPH α-titanium are (in order of ease of operation) the $\{10\bar{1}0\}$, $\{10\bar{1}1\}$ and (0001) planes with the $\vec{a}$ direction, $<11\bar{2}0>$, as the slip direction. This order seems to prevail over the entire temperature range of the CPH structure; however, slip on the $\{10\bar{1}1\}$ and the (0001) planes tends to become relatively more favorable compared to the $\{10\bar{1}0\}$ planes as the interstitial solute content or the temperature is increased. These three planes with the slip vector in the basal plane constitute a total of four independent slip systems.

Evidence for slip with a vector out of the basal plane has been obtained by electron transmission microscopy (15-17) and by slip trace analysis (15, 17).

The dislocation structure in deformed titanium single crystals was studied by Cass (16). For specimens deformed by prism slip and sectioned parallel to the prism plane the structure consisted of long lengths of single dislocation lines, most of them very near to the screw orientation. These screw dislocations trailed a large number of resolvable and, presumably, unresolvable dipoles. In one specimen, dislocations with a vector out of the basal plane (presumably $\vec{c} + \vec{a}$) were observed. For specimens deformed by basal glide and sectioned parallel to the basal plane, the dislocations were again near the screw orientation and as in the prism case, there was no tendency for pairing or tangling of long dislocations. However, the isolated debris appeared to be in the form of small, approximately circular loops rather than elongated ones.

Although the slip systems operative in unalloyed single crystals of BCC β-titanium have not been established, it is expected that they are the same as those for other metals of the BCC structure, i.e. {110}, {112} and {123} planes with <111> as the slip direction (18). For the stable BCC alloys of titanium with vanadium (20 to 50 at. %V) and with molybdenum (15 at. %) the slip direction was found to be <111> with {110}, {112} and {123} as possible slip planes (19, 20).

Twinning in titanium has been reported to occur on the $\{10\bar{1}x\}$ and $\{11\bar{2}x\}$ planes (11, 14, 17). The particular twinning plane which is operative in a given case is dependent on the direction of stressing with respect to the crystallographic axes and on the temperature. In general, the tendency for twinning decreases with increase in temperature and in interstitial solute content. However, Paton and Backofen (17) found that the stress for the $\{10\bar{1}1\}$ twinning mode decreased with increase in temperature. The magnitude of the shear associated with several twinning planes is given in Table III.

Table III. Twinning Shear in Titanium

		Magnitude of Shear			
Plane	Direction	Calculated	Ref.	Observed	Ref.
$\{10\bar{1}1\}$	$\langle 10\bar{1}2\rangle$	0.104	17	0.09 to 0.11	17
$\{11\bar{2}1\}$	$\langle 11\bar{2}6\rangle$	0.638	11	---	--
$\{11\bar{2}2\}$	$\langle 11\bar{2}3\rangle$	0.218	21	0.21	17
$\{11\bar{2}2\}$	$\langle 11\bar{2}3\rangle$	0.225	11	---	--

b. Effects of Temperature and Interstitial Solute Content on the Resolved Shear Stress for Slip and Twinning: The effects of temperature T and interstitial solute content C_i on the critical resolved shear stress τ_{crss} (resolved shear strain γ equal to $\sim 10^{-4}$) for {10$\bar{1}$0} <11$\bar{2}$0> prism slip are presented in Fig. 2. The total C + N + O interstitial solute content is given in terms of an oxygen equivalent (28), with C = 3/4 O and N = 2 O. The effect of temperature on the τ_{crss} for a given interstitial content is typical, permitting the separation of the flow stress into a thermal component τ^* and an athermal component τ_μ (29). The interstitial content influences both τ^* and τ_μ, the effect on τ^* increasing with decrease in temperature. The effect of the resolved shear strain γ on the flow stress is primarily through τ_μ (25, 27). Moreover, over the temperature range of 0°K to 650°K the strain hardening coefficient $\theta = d\tau/d\gamma$ divided by the shear modulus μ (C_{66}) is relatively independent of temperature (25, 27). Thus, one can write for the resolved shear stress for prism slip,

$$\tau = \tau^* (T, \dot{\gamma}, C_i) + \tau_\mu (\mu, \gamma, C_i); \qquad T_o (\tau^* = 0) = 600°K \qquad (1)$$

where $\dot{\gamma}$ is the resolved shear strain rate. τ^* decreases in an approximately linear fashion with the square root of the temperature (Fig. 3) yielding

$$\tau^* \approx \tau^*_o - a\, T^{1/2}; \qquad T'_o (\tau^* \approx 0.1\ Kg/mm^2) = 510°K \qquad (2)$$

The values of τ^*_o and a are proportional to the square root of the interstitial content with $\tau^*_o = 0.06\mu_o\, C_i^{1/2}$ and $a = 11\ C_i^{1/2} Kg/mm^2/°K^{1/2}$.

Using the etch pit technique, Tanaka and Conrad (30) investigated the motion of {10$\bar{1}$0}1/3 <1120> edge dislocations in zone refined single crystals with an interstitial content of ~0.1 at.% Oeq. The stress associated with a dislocation velocity of $\sim 10^{-3}$ cm sec^{-1} is plotted as a function of temperature in Fig. 4. There is good agreement between this stress and τ_{crss} for prism slip at a strain rate of about 10^{-4}sec^{-1}.

Also presented in Fig. 4 is the effect of temperature on the τ_{crss} for basal slip and on the shear stresses for twinning on the {10$\bar{1}$2}, {11$\bar{2}$1}, {11$\bar{2}$2} and {10$\bar{1}$1} planes. It is seen that the temperature dependence of τ_{crss} for basal slip is appreciably higher than that for prism slip, but the effect of interstitial content on the temperature dependence appears to be less. The stress for twinning on the {10$\bar{1}$2}, {11$\bar{2}$1} and {11$\bar{2}$2} planes increases with temperature while that for {10$\bar{1}$1} twinning decreases. The stress for twinning on the former three planes is less than that for prism slip below about 200°K, but it is above at 300°K and higher. The crossover for basal slip is somewhere between 300° and 400°K.

c. Deformation Kinetics: It has been shown (27) that the deformation kinetics for prism slip in unalloyed titanium are described by

$$\dot{\gamma} = \rho_m b \, \bar{v} = A \exp - \{G \, (\tau^*)/kT\} \tag{3}$$

$$= \frac{\rho_m}{\ell^*} A_s \, b \, \nu^* \exp \, (S/k) \exp \{-H(\tau^*)/kT\} \tag{3a}$$

$$= \nu \exp \{-H(\tau^*)/kT\} \tag{3b}$$

where $\dot{\gamma}$ is the shear strain rate, ρ_m the mobile dislocation density, b the Burgers vector, $\bar{v}$ the average dislocation velocity, G the Gibbs free activation energy, A a constant, ℓ^* the average spacing between obstacles along the mobile dislocations, A_s the area of the slip plane swept out per successful thermal fluctuations, ν^* the vibrational frequency of the dislocation segment involved in the thermal activation and k and T have their usual meaning. G = H–TS where H is the activation enthalpy and S is the activation entropy.

Plots of H versus T for prism and basal slip are given in Fig. 5. H was derived from constant strain rate type tests using the relationships developed by Conrad and Wiedersich (32). To be noted is that $H_{T'_o} = 1.25$ eV ($0.29 \, \mu b^3$) for prism slip ($\mu = C_{66}$) independent of interstitial content and 2.34 eV ($0.33 \, \mu b^3$) for basal slip ($\mu = C_{44}$). $H_{T'_o}$ for prism slip in titanium is similar in magnitude ($0.2 \, \mu b^3$–$0.3 \, \mu b^3$) to that observed for prism slip in the CPH metals Cd, Zn and Be (12, 33, 34). Considering Eqn. 3b, the slopes of the curves of Fig. 5 yield $\nu = 3.3 \times 10^8$ sec^{-1}and 5.2×10^{26}sec^{-1} respectively for prism and basal slip. The value of ν for prism slip is within the range of values (10^1 to 10^{15} sec^{-1}) generally obtained for the deformation of metals at low temperatures (33-35), while that for basal slip is considerably higher, casting some suspicion on the validity of the analysis for basal slip.

A consideration of the data in Figs. 2 to 5, and other results (27, 30), along with the results on polycrystalline specimens to be discussed below, leads to the conclusion that stress-assisted, thermally activated overcoming of interstitial solute obstacles by moving edge dislocations is the rate controlling mechanism for prism slip in titanium at low temperatures ($T < 0.3T_m$). In the cases of basal slip and twinning, there are at present insufficient data to permit an identification of the mechanisms which are controlling.

d. Strain Hardening: The strain hardening coefficient θ for prism slip was found to vary with temperature approximately as the shear modulus $\mu = C_{66}$ over the remperature range of 78° to 650°K with $\theta/\mu \approx 10^{-3}$ (25, 27). A slight increase in θ/μ did, however, occur at 4.2°K (27). The magnitude of the strain hardening in titanium is similar to that reported for prism slip in other CPH

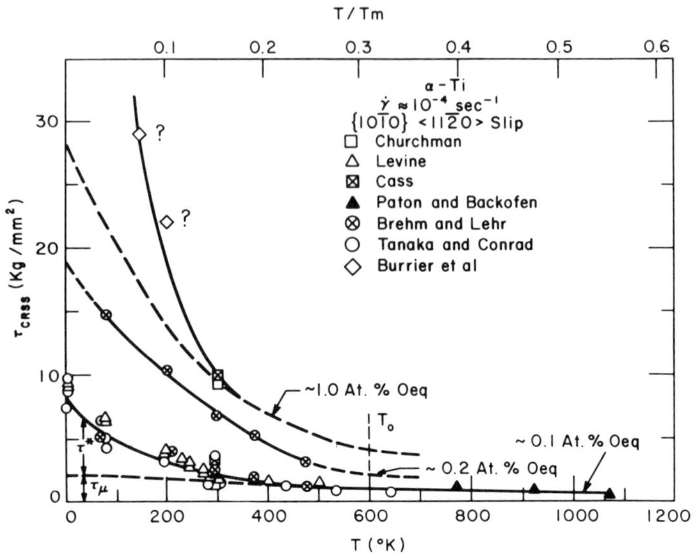

Fig. 2. The effects of temperature and interstitial content on the critical resolved shear stress for prism slip. Data from Refs. 16, 17, 22-27.

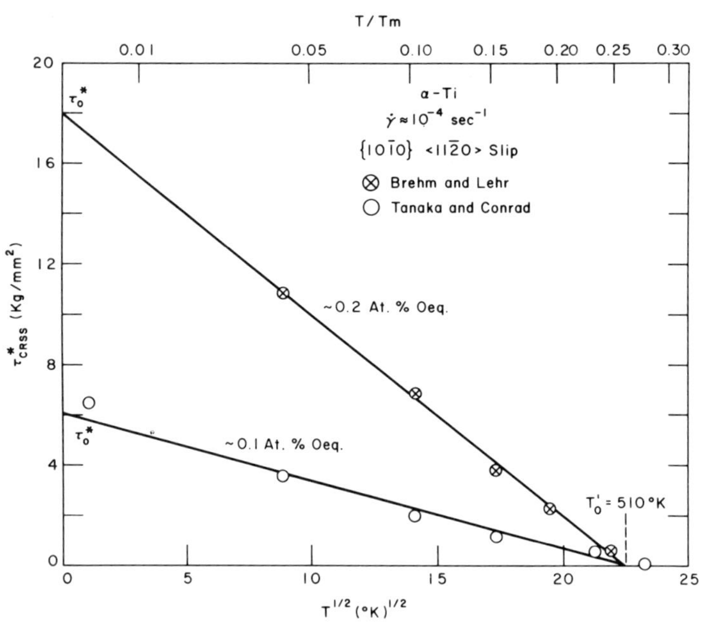

Fig. 3. τ_{crss} versus $T^{1/2}$ for prism slip in unalloyed titanium as a function of interstitial content. Data from Ref. 25 and 27.

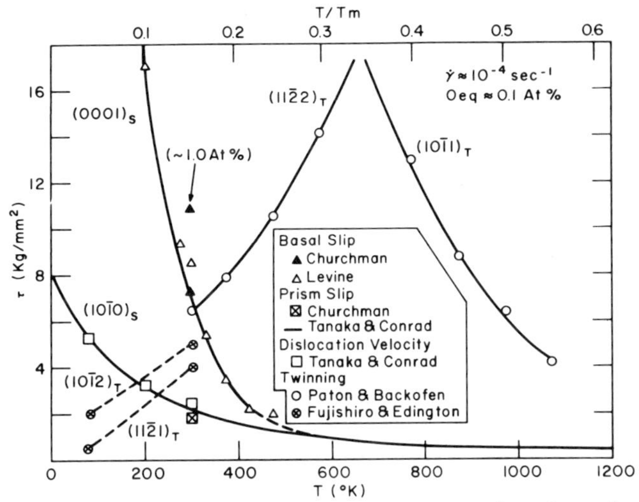

Fig. 4. The effect of temperature on τ_{crss} for basal slip, the shear stress for twinning on the $\{10\bar{1}2\}$, $\{11\bar{2}1\}$, $\{11\bar{2}2\}$, and $\{10\bar{1}1\}$ planes, τ_{crss} for prism slip and shear stress for $\{10\bar{1}0\}$ 1/3<$11\bar{2}0$> edge dislocation velocity of 10^{-3}cm sec^{-1}. Data from Refs. 16, 17, 22-27, 30, 31.

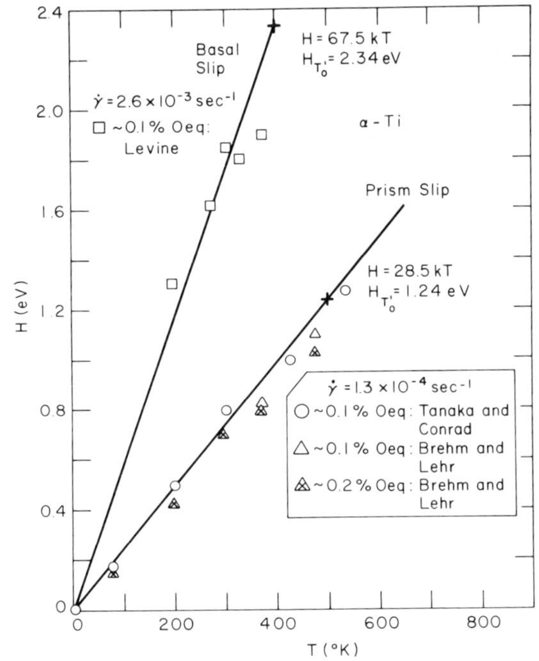

Fig. 5. The activation enthalpy for basal and prism slip versus the temperature. Data from Refs. 24-27.

metals (33, 34). It is about an order of magnitude larger than that generally found for Stage I of basal slip in CPH metals or easy glide in cubic metals, being more nearly that for Stage II of basal slip (27). A comparison of the dislocation density measurements on titanium single crystals (16) with predictions based on the observed strain hardening during prism slip suggests that cutting of forest dislocations may be the mechanism responsible for strain hardening on this slip system (27). No data are available on the strain hardening associated with basal slip in titanium.

2. Polycrystals

a. Characterization of Structure: Important to understanding the mechanical behavior of single-phase polycrystalline titanium is a knowledge of the "character" of the structure, i.e. the grain size, shape and distribution, the texture and the initial dislocation structure. A detailed characterization of the recrystallized structure of swaged and annealed titanium (0.1 at.% to 1.0 at.% Oeq) wires with 1 to 20 microns average grain size was carried out by Okazaki and Conrad (37, 38). In summary, quantitative optical microscopy (including serial sectioning) established that: (a) the grains are relatively equiaxed in all three dimensions, (b) the three dimensional grain size distribution is approximately log-normal, (c) the grains become more regular in shape with increase in grain size and (d) the grain size distribution for a given average grain size is independent of the combination of time and temperature used to obtain the grain size. Texture studies established that for grain sizes of about 1 to 2 microns there exists a strong texture of the $[10\bar{1}0]$ direction in the basal plane parallel to the wire axis. This splits into two $[10\bar{1}0]$ components, each about 15° from the wire axis, for grain sizes of 5 microns and larger, the intensity of these components tending to decrease with increase in grain size. The initial dislocation density increased with decrease in grain size and increase in interstitial content, ranging from 10^7 to 10^9 cm^{-2}.

The textures commonly observed in titanium sheet and rod or wire are illustrated in Fig. 6 (14, 18, 39-41). The texture of cold-rolled unalloyed sheet is most generally one with the $[10\bar{1}0]$ direction parallel to the rolling direction and the basal plane tilted about 20° to 40° to the plane of the sheet. Upon recrystallizing, the texture initially sharpens for fine grain sizes (1 to 2 microns). For larger grain sizes ($\gtrsim 5$ microns) there occurs a rotation of the basal plane so that the $\langle 11\bar{2}0\rangle$ direction becomes more nearly parallel to the rolling direction, although this direction may still be inclined to the rolling plane. For both the cold rolled and the recrystallized textures, interstitial and substitutional solutes tend to reduce the angle γ between the basal plane and the plane of the sheet, yielding in extreme cases a texture with the basal plane in the plane of the sheet. Hot working yields textures similar to those of the large grain size recrystallized material; it may also yield textures with the basal plane parallel to the plane of the sheet.

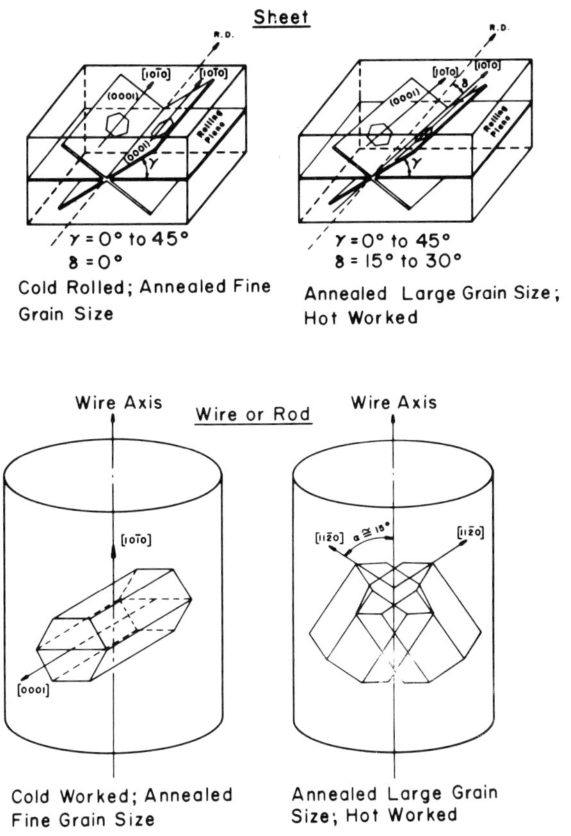

Fig. 6. Common textures in titanium sheet and rod or wire. Data from Refs. 14, 18, 39-41.

The texture of cold worked rod or wire consists of the $[10\bar{1}0]$ direction parallel to the rod or wire axis. Similar to sheet, a sharpening of cold worked texture occurs upon recrystallizing to a fine grain size. For the larger grain sizes, the basal plane becomes tilted about 10° to 15° to the wire axis and it rotates about the c-axis so that the $\langle 11\bar{2}0\rangle$ direction now becomes the projection of the wire axis on the basal plane.

Employing the method of Calnan and Clews (42), a theoretical deformation process which resulted in the experimentally observed deformation textures in titanium was developed by Williams and Eppelsheimer (43). This deformation process required the combined operation of three slip systems, $(0001)\ \langle 11\bar{2}0\rangle$, $\{10\bar{1}1\}\ \langle 11\bar{2}0\rangle$ and $\{10\bar{1}0\}\ \langle 11\bar{2}0\rangle$ and two types of twinning, $\{10\bar{1}2\}$ and $\{11\bar{2}2\}$. From theiranalysis the authors found that the critical shear stress for slip and for twinning at 300°K were related in the following manner: $\tau_S\ (0001) = 1.1\ \tau_S\ \{10\bar{1}1\} = 1.02\ \tau_S\ \{10\bar{1}0\} = \tau_T\ \{10\bar{1}2\} = \tau_T\ \{11\bar{2}2\}$. These ratios are in reasonable accord with the measured stresses for slip and twinning at 300°K depicted in Fig. 4.

b. Taylor Factor: To interpret the mechanical behavior of polycrystals in terms of slip in single crystals one needs to know the value of the Taylor factor $\bar{M}$ relating the tensile flow stress of polycrystalline specimens to the resolved shear stress for slip in single crystals. Armstrong et al (44) estimated $\bar{M}$ for CPH metals to be about 4 from the effect of grain size on the flow stress. Employing a Sachs-type (45) analysis they obtained a value of 6.5 for CPH metals with basal slip. Conrad and coworkers (27, 46) deduced a value of 5 for the Taylor factor in titanium from a comparison of the critical resolved shear stress for prism slip in single crystals with σ_i from Hall-Petch plots of the effect of grain size $\bar{d}$ on the flow stress σ of polycrystals, and from comparisons of the parameters associated with the deformation kinetics for polycrystals and for prism slip. An example of the agreement between σ_i and τ_{crss} for prism slip when a Taylor factor of 5 is used is given in Fig. 7. Also shown here is the agreement for the resolved shear stress at which $\{10\bar{1}0\}1/3\ \langle 11\bar{2}0\rangle$ edge dislocations move at a velocity of 10^{-3} cm sec^{-1} (30).

That the Taylor factor is relatively insensitive to the texture is indicated by the fact that those deformation kinetics parameters whose values include the Taylor factor (e.g. τ^* and the activation volume V^*) were found to be independent of grain size (and hence texture) in rod and wire (46). Furthermore, good agreement in flow stress and deformation kinetics parameters exists between polycrstalline wire or rod specimens (46) and sheet or plate specimens (47) of similar interstitial content.

c. Effects of Temperature, Interstitial Content and Grain Size on the Stress-Strain Behavior: Typical stress-strain curves over the temperature range of 4.2° to 1000°K for unalloyed titanium at a tensile strain rate $\dot{\varepsilon} = 10^{-4}$ sec^{-1} are given in Fig. 8. Serrations or

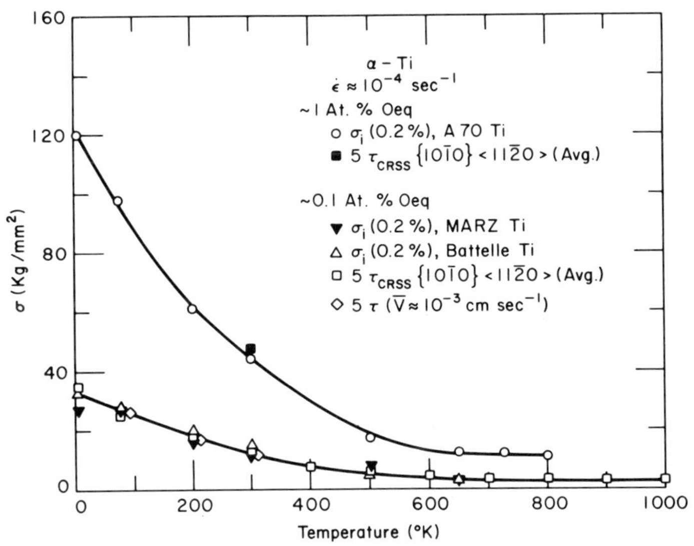

Fig. 7. Comparison of the effect of temperature on the flow stress of polycrystalline specimens with that for prism slip in single crystals and for the stress at which $\{10\bar{1}0\}$ $1/3\langle 11\bar{2}0\rangle$ edge dislocations move at a velocity of $\sim 10^3$ cm sec^{-1}. From Okazaki and Conrad (46).

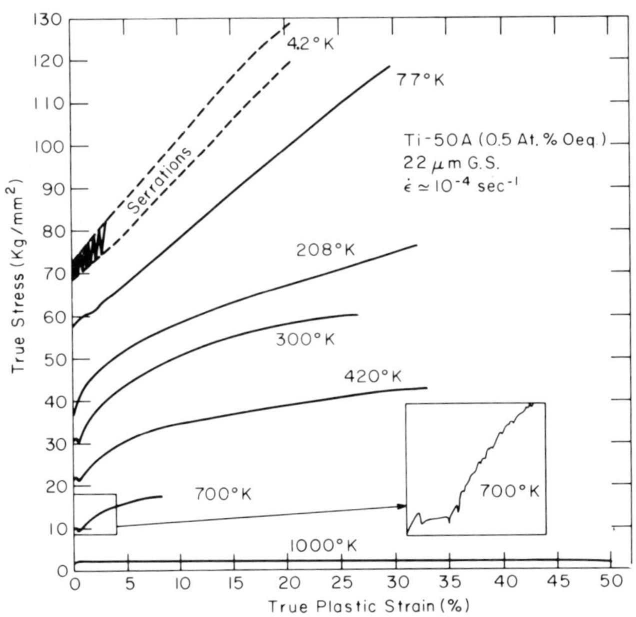

Fig. 8. Stress-strain curves for unalloyed titanium (0.5 at.% Oeq) with 22 microns grain size as a function of temperature. Data from Refs. 10 and 40.

irregularities occur at 4.2°K and in the temperature range of 700° to 800°K. Those at 4.2°K appear to be due to twinning and adiabatic heating effects (48), while those in the range of 700° to 800°K to dynamic strain aging (49). The stress-strain curves are essentially flat in the temperature range of 900° to 1000°K, parabolic between 200° and 800°K and nearly linear at 4.2° and 77°K.

The effects of temperature, grain size and interstitial content on the 0.2% yield strength are illustrated in Fig. 9. The yield stress decreases continuously from 4.2° to about 600°K, remains relatively flat, or increases slightly, between 600° and 650°K and then decreases again above 700°K. The effect of an increase in the interstitial content is to increase significantly the temperature dependence of the yield stress. The effect of grain size at temperatures below about 700°K is to increase the yield stress a relatively constant amount independent of temperature and can be described by the Hall-Petch relation

$$\sigma = \sigma_i (T, \dot{\varepsilon}, \varepsilon, C_i) + K (\mu) d^{-1/2} \qquad (4)$$

with $K/\mu = 2 \times 10^{-4}$ $mm^{-1/2}$, relatively independent of interstitial content and temperature (46). As mentioned above, good agreement is obtained between σ_i for 0.2% strain and τ_{crss} for prism slip and in turn with the shear stress for $\{10\bar{1}0\}1/3 <\bar{1}\bar{1}20>$ edge dislocation velocity of $\sim 10^{-3}$cm sec^{-1} for temperatures below about 700°K; see Fig. 7.

One means of describing strain hardening, which does not depend on a mathematical description of the stress-strain curve, is simply to take the difference in the flow stresses at two arbitrary values of the strain. This definition is used in Fig. 10, which is a plot of the flow stress at 5% strain minus that at 0.5% for commercial titanium plate (~0.5 at.% Oeq) with 16 micron grain size (50). Superimposed on the general decrease in $\Delta\sigma$ which occurs with increase in temperature are three maxima located at approximately 250°K, 550°K and 750°K for a strain rate of 2.7×10^{-4} sec^{-1}. The temperature of each maximum increases with increase in strain rate, yielding activation energies of 9, 32 and 57 Kcal/mole, respectively. It is difficult to relate these values directly to activation energies for interstitial or substitutional solute diffusion in titanium because of the scatter in the diffusion data and the limited number of strain rates investigated in the mechanical tests. However, the relative magnitudes of the activation energies for strain hardening correspond roughly to those for diffusion of the interstitials H, C and O or N, respectively. Of further note in Fig. 10 is that the strain hardening decreases very rapidly for temperatues above about 800°K.

Okazaki and Conrad (46) investigated the strain hardening of annealed titanium wire as a function of interstitial content and grain size over the temperature range of 4.2° to 650°K. For zone refined

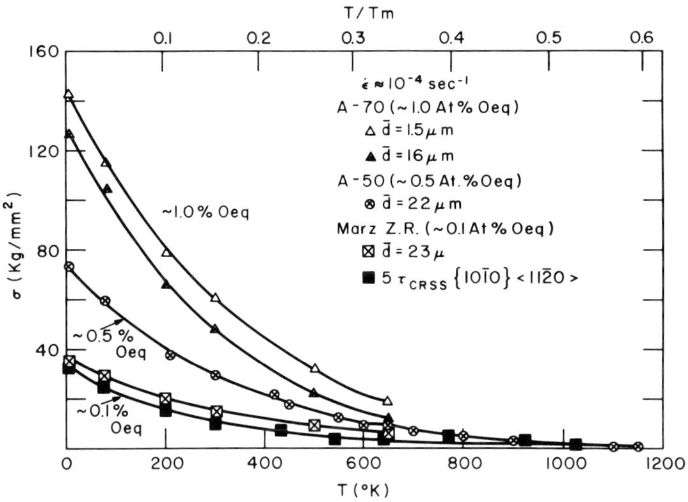

Fig. 9. Effects of temperature, grain size and interstitial content on the 0.2% yield stress. Data from Refs. 10, 27, 40, and 46.

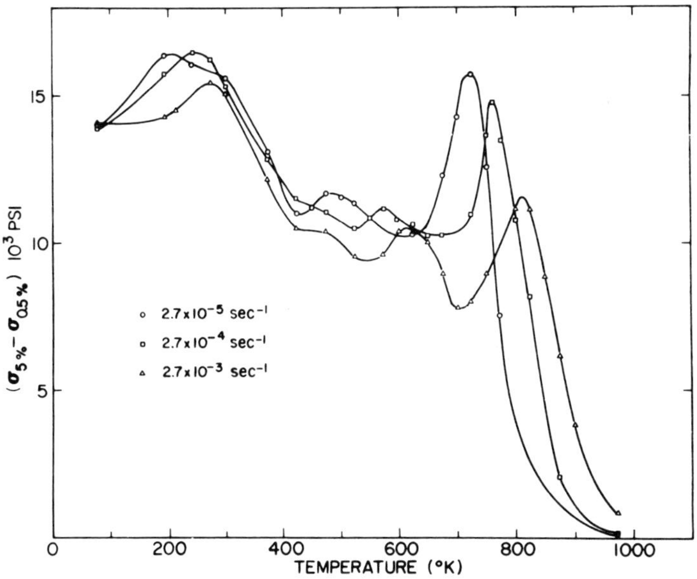

Fig. 10. Effect of temperature and strain rate on the strain hardening of titanium (0.5 at.% Oeq) with 16 micron grain size. From Santhanam and Reed-Hill (50).

material (~0.1 at.% Oeq) the strain hardening coefficient θ given by $\sigma_{.16} - \sigma_{.02}/0.14$ divided by the modulus $\mu = C_{66}$ decreased rapidly with temperature between 4.2° and 200°K, only slightly between 200°K and then rapidly again above 500°K, relatively independent of grain size. In the case of commerical A-70 material (~1.0 at.%Oeq) there occurred a slight increase in θ/μ between 4.2° and 300°K and thereafter a slight decrease, again independent of grain size. At 4.2° and 77°K the strain hardening for the zone refined material was greater than that for the A-70 while at higher temperatures the reverse was true. The values of θ/μ were between 7.2×10^{-2} and 1.5×10^{-3} for the temperature, purity and grain size ranges considered.

Wasilewski (51) and Conrad and coworkers (46, 52-54) found that the stress-strain curves in titanium could be approximated by the relation

$$\sigma \approx \sigma^{\circ}(T, \dot{\varepsilon}, \bar{d}, C_i) + h(\mu, T, \dot{\varepsilon}, \bar{d}, C_i)\, \varepsilon^{1/2} \tag{5}$$

σ° corresponded closely to the value of the proportional limit and h/μ decreased with temperature from values of the order of 10^{-2} at 77°K to near zero at 800°K.

The maxima in the strain hardening versus temperature curves of Fig. 10 suggest the occurrence of dynamic strain aging (49, 50). Other evidence for dynamic strain aging in titanium includes: (a) serrations in the stress-strain curves (Portevin-Le Chatelier effect) at 725° to 925°K (10, 52, 55), (b) a rise in the flow stress versus temperature curve at 675° to 750°K (47) and (c) a maximum in the creep strength in the vicinity of 475°K (56). Static strain aging in titanium has been observed by Rosi and Perkins (55) after heating in the temperature range of 473° to 673°K. Also, Kiessel and Sinott (56) found that the maximum creep resistance at room temperature occurred for cold-worked material aged for 100 hours at 477°K, compared to aging at lower or higher temperatures. Thus, there is little doubt that both static and dynamic strain aging occurs in titanium, being most pronounced in the temperature range of 650° to 800°K ($0.3 - 0.4T_m$) at normal testing rates. However, the solutes (presumably interstitials) which are responsible for the various phenomena have not yet been clearly defined.

d. <u>Deformation Kinetics at Low Temperatures ($T<0.3T_m$)</u>: The results presented in Fig. 9 indicate that the flow stress σ in polycrystals, similar to single crystals, can be separated into a thermal component σ^* and an athermal component σ_μ. Work by Conrad and coworkers (46, 53, 54, 57, 58) has shown that

$$\sigma \approx \sigma^*(T, \dot{\varepsilon}, C_i) + \sigma_\mu(\mu, \varepsilon, \bar{d}, C_i, C_s) \tag{6}$$

where T is the temperature, $\dot{\varepsilon}$ the tensile strain rate, C_i the interstitial solute content expressed as an oxygen equivalent, μ the modulus, ε the tensile strain, $\bar{d}$ the average grain size and C_s the substitutional solute content. To be noted is that C_i affects both σ^* and σ_μ while ε, $\bar{d}$ and C_s primarily influence σ_μ. σ^* was found to decrease in a linear fashion with the square root of the temperature and was proportional to the square root of the interstitial solute content. Using a Taylor factor of 5 there occurred good agreement between σ^* for polycrystals and τ^* for prism slip. Hence in the discussion to follow it will be assumed that $\tau = 1/5\ \sigma$ and $\dot{\gamma} = 5\dot{\varepsilon}$.

Plots of τ^* at 4.2°K and 300°K versus the square root of the interstitial content are presented in Fig. 11. The results indicate that τ^* is proportional to $C_i^{1/2}$ and that the proportionality constant decreases with temperature but is independent of the alloy content, crystal structure and whether the material is single-phase or two-phase. The value of τ^*_o is given by

$$\tau^*_o = 0.06\ \mu_o\ C_i^{1/2} \tag{7}$$

where μ_o is C_{66} extrapolated to 0°K. In Fleischer's (62) terminology this is a case of "rapid hardening".

A plot of the activation enthalphy H derived from constant strain rate type tests in the usual manner (32) versus the temperature is presented in Fig. 12. Good agreement occurs between prism slip in single crystals and plastic flow in unalloyed polycrystalline specimens and in polycrystalline alloys of the α(CPH) structure, the β(BCC) structure and $\alpha + \beta$ structures, independent of interstitial content over the range of 0.1 at.% to 1.0 at.% Oeq. The value of the activation energy at T_o' (510°K) is 1.24 eV (0.29 μ b^3), that at T_o (600°K) is 1.46 eV and the slope yields $\nu \approx 10 \times 10^8$ sec^{-1}. Values of H obtained from stress relaxation test (57, 63) and from creep tests (64) are in good accord with those in Fig. 12 derived from constant strain rate tests.

The results of Figs. 11 and 12 indicate that overcoming of interstitial solute obstacles is the rate controlling mechanism during the low temperature deformation of titanium and titanium alloys of both the CPH and BCC structures. In the case of unalloyed α-titanium, the good agreement between the results on polycrystals and those for prism slip in single crystals and the stress for the motion of edge dislocations on the prism plane indicates that the deformation kinetics reflect the motion of individual dislocations on the prism plane. It is presumed that this also holds for the α-alloys, although this still needs to be confirmed. The fact that similar values of τ^* and the thermal activation parameters are obtained for the all-β Ti-26Mo alloy suggests that the strong interaction between dislocations and interstitials in titanium is relatively independent of crystal structure.

Assuming that overcoming interstital solute atoms is rate controlling, a force-activation distance curve was derived (27, 46) from data on unalloyed titanium for the case of a random distribution of obstacles (65) and taking into consideration dislocation line flexibility (66); see Fig. 13. The approximately triangular shape of the F-d* curve is in accord with the experimentally observed dependence of τ^* on $T^{1/2}$. Moreover, the area under the F-d* curve between d*/b = 0 and

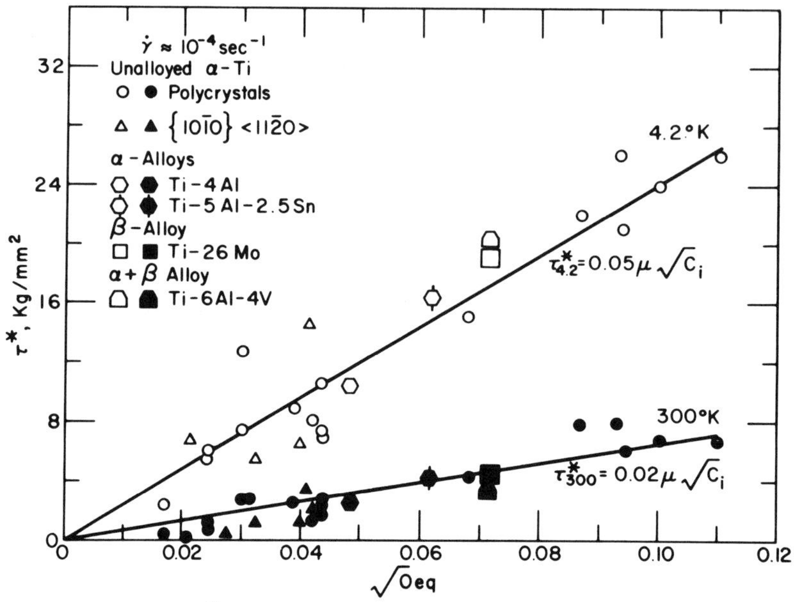

Fig. 11. τ^* at 4.2°K and 300°K versus the square root of the interstitial solute content for unalloyed titanium and some titanium alloys. Data from Refs. 27, 46, 58-61.

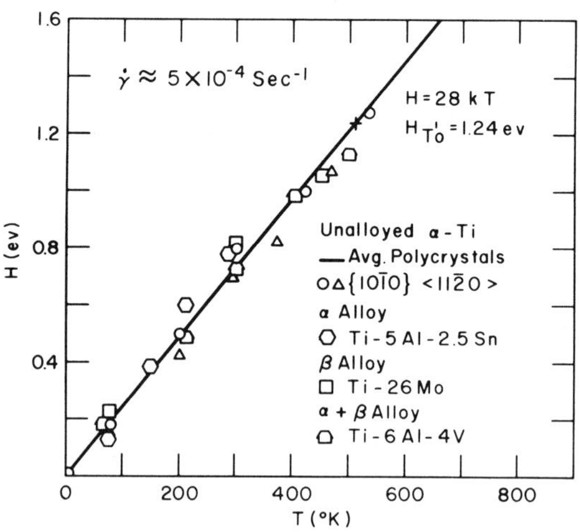

Fig. 12. Activation enthalpy versus temperature for the deformation of unalloyed titanium and some alloys. Data from Refs. 27, 46, 58, 60, and 61.

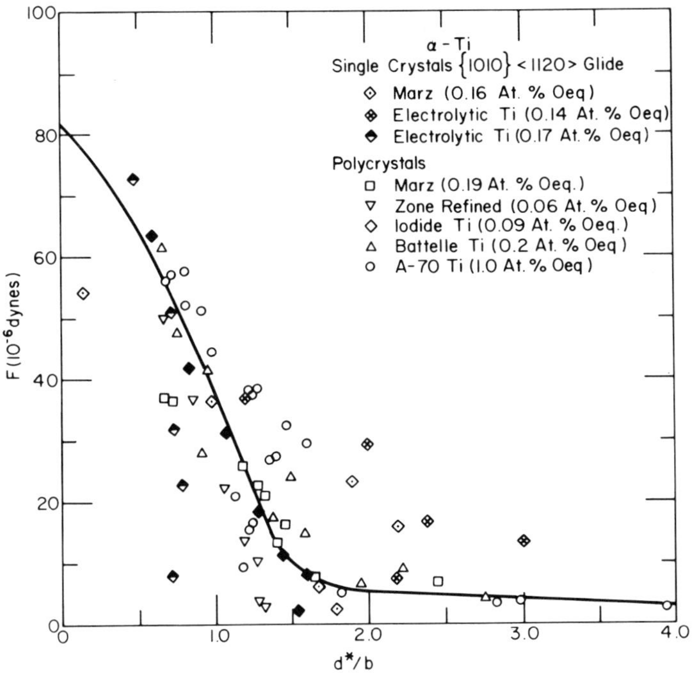

Fig. 13. Force-activation distance curve for the deformation of unalloyed titanium at low temperatures. From Okazaki and Conrad (46).

$d^*/b = 1.5$ yields $H = 1.3$ eV in good accord with the value of H_{T_o}. Of significance is that a rapid rise in the force first occurs at $d^* \approx 1.5b$. The maximum force $F_o = 82 \times 10^{-6}$ dynes ($0.21\ \mu_o\ b^2$). The detailed nature of the interaction between the moving dislocations and the interstitial solutes which leads to the F-d* curve of Fig. 13 is, however, still not clear (67).

e. Athermal Component at Low Temperatures ($T<0.3T_m$): It has been shown (54, 68-72) that there is reasonably good agreement in the value of the athermal component of the flow stress in titanium determined by a number of methods which include: (a) back-extrapolation of the flow stress at high temperatures, (b) the fully relaxed stress, (c) incremental unloading, (d) analysis of the stress relaxation curve, (e) single strain rate cycling, (f) double strain rate cycling, (g) the grain size method, (h) dislocation density versus flow stress and (i) measurements of the dislocation curvature.

Conrad and coworkers (54, 68, 69, 73) have established that the effect of grain size on the athermal component of the flow stress in unalloyed titanium is principally through its effect on the dislocation density ρ at a given strain and that flow stress increased in a linear fashion with $\rho^{1/2}$; see Fig. 14. Moreover, the effect of interstitial solute content on σ_μ was also primarily through the effect on the dislocation density. The dislocation density measurements on unalloyed titanium yield

$$\tau = \tau^* (T, \dot{\gamma}, C_i) + 0.5\ \mu\ b\ \{\rho\ (\gamma, \bar{d}, C_i)\}^{1/2} \tag{8}$$

Thus, the effects of temperature, interstitial content and grain size on strain hardening are through the accumulation of dislocations as a function of strain, i.e. through the multiplication and annihilation of dislocations. In fact, values of the strain hardening coefficient h and the Hall-Petch constant K derived from dislocation measurements are in good accord with those determined from stress-strain curves (54, 68, 73). Further interpretation of the effects of temperature, grain size and interstitial content on strain hardening in unalloyed titanium must await the development of a more detailed model for dislocation multiplication. The rapid increase in strain hardening at very low temperatures in the higher purity materials is probably due to twinning.

The value of 0.5 for the coefficient relating the flow stress to $\rho^{1/2}$ is in accord with that for most metals and with most strain hardening mechanisms (74) and hence does not permit an identification of the specific dislocation mechanism responsible for strain hardening in titanium. Considering the linear decrease in τ^* with $T^{1/2}$, Eqn. 8 can be expanded to

$$\tau \approx \{27-14.7\ T^{1/2} \log_{10} \left(\frac{5 \times 10^8}{\dot{\gamma}\ (\text{sec}^{-1})} \right)\}\ C_i^{1/2} + 0.5\mu b\rho^{1/2} (\text{Kg/mm}^2) \tag{9}$$

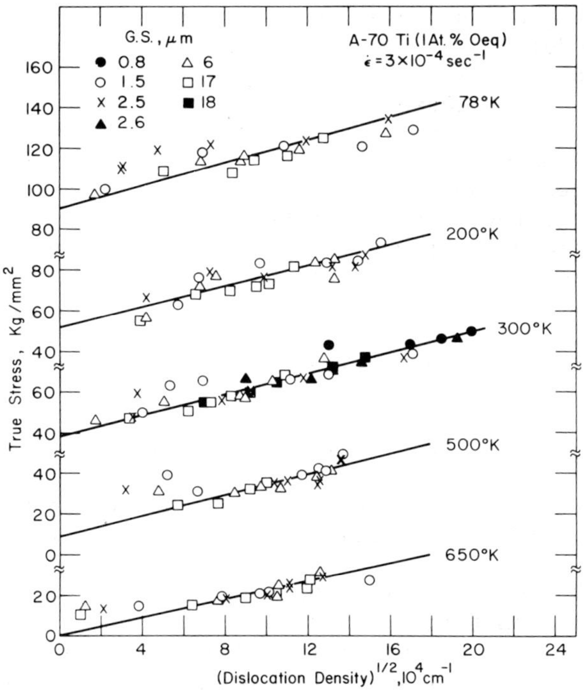

Fig. 14. Flow stress versus the square root of the dislocation density in A-70 titanium (1.0 at.% Oeq) as a function of grain size and temperature. From Conrad et al (54).

As pointed out above, the effect of substitutional solutes on the flow stress of titanium at low temperatures is primarily through the athermal component of the flow stress. Fig. 15 shows that the increase in tensile strength due to substitutional solutes which are α-stabilizers can be considered to be proportional to the square root of the atomic fraction of the solute content. Gegel and coworkers (75) have shown that the slopes of the straight lines of Fig. 15 are proportional to the interaction parameter given by the sum of the square roots of the absolute magnitude of the individual interaction coefficients Ω as defined by Fowler and Guggenheim (76). This proportionality is depicted in Fig. 16, where the individual interaction coefficients are expressed in electron volts per atom. The interaction parameter is similar to the Pauling electronegativity difference, which has been used by Kubaschewski and Sloman (77) in treating bonding mechanisms in intermetallic compounds. Of considerable importance regarding the data of Figs. 15 and 16 is that the strengthening results from the sum of all individual atom-atom interactions, including those between the individual solutes, as well as between titanium and each solute.

Considering the results of Figs. 15 and 16, we can expand Eqns. 6 and 8 for the α-stabilizing solutes to

$$\tau = \tau^* (T, \dot{\gamma}, C_i) + 0.5\mu b\rho^{1/2} + 34.1 \quad I\, C_\alpha^{1/2} \; (\mathrm{Kg/mm^2}) \qquad (10)$$

where I is the interaction parameter defined above and C_α is the concentration of α-stabilizing solutes.

An explanation for the effect of substitutional solutes on the strength of titanium in terms of dislocation-solute interactions is not yet available. However, Collings and coworkers (78) have developed a rationalization for the strengthening of α and β stabilizing solutes in terms of certain electronic factors. Their interpretation is the following: The significant solid solution strengthening of α-stabilizing solutes is due to the tendency for directional (partially covalent) bond formation by these solutes. The β-stabilizing solutes (transition elements) increase the conduction electron density and thereby there does not occur any tendency for directional bonding. Hence, the strengthening by the β-stabilizers which is largely due to d-electron bonding is considerably less than that of the α-stabilizers. Some correlation of the mechanical properties in th former alloys is obtained through the average group number or the electron-to-atom ratio.

f. Role of Twinning: The nature and role of twinning during the low temperature deformation of unalloyed polycrystalline titanium was investigated by Conrad and coworkers (40, 46, 54). Employing electron transmission microscopy they found that the tendency for twinning in wires deformed to a maximum strain of 10% increased with increase in purity, strain, grain size and temperature. Zone refined material (~0.1 at.% Oeq) exhibited twinning for all grain

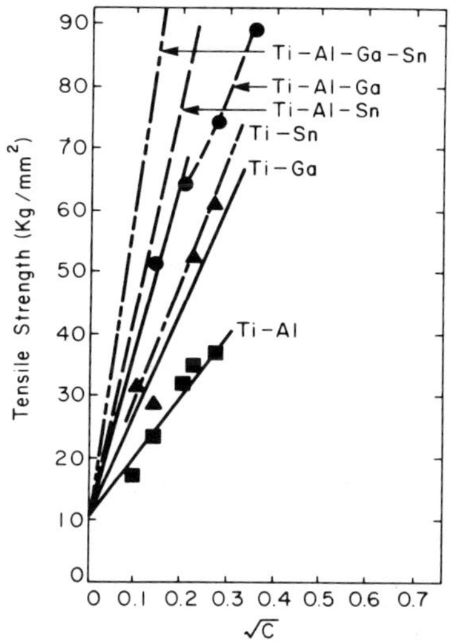

Fig. 15. Tensile strength at room temperature versus square root of the atomic fraction of α-stabilizing solutes. From Gegel (75).

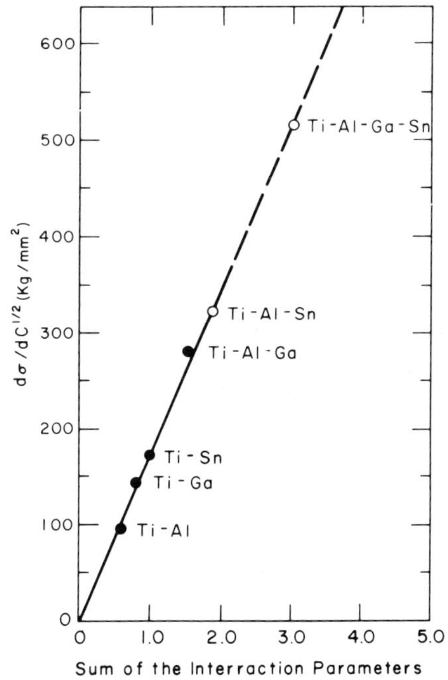

Fig. 16. d $\sigma/dc^{1/2}$ versus the interaction parameter defined as the sum of the square roots of the absolute values of the individual interaction coefficients expressed in eV per atom. From Gegel (75).

sizes (1 to 23 microns) at 78° and 200°K. At 300°K twinning was only seen in specimens with grain sizes in the range of 5 to 23 microns. No twins were observed for any grain size of this material at 500° and 650°K. Commercial A-70 material (1.0 at.% Oeq) exhibited some twinning for all grain sizes at 78°K but only the largest grain size (17 microns) at 200°K. No twins were observed in this material for any grain size (1.5 to 17 microns) at 300°K and above. When twinning occurred, the thermal component of the flow stress σ^* tended to be less than that in material where no twinning was evident, and in turn below that predicted from the single crystal results using a Taylor factor of 5. By correcting for the contribution of twinning to the total strain, good agreement in σ^* was obtained between material exhibiting twins and that with no evidence of twins (40).

Reed-Hill and coworkers (79, 80) found that the volume fraction of twins in polycrystalline titanium wire and sheet was approximately proportional to the strain,with the proportionality constant decreasing with increase in temperatue. Moreover, the amount of twinning increased with increase in purity. They attributed the linear stress-strain behavior at 78°K to the deformation twins and concluded that most of the temperature and strain dependence of the strain anistropy in titanium sheet can be accounted for by twinning.

g. <u>Deformation at High Temperatures ($T>0.4T_m$):</u> Weertman's (81) model for the high temperature deformation of metals based on dislocation climb leads to

$$\dot{\varepsilon}_s = \frac{3\pi^2 \sigma^2 D}{2\sqrt{2}\mu^2 b^2} \sinh\left(\frac{\sqrt{3}\,\sigma^{2.5}\, b^{1.5}}{8\mu^{1.5}\, N^{1/2}\, kT}\right) \tag{11}$$

where $\dot{\varepsilon}_s$ is the steady-state strain rate, σ is the applied tensile stress, μ the shear modulus and N the density of Frank-Read sources per cm^3. At low stresses Eqn. 11 reduces to the power law relation

$$\dot{\varepsilon}_s = \frac{3\sqrt{6}\,\pi^2\,\sigma^{4.5}\,D}{2^5\, b^{1/2}\, N^{1/2}\, \mu^{3.5}\, kT} \tag{12}$$

Hence, at low stresses a plot of log $\dot{\varepsilon}_s$ kT/Dμb versus log σ/μ should yield a straight line of slope 4.5. Such a plot is given in Fig. 17 for the available data on the high temperature deformation of unalloyed α-and β-titanium. The values of D which gave the good fit indicated in Fig. 17 were 55 exp {-60,000/RT} for α-titanium and 55 exp {-55,000/RT} for β-titanium. These values were chosen on the basis of the empirical correlation of the activation energy for diffusion with the melting temperature (82) and Zener's (83) theory for D_o. A slightly better fit was obtained using D = 1.0 exp {-56,500/RT} for α-titanium and D = 1.0 exp {-50,000/RT} for β-titanium. These latter values are in reasonable agreement with the "best" self-diffusion data for titanium, Table II.

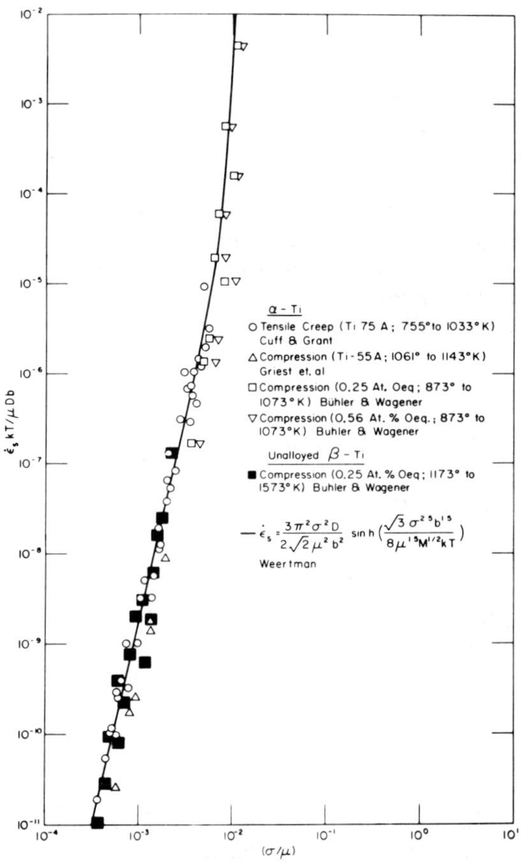

Fig. 17. Log $\dot{\varepsilon}_s$ kT.Dμb versus log σ/μ for the deformation of α- and β-titanium at high temperatures. Data from Refs. 84-86.

The correlation of Fig. 17 yields $N = 6.5 \times 10^{13}$ cm^{-3} and ρ $\{= (1.9\ N^{1/2})^{4/3}\} = 3.9 \times 10^{9}$ cm^{-2}. Using the second set of values for D gives $N = 2.8 \times 10^{12}$ cm^{-3} and $\rho = 4.7 \times 10^{8}$ cm^{-2}. Both sets of values for N and ρ are reasonable; however, the lower values are in better agreement with the dislocation densities measured in metals deformed at high temperatures (87). It is thus concluded that the high temperatue deformation of α- and β-titanium is controlled by dislocation climb.

The creep of binary Ti-Mo alloys has been studied by Collings et al (4). These investigators concluded that above 1073°K creep was glide controlled, i.e. dislocation glide limited by the drag of Cottrell atmospheres. The activation energies (28-44 Kcal/mole) obtained for creep were in the range of values reported for the diffusion of Mo in binary Ti-Mo alloys. However, one may question the creep activation energies, since there existed considerable curvature in the plots of log creep rate versus 1/T. Also, there is some question regarding the activation energy for the diffusion of Mo in Ti-Mo alloys (88).

Fracture

1. Unidirectional Stressing

a. Fracture Modes at Low Temperatures: In studies of fracture in large grain size titanium of commercial purity (~1.0 at.% Oeq), Amateau et al (89, 90) found that at low temperatures cracks developed at the interface between second-order twins (in primary $\{11\bar{2}2\}$ twins) and the matrix or first-order twins. Further, they found a correlation between ductility and the number of grains containing second-order twins. At temperatures above about 200°K the fracture mode was normal shear type.

b. Effects of Temperature, Grain Size and Solute Content on the Fracture Behavior at Low Temperatures ($T<0.3T_m$): The effects of temperatue, solute content and grain size on the fracture stress σ_F (= load at fracture/cross section area at fracture) is illustrated in Fig. 18. The fracture stress, similar to the flow stress, can be considered to consist of thermal and athermal components, i.e.

$$\sigma_F = \sigma_F^* (T, \dot{\varepsilon}, C_i) + \sigma_{\mu_F} (\mu, \bar{d}, C_i, C_s) \tag{13}$$

The temperature dependence of σ_F^* for the A-70 (1.0 at.% Oeq) material is only slightly higher than that of the flow stress for this material. The difference is essentially eliminated if the Bridgman's (92) correction for necking is applied to the fracture stress. However, there does not appear to be as strong an effect of interstitial content on the fracture stress as occurs for the flow stress. The reason for this is not clear.

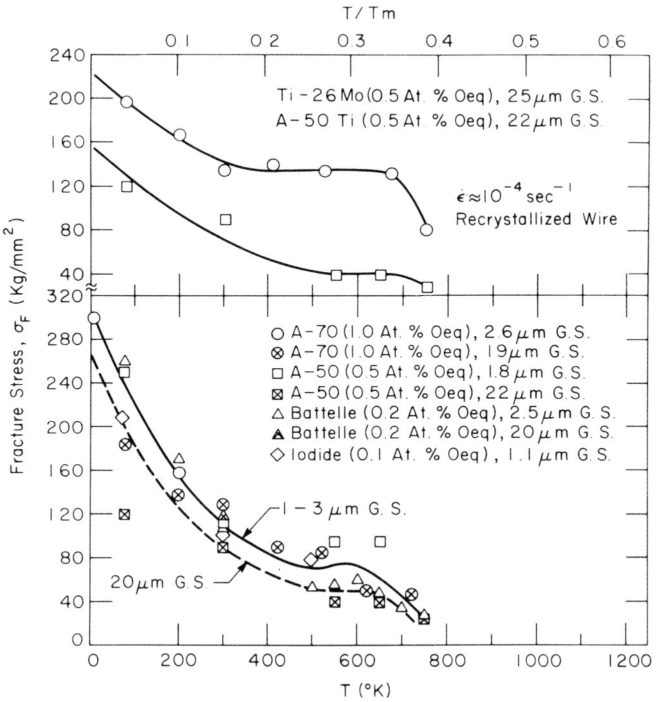

Fig. 18. Effects of temperature, solute content and grain size on the fracture stress of titanium. Data from Refs. 40, 58, 91.

The effect of grain size on the fracture stress (40, 91, 93) can be considered to obey the Hall-Petch relation yielding values of σ_{i_F} = 80 to 115 Kg/mm^2 and K_F = 1.0 - 1.4 Kg/mm$^{3/2}$ at room temperature, relatively independent of interstitial solute content. The value of K_F is about twice that of K for plastic flow. From the curves in the upper part of Fig. 18 it is seen that the effect of the substitutional solute Mo on the fracture stress appears to be mainly through the athermal component.

The effects of the interstitials C, N and O on the total elongation at fracture at 300°K are presented in Fig. 19. In each case the elongation decreases with increase in interstitial content, the rate being greatest for the lowest concentrations. The rate of decrease in elongation with interstitial content increases in the order C, O, N, which is the same order as their effect on the flow stress (28). The total elongation is relatively independent of grain size (91).

The effects of temperature and strain rate on the total elongation are presented in Fig. 20. To be noted is that a minimum in the elongation occurs in the vicinity of 750°K for a strain rate of 2.7×10^{-4} sec^{-1}, the temperature of which increases with increase in strain rate. The activation energy derived from the effect of strain rate on the temperature of this minimum is 48 Kcal/mole, in reasonable accord with the diffusion of O in titanium, Table II. The minimum in elongation thus appears to be related to the dynamic strain aging which occurs in this temperature range (49, 50).

The fracture mode in polycrystalline specimens at low temperatures is most generally of the cup-cone shear type.

c. Fracture at High Temperatures ($T>0.4T_m$): Orr et al (97) have shown that the stress-rupture data for titanium correlated very well through the relation

$$\sigma_F = f\,(t_r \exp\{-60{,}000/RT\}) \tag{14}$$

where σ_F is the fracture stress and t_r is the time to rupture. The value of the activation energy for rupture is thus similar to that for creep and suggests that the same mechanism is controlling, namely the climb of dislocations.

2. Fatigue

Unalloyed titanium exhibits a well defined fatigue limit between 77° and 673°K, the level of which increases with interstitial content (Fig. 21) and with decrease in grain size and temperature (98-103). From Fig. 22 it is seen that the ratio of the fatigue limit to the yield stress at 300°K is approximately constant, relatively independent of interstitial content and grain size, suggesting that the high cycle fatigue strength is controlled by the same dislocation mechanisms

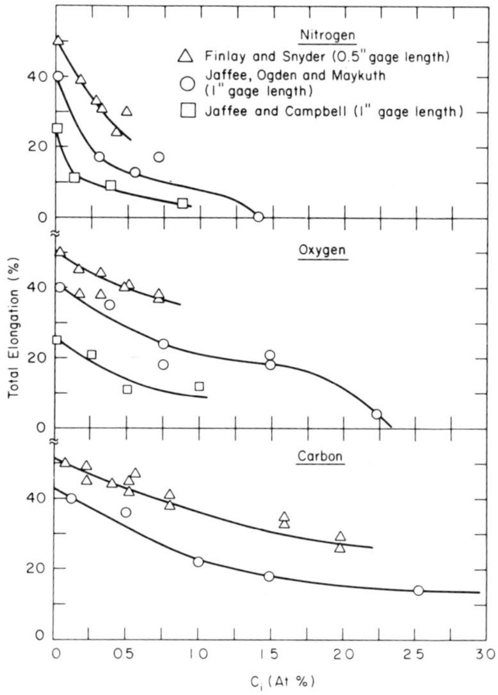

Fig. 19. Effects of the interstitials C, O, and N on the total elongation at fracture at 300°K. Data from Ref. 94-96.

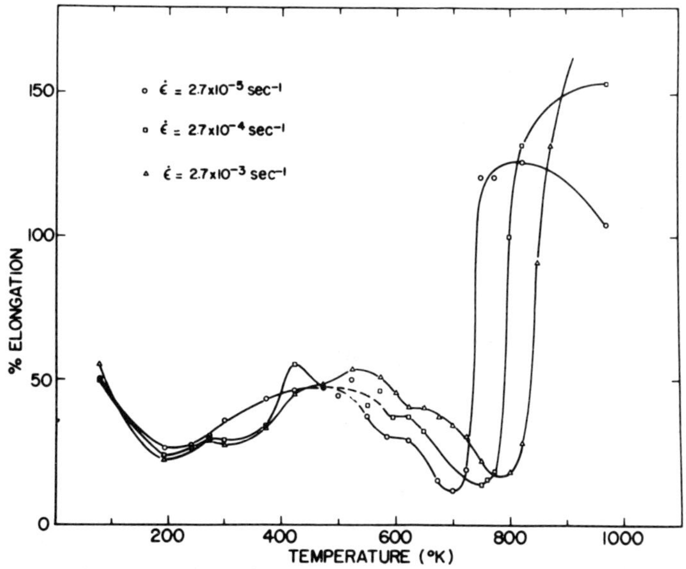

Fig. 20. Effects of temperature and strain rate on the total elongation at fracture in commercial purity titanium (~0.5 at.% Oeq) sheet of 16 micron grain size. From Santhanam and Reed-Hill (50).

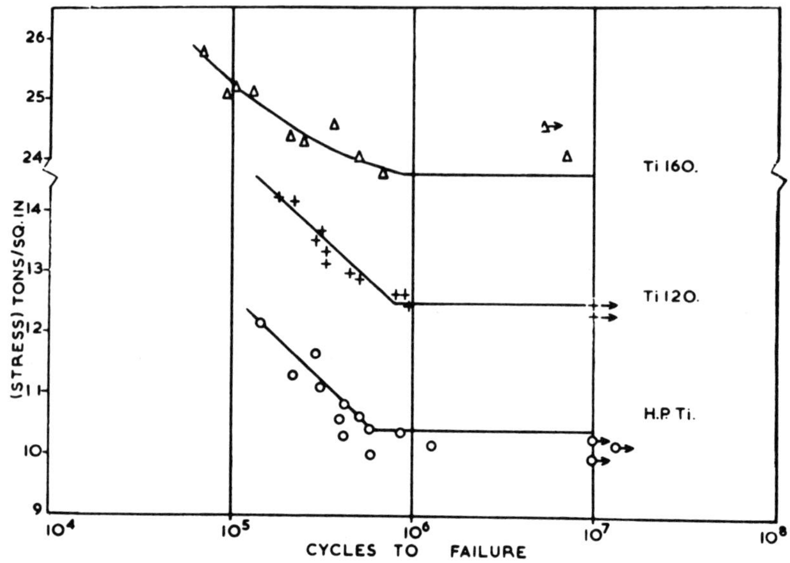

Fig. 21. Effect of interstitial content on S-N curves of unalloyed titanium with 32 micron grain size at 300°K. Interstitial contents in terms of Oeq are: Ti 160 (1.1 at.%), Ti 120 (0.4 at.%), and H.P.Ti (0.3 at.%). From Turner and Roberts (100).

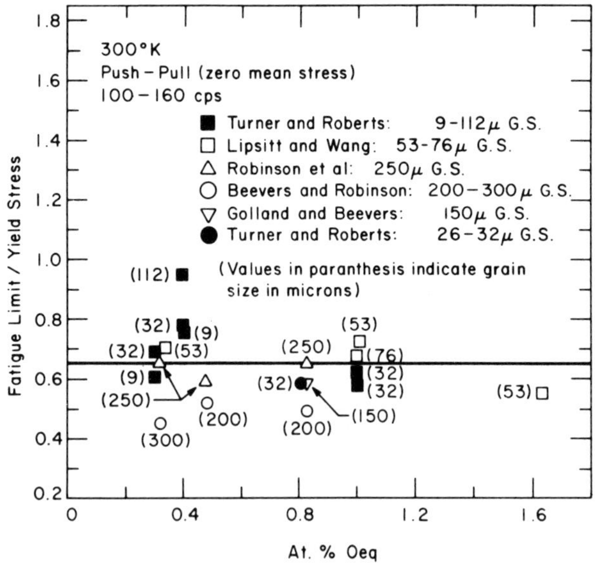

Fig. 22. Ratio of the fatigue limit to the yield stress in unalloyed titanium at 300°K as a function of interstitial content and grain size. Data from Refs. 98-104.

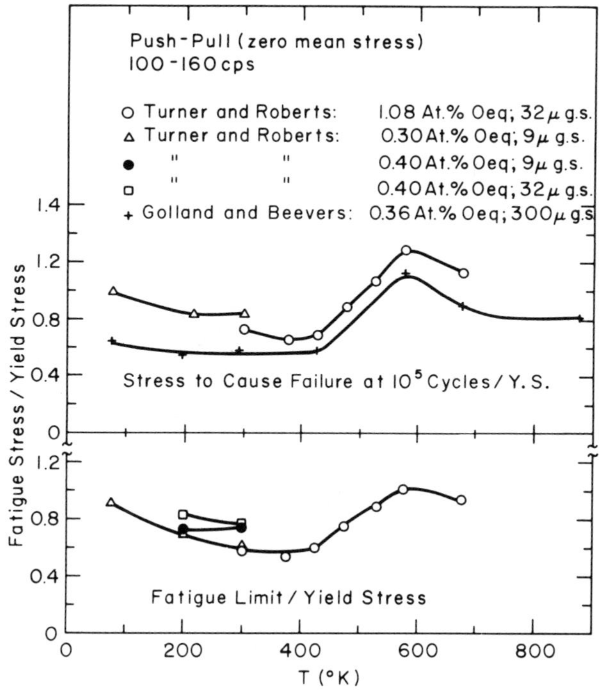

Fig. 23. Ratio of stress to cause failure at 10^5 cycles to the yield stress and ratio of the fatigue limit to the yield stress for unalloyed titanium as a function of temperature. Data from Refs. 99, 100, and 103.

as the flow stress. This conclusion is further supported by the results presented in Fig. 23, which contains plots of the ratio of the stress to cause failure at 10^5 cycles to the yield stress and the ratio of the fatigue limit (F.L.) to the yield stress (Y.S.) as a function of temperature. These ratios are relatively independent of temperature below about 450°K, but go through a maximum at higher temperatures. The maximum is probably related to the strain aging which occurs in this temperature range.

The observation that the F.L./Y.S. ratio and the ratio of the stress to cause failure at 10^5 cycles/Y.S. remain relatively constant with decreasing temperature below 300°K is not in accord with the commonly held belief that the low temperature fatigue behavior of unalloyed titanium is controlled by twinning (101-107).

The fracture modes under cyclic deformation in coarse-grained α-Ti have been investigated by Golland and Beevers (102). They reported that in the temperature range of 77° to 195°K fatigue cracks were associated with $\{11\bar{2}1\}$ type twins, slip bands and grain boundaries. The fracture surfaces exhibited twin-assisted transgranular and intergranular fracture paths. In the range of 473° to 773°K the failure occurred as a consequence of the transgranular propagation of surface nucleated cracks; between 873°K and 973°K the fatigue fracture was intergranular; above 973°K the internal fatigue damage existed in the form of grain boundary cavities.

Acknowledgement

The authors wish to express their appreciation to Drs. H. Gegel and E. Collings for permission to use their results on substitutional alloy strengthening prior to publication. They also wish to acknowledge financial support for this investigation under contract AF33615-68-C-1052, Dr. H. Gegel technical monitor.

References

1. Metals Handbook 8th Ed. Vol 1 (1961).

2. C. Kittell, Introduction to Solid State Physics, Wiley, N.Y. (1968).

3. E. W. Collings and J. C. Ho, The Science, Technology and Application of Titanium, Pergamon Press, N. Y. (1970) p. 331.

4. E. W. Collings, J. C. Ho and R. I. Jaffee, "Theory of Titanium Alloys for High Temperature Strength" Tech. Rept. AFML-TR-71-228 (Oct. 1971).

5. P. E. Armstrong and H. L. Brown, Trans. AIME 230 962 (1964).

6. E. S. Fisher and C. J. Renken, Phys. Rev. 135 No. 2A, A482 (1964).

7. H. R. Ogden, D. J. Maykuth, W. L. Finlay and R. I. Jaffee, Trans AIME 197 267 (1953).

8. E. S. Fisher and D. Dever, The Science, Technology and Application of Titanium, Pergamon Press, N.Y. (1970) p. 373.

9. N. Hsu and H. Conrad, Scripta Met. 5 905 (1971).

10. M. Doner and H. Conrad, in preparation.

11. P. G. Partridge, Met. Reviews 118 169 (1968).

12. W. R. Tyson, The Science, Technology and Application of Titanium, Pergamon Press, N.Y. (1970) p. 479.

13. L. J. Teutonico, Ibid, p. 425.

14. A. D. McQuillan and M. K. McQuillan, Titanium, Academic Press, N.Y. (1956).

15. J. C. Williams and M. J. Blackburn, Phys. Stat. Sol. 25 K1 (1968).

16. T. R. Cass, The Science, Technology and Application of Titanium, Pergamon Press, N.Y. (1970) p. 459.

17. N. E. Paton and W. A. Backofen, Met. Trans. 1 2839 (1970).

18. C. S. Barrett and T. B. Massalski, Structure of Metals, 3rd Ed., McGraw Hill, N.Y. (1966).

19. N. E. Paton and J. C. Williams, 2nd Int. Conf. Strength of Metals and Alloys, Vol I, ASM (1970) p. 108.

20. J. F. Breedis and M. K. Koul, "Relationship of Structure to Strength in Titanium Alloys" Rept. AFML-TR-69-236 (June 1969).

21. H. S. Rosenbaum, Deformation Twinning, Gordon and Breach, N.Y. (1964) p. 43.

22. A. T.Churchman, Proc. Roy. Soc. A226 216 (1954).

23. H. Burrier, M. Amateau and E. Steigerwald, "The Relationship Between Plastic Deformation and Fracture in Alpha Titanium", Rept. AFML-TR-65-239 (July 1965).

24. E. Levine, Trans. AIME 236 1558 (1964).

25. C. Brehm and P. Lehr, Mem. Sci. Rev. Metallurg. 68 277 (1971).

26. T. Tanaka and H. Conrad, 2nd. Int. Conf. Strength of Metals and Alloys, Vol I ASM (1970) p. 224.

27. T. Tanaka and H. Conrad, to be publ. Acta Met.

28. H. Conrad, Acta Met. 14 1631 (1966).

29. H. Conrad, AIME J. Met. 16 582 (1964).

30. T. Tanaka and H. Conrad, Acta Met. 19 1001 (1971).

31. S. Fujshiro and T. W. Edington, "Mechanical Twinning of Titanium Single Crystals", Tech. Rept. AFML-TR-70-176 (July 1970).

32. H. Conrad and H. Wiedersich, Acta Met. 8 128 (1960).

33. H. Conrad and I. Perlmutter, Conf. Int. Metallurgiè Beryllium, Presses Univ. de France, Paris (1965) p. 319.

34. H. Conrad, G. London and V. Damiano, Anisotropy in Single Crystal Refractory Compounds, Plenum, N.Y. (1968) p. 153.

35. H. Conrad, High-Strength Materials, Wiley, N.Y. (1965) p. 436.

36. D. B. Snow and J. F. Breedis, 2nd Int. Conf. Strength of Metals and Alloys Vol II ASM (1970) p. 418.

37. K. Okazaki and H. Conrad, to be publ. Japan Inst. Met.

38. K. Okazaki and H. Conrad, submitted to Met. Trans.

39. H. W. Babel and S. F. Frederick, AIME Jour. Met. (Oct. 1968) p. 32.

40. C. Yin, M. Doner and H. Conrad, in preparation.

41. H. Hu and R. S. Cline, TMS-AIME 242 1013 (1968).

42. E. A. Calnan and C. J. B. Clews, Phil Mag. 41 1085 (1950); ibid 42 616, 919 (1951); ibid 43 93 (1952).

43. D. N. Williams and D. S. Eppelsheimer, J. Inst. Met. 81 553 (1952-53).

44. R. Armstrong, I. Codd, R. Douthwaite and N. Petch, Phil. Mag. 7 45 (1962).

45. G. Sachs, Zeit. d. ver. deut. Ing. 72 734 (1928).

46. K. Okazaki and H. Conrad, submitted to Acta Met.

47. S. N. Monteiro, A. T. Santhanam and R. E. Reed-Hill, The Science Technology and Application of Titanium, Pergamon, N.Y. (1970) p. 503.

48. T. Tanaka and H. Conrad, in preparation.

49. A. M. Garde, A. T. Santhanam and R. E. Reed-Hill, Acta Met. 20 215 (1972).

50. A. T. Santhanam and R. E. Reed-Hill, Met. Trans. 2 2619 (1971).

51. R. J. Wasilewski, Trans. ASM 56 221 (1963).

52. R. N. Orava, G. Stone, and H. Conrad, Trans ASM 59 171 (1966).

53. H. Conrad and R. Jones, The Science, Technology and Application of Titanium, Pergamon, N.Y. (1970) p. 489.

54. H. Conrad, K. Okazaki, V. Gadgil, M. Jon, Electron Microscopy and Strength of Materials, Univ. of Calif. Press, Berkeley (1972) p. 438.

55. F. D. Rosi and F. C. Perkins, Trans ASM 45 972 (1953).

56. W. R. Kiessel and M. J. Sinott, Trans. AIME 197 331 (1953).

57. G. A. Sargent and H. Conrad, Scripta Met. 4 129 (1970).

58. R. Zeyfang and H. Conrad, Acta Met. 19 985 (1971).

59. K. Okazaki, T. Itoh and H. Conrad, in preparation.

60. B. de Meester, M. Doner and H. Conrad, in preparation.

61. S. F. Frederick and W. D. Hanna, "Kinetics of Cryogenic Plastic Deformation of Ti-5Al-2.5Sn" Douglas Missiles and Space Systems Div. DAC Rept. No 59595 (Apr. 30, 1968).

62. R. L. Fleischer, The Strengthening of Metals, Reinhold Pub. Corp., N.Y. (1964) p. 93.

63. G. Sargent and H. Conrad, Scripta Met. 3 43 (1969).

64. R. Zeyfang, R. Martin and H. Conrad, Met. Sci. Eng. 8 134 (1971).

65. U. F. Kocks, Canad. J. Phys. 45 737 (1967).

66. J. Friedel, Dislocations, Addison-Wesley, Reading, Mass. (1965) p. 224.

67. J. Kratochvil and H. Conrad, Scripta Met. 4 815 (1970).

68. R. Jones and H. Conrad, Trans. AIME 245 779 (1969).

69. H. Conrad and K. Okazaki, Scripta Met. 4 111 (1970).

70. H. Conrad and K. Okazaki, Scripta Met. 4 259 (1970).

71. H. Conrad, Mat. Sci. and Engr. 6 265 (1970).

72. K. Okazaki and H. Conrad, to be published Japan Inst. Met.

73. H. Conrad, Ultrafine Grain Size Metals, Syracuse Univ. Press, Syracuse, N.Y. (1970) p. 213.

74. H. Wiedersich, AIME J. Met. 16 425 (1964).

75. H. Gegel, private communication.

76. R. Fowler and E. A. Guggenheim, Statistical Thermodynamics, Cambridge Univ. Press, London (1939).

77. O. Kubaschewski and H. A. Sloman, The Physical Chemistry of Metallic Solid Solutions and Intermetallic Compounds, 1 HMSO, London (1959) paper 3B.

78. E. N. Collings, H. L. Gegel and J. C. Ho, "Fundamental Design of Titanium Alloys", private communication.

79. Y. Lii, V. Ramachandran and R. E. Reed-Hill, Met. Trans. 1 447 (1970).

80. A. M. Garde and R. E. Reed-Hill, Met. Trans. 2 2885 (1971).

81. J. Weertman, J. Appl. Phys. 28 1185 (1957).

82. G. V. Kidson and R. Ross, Radioisotopes in Scientific Research, Pergamon, N.Y. (1958) p. 185.

83. C. Zener, J. Appl. Phys. 22 372 (1951).

84. H. J. Griest, A. M. Sabroff and P. D. Frost, Trans. ASM, 51 9351 (1959).

85. H. Bühler and H. W. Wagener, Bänder Bleche Rohre 6 677 (1965).

86. F. B. Cuff and N. J. Grant, Iron Age 170 134 (1952).

87. F. Garofalo, L. Zewell, A. S. Keh and S. Weissmann, Acta Met. 9 721 (1961).

88. A. D. Le Claire, Diffusion in BCC Metals, ASM (1965) p. 13.

89. M. F. Amateau, H. I. Burrier and L. J. Ebert, Trans. ASM 59 920 (1966).

90. H. I. Burrier, M. F. Amateau and E. A. Steigerwald, "The Relationship Between Plastic Deformation and Fracture in Alpha Titanium", Tech. Rept. AFML-TR-65-239 (July 1965).

91. R. L. Jones, F. W. Cooke, H. Conrad and B. R. Banerjee, "Investigations to Understand the Deformation and Strengthening Mechanisms of Solid Solution Phases of Titanium", Tech. Rept. AFML-TR-68-28 (Feb. 1968).

92. P. W. Bridgman, Trans. ASM 32 553 (1944).

93. L. Sama, O. J. Opinski and L. L. Seigle, "Tensile and Impact Properties of Commercial Titanium over the Temperature Range -196°C to 500°C", WADC Tech. Rept. 54-422 (Sept. 1954).

94. W. L. Finlay and J. A. Synder, Trans. AIME 188 277 (1950).

95. R. I. Jaffee, H. R. Ogden and D. J. Maykuth, Trans. AIME 188 1261 (1950).

96. R. I. Jaffee and I. E. Campbell, Met. Progress 55 356 (1949).

97. R. L. Orr, O. D. Sherby and J. E. Dorn, Trans. ASM 45 113 (1954).

98. H. A. Lipsitt and D. Y. Wang, Trans. AIME 221 918 (1961).

99. N. G. Turner and W. T. Roberts, J. Less-Common Met. 16 37 (1968).

100. N. G. Turner and W. T. Roberts, Trans. AIME 242 1223 (1968).

101. J. Beevers and J. L. Robinson, J. Less-Common Met. 17 345 (1969).

102. D. I. Golland and C. J. Beevers, Met. Sci. Jnl. 5 174 (1971).

103. D. I. Golland and C. J. Beevers, J. Less-Common Met. 23 174 (1971).

104. S. L. Robinson, M. R. Warren and C. J. Beevers, J. Less-Common Metals, 19 73 (1969).

105. C. J. Beevers, The Science, Technology and Application of Titanium, Pergamon, N.Y. (1970) p. 535.

106. P. G. Partridge, Phil. Mag. 12 1043 (1965).

107. C. J. Beevers and M. D. Haliday, Met. Sci. Jnl. 3 74 (1969).

PART A

DEFORMATION

THE STABILITY OF <c + a > DISLOCATIONS

IN HEXAGONAL CRYSTALS*

L. J. Teutonico

Fairchild Industries, Fairchild Republic Division
Farmingdale, New York, U. S. A.

Abstract

Analytical expressions are given for the energies of dislocations with < a >, < c >, and < c + a > Burgers vectors in a hexagonal crystal. These expressions yield a criterion defining the orientations for which a <c + a> dislocation is stable against dissociation into an <a> dislocation and a <c> dislocation. The influence of elastic constant values on this criterion is discussed.

The stability considerations are relevant to the question of non-basal slip in hexagonal metals. It is shown that, in general, a non-basal slip plane will contain sectors for which a < c + a > dislocation is stable against dissociation, other sectors for which it is not. There is, however, one non-basal plane for which a glissile <c + a> dislocation can be stable for all orientations; this is the second-order pyramidal plane.

*This research was supported by the Air Force Office of Scientific Research (AFSC) under Contract No. F44620-71-C-0058.

Introduction

Dislocations in hexagonal crystals have been treated on the basis of anisotropic elasticity theory in a previous paper (1). Formulas were obtained for the displacement field, stress field, and energy of a dislocation of arbitrary orientation and arbitrary Burgers vector in a hexagonal crystal. The hexagonal system is the only crystal system for which a complete analytic solution exists.

The analysis contained in (1) is presently being applied to a number of topics relating to the mechanical behavior of hexagonal metals.* The present study is concerned with the stability of <c + a> dislocations against dissociation and bears directly on the problem of non-basal slip in hexagonal metals.

Energy Formulas

The energy of an infinitely long, straight dislocation in an anisotropic medium is given by

$$E = \frac{Kb_o^2}{4\pi} \log\left(\frac{R_o}{r_o}\right) \tag{1}$$

where K is the energy factor, b_o the Burgers vector of the dislocation, and (r_o, R_o) are the inner and outer radii, respectively, of the elastic region surrounding the dislocation, i.e., r_o is the core radius (the limit of applicability of linear theory) and R_o a dimension of the crystal.**

An analytical expression for the energy factor of a dislocation in a hexagonal crystal has been derived previously (1):

$$Kb_o^2 = K_{11}b_1^2 + K_{22}b_2^2 + K_{33}b_3^2 + 2K_{13}b_1b_3 \quad . \tag{2}$$

In this expression the quantities (b_1, b_2, b_3) are the components of the Burgers vector b_o with respect to a Cartesian coordinate frame (x_1, x_2, x_3) in which x_3 is parallel to the dislocation, x_2 lies in the basal

---

* An analytical study of the dilatational fields of dislocations in hexagonal crystals has already appeared (2).

** The logarithmic factor in Eqn. (1) is quite insensitive to the exact value of R_o/r_o; also, the core energy is expected to be a small fraction of the total energy. Hence the main effect of anisotropy is contained in the energy factor K.

plane, and x_1 makes an acute angle with the c-axis* (Cf. Fig. 1). The quantities (K_{11}, K_{22}, K_{33}, K_{13}) are functions only of the five elastic constants (c_{11}, c_{33}, c_{44}, c_{12}, c_{13}) and θ, the angle between the dislocation axis and the c-axis; the formulas necessary for their evaluation are given in the Appendix.

We now consider in turn dislocations with Burgers vectors $<a>$, $<c>$, and $<c+a>$. From geometrical considerations we are able to write the Burgers vector components of these dislocations in terms of two angles, θ and λ, where λ is defined as the angle between the prism plane containing the three Burgers vectors and the prism plane containing the dislocation (Cf. Fig. 1). Defining K_{b_0} as the energy factor of a dislocation with Burgers vector b_0, one obtains

$$K_a = K_{11}\cos^2\theta\cos^2\lambda + K_{22}\sin^2\lambda + K_{33}\sin^2\theta\cos^2\lambda - K_{13}\sin 2\theta\cos^2\lambda \quad (3)$$

$$K_c = K_{11}\sin^2\theta + K_{33}\cos^2\theta + K_{13}\sin 2\theta \quad (4)$$

$$K_{c+a}\,(c^2+a^2) = K_{11}\,(c\sin\theta - a\cos\theta\cos\lambda)^2 + K_{22}a^2\sin^2\lambda$$
$$+ K_{33}\,(a\sin\theta\cos\lambda + c\cos\theta)^2$$
$$+ K_{13}\,[(c^2 - a^2\cos^2\lambda)\sin 2\theta - 2ac\cos\lambda\cos 2\theta]\,. \quad (5)$$

Stability Criterion

The criterion for the stability of the $<c+a>$ dislocation against dissociation into an $<a>$ dislocation and a $<c>$ dislocation is

$$E_{c+a} < E_c + E_a \quad (6)$$

where E_{b_0} represents the energy of a dislocation of Burgers vector b_0 as given by Eqn. (1). The influence of the logarithmic factor in the energy is neglected;** the criterion for the stability of the $<c+a>$ dislocation can then be written as

$$K_{c+a}\,(c^2 + a^2) < K_a a^2 + K_c c^2\;. \quad (7)$$

---

* We neglected to note in (1) that this convention for the x_1 - axis had been adopted. We are indebted to W. R. Tyson for bringing this point to our attention.

** This is equivalent to assuming that the core radii of the three dislocations are the same.

This reduces to

$$f(c_{ij}, \theta) \sin 2\theta \cos \lambda > 0 \tag{8}$$

where we have set

$$f(c_{ij}, \theta) = K_{11} - K_{33} + 2K_{13} \operatorname{ctn} 2\theta \quad . \tag{9}$$

The factor $\sin 2\theta$ is positive in the half-space above the basal plane, negative in the half-space below the basal plane. The factor $\cos \lambda$ is positive in the half-space on that side of the second-order prism plane (normal to $<a>$) which contains the $<c+a>$ Burgers vector, negative in the other half-space (Cf. Fig. 2).

(i) If $f(c_{ij}, \theta)$ is positive for all θ (which is usually, but not always, the case) the stability criterion yields the following: the hexagonal crystal space can be divided into four quarter-spaces defined by the basal plane and the second-order prism plane normal to $<a>$; in two of these quarter-spaces (those containing the screw dislocation) the $<c+a>$ dislocation is stable against dissociation; in the other two quarter-spaces it is unstable. The latter quarter-spaces contain the orientations of the possible $<c+a>$ edge dislocations which are therefore unstable* (the only exception being the edge dislocation which lies in the basal plane and for which the criterion is ambiguous).

(ii) Should $f(c_{ij}, \theta)$ be negative for all θ, then results opposite to those in the above paragraph would obtain, i.e., the $<c+a>$ dislocation would be unstable in the two quarter-spaces containing the screw orientation, and stable in the two quarter-spaces containing the edge orientations.

(iii) It is elastically possible in an anisotropic** medium that f be positive for some values of θ, negative for other values of θ. The value of θ which satisfies $f(c_{ij}, \theta) = 0$ defines a cone*** about the c-axis, a cone which is bisected by the second-order prism plane (Cf. Fig. 3). For example, in the case $f < 0$ within the cone and positive elsewhere, a semicone of instability would be introduced into the (otherwise stable) quarter-spaces containing the screw orientation, and semicones of stability would correspondingly be introduced into the other two quarter-spaces.

---

* The analytically-derived stability criterion thus provides an explanation for the numerical results previously reported by Yoo (3).

** For an isotropic medium the function f reduces to $\mu\nu/(1-\nu)$, where μ, ν are the shear modulus and Poisson's ratio, respectively, i.e., f is independent of θ and has the sign of Poisson's ratio.

***The equation $f(c_{ij}, \theta) = 0$ reduces to a bicubic in $\sin\theta$.

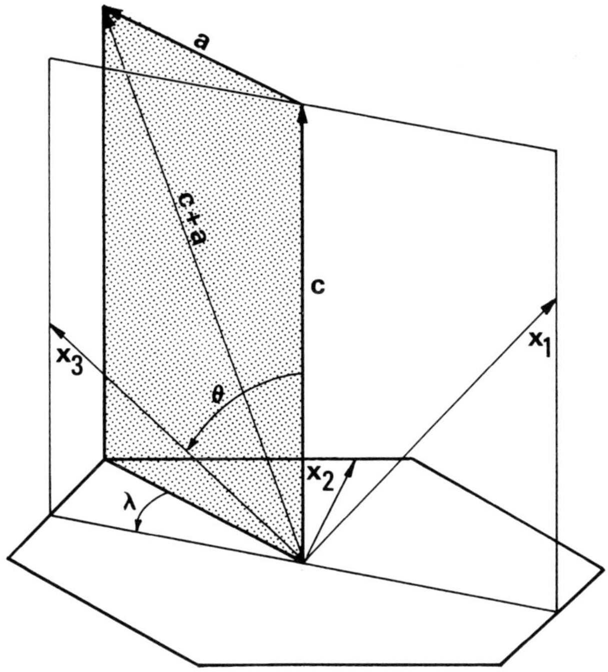
a
c+a
c
x3
x1
θ
x2
λ

Fig. 1

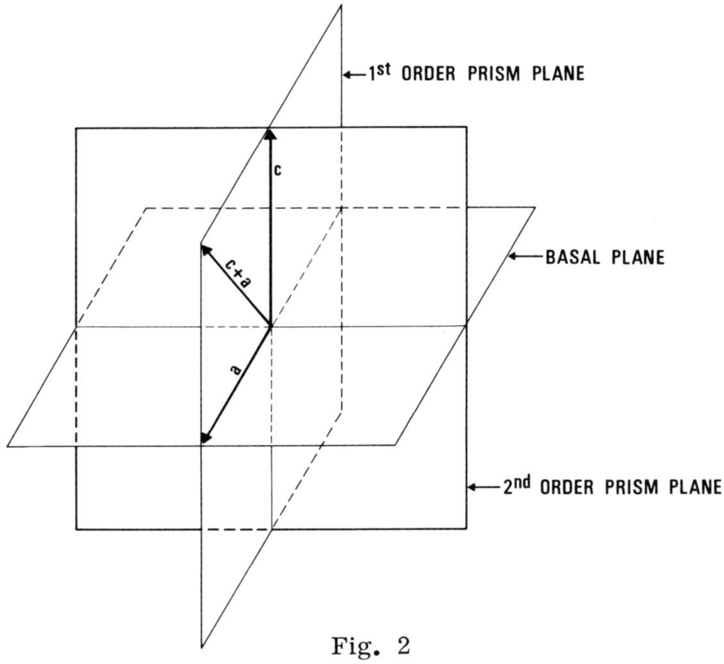
1st ORDER PRISM PLANE
c
BASAL PLANE
c+a
a
2nd ORDER PRISM PLANE

Fig. 2

Values of f have been computed as a function of orientation for the following seventeen hexagonal elements: beryllium, cadmium, cobalt, dysprosium, erbium, gadolinium, hafnium, holmium, magnesium, rhenium, ruthenium, scandium, thallium, titanium,* yttrium, zinc, and zirconium. For all materials except beryllium, it was found that f is positive for all θ. For beryllium it was found that f is negative within approximately eleven degrees of the c-axis (i.e., Fig. 3 applies to beryllium). This unique result for beryllium is directly attributable to the fact that c_{13} is negative in this material (5).**

Non-Basal Slip

The stability considerations of the previous section are relevant to the question of non-basal slip in hexagonal metals.

(i) Any non-basal plane, in general, will intersect all four quarter-spaces as defined above, i.e., on any given non-basal slip plane, there are sectors for which a <c+a> dislocation is stable against dissociation, other sectors for which it is not. As a stable <c+a> dislocation moves (under stress), its orientation may change sufficiently (due to pinning or interaction with other dislocations) so as to bring it into an unstable sector. Its dissociation into an <a> dislocation and a <c> dislocation obviously precludes the possibility of further <c+a> slip. There is, however, one plane which includes the <c+a> direction and lies wholly in two quarter-spaces of stability for positive f; this is the second-order pyramidal plane. In other words, the only non-basal slip plane for which a glissile <c+a> dislocation can be stable*** for all orientations is the second-order pyramidal plane.

---

* Computations for titanium were also performed as a function of temperature from 4°K to 1156°K, the transformation temperature. (Elastic constant data for this temperature range are given in (4).)

** By a limiting process, one can show that

$$f(c_{ij}, \theta) \xrightarrow[\theta \to 0]{} c_{13} c_{66} / c_{11} \quad .$$

***As noted earlier there is one orientation (the intercept with the basal plane) for which the energy criterion is ambiguous. This orientation corresponds to the <c+a> edge dislocation on the second-order pyramidal plane.

To demonstrate the statements in the previous paragraph we consider the three principal planes for <c+a> slip: the first-order prism plane $\{1\bar{1}00\}$, the first-order pyramidal plane $\{10\bar{1}\bar{1}\}$, and the second-order pyramidal plane $\{11\bar{2}\bar{2}\}$ (Cf. Fig. 4). The intercepts of these planes with the four quarter-spaces are shown in Figs. 5, 6, 7; in these figures I_B represents the intercept of the basal plane, I_{P2} the intercept of the second-order prism plane. The regions of instability are shown in the figures for the case in which f is positive for all θ. It is seen that there is no region of instability for the second-order pyramidal plane since $I_B = I_{P2}$; this is the only plane containing the <c+a> Burgers vector for which this is possible.

(ii) In the case that f is negative for all θ, the argument proceeds as in the previous section (i) except that the regions of stability and instability on any given non-basal plane are the reverse of those obtained for positive f [e.g., on the first-order prism plane (Cf. Fig. 5), the dislocation would be stable in the two quadrants containing the edge dislocation, unstable in the two quadrants containing the screw dislocation]. Thus, for negative f, the <c+a> dislocation is unstable for all orientations in the second-order pyramidal plane; there is no slip plane for which a glissile <c+a> dislocation is stable for all orientations.

(iii) To determine the stable and unstable sectors of a non-basal slip plane for a material in which the sign of f is a function of θ, one must consider not only the intercepts with the basal and second-order prism planes, but also the intercepts with the cones defining the zeros of f (Cf. Fig. 3). Each cone always intersects the first-order prism plane. Whether or not it intersects other possible slip planes will depend on the value of θ_c, the angle of the cone.

We have found that $\theta_c \sim 11°$ in beryllium. For a diagram of the first-order prism plane in beryllium one would have to add to that given in Fig. 5 a stable sector ($\sim 11°$ from the c-axis) in the shown regions of instability, and a corresponding unstable sector in the shown regions of stability. The results for the second-order pyramidal plane (Fig. 7) are unchanged, i.e., the <c+a> dislocation is stable for all orientations on the second-order pyramidal plane in beryllium also. This is due to the fact that the cone does not intersect this plane, since θ_c is less than $\psi \equiv \tan^{-1}(a/c)$.* This does bring out the point, however, that in a material for which θ_c were larger than ψ (and $f < 0$ within the cone), a sector of instability would be introduced on the second-order pyramidal plane for orientations about the screw orientation; in such a material there would thus be no plane for which the <c+a> dislocation would be stable for all orientations. It would be interesting to have the elastic constants of beryllium as a function of temperature in order to see to what extent θ_c does vary in this material.

* For beryllium ψ is equal to 32° 32′. In the case of ideal close-packing ($c/a = \sqrt{8/3}$), $\psi = 31° 29′$.

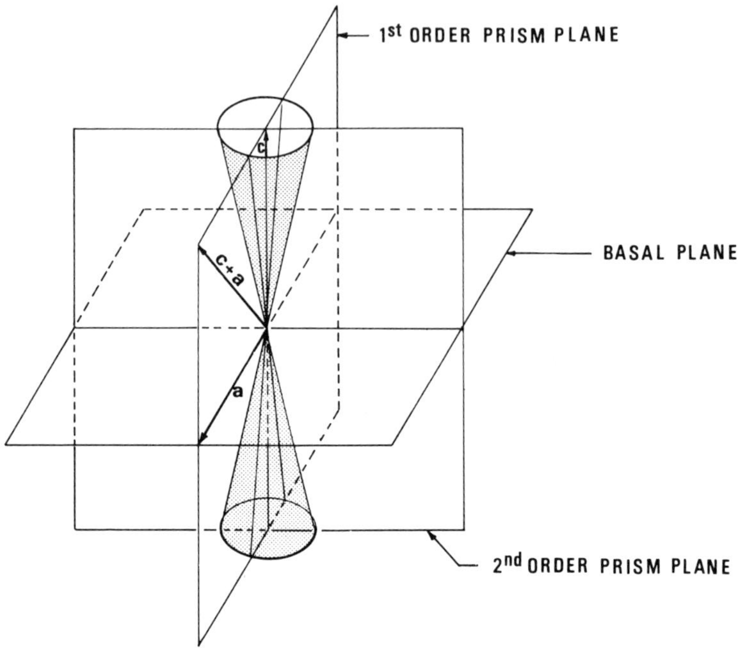
1st ORDER PRISM PLANE
BASAL PLANE
2nd ORDER PRISM PLANE
c
c+a
a

Fig. 3

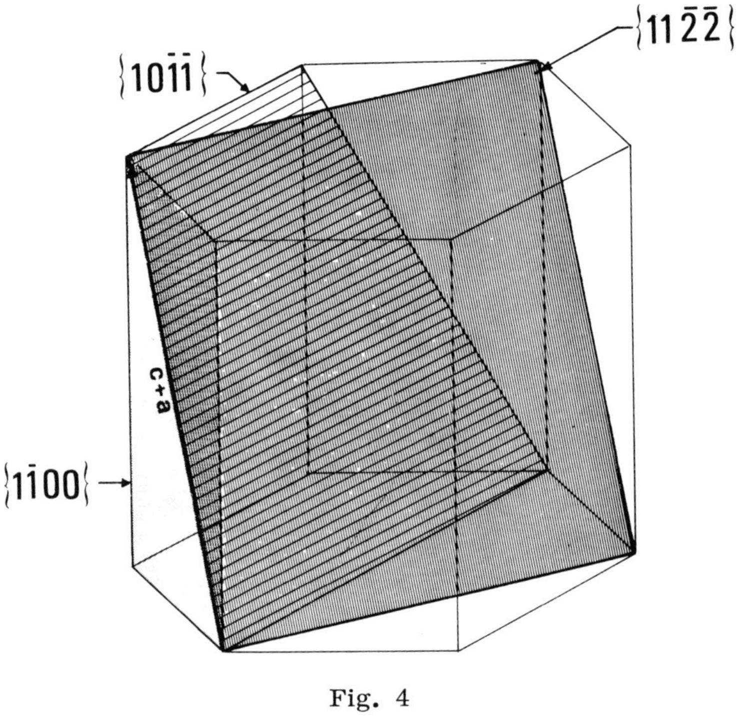
{10$\bar{1}\bar{1}$}
{11$\bar{2}\bar{2}$}
{1$\bar{1}$00}
c+a

Fig. 4

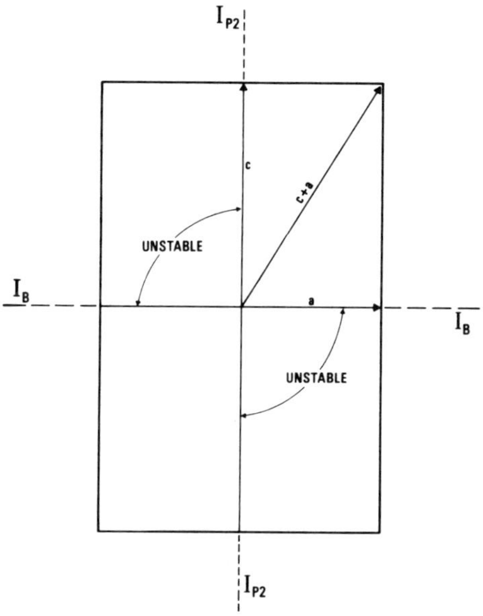

FIRST ORDER PRISM PLANE (f>o)

Fig. 5

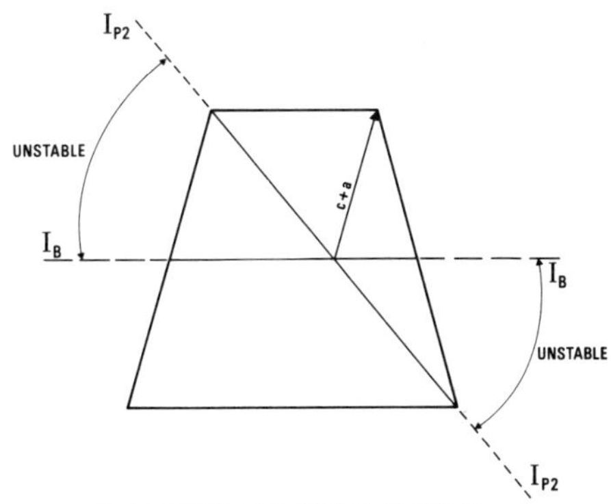

FIRST ORDER PYRAMIDAL PLANE (f>o)

Fig. 6

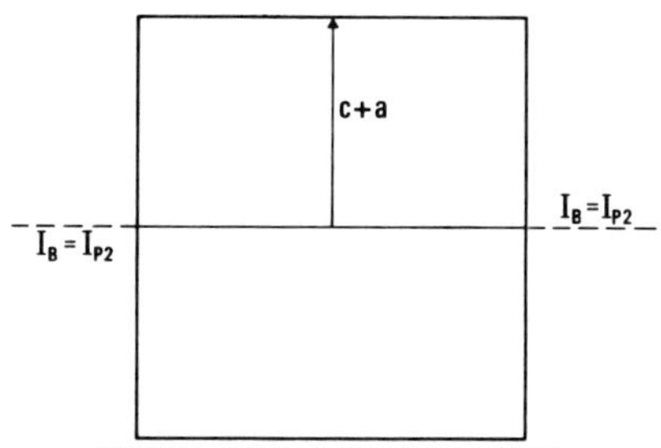

SECOND ORDER PYRAMIDAL PLANE

(No region of instability for f>o)

Fig. 7

References

1. Teutonico, L. J., "Dislocations in Hexagonal Crystals," Materials Science and Engineering, Vol. 6, 1970, pp. 27-47.

2. Teutonico, L. J., "The Dilatational Fields of Dislocations in Hexagonal Crystals," Proceedings of the Second International Conference on the Strength of Metals and Alloys, The Americal Society for Metals, 1970, pp. 394-398.

3. Yoo, M. H., "$\vec{c} + \vec{a}$ Dislocation Reactions in H. C. P. Metals," Scripta Metallurgica, Vol. 2, 1968, pp. 537-540.

4. Fisher, E. S. and Renken, C. J., "Single-Crystal Elastic Moduli and the hcp → bcc Transformation in Ti, Zr, and Hf," The Physical Review, Vol. 135, 1964, pp. A482-A494.

5. Silversmith, Donald J. and Averbach, B. L., "Pressure Dependence of the Elastic Constants of Beryllium and Beryllium-Copper Alloys," The Physical Review B, Vol. 1, 1970, pp. 567-571.

Appendix

Evaluation of Energy Factors

The energy factor of a dislocation in a hexagonal crystal involves the four quantities K_{11}, K_{22}, K_{33}, and K_{13} [Cf. Eqn. (2)]; they are determined as follows.

(i) One defines five quantities (b, ρ, k, ℓ, m) which are functions only of the c_{ij}:

$$b = \frac{c_{11} - c_{12}}{2} = c_{66}, \qquad \rho = c_{44} - c_{66}, \qquad k = c_{11} c_{44}$$

$$\ell = (c_{11} c_{33} - c_{13}^{2}) - 2 c_{44} (c_{11} + c_{13})$$

$$m = (c_{11} c_{33} - c_{13}^{2}) - c_{44} (c_{11} + c_{33} + 2 c_{13})$$

(ii) One next defines four quantities (δ_1, δ_2, δ_3, d) given by

$$\delta_1 = \left(\frac{k + \ell \sin^2\theta - m \sin^4\theta}{k}\right)^{1/2} \qquad \delta_3 = \left(\frac{b + \rho \sin^2\theta}{b}\right)^{1/2}$$

$$\delta_2 = \frac{2k + \ell \sin^2\theta}{k} \qquad d = \frac{1}{2}\left(\delta_2 + 2\delta_1\right)^{1/2}$$

(iii) Six quantities s_i (i = 1, ... , 6) are defined as

$$s_1 = \frac{b^2 \left(g \cos^4\theta + \delta_{13}\delta_3^{\,4}\right)}{2k\,(b+\rho)^2} \qquad s_4 = \frac{2b \operatorname{ctn}\theta\,(b - \rho\cos 2\theta)}{(b+\rho)^2}$$

$$s_2 = \frac{4b^2\delta_3^{\,2}\operatorname{ctn}^2\theta}{(b+\rho)^2} \qquad s_5 = \frac{(g b^2 + \delta_{13}\rho^2)\sin^2 2\theta}{8k\,(b+\rho)^2}$$

$$s_3 = \frac{b \sin 2\theta\,(\delta_{13}\rho\delta_3^{\,2} - g b \cos^2\theta)}{4k\,(b+\rho)^2} \qquad s_6 = \frac{4\rho b \cos^2\theta}{(b+\rho)^2}$$

where

$$\delta_{13} = c_{11}c_{33} - c_{13}^{\,2}, \quad g = c_{11}c_{33} - (c_{13} + 2c_{44})^2 .$$

(iv) Finally, the quantities K_{11}, K_{22}, K_{33}, K_{13} are given by*

$$K_{11} = \frac{(b+\rho)}{d}\left[s_1 + s_2 d\,(\delta_3 - d)\right], \quad K_{22} = \frac{(b+\rho)}{d\delta_3\delta_1}\left[\delta_3 s_1 + s_2 d\,(d\delta_3 - \delta_1)\right]$$

$$K_{33} = (b+\rho)\left[\frac{s_5 + s_6 d\,(d - \delta_3)}{d} + \frac{1}{\delta_3}\right]$$

$$K_{13} = \frac{(b+\rho)}{d}\left[s_3 + s_4 d\,(d - \delta_3)\right] .$$

---

*In Eqn. (74) of Reference (1) the last quantity in K_{22} was incorrectly given as δ_2 , instead of δ_1.

TENSILE STRESS-STRAIN CURVES OF COLD WORKED POLYCRYSTALLINE TITANIUM

A W Bowen and P G Partridge
Materials Department
Royal Aircraft Establishment
Farnborough, Hants, UK

Introduction

High strengths are expected in fcc and hcp metals after working at subzero temperatures (1). Titanium is particularly attractive for this purpose since in addition to a high dislocation density, deformation twinning may reduce the effective grain size and increase the ductility (2,3). The present paper describes some preliminary results on the effect of working titanium of different initial grain sizes at both 20°C and -196°C. The use of empirical equations in analysing the stress-strain curves is discussed briefly.

Experimental Procedure

Commercial purity sheet (containing 0.15 wt % O_2) was annealed to give grain sizes of 0.018, 0.036, 0.11 and 0.90 mm diameter. Each grain size was reduced in thickness 50% by rolling at 20°C and at about liquid nitrogen temperature, i.e. transferred rapidly from the liquid nitrogen bath to the rolls (see ref 2). Electropolished test pieces with dimensions of 0.5 in gauge length, 0.0125 in wide and 0.035-0.050 in thick were tested in an Instron machine at 20°C at nominal strain rates of 1.6×10^{-2}, 1.6×10^{-3} and 1.6×10^{-4} sec^{-1}. Load-elongation curves were obtained by means of an X-Y plotter linked to the load cell and an extensometer attached to the gauge length. The data was converted to true stress and true strain using the relationships $\bar{\sigma} = \sigma\ (1 + e)$ and $\bar{\varepsilon} = \ell n\ (1 + e)$.

Results and Discussion

Tensile Properties

The effect of working on the true stress - true plastic strain curves is shown in Fig 1 for the smallest grain size. Smooth curves were obtained in all the tests. The large increase in strength obtained by working coincides with a large reduction in the uniform strain at maximum load. Clearly the successful exploitation of these strength levels must depend on the associated ductility values and this aspect will be emphasised in the following discussion.

Following the work of Geil and Carwile (4), DeSisto and Carr (2) studied the effect on the room temperature strength and ductility of subzero working temperatures and percentage rolling reduction. They found the strength increased with increased rolling reduction (maximum reduction was 40%) and decreasing rolling temperature (maximum temperature was $-196^{\circ}C$). The total and uniform elongation decreased with increasing rolling reduction but was independent of rolling temperature. Some important features of their results were firstly, that although the strength increased continuously with increasing reduction, the ductility values either decreased markedly after only 10% reduction and thereafter changed very little up to 40% reduction (i.e. elongation and maximum uniform strain) or were almost unchanged by working (i.e. R of A and true fracture strain). Secondly, working at $-196^{\circ}C$ resulted in the same trend with rolling reduction but both strength and ductility values were higher than in material worked at room temperature.

For comparable grain sizes (0.034 mm = $5.5^{-\frac{1}{2}}mm^{-\frac{1}{2}}$) there is good agreement with the present results. The strength of the annealed material was not dependent on grain size in the range studied (Fig 2) although Jones and Conrad (5) have reported an increase in strength with smaller grain sizes ($d^{-\frac{1}{2}}mm^{-\frac{1}{2}}$ = 20-25). However the strength of worked material was dependent on the initial grain size and this dependence was greater in material worked at $-196^{\circ}C$ (Fig 2).

The effect of initial grain size on the ductility parameters is shown in Fig 3. All the ductility parameters fell sharply in the largest grain size material. The elongation and uniform strain values were almost independent of initial grain size when $d < 0.1$ mm, but the true fracture strain in annealed and worked material increased rapidly with decreasing grain size. The effect of working on the relative strength and ductility parameters is summarised in Table I. The subzero working temperature increases the tensile strength significantly compared with working at $20^{\circ}C$ (e.g. 25% increase) and slightly increases the ductility parameters.

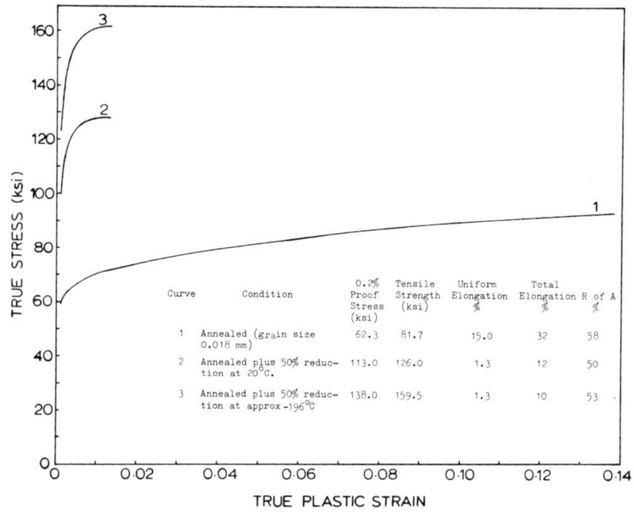

Curve	Condition	0.2% Proof Stress (ksi)	Tensile Strength (ksi)	Uniform Elongation %	Total Elongation %	R of A %
1	Annealed (grain size 0.018 mm)	62.3	81.7	15.0	32	58
2	Annealed plus 50% reduction at 20°C.	113.0	126.0	1.3	12	50
3	Annealed plus 50% reduction at approx -196°C	138.0	159.5	1.3	10	53

Fig. 1. True stress - true plastic strain curves to maximum load for commerical purity titanium. $\dot{\varepsilon} = 1.6 \times 10^{-3}$ sec^{-1}. Also shown are nominal strength and ductility values for these curves.

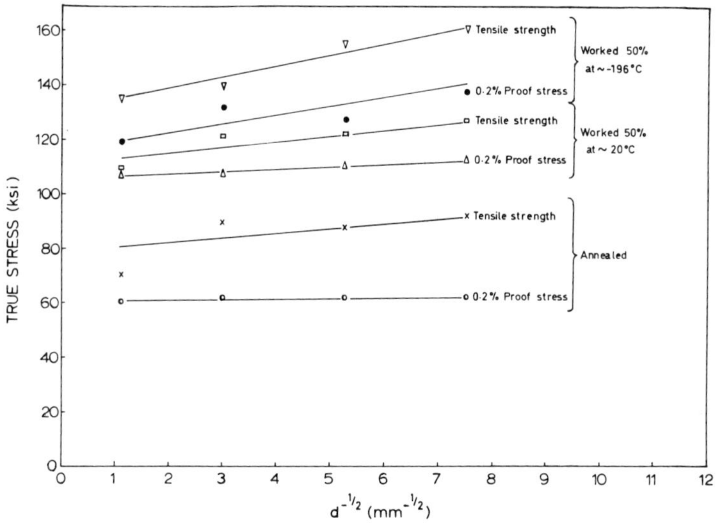

Fig. 2. Effect of initial grain size on the proof stress and tensile strength of commercial purity titanium after working $\dot{\varepsilon} = 1.6 \times 10^{-3}$ sec^{-1}.

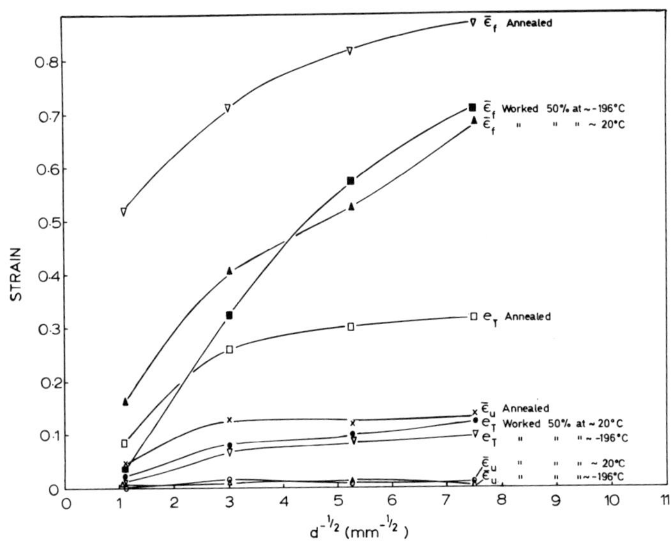

Fig. 3. Effect of initial grain size on the ductility of commercial purity titanium after working $\dot{\epsilon}$ = 1.6 x 10^{-3} sec^{-1}. $\bar{\epsilon}_f$ - true fracture strain; $\bar{\epsilon}_u$ - true uniform strain e_T - total elongation.

Table I

Grain Dia. (mm)	Treatment		True Tensile strength (ksi)		True Fracture Stress (ksi)		True Uniform plastic strain at maximum load		True Plastic strain at fracture		R of A (%)	
0.036	Annealed		89		138		0.13		0.82		56	
	Rolling reduction %		40 *	50	40 *	50	40 *	50	40 *	50	40*	50
	Rolling temperature											
		$20^{\circ}C$	103	123	130	165	0.03	0.011	0.68	0.53	50	41
		$-196^{\circ}C$	128	156	138	202	0.04	0.015	0.75	0.58	53	44
0.018	Annealed		93		144		0.14		0.87		59	
	Rolling temperature											
		$20^{\circ}C$		128		186		0.014		0.69		50
		$-196^{\circ}C$		162		220		0.014		0.71		53

* from ref. 2

Increased rolling reduction from 40% to 50% also increased the tensile strength by ~ 25% but reduced the ductility parameters. A slight increase in strength after working was obtained by decreasing the grain size to 0.018 mm; the uniform strain was not affected but the fracture strain increased by 25%. These results suggest that either decreasing initial grain size or increasing rolling reduction is effective in raising the tensile strength. Some indication of the effect of reducing the grain size is provided by the work of Jones (6) on titanium wire (Table II).

Table II from Ref 6

Grain Diameter mm	True Tensile Strength ksi	Uniform Elongation	% Reduction of Area
0.020	105	0.16	52
0.008	125	0.080	62
0.008 +44% reduction at room temperature	165	0.040	68
0.0035	160	0.080	72
0.0035 +50% reduction at room temperature	182	0.025	72

Much higher strength combined with higher ductility is obtained with 0.0035 mm initial grain size after 50% reduction at room temperature. However, the strength levels in sheet form were much less than those obtained in the present work after deformation at -196°C. Thus the optimum combination of strength and ductility might be rolling reductions < 40%, grain size < 0.02 mm and sub-zero rolling temperature. It is interesting to note that the highest strength in the present work, of 162 ksi nominal tensile strength, was obtained in sheet material with 8% total elongation, 53% reduction of area, 0.014 maximum uniform strain and 0.71 true fracture strain. This strength and elongation is typical of some titanium alloys (7) but the reduction of area is exceptionally high for this strength level. A combination of high strength, high reduction of area and low uniform elongation is also characteristic of maraging steels of 193 ksi tensile strength (7).

Annealed and worked material exhibited a positive strain rate sensitivity. The yield stress (0.2% proof stress) data gave values of m, in the equation $\sigma = A\dot{\varepsilon}^{m}$, of 0.024–0.029. This agrees with previous work (8). Tensile strength data produced values of $m \approx 0.017$. Values of m in the worked material were $m_{Y.S.} \approx 0.01 - 0.022$, $m_{T.S.} \approx 0.011 - 0.03$. The ductility parameters were not significantly affected by strain rate.

Validity of Strain Hardening Equations

Two empirical equations are widely used to describe tensile stress-strain curves (9), e.g. the Hollomon equation

$$\bar{\sigma} = K\bar{\varepsilon}_T^{\,n} \quad \dots \quad (1)$$

where $\bar{\sigma}$ = true stress

$\bar{\varepsilon}_T$ = true total strain, i.e. elastic + plastic

K, n = constants,

and the Ludwik equation

$$\bar{\sigma} = \bar{\sigma}_o + h\bar{\varepsilon}_p^{\,n^1} \quad \dots \quad (2)$$

where $\bar{\sigma}$ = true stress

$\bar{\varepsilon}_p$ = true plastic strain

$\bar{\sigma}_o$, h, n^1 = constants.

The validity of these equations when applied to non-ferrous metals has been questioned. This is discussed more fully eslewhere (9). The data from the present study has been used to examine the validity of equations (1) and (2) when applied to titanium. The effect of using $\bar{\varepsilon}_p$ in place of $\bar{\varepsilon}_T$ is examined for equation (1). In equation (2) the term σ_o is equated either to the proportional limit ($\bar{\sigma}_{\ell p}$) or to the 0.2% proof stress ($\bar{\sigma}_{0.2}$) and the exponent n^1 determined in each case. In titanium (10-13) if a sufficient strain range is considered plots of log $\bar{\sigma}$ v log $\bar{\varepsilon}$ produce at least two linear regions of different slopes n_1 and n_2. The strain at which the change in slope occurs is denoted ε_1. In all the titanium data reported the curves are concave upwards at high strains, i.e. $n_1 < n_2$.

Typical log $\bar{\sigma}$ v log $\bar{\varepsilon}$ curves for the Hollomon equation are shown in Fig 4 for annealed titanium. When either $\bar{\varepsilon}_T$ or $\bar{\varepsilon}_p$ terms are used two linear regions are apparent. At strains greater than $\varepsilon_1 \approx 0.025$ both curves are almost parallel with slopes $n_2 = 0.135$ and 0.13 respectively; these values agree with data of (2) and (11). The n values equal approximately the uniform strain at maximum load. For strains $< \varepsilon_1$ there exists a linear region with $n \simeq 0.05$. When $\bar{\varepsilon}_T$ is used, a third linear region is obtained at strains < 0.004, with n values approaching unity. The transition strain ε_1 did not appear to be related either to the grain size or the strain rate. However in the region of high n the elastic strain was > 80% of the total strain term, whilst the slope n_1 coincided with a gradual decrease in the percentage elastic from 80% to 20%, (Fig 4). Because of the high stresses and small plastic strains involved in the worked material the whole stress-strain curve is associated with high elastic strains and it is difficult to interpret these curves in terms of either equation (1) or (2) (Fig 5).

The Ludwik equation also produced two linear regions in curves of log $(\bar{\sigma} - \bar{\sigma}_{\ell p})$ and log $(\bar{\sigma} - \bar{\sigma}_{0.2})$ versus log $\bar{\varepsilon}_p$ (fig 4). When $\bar{\sigma} = \bar{\sigma}_{0.2}$, n values of 0.5 - 0.7 were obtained. When $\bar{\sigma}_o = \bar{\sigma}_{\ell p}$, n $\simeq$ 0.17 - 0.3. The transition strain ε_1 was similar to that found in the Hollomon curves; below this strain n $\simeq$ 0.05 - 0.20. These results differ from those of (14) and (15) who found $\bar{\sigma}_o$ equal to the proportional limit when n = 0.5; in the present work the $\bar{\sigma}_{\ell p}$ was significantly lower than in (15). Values for a Ti-Al-Mn alloy (13) were n = 0.23 - 0.45; again the $\bar{\sigma}_{\ell p}$ was close to the $\bar{\sigma}_{0.2}$ value.

The results show that the Hollomon equation is a good description of the experimental stress-strain curve at strains > 0.025. The Ludwik equation is in reasonable agreement if n $\simeq$ 0.6 and $\bar{\sigma}_o = \bar{\sigma}_{0.2}$. The results also suggest that the different slopes in the log-log plots at low strains are related to the elastic behaviour. Other evidence is provided by the log $\bar{\sigma}$ v log $\bar{\varepsilon}_T$ curves for a dilute Ti-Al-Mn alloy (13). Two linear regions are obtained, one completely elastic, the other with n = 0.16. The same data plotted as log $(\bar{\sigma} - \sigma_{\ell p})$ v log $\bar{\varepsilon}_p$ produced three linear regions with $n_1 = 0.90$, $n_2 = 0.17$, $n_3 = 0.4$ i.e. the region n_2 represents the transition between the elastic and plastic regions. Similarly, in high strength Ti-alloy (7), the log $\bar{\sigma}$ v log $\bar{\varepsilon}_p$ curve could be divided into many regions with decreasing n and increasing strain; at the stress levels involved these regions all involved high elastic strains, analogous to present worked material (Fig 5). It is therefore concluded that no relationship can be expected between 'n' values and the true plastic strain hardening behaviour in these transition regions. Wasilewski (14) reached similar conclusions for titanium deformed at low temperatures.

Conclusions

1. High strengths (~ 160 ksi tensile strength) and good ductility (~ 10% Elongation, 53% reduction of area) are obtained in unalloyed

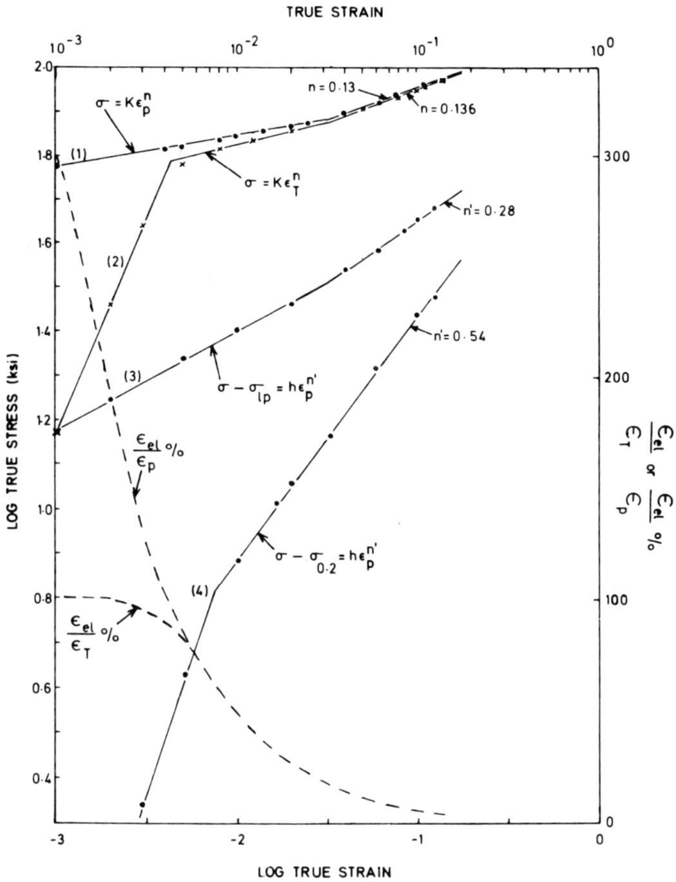

Fig. 4. Plot of strain hardening equations for annealed commercial purity titanium. Grain size 0.018 mm $\dot{\varepsilon}$ = 1.6 x 10^{-3} sec^{-1}.
Curve (1) Hollomon equation log true stress vs log true plastic strain ($\bar{\varepsilon}_p$)
(2) Hollomon equation log true stress vs log true total strain ($\bar{\varepsilon}_t$)
(3) Ludwik equation log ($\bar{\sigma}$ - $\bar{\sigma}_{\ell p}$) vs log true plastic strain
(4) Ludwik equation log ($\bar{\sigma}$ - $\bar{\sigma}_{0.2}$) vs log true plastic strain

Also shown are curves of percentage elastic strain ($\bar{\varepsilon}_{el}$) in the total ($\bar{\varepsilon}_T$) and plastic ($\bar{\varepsilon}_p$) strain terms.

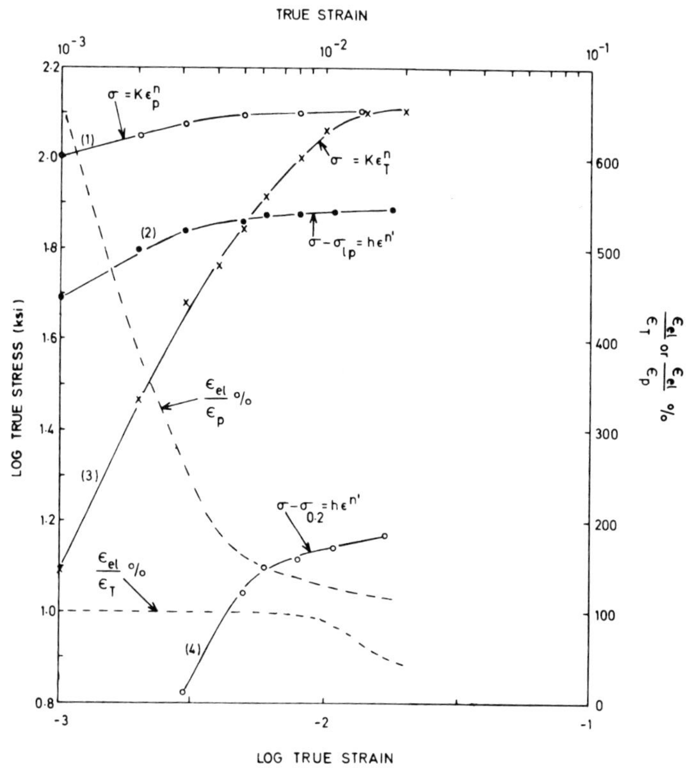

Fig. 5. Plot of strain hardening equations for commercial purity titanium worked at 20°C. Grain size 0.018 mm $\dot{\varepsilon} = 1.6 \times 10^{-3}$ sec^{-1}.

Curve (1) Hollomon equation log true stress vs log true plastic strain ($\bar{\varepsilon}_p$)

(2) Hollomon equation log true stress vs log true total strain $\bar{\varepsilon}_t$

(3) Ludwik equation log $(\bar{\sigma} - \bar{\sigma}_{\ell p})$ vs log $\bar{\varepsilon}_p$.

(4) Ludwik equation log $(\bar{\sigma} - \bar{\sigma}_{0.2})$ vs log $\bar{\varepsilon}_p$.

Also shown are curves of percentage elastic strain ($\bar{\varepsilon}_{el}$) in the total ($\bar{\varepsilon}_T$) and plastic ($\bar{\varepsilon}_p$) strain terms. Similar shaped curves were obtained for material worked at -196°C.

titanium after 50% reduction by rolling at -196°C. It is concluded that strength and ductility similar to that found in titanium alloys could be produced by subzero reductions of ~40% in material with a gram size < 0.02 mm.

2. For annealed titanium both the Hollomon and the Ludwik equations adequately describe the stress-strain curves at high strains (> 0.025). Anomalies occur at lower strains where the elastic strains are significant. The strain hardening exponent n^1 in the Ludwik equation is very dependent on the value of σ_o. For worked titanium neither equation successfully describes the stress-strain curves. This is attributed to the very high elastic strain contribution at all strains at these high stress levels.

References

1. Conrad, H., "The Cryogenic Properties of Metals" High Strength Materials (ed. V.F. Zackay) Wiley New York 1965 pp 436 - 505

2. De Sisto, T. S., Carr, F. L. "Influence of Subzero Rolling on the Tensile Properties of Commercially Pure Titanium" Proc ASTM. Vol 64, 1964, pp 636 - 648.

3. Partridge, P. G. "Crystallography and Deformation Modes of Hexagonal Close Packed Metals." Metallurgical Reviews, Vol 12, No 118, Nov 1967, pp 169 - 194.

4. Geil, G. W., Carwile, N. L. "Effect of Strain-Temperature History on The Tensile Behaviour of Titanium and a Titanium Alloy." J.Res. Nat. Bureau Stds., Vol 61, Sept 1958, pp 173 - 186.

5. Jones, R. L., Conrad, H. "Effect of Grain Size on the Strength of α-Titanium at Room Temperature." Trans Met Soc AIME, Vol 245, April 1969, pp 779 - 789.

6. Jones, R. L. "Investigations to Develop Titanium with Improved Mechanical Properties." Franklin Inst Rep F-C2470, Jan 1970.

7. Fitzpatrick, J. M. "Strain Hardening Characteristics of Four New Steels and a Titanium Alloy." J. Mater. Vol 3, 1968, pp 977 - 982.

8. Monteiro, S. N., Santhanam, A. T., Reed-Hill, R. E. "A thermal Plastic Deformation in Commercial Purity Titanium." The Science Technology and Application of Titanium. (ed R. I. Jaffee and N.E. Promisel) Pergamon Oxford 1970, pp 503 - 516.

9. Bowen, A. W., Partridge, P. G. "Analysis of Tensile Stress-Strain Curves of Polycrystalline Metals." To be published.

10. Holden, F. C., Williams O. N., Riley, W. E., Jaffee, R. I. "Flow Properties of Titanium and Titanium Alloys." T.M.L. Rep. No.30, Battelle, Jan 1956.

11. Holden, F. C., Ogden, H. R., Jaffee, R. I. "Effects of Temperature on the Uniform Elongation of Titanium Alloys" American Society for Testing and Materials, STP204, 1956, pp 14 - 31.

12. Klier, E. P., Gazzara, C. P. "Stress-Strain Properties of Selected Titanium Alloys". Proc ASTM, Vol 59, 1959, pp 1015 - 1028.

13. Haynes, R., Maddocks, P. J. "Effect of Hydrogen on Work Hardening and Fracture Behaviour of Ti-Al-Mn Alloys." Met Sci.J. Vol 3, Sept 1969, pp 190 - 195.

14. Wasilewski, R. J. "Discontinuous Yield and Plastic Flow in α - Titanium." ASM Trans, Vol 56, 1963, pp 221 - 235.

15. Orava, R. N., Stone, G., Conrad, H. "Effects of Temperature and Strain Rate on the Yield and Flow Stresses of α - Titanium." ASM Trans Quart, Vol 59, June 1966, pp 171 - 184.

THE EFFECTS OF GRAIN SIZE REFINEMENT ON THE MECHANICAL PROPERTIES OF UNALLOYED TITANIUM

Robin L. Jones*
Senior Research Metallurgist
Franklin Institute Research Laboratories
Philadelphia, Pa.

Introduction

In the course of an extensive study of the interrelated effects of thermomechanical processing history, impurity content and substitutional alloying on the mechanical properties of α-titanium (1-4), the grain size dependence of the uniaxial tensile behavior and toughness of unalloyed titanium sheet has been investigated. Wire test samples, made by recrystallization of heavily cold swaged material, were used exclusively in our previous investigations of the low temperature deformation mechanism (5-8) and of the origin of the grain size dependence of the strength (8,9) of unalloyed titanium. The question then arises as to the extent to which the observations reported and the conclusions drawn from them were influenced by the crystallographic texture of the wire samples. One objective of the present work was to determine whether or not unalloyed titanium sheet behaved in a similar manner to that reported previously for wire (5-9). The second objective of the program was to investigate, using a modified Kahn-type tear test (10), the effect of grain refinement on the room temperature toughness of the sheet material, since the increase of strength which results from grain refinement is of little utility if it is accompanied by a substantial toughness decrease.

Sample Fabrication and Testing Procedures

The material used in this program was obtained from General Aerospace Materials Corp. in the form of a billet. The billet conforms to ASTM Spec. #B265-58T Grade 3, and the as-received chemical analysis is given in Table I. The billet was received in the

*Now associated with the Stanford Research Institute, Menlo Park, Cal.

mill annealed condition.

TABLE I: Chemical Analysis of Grade 3 Titanium

O	C	N	H	Fe	Ti
0.12	0.02	0.005	<0.01	0.11	Balance

Sheet material with a thickness of 0.06 in. was prepared by unidirectional cold rolling without intermediate annealing of 0.6 in. thick slabs cut from the billet. This rather drastic reduction, which caused considerable amounts of edge cracking, was used so as to permit the attainment of the widest possible range of recrystallized grain sizes. Test samples were machined from the as-rolled sheet. Recrystallized samples with grain sizes in the range 0.5-20 microns were prepared by isothermal annealing in the range 823°K - 973°K. Grain sizes less than about 2 microns were produced by short annealing treatments in a molten lead bath, while the longer anneals required for coarser grain sizes were performed in a dynamic vacuum of $\sim 5 \times 10^{-6}$ torr. Test samples were surface ground following annealing to a 600 grit final finish, the grinding direction being parallel to the tensile axis of the sample.

Grain sizes were equated to the mean distance between intercepts with grain boundaries on a linear traverse. Three different quantitative metallographic techniques were employed which are described elsewhere (4). The three techniques were found to give mutually consistent estimates of grain size to an accuracy of about $\pm$ 5%. Grain shapes were essentially equiaxed and the grain sizes were quite uniform with no marked tendency for any systematic through-thickness grain size variation.

Quantitative pole figures were generated for selected representative samples using a semi-automated, modified Schultz technique. The textures observed were similar to those reported in the literature (11-13) for the cold rolled and intermediate annealed state and no systematic variation with grain size was observed. A typical example of the characteristic anisotropic basal texture is shown in Fig. 1. Williams and Eppelsheimer (14) have shown that the anisotropy can be explained by supposing that slip occurs during rolling on basal, prism and pyramidal systems in addition to operation of compression twinning modes. The operation of all of the three major slip systems appears reasonable since Churchman (15) has shown that the critical resolved shear stresses differ by only $\sim$ 15% at room temperature in commercially pure titanium.

Tensile tests were performed using an Instron machine at a nominal strain rate of $3.3 \times 10^{-4}sec^{-1}$. Change-of-strain rate information was generated by instantaneous changes of nominal strain rate between $3.3 \times 10^{-4}sec^{-1}$ and $3.3 \times 10^{-5}sec^{-1}$. Load sensitivity was improved to about 0.1% of the applied load using a zero sup-

suppression device and plastic elongations were obtained as deviations from the extrapolated 'elastic' loading line on the load-displacement chart. Test temperatures below ambient were obtained by immersion in liquid coolants while elevated temperature tests were performed in air.

Samples were prepared for tear testing by machining a 5/16 inch deep 60° edge notch in one long side of a 1.5 x 0.8 inch blank. For most samples the root radius of the notch was sharpened to about 0.0005 in. by drawing a razor blade across it. A few samples were tested with as-machined notches (radius ∿ .005 in.) to assess the effect of notch sharpness. The samples were pin-loaded in the Instron at a nominal strain rate of $1.1 \times 10^{-3}sec^{-1}$. The line joining the centers of the loading pins was within 0.0005 in. of the tip of the starting notch and the behavior of the sample during tear failure was followed by a conventional load-deflection chart recorder.

Experimental Results - Tensile Tests

General Features

Unlike the behavior observed previously for fine grain size wire samples of commercial purity titanium (1,9), the present sheet samples showed no evidence of discontinuous yielding at any test temperature, the transition from elastic to plastic behavior being accomplished smoothly without yield drops or yield plateaux. As observed in wire experiments (1,7,8,9), the relation between true stress σ and true strain ε in the uniform flow portion of the stress strain curve obeyed the relation

$$\sigma = \sigma(0) + h\varepsilon^{1/2} \qquad \ldots 1$$

The hardening coefficient, h, typically showed values of ∿ 70 ksi at room temperature and depended slightly, if at all, on grain size. The temperature dependence of h was similar to that observed previously for commercial purity wire samples (7). The values of the intercept stress $\sigma(0)$ were generally in the vicinity of the observed proportional limits. Scanning electron fractography indicated that fracture was entirely ductile in nature at all testing temperatures and for all grain sizes (3).

Effects of Grain Size at Room Temperature

Selected tensile data obtained in room temperature tests are tabulated in Table II to illustrate the effects of grain refinement. All the data in Table II were obtained using longitudinal samples (tensile axis parallel to the rolling direction). The yield stresses tabulated are 0.2% plastic offset flow stresses and the UTS values are true stresses. It is evident in Table II that grain refinement leads to an increase in yield stress, a smaller increase in UTS, a

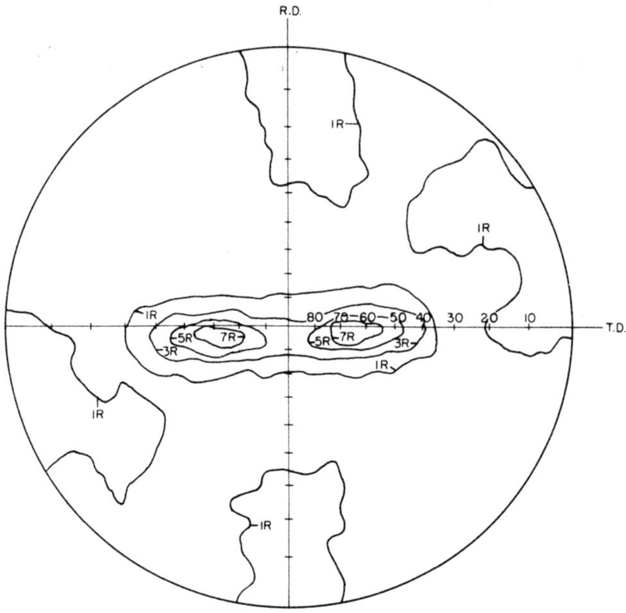

Fig. 1. Typical basal texture for unidirectionally cold rolled and recrystallized Grade 3 titanium sheet.

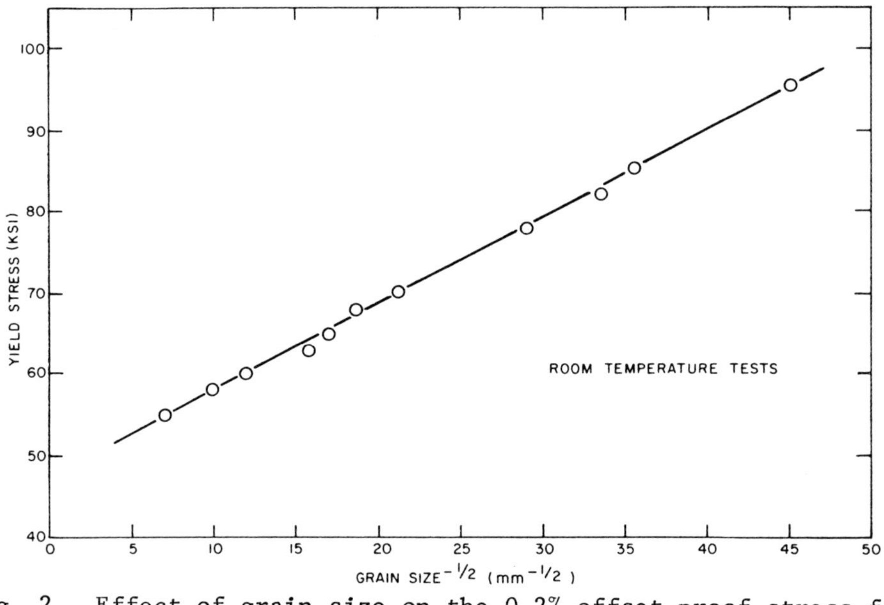

Fig. 2. Effect of grain size on the 0.2% offset proof stress for Grade 3 titanium sheet. Tests at room temperature at a strain rate of 3.3 x 10^{-4}sec^{-1}.

slight increase in reduction in area, a systematic decrease in uniform elongation and little effect on total elongation. The complete data for the yield stress are plotted against the inverse square root of the grain size in Fig. 2 and show excellent support for a linear relationship.

TABLE II. Effect of Grain Size on the Room Temperature Tensile Behavior of Grade 3 Titanium Sheet

Grain Size (Microns)	Yield (ksi)	UTS (ksi)	Uniform Elongation (%)	Total Elongation (%)	Reduction of Area (%)
20	55	89	18	27	50
5.0	62	95	17	30	52
1.2	78	107	16	26	53
0.5	96	118	10	25	57

Effect of Test Temperature and Strain Rate

The effect of test temperature on the tensile properties was established for samples with 2 grain sizes, 0.5 micron and 5 microns. The data obtained are summarized in Table III and values of the yield stress are plotted against temperature in Fig. 3. Below about 500°K the increase in yield stress due to grain refinement is seen to be athermal in nature i.e. the strength increase is temperature independent. Above 573°K the curves converge and metallographic examination indicated that some grain growth occurred in the 0.5 micron grain size samples during the tests. Total elongations and

TABLE III: Effect of Temperature on the Tensile Behavior of Grade 3 Titanium Sheet

Grain Size (Microns)	Test Temp. (°K)	Y. Stress (ksi)	UTS (ksi) (ksi)	Uniform Elong. (%)	Total Elong. (%)	R.A. (%)
0.5	77	182	192	2	18	49
	200	123	147	5	22	55
	300	96	118	10	25	57
	373	89	107	6	28	53
	473	71	85	5	28	55
	573	65	72	3.5	27	65
	673	45	55	2.5	36	>90
5.0	77	146	163	6	23	42
	200	90	115	10	24	45
	300	62	95	17	30	52
	373	50	74	16	28	53
	473	38	54	10	27	52
	573	33	42	6	27	69
	673	30	38	3	38	>90

reductions in area show a general tendency to increase with increasing test temperature, with the most marked increase occurring at the highest testing temperatures, while the uniform elongations tend to peak at around ambient temperature.

In addition to constant strain rate tests, one sample at each grain size was subjected to change-of-strain rate tests at each test temperature. Values for the change in flow stress resulting from order of magnitude changes in strain rate, which were found to be independent of grain size and strain, increased with test temperature up to 373°K, decreased rapidly at higher test temperatures and reached small values at test temperatures $\geq$ 573°K.

Experimental Results - Tear Tests

General Features

Longitudinal samples of Grade 3 titanium covering the entire grain size range were tear tested at room temperature. Samples of 2024 aluminum alloy (in the T3 condition) and Ti-6Al-4V (in both the annealed and the solution treated and aged (STA) conditions) were tested to provide comparative data. With the exception of Ti-6Al-4V samples in the STA condition, qualitatively similar load-deflection curves were obtained, typified by the example shown in Fig. 4. The area under the rising portion of the curve is considered to be associated with the energy required for tear initiation while the area under the falling part of the curve is taken to represent the energy required to propagate the crack (16). Also defined on Fig. 4 are the two most useful parameters which can be obtained from this test - the tear strength (maximum nominal combined tensile and bending stress) and the unit propagation energy (UPE). Values of these parameters were calculated for all samples tested, the areas under the load-deflection curves being measured directly with the aid of a planimeter. Samples of Ti-6Al-4V in the STA condition behaved in a rather different manner, in that propagation occurred catastrophically following a similar initiation curve. For these samples an upper limit estimate of the value of the UPE was made by measuring the elastic energy stored in the loading rig at the maximum load, P_{max}. In all cases the tear propagation direction was normal to the loading axis. The root radius of the starting notch was not found to have any effect on the tear properties observed for Grade 3 titanium, probably because extensive plastic deformation in the notch root region occurs prior to the initiation of the tear failure.

The Tear Strength and Unit Propagation Energy

Fig. 5 shows the tear strength data plotted as a function of yield stress. The tear strength is linearly related to yield stress for the unalloyed titanium samples. The tear strength for annealed Ti-6Al-4V lies close to the line defined by the Grade 3 data while the data points for 2024-T3 aluminum and Ti-6Al-4V STA lie far below

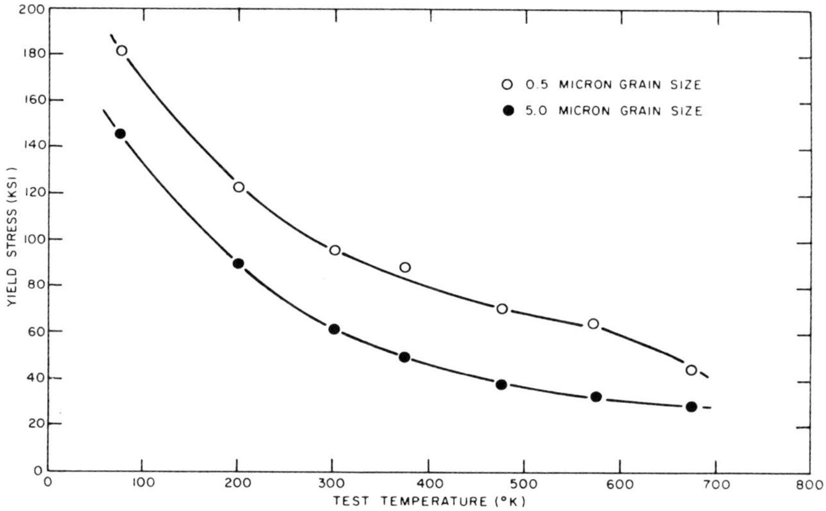

Fig. 3. Effect of temperature on the 0.2% offset proof stress for Grade 3 titanium sheet samples of two grain sizes. Nominal strain rate 3.3 x $10^{-4}sec^{-1}$.

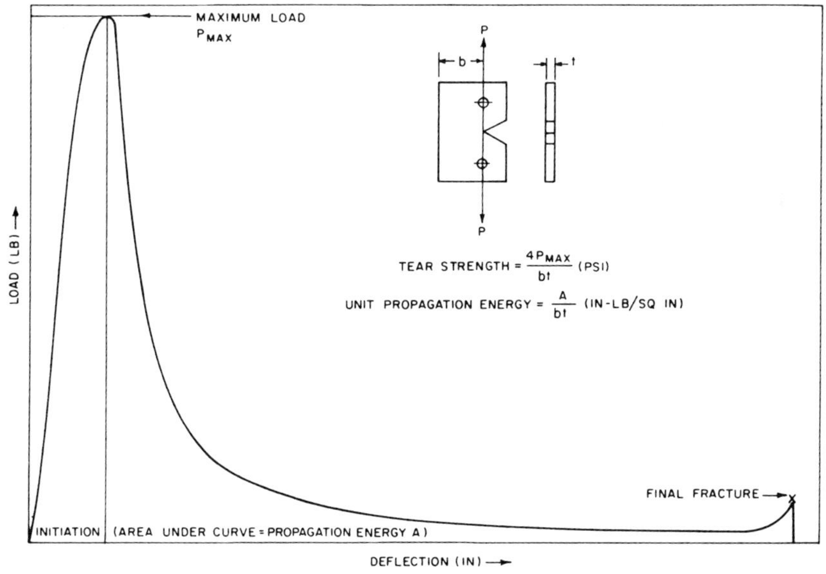

Fig. 4. Typical load-deflection curve observed in modified Kahn-type tearing tests. The rising portion preceding final fracture was observed for all Grade 3 titanium samples and indicates that final fracture is tensile rather than tearing in nature.

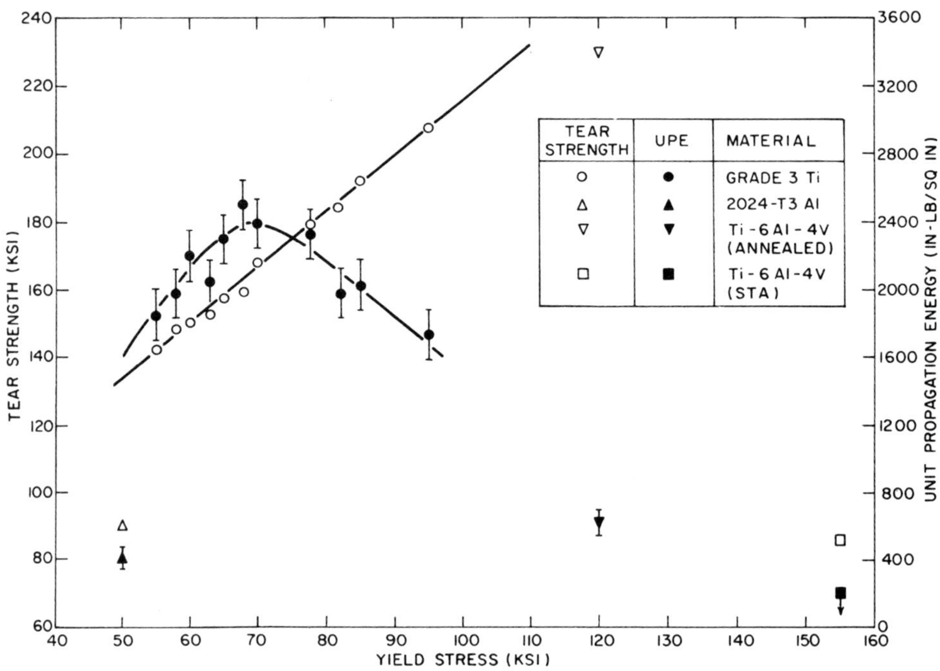

Fig. 5. Variation of tear strength and unit propagation energy with yield stress for three materials tear tested at room temperature.

the line. Mean values of the unit propagation energy (UPE) obtained are also plotted in Fig. 5. The error bars indicate the maximum experimentally observed scatter for nominally identical samples and the mean values are for two or more tests. Although the data show considerable scatter, there appears to be a peak in UPE for Grade 3 titanium at a yield strength of $\sim$ 70 ksi. The aluminum and titanium alloys both show very much lower values of UPE than the unalloyed material.

Discussion - Tensile Tests

Apart from the absence of yield points, the effects of grain size, testing temperature and strain rate on tensile properties of longitudinal sheet samples are qualitatively identical to the effects previously reported for wire specimens (5-9). Fig. 2 shows that the effect of grain size on the room temperature yield strength obeys the commonly observed Hall-Petch (17,18) relation, namely:

$$\sigma_y = \sigma_i + kd^{-1/2} \qquad \ldots.2$$

where σ_y is the yield stress, σ_i is the 'friction stress', d is the grain size and k is the Hall-Petch constant. The previous results (1,8,9) showed that for unalloyed titanium both σ_i and k increased with increasing interstitial content. The data plotted in Fig. 2 indicate values of σ_i and k for Grade 3 sheet of 47ksi and 1.1 ksi-mm$^{1/2}$. Both values lie within the range reported previously (1,8,9) and are close to those expected on the basis of the interstitial content of the present material given in Table I.

The reduction in uniform elongation which accompanies grain refinement can be explained as follows. In well-behaved polycrystalline materials the strain hardening rate decreases as the flow stress increases with increasing strain, and it is easy to show (17) that since deformation occurs at almost constant volume, necking will begin when the values of the strain hardening rate and flow stress are equal, i.e.

$$(d\sigma/d\varepsilon)_u = \sigma_u \qquad \ldots.3$$

where $(d\sigma/d\varepsilon)_u$ and σ_u are respectively the strain hardening rate and flow stress at the UTS, and σ and ε are true stress and strain. If equation 1 is an accurate representation of the uniform flow behavior, we can use it to substitute for σ_u and, after differentiating, for $(d\sigma/d\varepsilon)_u$ in equation 3, obtaining a relation for ε_u, the true strain at the UTS which can be solved obtaining:

$$\varepsilon_u = \left[\frac{h}{\sigma(0) + \sqrt{(\sigma(0))^2 + 2h^2}}\right]^2 \qquad \ldots.4$$

We can then use the observed values of $\sigma(0)$ and h to predict expected values of ε_u as a function of grain size. This is done in Table IV, the value of h used (70 ksi) being the mean value observed in room temperature tests. The agreement between the calculated and observed values of ε_u is surprisingly good and serves to show that the decrease in uniform elongation which accompanies grain refinement arises because decreasing the grain size increases the strength of α-titanium while having little effect on the strain hardening behavior. A similar argument can be used to explain the temperature variation of ε_u.

TABLE IV: Predicted and Observed Values of Uniform Strain at Room Temperature

Grain Size (Microns)	$\sigma(0)$ (ksi)	h (ksi)	ε_u(calculated)	ε_u(observed)
20	52	70	.18	.17
5	59	70	.16	.16
1.2	75	70	.12	.15
0.5	93	70	.09	.095

The increase in reduction in area which accompanies grain refinement probably occurs because the fracture stress is more sensitive to grain size than are the yield and flow stresses. Larger values of k for fracture than for yield have been reported for unalloyed titanium (9) as well as zirconium (20) and mild steel (21). The almost constant total elongation observed for specimens of differing grain sizes is not believed to have any fundamental significance. The elongation during necking is very sensitive to specimen geometry and its observed insensitivity to grain size is most probably coincidental.

In order to determine whether or not the deformation mechanism in the present sheet samples is the same as that deduced for wire (1,8), the change-of-strain rate and yield strength vs. temperature data, have been subjected to a brief thermal activation analysis. Following the same approach as before (1,8) the present data for the yield stress have been separated into two components, the short range (strain rate and temperature dependent) thermal component σ^* and the long range (strain rate and temperature independent) component σ_μ. As noted earlier the effect of grain size primarily resides in σ_μ. It has been confirmed that σ^* varies linearly with the square root of the absolute temperature (1,5,8) and that σ^* (the extrapolated value of σ^* at 0°K) is in accordance with the previously published relation between this parameter and the interstitial content (1,8). The change-of-strain rate data have been used in conjunction with the temperature dependence data to deduce values of H, the activation enthalpy for the operative, rate - controlling, thermally activated deformation process, at each testing temperature. The values obtained are independent of strain and grain size and are plotted against temperature in Fig. 6. The line plotted through the data shows the relation previously obtained for wire samples of

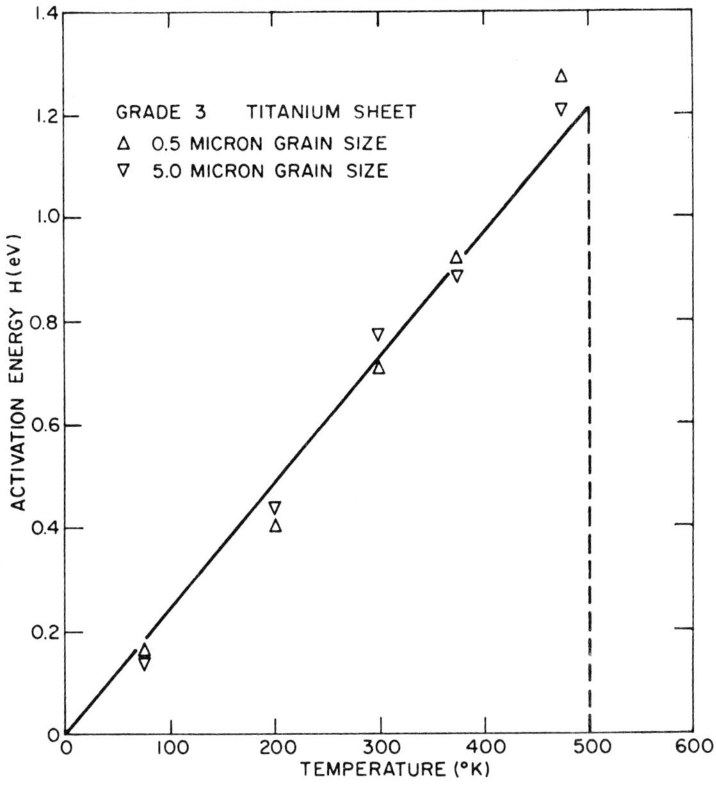

Fig. 6. Temperature dependence of the activation enthalpy for deformation of Grade 3 titanium sheet. The line through the data is replotted from reference 1.

a range of purities (1,8). The linear relation indicates a single rate controlling process and the value of the activation enthalpy at 500°K (the temperature at which the relation between σ^* and T extrapolates to $\sigma^* = 0$) is about 1.2 ev.

The analysis indicates that the rate controlling deformation mechanism in fine grained unalloyed titanium sheet is the same as that previously observed in wire samples. On the basis of the more detailed analysis for wire samples (1,8) it is believed that this mechanism is dislocation glide on first order prism planes opposed by obstacles associated with individual interstitial atoms. The magnitude of the activation energy suggests that the interaction is not due to elastic repulsion between the strain fields of the dislocation and interstitial (22), but rather arises due to interference by the interstitials with the smooth motion of atoms in the dislocation core (23).

It should be noted that the test samples used here were all tested in the longitudinal direction i.e. with the tensile axis parallel to the rolling direction. The anisotropy of the basal texture shown in Fig. 1 is such that the longitudinal test direction favors prism slip at the expense of basal in the same kind of way as occurs in swaged and recrystallized wire samples. Sheet samples tested transverse to the rolling direction are found to show somewhat different tensile behavior (3). The most striking difference is that transverse samples show much lower strain hardening rates than longitudinal samples, which leads to low values of uniform elongation and, more indirectly, to a lower value of k, the Hall-Petch constant. The origin of this low strain hardening rate has not yet been investigated. The increased importance of basal slip may be significant, or possibly deformation twinning (which was not observed in deformed longitudinal samples) may occur. Interestingly, strain rate change experiments lead to values of the stress activation volume which are indistinguishable for the two test directions (3), suggesting that the marked difference in strain hardening behavior does not reflect any change in the rate controlling deformation mechanism.

Discussion - Tear Tests

The objective of the tear testing portion of the program was to establish the effect of grain size on the toughness of unalloyed titanium. Undoubtedly the best method of doing this would be to measure the plane strain fracture toughness K_{Ic} as a function of grain size. Unfortunately the specimen size required for a valid K_{Ic} test on a material as tough as unalloyed titanium makes the performance of such a program extremely difficult, if not impossible. In ASTM Method E399-70T the plastic enclave factor R, defined as $(K_{Ic}/\sigma_y)^2$, where σ_y is the 0.2% proof stress, must, for a valid test, be less than the smaller of $0.4a_o$ or 0.4B where a_o is the initial crack size and B is the specimen thickness. Taking typical handbook values for Grade 3 unalloyed titanium of 80 ksi $\sqrt{in}$ and 60 ksi we obtain R = 1.78 inches, and therefore for a valid test B $\geq$ 4.45 inches. To investigate the effect of grain size on K_{Ic}, we are thus faced with the necessity for obtaining a wide range of equiaxed, uniform grain sizes in material $\geq$ 4.45 inches thick - a daunting prospect indeed.

In view of this problem it seemed most reasonable to use instead of K_{Ic} some other toughness indicator with less stringent size requirements, preferably one which previous work has shown to give a reasonable correlation with K_{Ic}. Such a parameter is the unit propagation energy (UPE) obtained from the modified Kahn tear test. Kaufman and Hunsicker (16) have made extensive use of this parameter as a toughness indicator for aluminum alloys in sheet form, and have reported a correlation between their measurements of K_{Ic} and UPE. Although the material available dictated the use of a smaller sample in the present work than that used by Kaufman and Hunsicker, a qualitative relation between UPE and K_{Ic} is evidenced by the data in Table V where the present values of UPE are compared with typical values of K_{Ic} obtained from the literature. Fig. 5 shows that the UPE for Grade 3 titanium rises initially as the yield strength is increased by grain refinement, and subsequently falls as grain refinement is continued. However the value of the UPE remains very high over the entire grain size range investigated and is probably equivalent to K_{Ic} values of $\geq$ 80 ksi $\sqrt{in}$ at all grain sizes. The highest values of UPE are observed at a yield stress of $\sim$ 70 ksi, equivalent to a grain size of about 2 microns.

TABLE V: Correlation of UPE and K_{Ic}

Material	Condition	UPE (in-lb/sq inch)	K_{Ic} (ksi $\sqrt{in}$)
Grade 3 Ti	Recrystallized (10-20μ)	1850	>80
2024 Al	T3	400	40
Ti-6Al-4V	Annealed	510	50
Ti-6A-4V	STA	<200	30

The other parameter obtained from the tear test, the tear strength, is generally used as an indicator of notch sensitivity in much the same way as the notched tensile strength is used in the notched tensile test (16). It is seen in Fig. 5 that the tear strength is a linear function of yield strength for Grade 3 titanium samples, indicative of notch insensitive behavior. This conclusion is supported by the absence of any effect of notch root radius on tear strength and by the notch insensitive behavior shown by Grade 3 titanium in notched tensile tests (3). The other materials tested fall below the relation between tear strength and yield strength defined by the Grade 3 data, indicating varying degrees of notch sensitivity.

Conclusions

Grain size refinement increases the yield strength of Grade 3 titanium sheet according to the Hall-Petch relation. The magnitude of the grain size effect in sheet samples is similar to that expected on the basis of the results of previous work using wire specimens and the interstitial content of the Grade 3 material. The decrease in uniform elongation which accompanies grain refinement can be explained in terms of the interrelation between flow stress and strain hardening rate at the onset of necking. Thermal activation analysis of the effects of temperature and strain rate on the yield and flow stresses indicates that the rate controlling, thermally activated deformation process at temperatures below 700°K is identical for both sheet and wire test specimens.

In modified Kahn-type tear tests, Grade 3 unalloyed titanium sheet behaves in a notch insensitive manner at all grain sizes, and consequently shows high values of tear strength which depend on grain size in a similar manner to the yield stress. The unit propagation energy for tear propagation is somewhat dependent on grain size, showing maximum values for grain sizes around 2 microns. Grade 3 titanium of any grain size is very resistant to tear propagation; values of the UPE observed were much larger than those for tear failure of 2024-T3 aluminum and Ti-6Al-4V, which were tested for comparison. The results show that substantial increases in strength can be obtained by grain refinement in unalloyed titanium without sacrificing the high toughness of the material. It is thus possible to fabricate unalloyed titanium sheet which combines a yield strength close to 100 ksi with high toughness, probably equivalent to a $K_{Ic} \geq 80$ ksi $\sqrt{in}$.

Acknowledgements

The work described was supported by the Department of the Navy, Naval Air Systems Command under contract number N00019-71-C-0133. The author is indebted to Dr. J. D. Meakin of FIRL, Mr. R. Schmidt of NASC and Dr. J. M. Krafft of NRL for helpful comments regarding various aspects of the program.

References

1. R. L. Jones, F. W. Cooke, H. Conrad and B. R. Banerjee, AFML-TR-68-28 (Feb. 1969).

2. R. L. Jones, FIRL Rept. F-C2470, NASC Contract N00019-69-0437 (Jan. 1970).

3. R. L. Jones, FIRL Rept. F-C2739, NASC Contract N00019-70-C-0206 (March 1971).

4. R. L. Jones, FIRL Rept. F-C3013, NASC Contract N00019-71-C-0133 (March 1972).

5. R. L. Jones and H. Conrad, Acta Met. 15, 4 (1967), pp. 649-650.

6. R. L. Jones and H. Conrad, Scripta Met. 2, (1968), pp. 239-242.

7. R. L. Jones, Scripta Met. 2, (1968), pp. 345-349.

8. H. Conrad and R. L. Jones, The Science, Technology and Application of Titanium, Eds. Jaffee and Promisel, Pergamon Press, N. York (1970), pp. 489-501.

9. R. L. Jones and H. Conrad, Trans. TMS-AIME, 245 (1969), pp. 779-789.

10. N. A. Kahn and E. A. Imbembo, Welding Journal 27, (1948), pp. 169s-182s.

11. C. J. McHargue and J. P. Hammond, Trans. TMS-AIME 197, (1953), p. 57.

12. H. T. Clark, Trans. TMS-AIME 188, (1950), p. 1154.

13. D. N. Williams and D. S. Eppelsheimer, Trans. TMS-AIME 197, (1953), p. 1378.

14. D. N. Williams and D. S. Eppelsheimer, J. Inst. Met. 81, (1953), p. 533.

15. A. T. Churchman, Proc. Roy. Soc. A226, (1954), p. 216.

16. J. G. Kaufman and H. Y. Hunsicker, ASTM STP 381 (1965), pp. 290-307.

17. E. O. Hall, Proc. Phys. Soc. (London) B64 (1951) p. 747.

18. N. J. Petch, J. Iron and Steel Inst., 174 (1953) p. 25.

19. A. H. Cottrell, Introduction to Metallurgy, St. Martin's, N.Y. (1967), pp. 388-389.

20. C. E. Coleman and D. Hardie, J. Inst. Met. 94 (1966), p. 387.

21. N. J. Petch, Phil. Mag. 1 (1956), p. 186.

22. R. Tang, J. Kratochvil and H. Conrad, Scripta Met. 2 (1969), p. 485.

23. W. R. Tyson, Canad. Met. Quarterly 6 (1968), p. 307.

THE DEFORMATION OF α-PHASE TITANIUM

N. E. Paton, J. C. Williams, and G. P. Rauscher

North American Rockwell Science Center

Thousand Oaks, California

Introduction

The hexagonal close-packed α-phase of pure titanium is stable below about 882°C. The deformation of the α-phase has been the subject of several intensive studies over the last 20 years, and is now fairly well understood and documented. The deformation of α-phase titanium of a composition approaching that of the α-phase constituent in commercial alloys has received very limited attention, however. Since the available deformation mechanisms in hexagonal metals can be extremely numerous and complex (1) it can be expected that binary and ternary alloying additions might have a fairly complicated effect on deformation. An understanding of the deformation of α-phase titanium alloys is essential for the solution of such problems as the development of rolling textures, creep, and fracture toughness. The lack of data presently available is probably a result of the difficulty of growing α-phase alloy single crystals of a suitable size. This problem has now been partially overcome and crystals of Ti containing up to 9 wt.% Al (2) can be grown fairly readily.

Deformation in α-phase titanium can occur by either slip or twinning. The slip and twin modes which have been reported in the literature are listed in Table I, along with the appropriate references. A complete review of deformation mechanisms is beyond the scope of this paper, but reference should be made to the recent literature and important findings. Historically basal slip ({0001} <11$\bar{2}$0>) and prismatic slip ({10$\bar{1}$0}<11$\bar{2}$0>) have been considered the most important deformation modes in pure α titanium (3). Levine (4) concluded that prismatic slip in the temperature range 0-210°K

Table I. Slip and Twin Modes in α-Phase Titanium

Slip Modes		*CRSS (kg/mm^2) Room Temp. Ti-250 ppm O_2	Reference
$\bar{a}$ slip	$\{0001\}\langle 11\bar{2}0\rangle$	8.5	3, 6, 4
	$\{10\bar{1}0\}\langle 11\bar{2}0\rangle$	1.4 to 2.4	3, 4
	$\{10\bar{1}1\}\langle 11\bar{2}0\rangle$	---	6
$\bar{c}+\bar{a}$ slip	$\{11\bar{2}2\}\langle 11\bar{2}3\rangle$	---	5, 2, 11
	$\{10\bar{1}1\}\langle 11\bar{2}3\rangle$	---	5, 2, 11
Twin Modes			
	$\{10\bar{1}2\}$		3, 6
	$\{11\bar{2}1\}$		6
	$\{11\bar{2}2\}$		3, 6
	$\{11\bar{2}3\}$		6
	$\{11\bar{2}4\}$		6
	$\{10\bar{1}1\}$		5

*Critical Resolved Shear Stress

was controlled by overcoming the Peierls barrier, while Conrad (7), Evans (8) and Tyson (9) observe that flow stress is controlled by a Fleischer-type hardening mechanism. This latter interpretation seems to be more generally accepted, although recently Sastry and Vasu (10) have revived the concept of the Peierls stress as being the hardening mechanism. An alternate approach which combines some features of both models has been forwarded by Ono and Sommer (12).

Deformation modes other than prism slip have been less thoroughly investigated, but the importance of non-basal $\bar{c}+\bar{a}$ slip has recently received considerable attention (2,5,11) while Cass (2) has shown that $\bar{c}+\bar{a}$ slip is quite common in Ti-Al alloys. A recent review of the role of twinning in the deformation of metals has demonstrated that twinning can have a pronounced effect on the anisotropy of yield stress and work hardening rate. Since pure titanium twins at almost any temperature (5) ignoring its importance as a deformation mode may invoke serious errors in analysis of the deformation of α-phase Ti.

The purpose of this research was to determine the operative deformation modes as a function of stress, alloy content and temperature, and then apply this knowledge to a possible solution of the hardening mechanisms. The great variety of available deformation modes listed in Table I makes characterization of the deformation of polycrystals extremely complicated. Interpretation of the deformation of polycrystals in terms of a single, thermally activated mechanism, could likely be in error, unless polycrystals with a strong texture, or even better, single crystals, are used to ensure that a single mechanism is operating.

Experimental

Single crystals were prepared by electron-beam zone melting from rods of a suitable composition. Some allowance had to be made for loss of Al by evaporation. The rods were fabricated by swaging arc melted and homogenized ingots. Except where noted, the starting materials were electrolytic grade EL-60 titanium supplied by Titanium Metals Corporation of America, and 99.999% aluminum supplied by United Mineral Corporation. Crystals suitable for testing were centerless ground, either chemically or electrolytically polished and the ends of the crystals were carefully lapped flat and parallel to form a finished specimen size of approximately 4 mm diameter by 8 mm long. The crystals were then annealed in a dynamic vacuum of $1x10^{-6}$ torr for 8 hours at a temperature 50°K below the $\alpha/\alpha+\beta$ transus temperature. Crystal orientation and perfection were determined using the Laue back-reflection x-ray diffraction technique.

Polycrystalline samples were cut from material supplied by Titanium Metals Corporation of America in hot rolled sheet form. The final chemical compositions of all materials - both single crystals and polycrystals-are given in Table II. Rolling procedures

Table II. Chemical Compositions of Alloys Tested

Al (at.%)	Al (wt.%)	PPM by Weight		
		O	N	H
3.5	2.1	691	21	37
6.9	4.0	749	32	49
7.0*	4.1	316	19	5
10.2	6.1	697	26	55

* single crystals

used to fabricate the polycrystalline material into sheet form were such that a strong basal pole texture normal to the rolling plane was developed in all compositions. The texture of the Ti - 10.2 at. % Al† material is illustrated in Fig. 1 and the textures of sheets of the other compositions were essentially identical. This strong texture encouraged a comparison of through-thickness compression properties and uniaxial tension properties since the former enforces c-axis compression of many grains, while the latter requires prism slip only. Through-thickness compression tests were achieved on the thin sheet by diffusion bonding a stack of several layers of sheet together to form a compression specimen (subsequently machined from the diffusion bonded block) approximately 14 mm high x 6.5 mm square. All materials were heat treated in a dynamic vacuum of $1x10^{-6}$ torr to provide a uniform final grain size of 35μ.

†All alloy compositions are given in at.% unless otherwise stated.

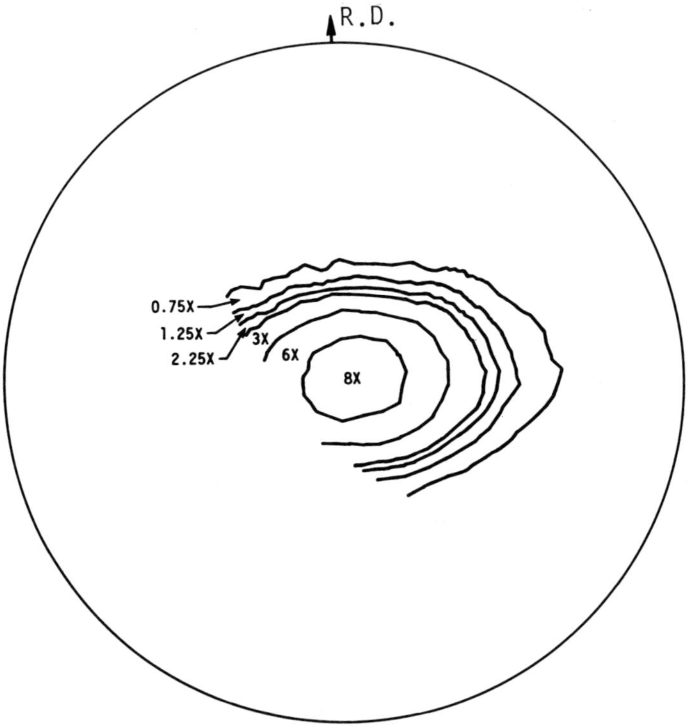

Fig. 1. Basal plane (0001) pole figure of Ti-10.2% Al sheet. The intensity contours are numbered in relation to 1 ≡ random.

Specimens were tested at temperatures from 77°K to 1100°K, those tests above 500°K being conducted in a vacuum of $5x10^{-5}$ torr or better. The high temperature tests were followed by a helium gas quench within a few seconds after completion, in order to minimize recovery and retain as much of the dislocation structure present during straining as possible. The surfaces were sufficiently well protected from oxidation by this technique that slip line analyses could be performed without difficulty after testing.

A two surface trace analysis method was used on single crystals for identification of slip and twin planes. This was done by mounting the crystals in a fixture which permitted simultaneous optical examination of the surface, and rotation of the crystal about its axis.

Thin foils were prepared by a method described previously (13) and examined in a Philips 300 Electron Microscope. Sections were generally taken parallel to the plane of the sheet in the case of polycrystals, and both parallel to the slip plane, and normal to the slip plane for single crystals. Burgers vector determinations were accomplished by standard extinction criteria, usually in a $11\bar{2}0$ zone. Occasionally a section was cut specifically for this purpose to avoid excessive tilting of the foil.

Results

The results described below have been obtained on polycrystalline Ti-Al alloys listed in Table II, on Ti-O alloys containing 250 and 2500 weight ppm O_2, and on Ti-7.02 Al single crystals. These results demonstrate the diversity of deformation modes which have been encountered during this investigation; the results pertinent to each of several deformation modes are discussed separately for clarity.

$\bar{a}$ Slip

The temperature dependence of the resolved shear stress for prism slip with an $\bar{a}$ vector, i.e. $\{10\bar{1}0\}$ $\langle 11\bar{2}0\rangle$ slip, is shown in Fig. 2 for each of the three polycrystalline Ti-Al alloys and for iodide Ti obtained from single crystal experiments, which is included for comparison. The resolved stress for the polycrystalline sheet was calculated by computing the appropriate Schmidt factor for prism slip from the pole figures. The data from Fig. 2 have been re-plotted as resolved shear stress vs Al concentration for each of six deformation temperatures and these results are shown in Fig. 3. Of particular interest in this figure is the increasing concentration dependence of the resolved shear stress as the temperature is decreased, i.e. the slopes of the curves increase with decreasing temperature. The resolved shear stress vs temperature curves for these alloys show the characteristic shape which has been shown previously (14).

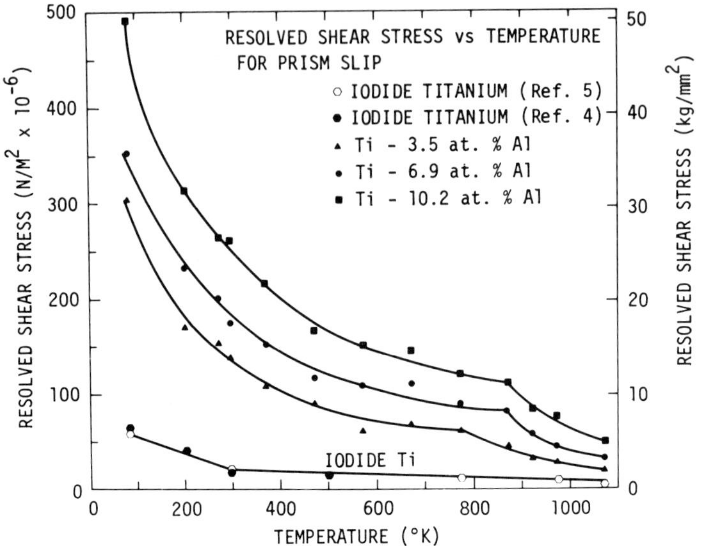

Fig. 2. Resolved shear stress for prism slip in textured polycrystalline Ti-Al sheet with a 35μ grain size and for iodide Ti single crystals.

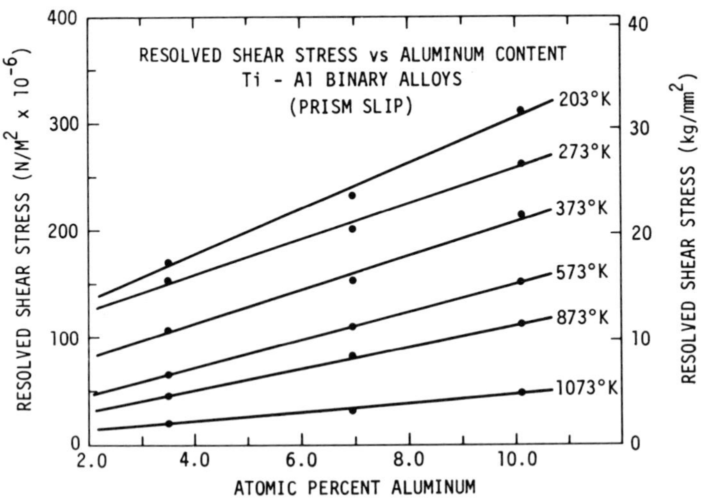

Fig. 3. Resolved shear stress versus Al concentration from the data in Fig. 2.

Each of the curves in Fig. 2 exhibits a discontinuity in resolved stress in the region 750-900°K. The origin of this discontinuity has been previously ascribed to strain aging; however, our observations on Ti-Al substitutional solid solutions suggests that the temperature dependence of the dislocation arrangements which occur during deformation may also contribute to such discontinuities. Thin foil electron microscopy of tensile specimens oriented for $\bar{a}$ slip has shown that planar slip occurs at low temperatures in all the Ti-Al alloys examined (Fig. 4), but that the slip becomes increasingly random as the temperature is increased; the temperature at which planar slip disappears altogether increases with Al content. From Fig. 4 it can be seen that a suggestion of planar slip is present in basal sections even in the 3.5% Al alloy. The principal effect of increasing Al content is to cause the bands of planar slip to become more constricted and better defined, e.g. 4(b) and (c). The detection of banded slip is strongly orientation dependent in the polycrystalline specimens examined in this study, particularly at the lower Al contents. Thus, prism and pyramidal sections do not show the well-defined bands that can be seen in basal sections, presumably because the principal slip plane lies in the foil plane, rather than normal to it.

The influence of temperature on deformation mode at a given Al content is shown in Fig. 5 for the 10.2% Al alloy. In 5(a) the well defined planar bands can be seen. These bands lie parallel to the prism planes in most cases. In 5(b) the random dislocation arrangements can be seen which are typical of this alloy deformed at 925 or 975°K. Figure 5(c) shows the well-defined hexagonal networks which form at deformation temperatures of 1025°K or greater. Each of the adjacent sides of the hexagonal net are composed of different $\bar{a}$ type Burgers vectors; this point has been verified by extinction experiments in the microscope. A significant amount of serrated yielding was observed in all three alloys in the 775-875°K temperature range, suggesting that dynamic strain aging was contributing to the flow stress of the alloys in this temperature range. This observation has also been made previously by Reed-Hill (15).

Non-Basal Slip

As mentioned in the experimental section, diffusion bonded through-thickness compression specimens have been prepared from the 10.2% Al alloy and tested at temperatures between 77 and 1075°K. Figure 6 shows, as has been reported elsewhere (16), that at temperatures $<900°K$, c-axis compression specimens are significantly stronger than the specimens oriented for $\bar{a}$ slip. Thin foil electron microscopy of the c-axis compression specimens showed that extensive slip with a non-basal vector had occurred at temperatures up to and including 575°K. Contrast experiments performed in the microscope have verified that this

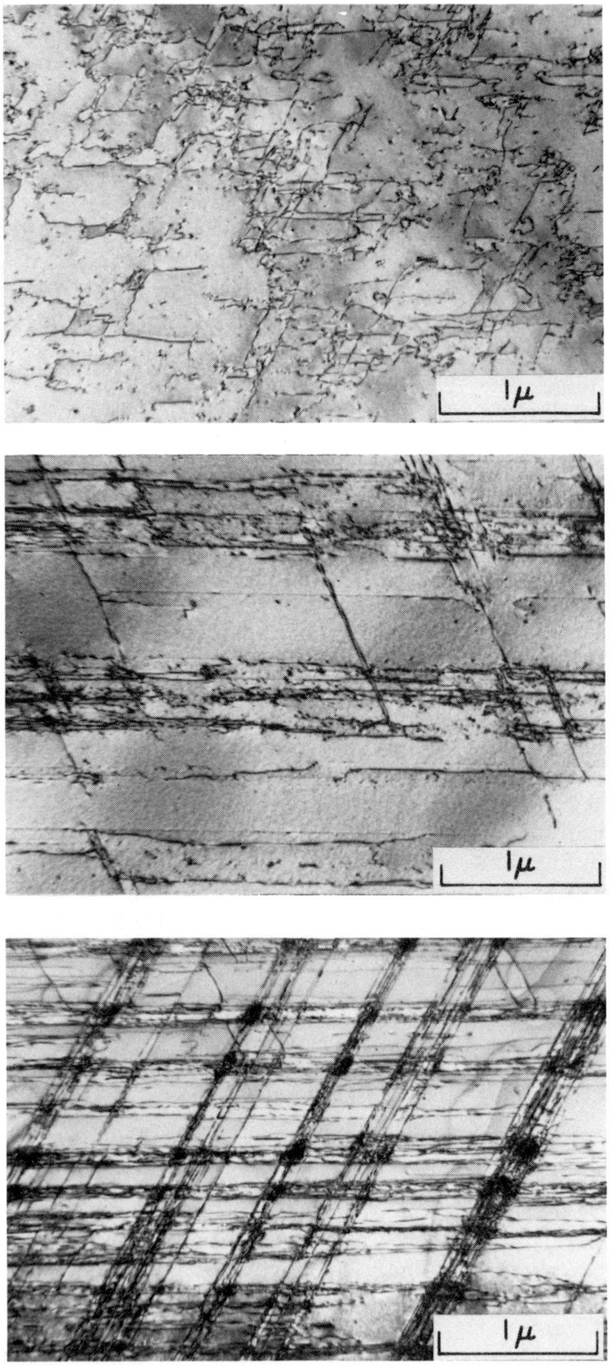

a) 3.5% Al

b) 6.9% Al

c) 10.2% Al

Fig. 4. Thin foil micrographs of Ti-Al alloys deformed ~3% in tension at 300°K showing the increasing confinement of $\bar{a}$ dislocations to planar bands with increasing Al content.

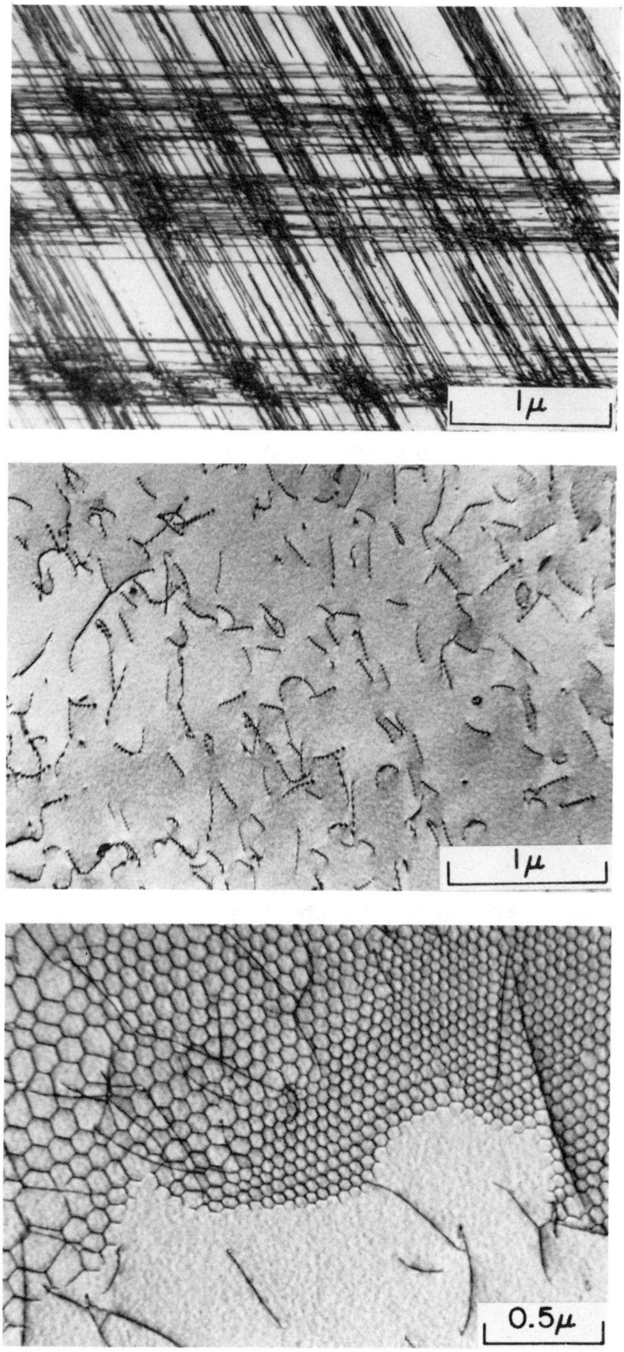

a) 77°K

b) 975°K

c) 1075°K

Fig. 5. Thin foil micrographs of a Ti-10.2% Al deformed ~3% in tension, showing the variation in dislocation arrangements as a function of deformation temperature.

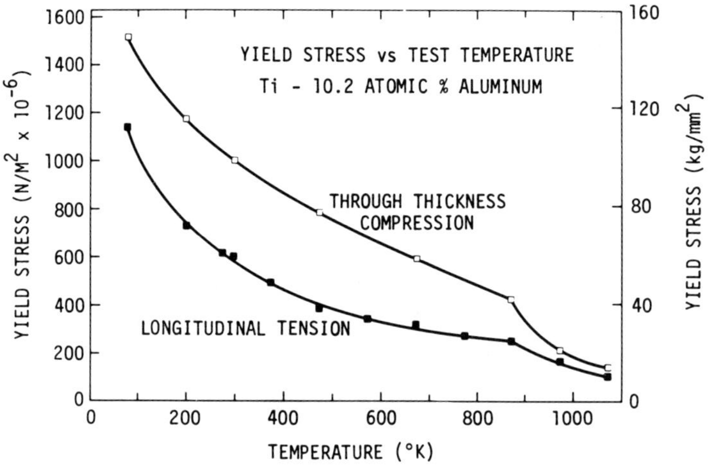

Fig. 6. Temperature dependence of the yield stress of textured Ti-10.2% Al polycrystals for two loading directions.

non-basal slip vector is $\bar{c}+\bar{a}$; an example of $\bar{c}+\bar{a}$ dislocations is shown in Fig. 7. In Ti-Al alloys containing 7.0 and 10.2% Al the $\bar{c}+\bar{a}$ dislocations are banded (Fig. 7(a)), whereas in the lower Al alloys and in unalloyed Ti the $\bar{c}+\bar{a}$ dislocations are often distributed in a more random fashion (Fig. 7(b)). At temperatures >575°K the extent of $\bar{c}+\bar{a}$ slip which could be detected in thin foil specimens decreased significantly, although the specimens had undergone ~ 3% plastic strain.

Single crystals oriented with their c-axis 13-16° off the loading direction have been tested to determine the temperature dependence of the resolved shear stress for $\bar{c}+\bar{a}$ slip. These results are shown in Fig. 8 together with data for two other deformation modes which also occur in these crystals; e.g. at temperatures >575°K basal slip occurs although the Schmidt factor for slip on this system is extremely low. At elevated temperatures the crystals do not undergo $\bar{c}+\bar{a}$ slip to a detectable extent; rather, they deform by a catastrophic unstable shear process, the habit plane of the unstable shear being either $\{10\bar{1}1\}$ or $\{11\bar{2}2\}$. The resolved stress at which this process occurs is also shown in Fig. 8. The nature of this process is best illustrated by the macro-photograph in Fig. 9 showing a shear band in a crystal deformed at 800°K. Extensive plastic deformation accompanies this process as shown in the scanning electron micrograph, Fig. 9(b). The occurrence of this catastrophic shear process is in agreement with the observation that $\bar{c}+\bar{a}$ slip is virtually absent at deformation temperatures above about 600°K. A similar unstable shear process has previously been observed in Mg single crystals (17). Light metallography of a polished section containing the shear band indicates that once the unstable shear occurs, all subsequent compressive strain is concentrated in this region. In view of this and the rapidity at which this unstable shear process occurs, it is possible that adiabatic heating causes rapid recrystallization of this region once the load drop has occurred.

In addition to the occurrence of $\bar{c}+\bar{a}$ slip at temperatures up to 575°K and unstable shear at temperatures between 575 and 1100°K, a third non-basal slip mode has been observed. This mode is characterized by straight dislocation segments which lie parallel to the basal plane trace, as shown in Fig. 10(a). If the foil is tilted about an <hko> axis it can be seen that these dislocations have a stacking fault associated with them, as shown in Fig. 10(b). The fringe contrast of the fault is symmetrical in bright field and the faults exhibit contrast features characteristic of basal plane stacking faults. This deformation mode is concluded to be $\bar{c}/2$ slip. Its occurrence is apparently limited to a small temperature regime and restrictive loading conditions and is thus probably not an important deformation mode per se.

Twinning

Extensive twinning was observed at low temperatures in the

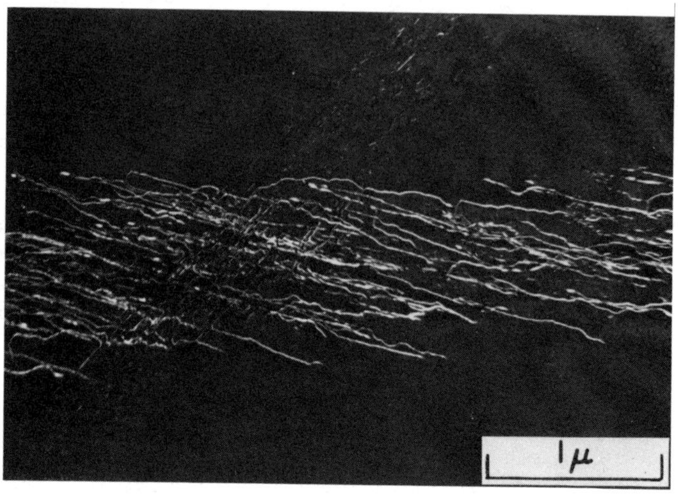

A

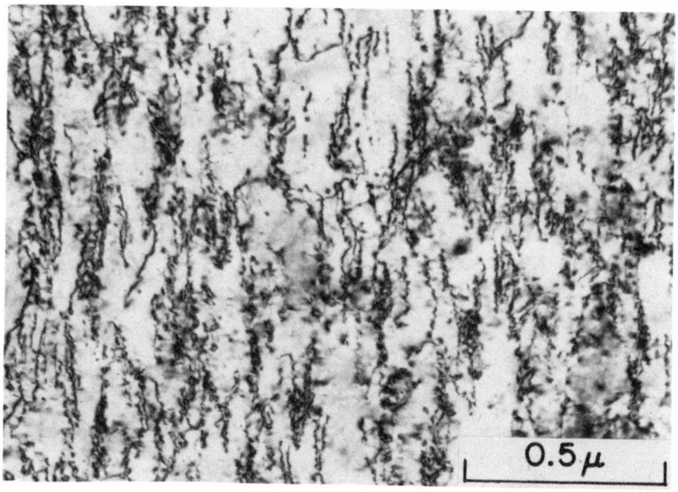

B

Fig. 7. Thin foil micrographs showing the variation in $\bar{c}+\bar{a}$ dislocation arrangements as a function of Al content.
a) Ti - 7.0% single crystal deformed at $200^{\circ}K$, $\bar{g}_{0002}$.
b) Commercially pure Ti containing 2500 wt. ppm oxygen deformed at $475^{\circ}K$, $\bar{g}_{0002}$.

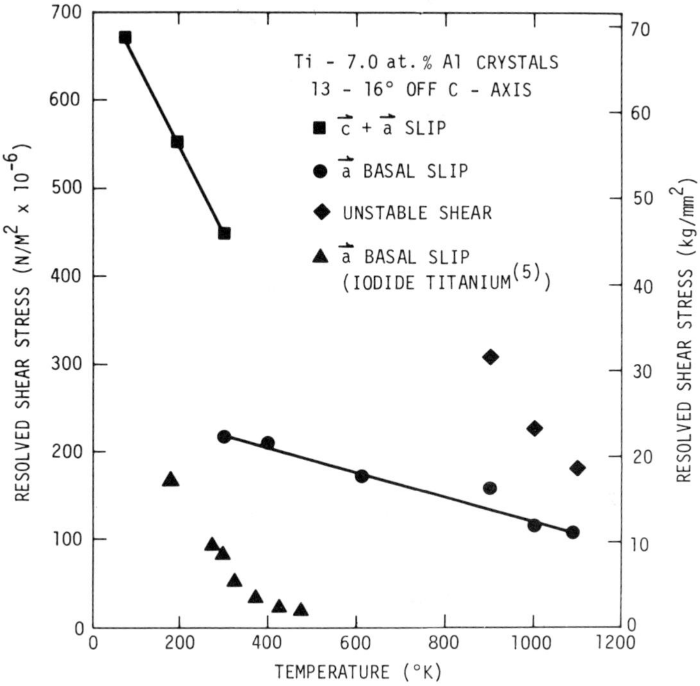

Fig. 8. Resolved shear stress vs temperature of iodide Ti and Ti-7.0% Al single crystals for the deformation modes indicated in the legend.

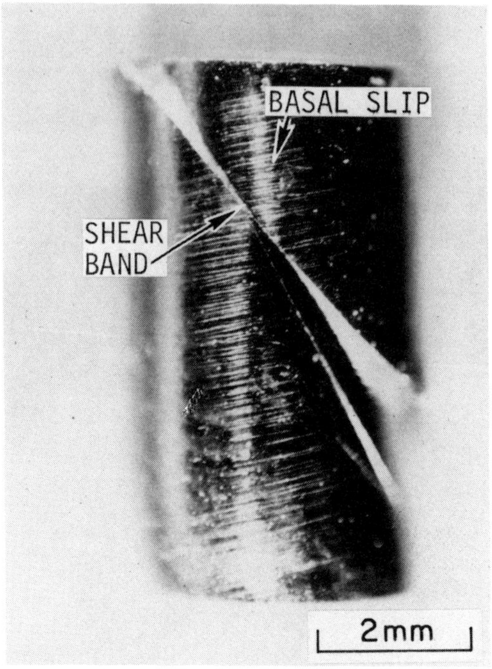

A

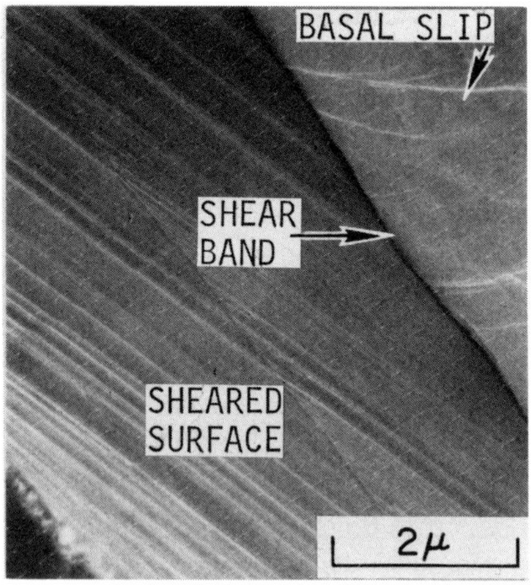

B

Fig. 9. (a) Macro-photograph of Ti-7.0% Al single crystals after unstable shear. (b) Scanning electron micrograph showing extensive plastic deformation which has occurred during unstable shear.

3.5% Al alloy; however, increasing Al tends to suppress the occurrence of twinning rather rapidly. In the 3.5% Al tensile specimens, the twins were usually of a type $\{10\bar{1}2\}$. Through-thickness compression of 6% Al alloys tested at 77°K resulted in a limited amount of twinning of the type $\{11\bar{2}2\}$. In connection with the occurrence of twinning, extensive $\bar{c}+\bar{a}$ slip is always seen in the matrix surrounding the twins and thus is considered to be an accommodation mode, an example of such accommodation is shown in Fig. 11. The material contained within the twinned region is often reoriented more suitably for slip and in such cases the twins frequently contain a higher density of $\bar{a}$ dislocations than the matrix. The amount of strain which is accommodated by twinning depends on the relative values of the twinning stress and the resolved shear stress for the various slip systems. Such dependencies give rise to a variable volume percent of twins at a constant macroscopic strain. In Fig. 12 it can be seen that alloy composition or impurity content has an important influence on twinning frequency.

Discussion

The results shown in Figs. 2 and 3 for $\bar{a}$ vector prism slip in Ti-Al alloys show that the composition dependence of the resolved shear stress varies linearly with the Al concentration. Such a dependence is typical of hardening by solute atoms which produce a symmetrical distortion (18). The variation in slope of the resolved shear stress vs composition curves is similar to the temperature dependence of solid solution strengthening in other systems. However, two additional factors which may influence the temperature dependence of the solution hardening in Ti-Al alloys should be mentioned. First, at low temperatures, e.g. 203°K, increasing Al to 10.2 at.% results in suppression of twinning which may cause the slope of the 203°K curve to increase. Second, the random to planar slip transition which occurs with increasing Al content also complicates the analysis of the data. In this connection, it has recently been shown that a comparable slip mode transition in Ti-O alloys is accompanied by an increase in the internal component of the flow stress as measured by stress relaxation techniques (19).

The single crystal results (Fig. 8) show also that basal slip can operate at a relatively low resolved stress in Ti-Al alloys, particularly at elevated temperatures, and it is possible that the resolved shear stress for prism, pyramidal, and basal slip may converge at these temperatures. Such convergence of resolved shear stress would also allow the banded to random slip mode transition to be explained. That is to say, if the relative values of resolved shear stress approach one another, then slip on secondary or cross-slip systems will become more favorable. The observation of planar slip demonstrates that cross-slip is difficult at low temperatures and high Al content. The random dislocation

A

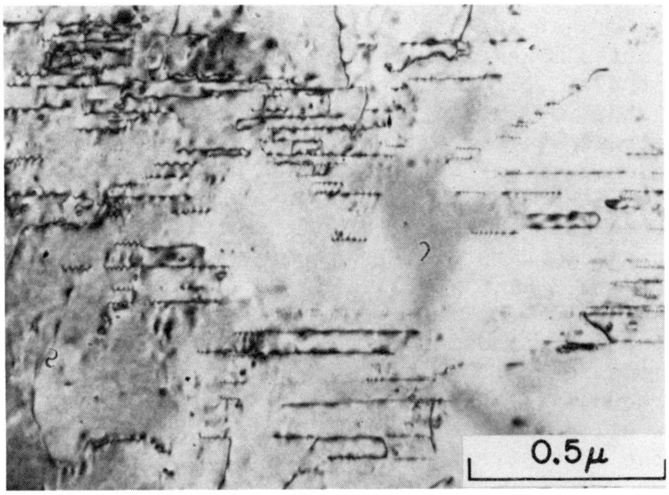

B

Fig. 10. Thin foil micrographs showing $\bar{c}/2$ slip in commercially pure Ti containing 2500 wt. ppm oxygen deformed at 750°K. a) Basal plane stacking faults, $\bar{g}_{10\bar{1}1}$. b) $\bar{c}/2$ dislocations with faults parallel to electron beam, $\bar{g}_{0002}$.

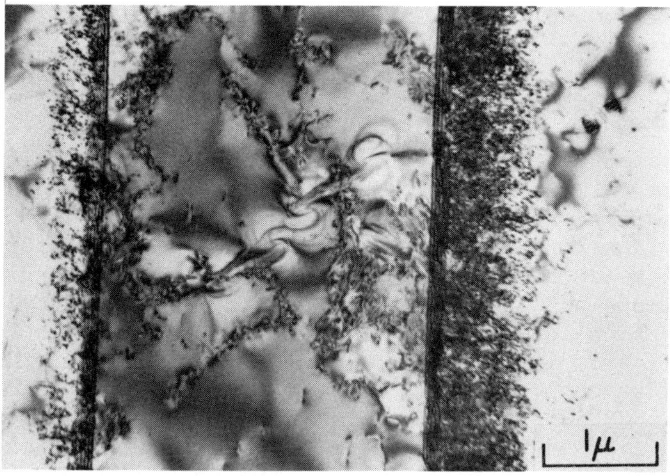

Fig. 11. Zone of $\bar{c}+\bar{a}$ accommodation slip adjacent to a $(11\bar{2}2)$ twin in iodide Ti deformed at 77°K.

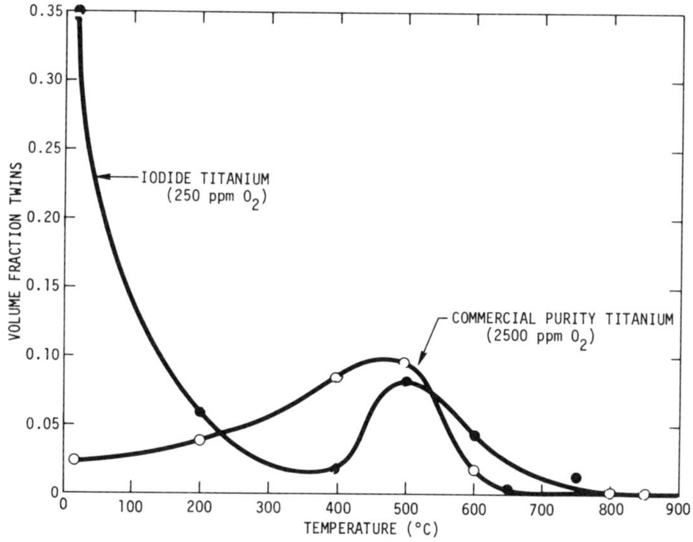

Fig. 12. Influence of deformation temperature on the volume fraction of twins after 5% strain in Ti of two different purities. (Oxygen concentration in wt. ppm.)

arrangements shown in Fig. 5 indicate that with increasing temperature the stress for dislocation motion on the various slip planes becomes comparable, permitting easy cross-slip which results in random arrangements of dislocations.

An increased resolved shear stress for $\bar{c}+\bar{a}$ slip is implied by the difference in flow stress between uniaxial tension and through-thickness compression, shown in Fig. 6. At high temperatures, however, the yield stress drops more dramatically for the through-thickness compression specimens than it does for the longitudinal tension specimens and in this figure, as in Fig. 2, the yield stress-temperature curves converge at temperatures greater than 1000°K. The reason for this is not obvious since $\bar{c}+\bar{a}$ slip is virtually absent in the through-thickness compression specimens deformed at temperatures >575°K. It appears, however, that yield stress measured is actually the stress required to move $\bar{a}$ dislocations on pyramidal, prismatic, or basal planes, and so the stress would be expected to converge if the resolved stress for $\bar{a}$ slip on all slip systems converges. The single crystal results show that the resolved stress for basal slip does in fact fall to low values at the elevated temperatures in question. Confirmation of this point must await additional resolved shear stress data for the slip systems involved. Some of these data should be forthcoming from a study currently in progress (20).

The occurrence of the unstable shear process was an unexpected result. It is possible that the unstable shear process is a combination of twinning followed by concentrated slip within the twin due to localized reduction in flow stress. This reduction in flow stress is a consequence of the more favorable orientation for $\bar{a}$ slip in the twin, and adiabatic heating. The evidence for twinning is incomplete since the large shears in the twinned volume preclude identification by the standard twin orientation relation methods, but the habit plane of the unstable shear coincides with the composition plane for $\{10\bar{1}1\}$ and $\{11\bar{2}2\}$ twinning. A comparable situation has been observed in Mg (17) single crystals, and in this case sufficient adiabatic heating was incurred to promote recrystallization. In this connection, Fig. 2 shows that the temperature dependence of the flow stress of Ti-Al alloys is large enough that adiabatic heating could in fact promote a significant reduction in the local flow stress and account for the concentration of strain in the region of unstable shear. It is possible that the observed unstable shear is directly related to the well-known internal cracking problem, which is encountered during deformation processing of Ti-Al and complex commercial Ti alloys which contain Al. This internal cracking process has been the source of considerable concern in the metal processing industry because of its severe restraint on the amount of work which can be put into a piece of sheet during hot rolling.

At very high temperatures the formation of a dislocation network can occur by the following reaction of $\bar{a}$ dislocation contained in the basal plane;

$$\frac{1}{3}[11\bar{2}0] + \frac{1}{3}[1\bar{2}10] \rightarrow \frac{1}{3}[2\bar{1}\bar{1}0] \ . \qquad (1)$$

Occurrence of this reaction is favored because the long range stress associated with such a network is low. Observed changes in work hardening rates which accompany these various deformation modes are consistent with the fact that relatively high work hardening occurs in the banded slip regime when the mean spacing between dislocations in the active slip bands is small; somewhat less work hardening occurs in the random slip regime. In the temperature regime where networks are formed, the work hardening rate is either zero or negative, indicating in fact that the build-up of internal stress in the crystals is negligible. In view of the rather complicated changes in deformation mode in Ti-Al alloys, detailed analysis of the deformation dynamics in polycrystalline specimens should be approached with caution because of the complications introduced by the varying dislocation arrangements on the many slip systems. The most desirable way to circumvent this difficulty is to use single crystals oriented for slip on a single system, or at least to use strongly textured sheet so that one slip mode predominates.

The $\bar{c}/2$ slip observed in α-Ti alloys between 750°K and the α/α+β transus temperature is probably relatively unimportant as a deformation mode for accommodating significant amounts of plastic strain. However, the frequently observed stacking faults in Ti martensites can be explained by this result. The origin of various types of martensitic substructure has led to some rather detailed analysis and extensive speculation about the mode of lattice invariant shear which accompanies the martensitic transformation in titanium alloys (21). Thus, the observation of $\bar{c}/2$ slip as a deformation mode strengthens arguments that the faults observed in Ti martensites are the result of accommodation slip rather than lattice invariant shear. The alloys in which the faults are most often seen in the martensite have characteristically high M_s temperatures and thus the operation of $\bar{c}/2$ accommodation slip may be fortuitous in that the martensite forms in the temperature range for which this deformation mode is favored.

Twinning has been shown to occur at virtually all temperatures studied in this investigation, and thus it should be re-emphasized that any attempt to analyze the deformation behavior of Ti-Al alloys without accounting for the occurrence of twinning must be regarded with skepticism. The observed tension twin mode in our polycrystalline samples appears to be predominantly $\{10\bar{1}2\}$ twinning. Considering the texture of the sheet, this result is not in complete agreement with the result of Reed-Hill (22). The source of this possible discrepancy has not been elucidated as yet; however, it is the source of continuing consideration. The observation of $\{11\bar{2}2\}$ twinning in the through-thickness compression specimens is in agreement with other worker's results (22). The accommodation dislocations which accumulate in the matrix adjacent to the twin are largely $\bar{c}+\bar{a}$ dislocations because the shears required to accommodate twins make this most favorable. The width of the band of accommodation slip decreases with decreasing temperature which apparently results from the increased resistance of motion of the $\bar{c}+\bar{a}$ dislocations out into the matrix away from the twin-matrix interface. The importance of this layer of accommodation slip has recently been pointed out by Partridge (23) who shows that such a layer can result in the initiation of cracks during cyclic loading. It thus appears that the occurrence of twinning and accommodation slip during cyclic deformation in low Al alloys, or at low temperatures, should have undesirable effects on the fatigue life of these alloys.

Summary and Conclusions

A discontinuity in yield stress versus temperature plot for α-Ti alloys (Fig. 6) coincides with a transition from planar to random slip with increasing temperature. The temperature at which this transition occurs increases with increasing solute content. Through-thickness compression of textured Ti-Al or Ti-O polycrystals or near c-axis loading of Ti-Al single crystals below about 575°K occurs by a combination of $\bar{c}+\bar{a}$ slip and twinning. Above 575°K $\bar{c}+\bar{a}$ slip is absent in the single crystals and replaced by a catastrophic, unstable shear process. Twinning can occur at virtually all temperatures and the twinning frequency is strongly dependent on alloy composition, orientation and temperature. The results suggest that analysis of deformation dynamics of α-phase Ti alloys should be approached with considerable caution because of the complications introduced by the varying dislocation arrangements and the variety of slip and twin systems which may occur during deformation.

Acknowledgements

The authors wish to acknowledge the experimental assistance of Mr. H. Nadler, R. A. Spurling, F. Nevarez, E. Wright, and P. Sauers. Mr. R. Olsen, of the Boeing Company, determined the texture of the polycrystals. The single crystal studies were supported by the Air Force Office of Scientific Research under Contract F44620-72-C-0043.

References

1. Partridge, P. G., Met. Rev. No. 118, 1967.

2. Cass, T. R., "The Science, Technology and Application of Titanium," R. I. Jaffee and N. Promisel, eds., Pergamon Press, 1970, p 459.

3. Churchman, A. T., Proc. Roy. Soc., Vol. A226, 1954, p 216.

4. Levine, E. D., Trans. AIME, Vol. 236, 1956, p 1558.

5. Paton, N. E. and Backofen, W. A., Met. Trans., Vol. 1, 1970, p 2839.

6. Rosi, F. D., Perkins, F. C., and Seigle, L. L., Trans. AIME, Vol. 197, 1953, p 257.

7. Conrad, H., Can. J. Phys., Vol. 45, 1967, p 581.

8. Evans, K. R., Trans. AIME, Vol. 242, 1968, p 648.

9. Tyson, W. R., Can. Metals Quart., Vol. 6, 1968, p 301.

10. Sastry, D. H. and Vasu, K. I., Acta Met., Vol. 20, 1972, p 399.

11. Williams, J. C. and Blackburn, M. J., Phys. Stat. Sol., Vol. 25, 1968, K1.

12. Ono, K. and Sommer, A. W., Met. Trans., Vol. 1, 1970, p 877.

13. Blackburn, M. J. and Williams, J. C., Trans. AIME, Vol. 239, 1967, p 287.

14. Lee, D. and Backofen, W. A., Trans. AIME, Vol. 236, 1966, p 1696.

15. Monteiro, S. N., Santhanam, A. T., and Reed-Hill, R. E., same as ref. 2, p 503.

16. Paton, N. E., Ph.D Thesis, M.I.T., 1969.

17. Wonsiewicz, B. C. and Backofen, W. A., Trans. AIME, Vol. 239, 1967, p 1422.

18. Fleischer, R. L., "The Strengthening of Metals," D. Peckner, ed., Chapman and Hall, New York, 1964, p 93.

19. Williams, J. C., Sommer, A. W., and Tung, P. P., unpublished research, N. R. Science Center, 1969.

20. Sakai, T. and Fine, M. E., research in progress, Northwestern University, 1972.

21. Otte, H. M., same as ref. 2, p 645.

22. Reed-Hill, R. E., paper presented at the 1971 ASM Seminar, Detroit. To be published in the Proceedings of the Seminar.

23. Partridge, P. G. and Peel, C. J., same as ref. 2, p 517.

SLIP IN TITANIUM-ALUMINUM ALLOYS CONTAINING SMALL Ti_3Al PRECIPITATES

J.D. Boyd and R.G. Hoagland
Battelle
Columbus Laboratories
505 King Avenue
Columbus, Ohio 43201

Introduction

In titanium-aluminum alloys containing more than approximately 12 atomic percent aluminum, (7.2 wt. pct.) and less than approximately 20 at. pct. aluminum, (12.4 wt. pct.), the compound Ti_3Al forms as discrete, ordered precipitates in an alpha solid-solution matrix.(1) The size distribution and morphology of the Ti_3Al precipitates can vary widely depending on the aluminum concentration and the details of heat treatment. (1,2) The precipitation of Ti_3Al produces a relatively small increase in the yield stress, but has profound effects on the fracture properties. Generally, Ti_3Al precipitation decreases the fracture toughness, (3,4) but there is some evidence that the degree of embrittlement is sensitive to the actual size distribution of the Ti_3Al particles.(2) The formation of Ti_3Al also increases an alloy's susceptibility to stress-corrosion cracking in saltwater, (5-7) and the fracture toughness in saltwater also depends on the size distribution of the Ti_3Al particles.

It is suspected that the effects of Ti_3Al precipitates on the fracture toughness and stress-corrosion susceptibility of titanium-aluminum-base alloys is due to the character of slip. Specifically, it has been suggested that in alloys containing Ti_3Al precipitates which are sheared by glide dislocations, slip is more planar and nonhomogeneous in distribution than for the single-phase solid solution.(2,3) In support of this view it has been demonstrated that when the size distribution of the Ti_3Al particles is such that they are bypassed by the glide dislocations, slip is much more homogeneous than when the particles are sheared.(2) The investigation reported here was a study of the relation between the size distribution of Ti_3Al particles, slip character, and mechanical properties in a titanium-aluminum alloy. The ultimate objective of this work, (which is ongoing), is to devise a size distribution of Ti_3Al particles which gives improved fracture properties, particularly with regard to stress-corrosion cracking.

Experimental

The alloy studied throughout this program was titanium-13.4 at. pct., (8.03 wt. pct.), aluminum. The complete chemical analysis of this material and the details of specimen preparation are given elsewhere.(7) All specimens were annealed for 24 hours at 900°C and water quenched. This heat treatment produced a microstructure consisting of dislocation-free equiaxed alpha grains, (average grain diameter ≃60 μm), and there was no evidence of ordering in the electron diffraction patterns. Various size distributions of Ti_3Al precipitates were obtained by giving the solution-annealed-and-quenched material different aging treatments. Most of the results reported in this paper refer to material aged 100 hours at 500°C. This heat treatment produces a very fine dispersion of spherical Ti_3Al particles 25-50 Å in diameter.

Strip tensile specimens 0.025-in. thick with 1.0 in. by 0.25 in. gauge sections were tested in air and saltwater, (3.5 wt. pct. NaCl), at a strain rate of 0.005 min.$^{-1}$. The surface topography of the deformed specimens was characterized by a two-stage replica technique, which is described in detail in Reference 7. Specimens were thinned for transmission electron microscopy by chemical polishing in nitric-hydroflouric acid, (60% HNO_3-40%HF) to approximately 0.005-in. thickness, followed by electropolishing in a 6% perchloric acid-35% butanol-59% methanol solution at -30°C.

Results

The stress-strain behaviors of the aged and unaged material are illustrated in Figure 1. Aging for 100 hours at 500°C does not significantly change the yield stress or elongation-to-fracture in air, but it does alter the shape of the stress-strain curve. The aged specimens exhibit a yield drop and Lüders extension, whereas these effects are absent in the unaged specimens. When tested in salt water, the aged specimens fail by subcritical crack growth after approximately 10 percent strain, whereas the tensile properties of unaged specimens are unaffected by a salt-water environment.

The surface slip lines on deformed specimens also show an effect of heat treatment. In the single-phase material, (unaged),

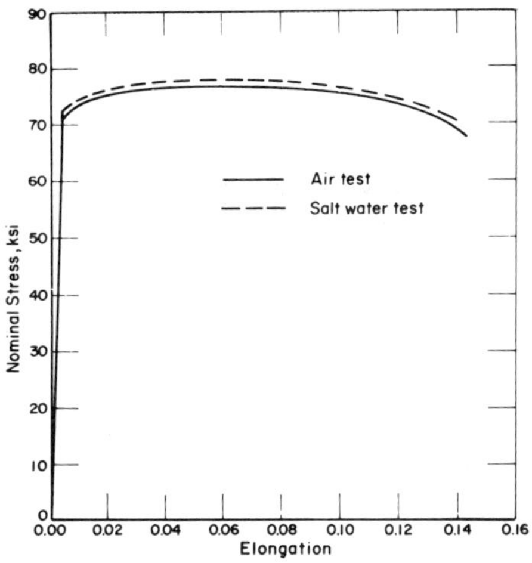

A

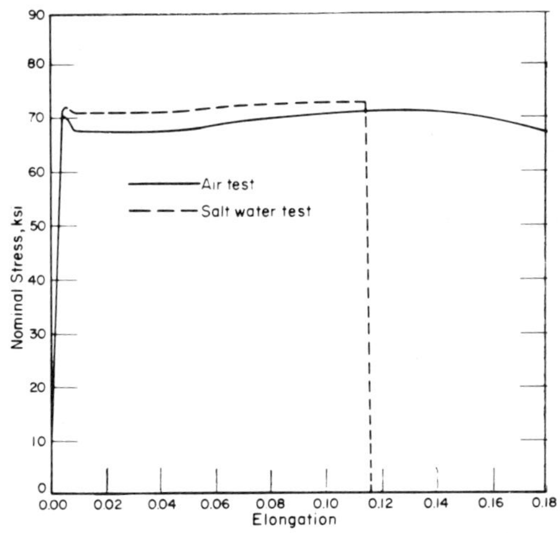

B

Fig. 1. Stress-strain curves for Ti-8Al, ($\dot{\varepsilon}$ = .005 min^{-1}); (a) 24 hr at 900°C, W.Q. (b) 24 hr at 900°C, W.Q. + 100 hr at 500°C.

slip is nonhomogeneous in that it is generally confined to bands which have an average spacing of 0.2-2 μm, (~10^3-10^4b). However, there is much evidence of bands which are wavy on a fine scale, (Figure 2a), and often, in the vicinity of an obstacle such as a grain boundary or another slip band, a coarse slip band will disintegrate into a region of fine wavy slip, (Figure 2b). By constrast, the slip bands in the aged material are exceedingly straight or are composed of perfectly straight segments, and this slip character persists up to the point of fracture, (Figure 3). There is rarely any fine, wavy slip such as that which occurs in the single-phase material. Instead, cracks often form where slip bands intersect grain boundaries or other slip bands, (e.g., Figure 3b at A).

Another feature characteristic of the slip bands in the aged material is the existence of trenches along the slip steps. This feature appears in the two-stage replicas as a black line adjacent to a white shadow, (Figures 4b and 4c). The shadows from the trenches are usually very ragged, suggesting the contours of the trenches are quite irregular. The trenches occur in aged specimens deformed in air as well as in those deformed in salt-water, provided the strain is more than about 5 percent. Trenches are found in approximately half of the grains, and they are specific to a particular slip system in a given grain. For example, in grains deforming by multiple slip, trenches can be found on nearly every slip band belonging to only one or perhaps two of the systems, but are not evident on the remaining slip systems in the grain, (Figure 4b).

The dislocation substructures, as revealed by transmission electron microscopy, are quite similar in both the aged and unaged specimens. At the low strains which have been investigated, ($\varepsilon \approx 0.02$), both alloy conditions contain coplanar arrays of $\frac{a}{3}\langle 11\bar{2}0\rangle$ dislocations lying on $\{1\bar{1}00\}$ and (0001). There is more evidence of cross slip between (0001) and $\{1\bar{1}00\}$ in the single-phase material than in the alloy containing Ti_3Al precipitates, while in the latter the dislocation arrays are somewhat narrower and more rigorously planar. However, dark-field microscopy using a Ti_3Al superlattic reflection reveals some interesting features of the slip bands in the aged material. Figure 5 shows that the density of Ti_3Al particles is much less in the slip bands than throughout the rest of the grain. It is not possible to determine the shear in a slip plane from the dislocation arrangements, but it appears that the density of Ti_3Al particles is less for the more active slip bands (cf., areas A and B in Figure 5). The particles which are visible in the slip bands have much weaker images than the others, and in those areas where the shear has presumably been the greatest there are no particles whatsoever. These effects can be seen more clearly in Figure 6 where the mean particle size is much larger, (100-300 Å). Many sheared particles can also be seen in the slip bands.

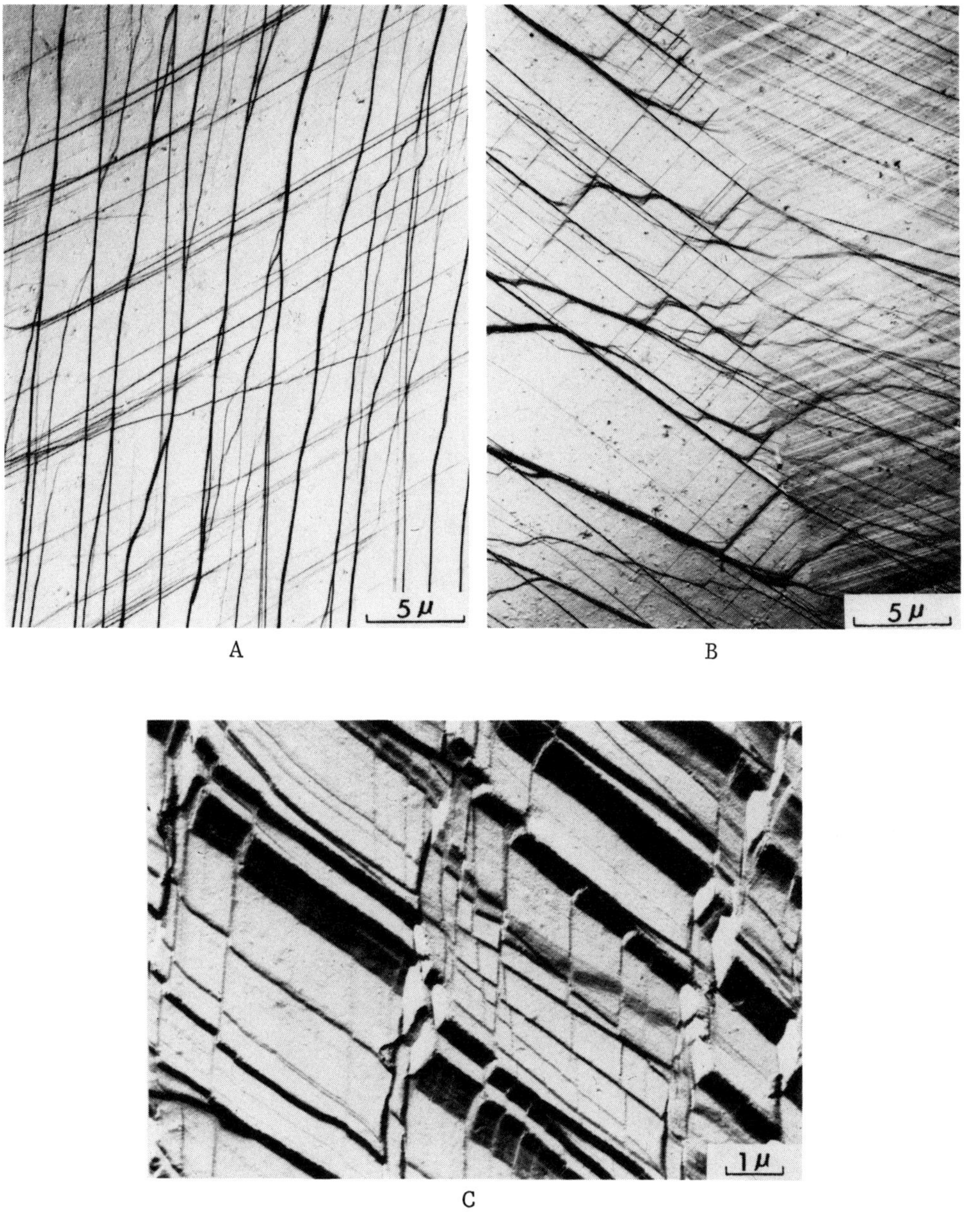

Fig. 2. Surface slip lines for the single-phase alloy, (24 hr at 900°C, W.Q.). (a) ε = .02; (b) ε = .05; (c) adjacent to fracture.

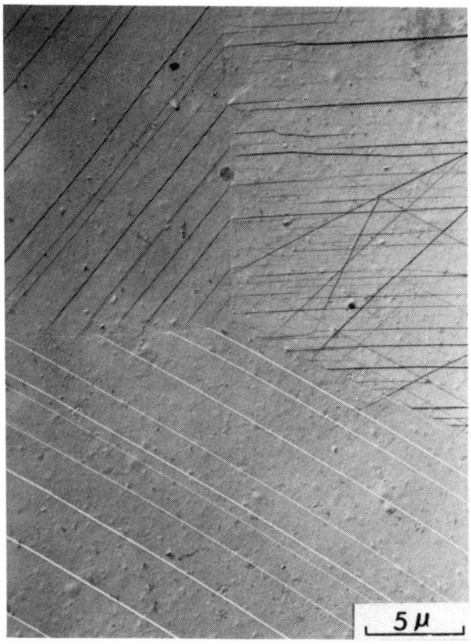

A

B

C

Fig. 3. Surface slip lines for the two-phase alloy, (24 hr at 900°C, W.Q., + 100 hr at 500°C). (a) $\varepsilon = .02$; (b) $\varepsilon = .05$; (c) adjacent to fracture.

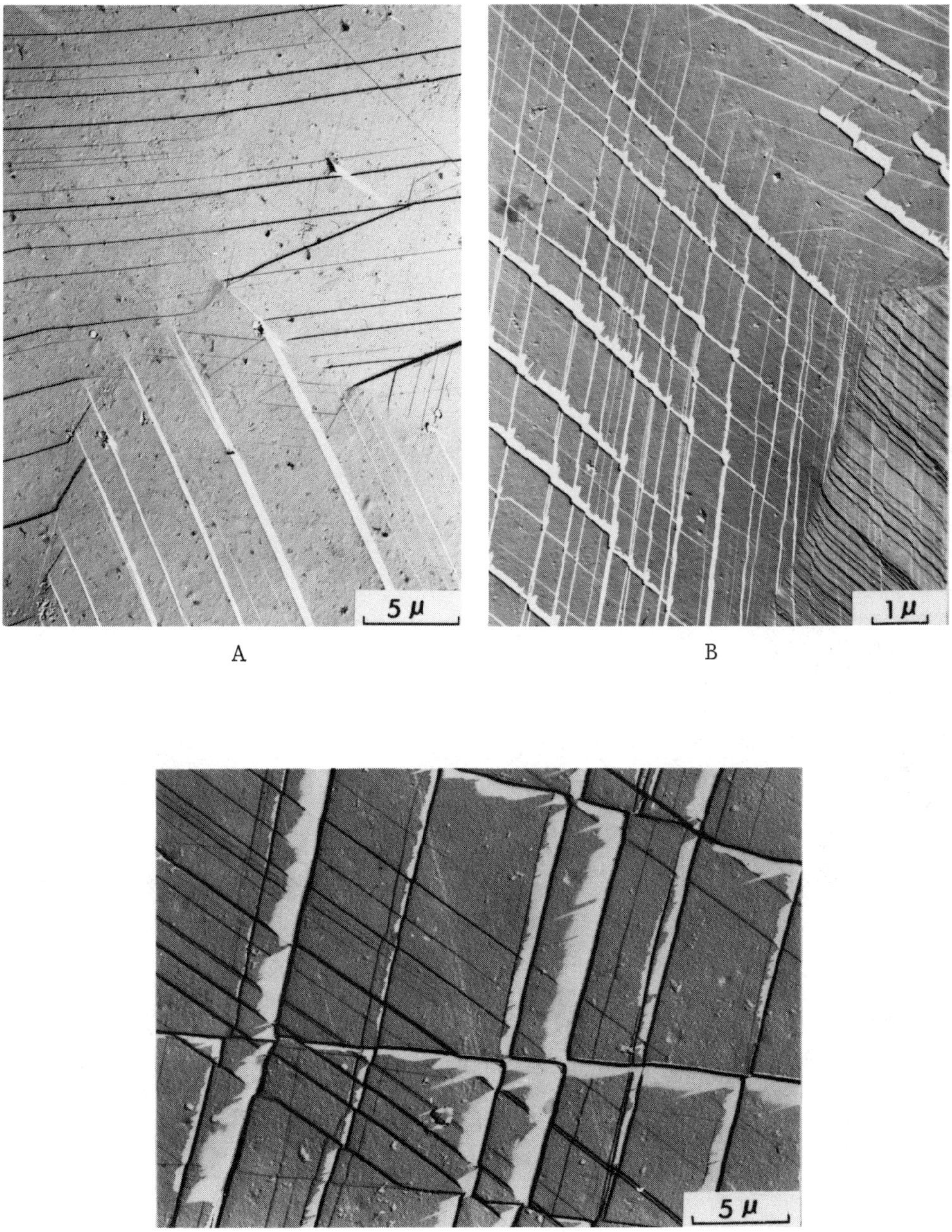

A B

C

Fig. 4. Surface slip lines for the two-phase alloy deformed in salt water, (24 hr at 900°C, W.Q., + 100 hr at 500°C). (a) ε = .02; (b) ε = .05; (c) adjacent to a stress-corrosion crack.

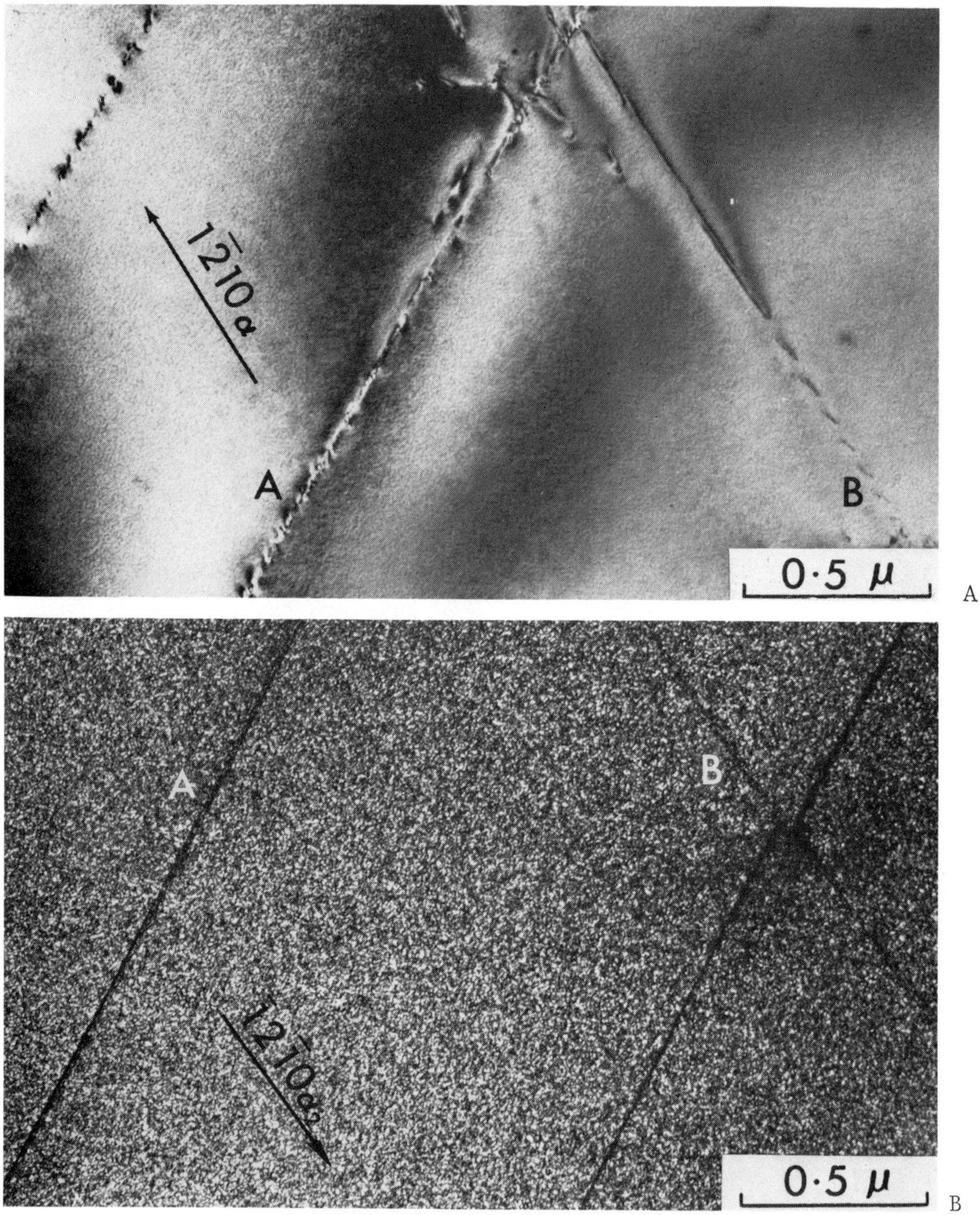

Fig. 5. $\{10\bar{1}0\}$ slip bands in Ti-8Al aged 100 hours at 500°C, and strained ~1 percent. (Arrows indicate diffraction vectors, zone axis near [0001]). (a) bright field; (b) same area, dark field, showing reduced density of Ti_3Al precipitates in slip bands.

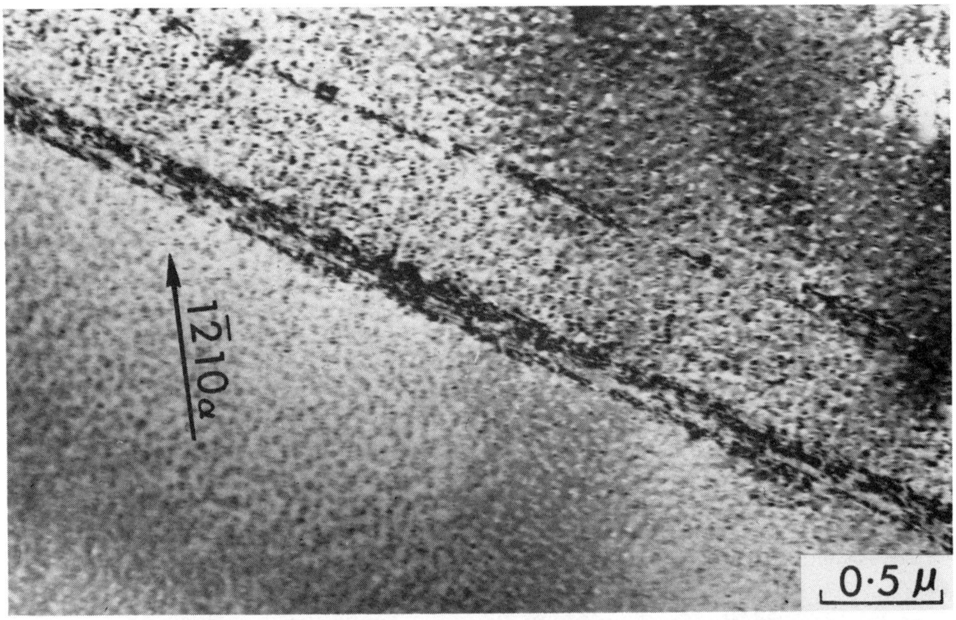

A

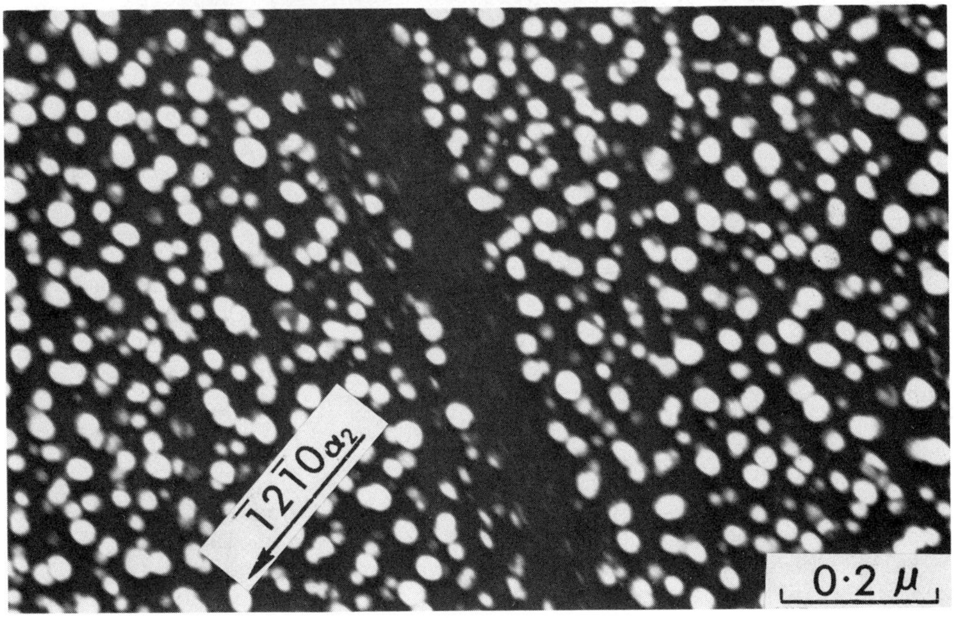

B

Fig. 6. $\{10\bar{1}0\}$ slip bands in Ti-8Al aged 500 hours at 650°C, and strained ~1 percent. (Arrows indicate diffraction vectors, zone axis near [0001]). (a) bright field; (b) same area, dark field, showing reduced density of Ti_3Al precipitates in a slip band, and sheared particles.

Discussion

The values of yield stress of titanium-aluminum alloys reported in the literature range from 75 ksi (2) to 110 ksi (3,9) for compositions and heat treatments similar to that employed in the present work. The values of yield stress reported here are all at the bottom of this range. It is felt that this discrepancy is due to contamination, (probably by oxygen), which occurs during the high-temperature processing and heat treatments. In the work reported in References 3 and 9, the specimens were sealed in evacuated tubes for the high-temperature heat treatments, whereas in the work reported in Reference 2, the specimens were sealed in tubes and heat treated in a dynamic vacuum. In the present work the specimens were just sealed in evacuated tubes for all heat treatments, but at least 0.020 in. was removed (by surface grinding) from all surfaces of each specimen after heat treatment and prior to testing. To check this effect some specimens were tested after only 0.010 in. had been removed from the surfaces, and the yield stresses were significantly higher than for the standard specimens, (~90 ksi). Therefore, it is considered that the tensile data and the descriptions of surface slip lines given in this paper are accurate representations of the slip behavior, with no effects due to a contaminated or deformed surface layer.

The transmission electron microscopy observations indicate that the Ti_3Al particles are sheared by the glide dislocations. Furthermore, the weak dark-field images of the sheared particles and the sharply reduced density of visible particles in slip bands suggests that the long-range-ordered structure of the particles is destroyed by the slip process. This could occur by the dissolution of fragmented Ti_3Al particles in the slip band, or by an increase in the density of anti-phase boundary, (APB), in the particles. Alternatively, the apparent low density of particles in the slip bands could be entirely a diffraction-contrast effect. The contrast from the particles is extremely sensitive to the orientation of the specimen relative to the electron beam. Local lattice rotations due to the residual dislocations in the slip band could make contrast conditions poor for particles in a slip band when they are optimum for matrix precipitates. However, this does not appear to be the cause of the observed effects, because it was not possible to improve the contrast from the particles in a slip band by tilting the specimen. Regardless of the precise description of the precipitates in the slip bands, two important conclusions can be drawn from the observations:

(1) Slip should tend to concentrate in preexisting slip bands because the areal fraction of precipitates per slip plane decreases with the passage of each glide dislocation. Therefore, slip should be nonhomogeneous and planar because it will be difficult for a dislocation to cross slip into an unsheared region.

(2) The chemical potential of the volume of material in a slip band should be higher than the adjacent matrix if the sheared particles redissolve, if the particles in the slip bands have a higher density of APB's, or if they are smaller than the matrix particles, (Gibbs-Thompson effect).

Thus, the nonhomogeneous, planar slip character observed for the aged material is consistent with the particle-shearing mechanism. The slip character in the single-phase material is also nonhomogeneous, probably due to short-range order, but in this material cross slip can occur readily on a fine scale, (<25 Å). By contrast the slip bands in the aged material are rigorously planar, and there is little evidence of fine, wavy slip. In both alloy conditions the rates of work hardening are low, but only the aged alloy shows a yield drop and Lüders extension. This is also consistent with the particle-shearing mechanism.

The particle-shearing mechanism also accounts for the occurrence of trenches in the aged alloy. The irregular contours of the trenches and their appearance in the early stages of formation as discrete holes along a slip band suggests that they form by localized chemical attack, (7) which is promoted by the enhanced chemical potential of the slip bands and the larger surface slip offsets in the aged material. Conversely, in the single-phase material a passivating reaction should dominate when a fresh slip step is produced because the slip steps are smaller and more closely spaced, and there is no gradient in chemical potential between a slip band and the adjacent matrix, other than the small effect due to the elastic strain energy of the residual dislocations in the slip band. For all of the mechanisms discussed above the chemical potential of a slip band should increase with increasing shear strain. The observations indicate that conditions are not suitable for trench formation until the average strain is approximately 5 percent. It is likely that this minimum strain for trench formation also depends on the size distribution of the Ti_3Al particles. For instance if the enhanced chemical potential of the slip bands is due to particle dissolution or the Gibbs-Thompson effect, then it is expected that the minimum strain required before trenches are observed will increase rapidly with increasing mean-particle diameter.

These observations on slip give some indication of the specific effects of Ti_3Al particles on the fracture toughness and stress-corrosion susceptibility of titanium-aluminum alloys. The nonhomogeneous, planar slip character which obtains in alloys containing deformable Ti_3Al particles causes large shear offsets and microvoid nucleation at internal boundaries, and coplanar pileups of dislocations at obstacles. Both of these microstructural conditions are conducive to macroscopic crack propagation by microvoid coalescence or by cleavage. The fracture toughness should be improved when the size distribution of Ti_3Al is such that the particles are bypassed by glide dislocations, and this has been shown to be the case by Lütjering and Weissmann.(2) It is not clear how, or indeed, if the trenches contribute to subcritical crack propagation in an aqueous environment, but their appearance correlates with a susceptible alloy condition, and indicates that the stress-corrosion susceptibility is related to the high chemical potential of the slip steps in alloys containing deformable Ti_3Al particles. Since the rate of increase of chemical potential with strain is expected to vary inversely with mean particle size, highly overaged alloys should show a decreased stress-corrosion susceptibility, as found by Cavallaro and Wilcox.(8)

Acknowledgements

The authors would like to thank Dr. R.I. Jaffee for helpful discussions, and acknowledge the assistance of Mr. Dick Barnes and Mr. Chris Martin with the experimental work. This research was supported financially by the National Aeronautics and Space Administration, (Contract No. NASW-2242), and by Battelle, Columbus Laboratories.

References

1. M.J. Blackburn, "The Ordering Transformation in Titanium: Aluminum Alloys Containing up to 25 at. pct. Aluminum", Trans. AIME, Vol. 239, 1200 (1967).

2. G. Lütjering and S. Weissmann, "Mechanical Properties of Age-Hardened Titanium-Aluminum Alloys", Acta Met., Vol. 18, 785 (1970).

3. M.J. Blackburn and J.C. Williams, "Strength, Deformation Modes and Fracture in Titanium-Aluminum Alloys", ASM Trans. Quart., Vol. 62, 398 (1969).

4. K.R. Evans, "The Embrittlement and Fracture of Ti-8 pct. Al. Alloys", Trans. AIME, Vol. 245, 1297 (1969).

5. M.J. Blackburn and J.C. Williams, "Metallurgical Aspects of the Stress-Corrosion Cracking of Titanium Alloys", in Fundamental Aspects of Stress Corrosion Cracking, eds. R.W. Staehle, et al., NACE (Houston), 1969, 620.

6. I.R. Lane and J.L. Cavallaro, "Metallurgical and Mechanical Aspects of the Sea-Water Stress Corrosion of Titanium", in Applications Related Phenomena in Titanium Alloys, ASTM STP 432, 1968, 147.

7. J.D. Boyd and R.G. Hoagland, "The Relation Between Surface Slip Topography and Stress-Corrosion Cracking in Ti-8wt.%Al", presented at the International Symposium on Stress-Corrosion Mechanisms, Atlanta, Georgia, 1971, to be published by NACE.

8. J.L. Cavallaro and R.C. Wilcox, "Embrittlement of Ti-7Al Binary Alloy in Sea Water", Corrosion, Vol. 27, 157 (1971).

9. K.R. Evans, "Deformation Mechanisms in Titanium and Titanium-Aluminum Alloys", Trans. AIME, Vol. 242, 648 (1968).

THE LOW TEMPERATURE DEFORMATION BEHAVIOUR OF THE INTERMETALLIC COMPOUND Ti_3Sn

P.J. Jones and J.W. Edington
Department of Metallurgy and Materials Science
University of Cambridge
England CB2 3QZ

Abstract

The mechanical properties of the intermetallic compound Ti_3Sn (DO_{19} structure) have been measured for polycrystalline material in the temperature range -196 to 600°C. The temperature dependence of the 0.2% flow stress is similar to that of pure titanium and the material exhibits ~2% strain in compression at -196°C increasing to ~12% at 600°C. Two surface analysis of individual grains in deformed material has led to the identification of $\{1\bar{1}00\}$, $\{1\bar{1}01\}$, (0001) and $\{11\bar{2}0\}$ as slip planes. Transmission electron microscopy of deformed material has shown the existence of dislocations with Burgers vectors 1/3 $\langle 11\bar{2}0\rangle$ and $\langle 0001\rangle$ using the technique of matching experimental and computed micrographs. The lack of ductility of Ti_3Sn is attributed to the planar nature of the slip and the operation of insufficient independent slip systems.

Introduction

The addition of greater than 6 at.% (13wt%) Sn to titanium leads to embrittlement of the α-phase. This embrittlement has been attributed to long range ordering in the lattice resulting in the formation of the intermetallic compound Ti_3Sn (1). (DO_{19} structure - isostructural with Ti_3Al and Mg_3Cd.) However, no details of the mechanical properties of Ti_3Sn have been reported and it is not clear whether the reported lack of ductility of these alloys is caused by an inherent lack of ductility in the γ (Ti_3Sn) phase.

The purpose of this study was to establish the mechanical properties of Ti_3Sn at ambient temperatures and to study in particular the fracture properties of this phase.

Experimental

The Ti_3Sn was prepared from Osaka 160 titanium (700 ppm by wt. oxygen, nitrogen and carbon, 20 ppm hydrogen) and (99.98%) Pass No.1 tin. The constituent elements were melted together four times in a water cooled copper boat using an R.F. induction heating coil under a positive pressure of high purity argon. The final composition of the ingot was 24.8 at.% (45.0 wt%) Sn and analysis showed almost no difference in impurity level compared with the constituent elements.

Specimens with dimensions 7.5 x 3 x 3 mm were cut by diamond saw from the ingots (coarse columnar grains approx. $\frac{1}{2} \times \frac{1}{2} \times 1$ mm), wrapped in tantalum sheet, sealed in quartz tube and homogenised in argon for 24 hrs at 1200°C. Compression tests were performed at a strain rate of 10^{-4} sec^{-1} in the temperature range -196°C to 600°C (above room temperature under vacuum). Two series of differently oriented specimens were tested, series A with the (0001) planes parallel to the compression axis and series B with the (0001) planes at ~ 45° to the compression axis. Those specimens required for two surface trace analysis were electropolished prior to deformation and then lightly deformed 1 - 2%. The remaining specimens were tested to failure.

Thin foils for transmission electron microscopy studies were prepared from deformed material by standard spark erosion and electropolishing techniques (2).

Mechanical properties

The transition from elastic to plastic behaviour was smooth with no well defined yield point, over the whole of the temperature range investigated. The temperature dependence of the 0.2% proof stress for both series of specimens, A and B, was similar to that of pure (iodide) titanium (3) (fig. 1). However, the proof stress of specimens B was twice that of specimens A.

For both orientations the plastic deformation was inhomogeneous and a high density of small cracks were present in most grains at strains of 1-2%. The temperature variation of the average compressive engineering strain to fracture (fig. 2) was independent of the specimen orientation. The inhomogeneous distribution of strain (slip and cracks) from grain to grain was reflected in the work hardening rate which varied considerably even for specimens of the same orientation tested at the same temperature. Indeed, in some cases the work hardening rate was linear while in other cases the rate was parabolic.

Surface observations of the deformation and fracture modes

Deformation modes

Fig. 3 shows a typical two surface analysis of a deformed grain from a specimen tested at 23°C. The planar nature of the slip is apparent with occasional well defined cross slip events. There was no evidence for mechanical twinning.

The operative slip planes were identified for deformation at -196°, 23° and 400°C, see table I. At all temperatures the most common slip plane was $\{10\bar{1}0\}$, independent of specimen orientation. Slip on the remaining planes generally occurred adjacent to grain boundaries or in short cross slip segments. Slip on the (0001) plane was only observed in specimens B because the (0001) planes in specimens A were parallel to the compression axis. Although the (0001) planes were favourably oriented for slip in specimens B, this slip mode was limited to grain boundary regions. In specimens tested at -196°C the slip lines were more finely spaced with less cross slip than in specimens tested at 23°C and above.

It was possible to identify the $\langle 1\bar{2}10\rangle$ direction as slip direction for the $\{10\bar{1}0\}$, $\{10\bar{1}1\}$ and (0001) planes from cross slip events for deformation at 23° and 400°C. Cross slip between $\{11\bar{2}0\}$ and $\{10\bar{1}0\}$ planes indicated that an additional slip direction $\langle 0001\rangle$ was probably operative.

Fracture modes

There was no evidence of crack nucleation at the grain boundaries and the fracture mode was predominantly transgranular at all temperatures. Similar behaviour has been observed in other titan-

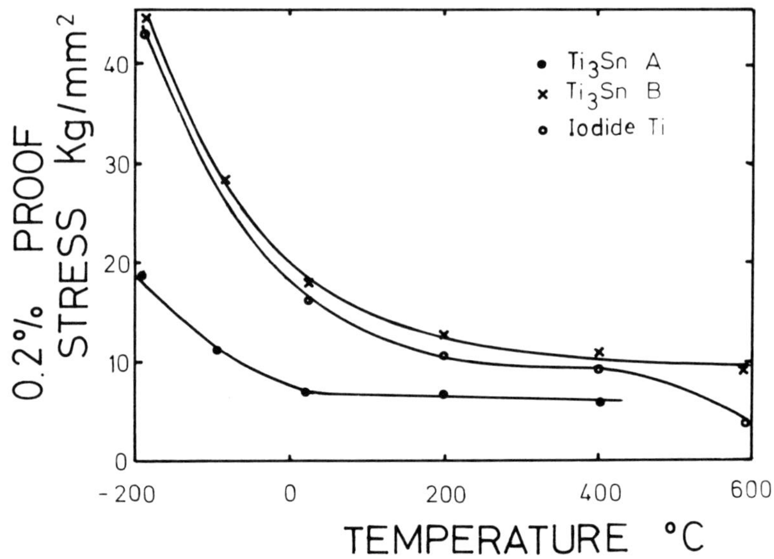

Fig. 1. Variations of 0.2% proof stress of polycrystalline Ti_3Sn and iodide α-Ti (3) with temperature.

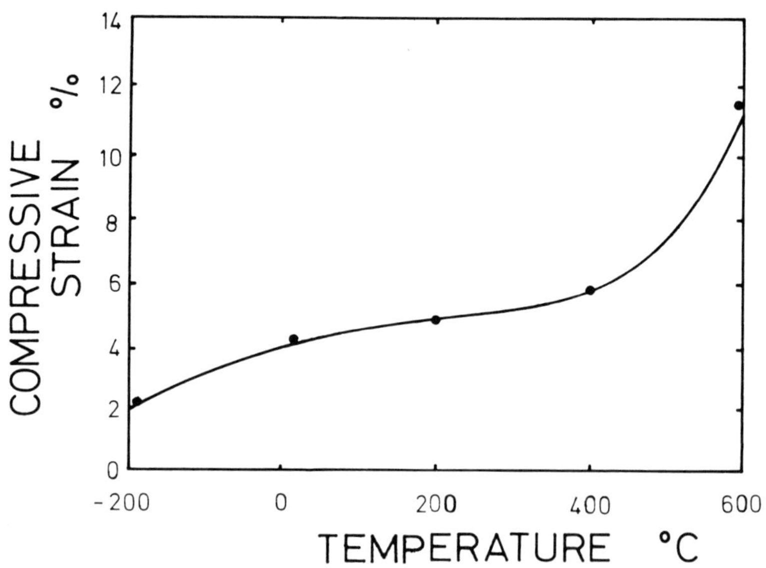

Fig. 2. Variation of compressive strain with temperature for polycrystalline Ti_3Sn.

Table I

Operative slip planes at different deformation temperatures

Temperature	Slip plane: Specimens A	Slip plane: Specimens B	Total number of grains analysed
$400^{\circ}C$	$\{10\bar{1}0\}$	$\{10\bar{1}0\}$	6
	$\{10\bar{1}1\}$	$\{10\bar{1}1\}$	
	$\{11\bar{2}0\}$ S	(0001)	
$23^{\circ}C$	$\{10\bar{1}0\}$	$\{10\bar{1}0\}$	15
	$\{10\bar{1}1\}$	$\{10\bar{1}1\}$	
	$\{11\bar{2}0\}$	(0001)	
		$\{11\bar{2}0\}$ S	
$-196^{\circ}C$	$\{10\bar{1}0\}$	None Analysed	6
	$\{10\bar{1}1\}$		

S single surface analysis only
others by two surface analysis

ium-base DO_{19} compounds (4). Although small cracks were nucleated in many grains at strains of 1-2% in all the specimens tested (fig. 3), macroscopic cleavage failure only occurred at $-196^{\circ}C$. Thus it was concluded that most cracks grow slowly above $\sim -196^{\circ}C$.

Fig. 4 is a scanning electron micrograph which shows the method of crack linkage. Considerable plastic bending and tearing near the cracks is evident. It is clear that heavy localised plastic deformation takes place in the region near the crack and thus progress of the crack across the grain is impeded. This is similar to the fracture behaviour of ordered Mg_3Cd where cleavage failure was preceded by considerable plastic work associated with crack propagation (5).

In order to study fracture surfaces a few unnotched specimens were subjected to impact loading at 23° and $-196^{\circ}C$. Those specimens tested at $-196^{\circ}C$ failed completely and exhibited large primary cleavage facets which were identified as (0001) planes by X-ray Laue back reflection. However, even in impact at $23^{\circ}C$, simple cleavage only occurred across the first grain with branching and secondary cleavage becoming common as the crack propagated further into the specimen.

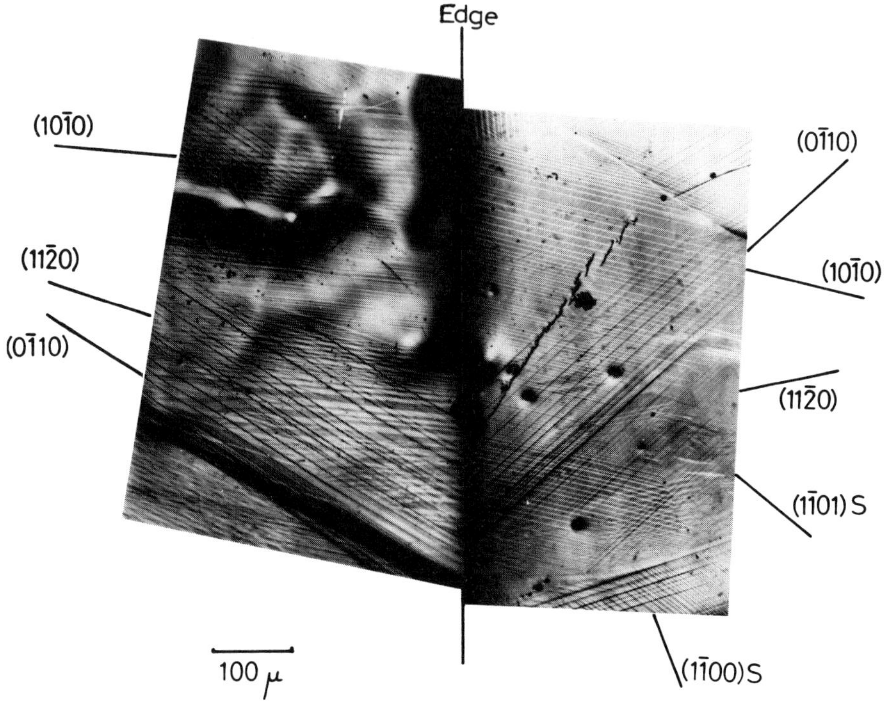

Fig. 3. Two surface analysis of a grain deformed at 23°C.

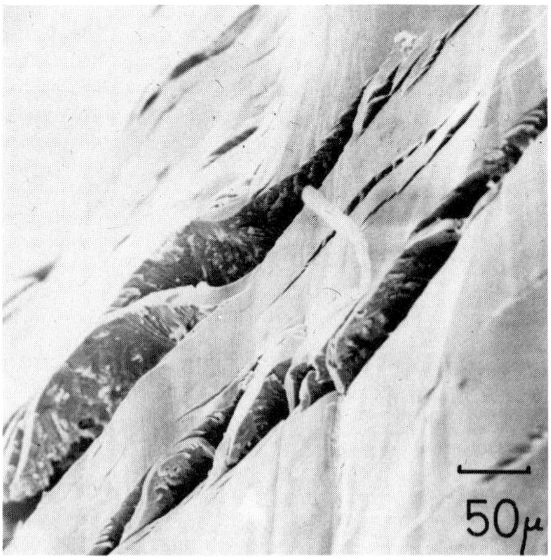

Fig. 4. Method of crack linkage in a specimen tested at 23°C.

Transmission electron microscopy observations

Identification of Burgers vectors in h.c.p. metals by contrast experiments in the electron microscope involves the use of computer simulation techniques (6). The details of the identification of the Burgers vectors of slip dislocations in Ti_3Sn using both 'weak beam' (7) transmission electron microscopy and computer simulation techniques will be reported elsewhere (8). However, the importance of computational techniques for Burgers vectors determination is shown in fig. 5.

The dislocations in fig. 5 lie on a high A.P.B. energy type 2 $(10\bar{1}0)$ plane (9) and are predominantly edge in character. Computed micrographs for undissociated dislocations* with the possible Burgers vectors, $\underline{b} = {}^1/_3[\bar{1}2\bar{1}0]$, ${}^1/_3[\bar{1}2\bar{1}3]$, ${}^1/3[\bar{1}2\bar{1}\bar{3}]$ and $[0001]$ for the experimental diffracting conditions are shown in fig. 6. Comparison of the computed and experimental micrographs confirms that the dislocations have $\underline{b} = {}^1/_3[\bar{1}2\bar{1}0]$ and this has been supported by further computer matching for a number of different reflections. In addition, fig. 5 demonstrates some of the difficulties associated with Burgers vectors determination in DO_{19} structures. All images are strong but the $\underline{g}.\underline{b} = 0$ case is not the weakest image because $\underline{g}.\underline{b} \times \underline{u} \simeq 2.2$. Furthermore, although the image is double (i.e. two parallel dotted lines), it cannot be concluded that the dislocations are dissociated because there is good agreement in the separation of the two parts of the computed (~90Å) and experimental (~100Å) images.

Dislocations lying on the $\{10\bar{1}0\}$, $\{10\bar{1}1\}$ and (0001) planes have been shown to have Burgers vectors, $\underline{b} = {}^1/_3\langle 11\bar{2}0\rangle$ using image matching techniques. The majority of the dislocations, $\underline{b} = \langle 0001\rangle$ detected, were edge in character and lay on $\{10\bar{1}0\}$ or $\{11\bar{2}0\}$ planes. These dislocations were uncommon and many foils contained none apart from those in low angle tilt boundaries.

---

* The A.P.B. energy and equilibrium spacings of superlattice pairs for Ti_3Sn estimated by the method of Bushnev and Kitaeva (10) are shown in table II. The small predicted separations justify the use of computing the contrast for undissociated dislocations.

Table II

A.P.B. plane	E ergs cm^{-2}	r Å
(0001)	216	28–42
$\{10\bar{1}0\}$	0 or 232	26–39
$\{10\bar{1}1\}$	65 or 159 or 191	32–136

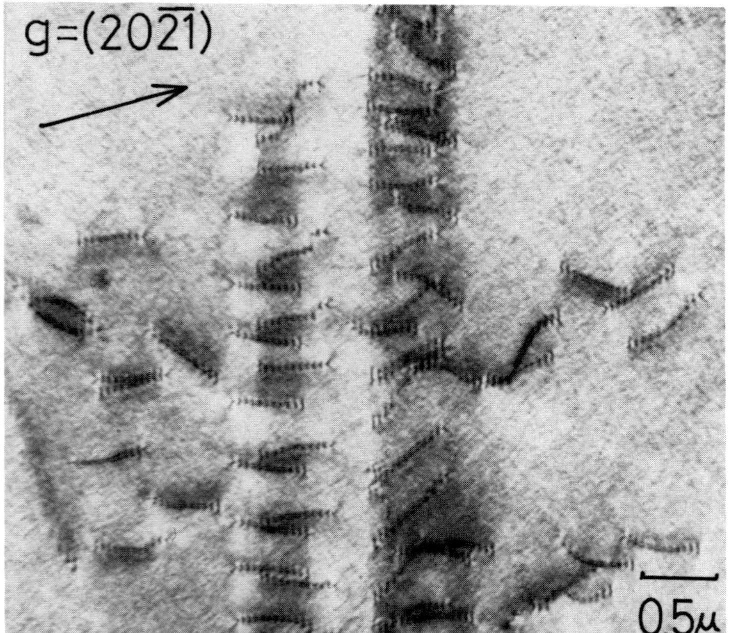

Fig. 5. Dislocations on a type 2 (10$\bar{1}$0) plane.

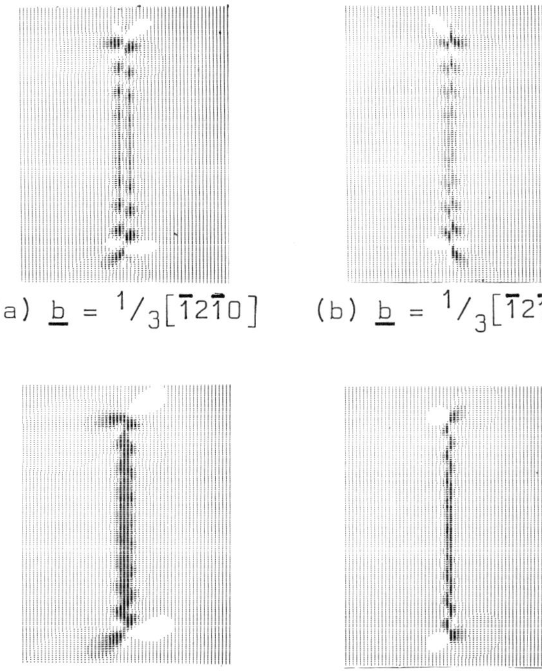

(a) $\underline{b} = {}^1/_3[\bar{1}2\bar{1}0]$ (b) $\underline{b} = {}^1/_3[\bar{1}2\bar{1}3]$

(c) $\underline{b} = {}^1/_3[\bar{1}2\bar{1}\bar{3}]$ (d) $\underline{b} = [0001]$

Fig. 6. Computed images for Fig. 5. Line direction, u = [12$\bar{1}$ 25] , g = (20$\bar{2}\bar{1}$); beam direction = [10$\bar{1}$4] ; foil normal = [11$\bar{2}$8]; deviation parameter, w = 0.1.

Discussion

The higher strength of specimens B compared with specimens A implies that the critical resolved shear stress (τ_{crit}) for slip is different on different planes. Furthermore, the scarcity of slip on the basal plane even in favourably oriented grains indicates that in Ti_3Sn there is considerable difference between τ_{crit} for slip on basal and $\{10\bar{1}0\}$ planes. In pure titanium (1000 ppm by wt. oxygen and nitrogen) a similar effect is seen as τ_{crit} for basal slip is three times that for slip on $\{10\bar{1}0\}$ planes (11). In contrast τ_{crit} for slip on these planes is similar in ordered Mg_3Cd (12).

Slip on the following systems has been identified in Ti_3Sn:

$\langle 11\bar{2}0 \rangle$	$\{1\bar{1}00\}$	primary
$\langle 11\bar{2}0 \rangle$	$\{1\bar{1}01\}$	secondary
$\langle 11\bar{2}0 \rangle$	(0001)	secondary
$\langle 0001 \rangle$	$\{1\bar{1}00\}$	very occasionally
$\langle 0001 \rangle$	$\{11\bar{2}0\}$	very occasionally

The slip is planar with occasional well defined cross slip events.

The combination of the primary and secondary slip modes give only four independent slip systems. Thus the von Mises criterion (13) is not satisfied. Indeed, even if extensive slip took place on the $\langle 0001 \rangle$ $\{1\bar{1}00\}$ and $\langle 0001 \rangle$ $\{11\bar{2}0\}$ systems there would still only be four independent slip systems. Hence the cracking observed in the grains is to be expected as no relaxation of the strain in the $\langle 0001 \rangle$ direction is possible.

A number of nucleation mechanisms of cleavage cracks have been suggested (for a review see Petch (14)). In the present case nucleation caused by twin intersection and dislocation pile-up mechanisms can be ruled out because no twinning or fracture at grain boundaries is observed. The most likely explanation of crack nucleation involves the interaction between slip and low angle boundaries and intersecting slip planes themselves. This is consistent with the observed uniform distribution of cracks and the complex dislocation tangles frequently seen at the intersection of slip planes and slip planes with low angle boundaries.

The principal cause of the lack of ductility of Ti_3Sn is the lack of available independent slip systems. The planarity of the slip increases the tendency for cleavage failure, together with differences in τ_{crit} for different slip systems and the coarse grain size of the material. The ease of dislocation production and movement indicated by the low flow stress of the material suggests that interstitial impurities have little effect on the ductility of Ti_3Sn. The combination of four independent slip systems with a very low flow stress appears to be effective in preventing the rapid growth of cracks formed as a result of failure to satisfy the von Mises criterion. Thus the material, though brittle, exhibits a degree of ductility.

Acknowledgements

The authors wish to gratefully acknowledge the support of this work by the Tin Research Institute, Greenford, Middlesex, England.

References

1. Smart, R. F., "Alloys of titanium and zirconium containing tin," Metallurgia, Vol. 57, April 1958, pp. 181-188.

2. Blackburn, M. J. and Williams, J. C., "The preparation of thin foils of titanium alloys," Transactions of the Metallurgical Society of AIME, Vol. 239, Feb. 1967, pp. 287-288.

3. Savitsky, E. M., The influence of temperature on the mechanical properties of metals and alloys, Oxford University Press, London, 1962, pp. 143-149.

4. Williams, J. C. and Blackburn, M. J., "The structure, mechanical properties and deformation behaviour of Ti-Al and Ti-Al-X alloys," Ordered Alloys, Claitors Pub. Div., Baton Rouge, 1970, pp. 425-445.

5. Stoloff, N. S. and Davies, R. G., "The effect of ordering on the plastic deformation of Mg_3Cd," Transactions of the ASM, Vol. 57, March 1964, pp. 247-260.

6. Jones, P. J. and Edington, J. W., "Computer simulation of images of undissociated dislocations in hexagonal metals," Philosophical Magazine, in press.

7. Cockayne, D. J. H., Ray, I. L. F. and Whelan, M. J., "Investigation of dislocation strain fields using weak beams," Philosophical Magazine, Vol. 20, Dec. 1969, pp. 1265-1270.

8. Jones, P. J. and Edington, J. W., Philosophical Magazine, to be published.

9. Blackburn, M. J., "Ordering and its effect on dislocation arrangements in Mg_3Cd," Transactions of the Metallurgical Society of AIME, Vol. 239, May 1967, pp. 660-668.

10. Bushnev, L. S. and Kitaeva, L. P., "Electron microscope study of the domain and dislocation structure in the ordered alloy Mg_3Cd," Soviet Physics - Crystallography, Vol. 9, May 1965, pp. 740-745.

11. Churchman, A. T., "The slip modes of titanium and the effect of purity on their occurrence during tensile deformation of single crystals," Proceedings of the Royal Society, Vol. A226, 1954, pp. 216-226.

12. Kirby, J. H. and Noble, F. W., "The effect of degree of order on the critical resolved shear stress for slip in Mg_3Cd," Philosophical Magazine, Vol. 16, Nov. 1967, pp. 1009-1019.

13. von Mises, R., Zeitschrift für angewandte Mathematik und Mechanik, Vol. 8, 1928, p. 161.

14. Petch, N. J., "Metallographic aspects of fracture," Fracture, Vol. 1, Academic Press, 1968, pp. 351-393.

ASYMMETRY OF SLIP IN BETA Ti-V SINGLE CRYSTALS**

D. A. Koss* and J. C. Chesnutt†

*Department of Metallurgical Engineering
Michigan Technological University
Houghton, Michigan

†M.E.R.L., Pratt and Whitney Aircraft
Middletown, Connecticut

The macroscopic slip plane and the yield stress were determined for Ti-40 a/o V (Ti-42 w/o V) single crystals deformed in compression at 77° and 297°K. The orientation dependence of the active slip plane indicates a deviation from the maximum resolved shear stress plane to the {112} plane, with {112} slip being predominant at 77°K. An asymmetry of the yield stress exists such that if [111] is the slip direction, then slip on the ($\bar{2}11$) plane is easier than on the ($\bar{1}\bar{1}2$) plane. The degree of asymmetry is larger at 77° than at 297°K. Many jogged screw dislocations are observed in lightly deformed samples. The results indicate that the flow stress is influenced by the core structure of screw dislocations.

**Supported in part by the National Science Foundation Science Development Grant, Department of Metallurgical Engineering, Michigan Technological University.

Introduction

Plastic deformation and solute strengthening of dilute substitutional bcc alloys has been the subject of several studies in recent years. A characteristic feature of plastic deformation of both bcc metals and dilute alloys is the complex slip geometry (for a review, see Christian (1) or Stein *et al.* (2)). A unique slip direction, the {111} direction, is observed, but there is considerable evidence that there is no unique slip plane. In many cases, the observed macroscopic slip plane is an irrational plane which depends on crystal orientation and is believed to reflect a varying degree of microscopic slip on several {110} and {112} systems (3). In addition there is an orientation dependence of the critical resolved shear stress such that resistance to {112} slip depends on whether the crystal is being stressed in the twinning or anti-twinning sense (4). The resultant asymmetry of slip is generally interpreted as being a consequence of the core structure of screw dislocations. For a solid solution alloy, the existence of slip and yield stress asymmetry indicates whether or not the dislocation core structure is important in controlling the flow stress. In order to determine the mechanism of strengthening in concentrated bcc alloys, this investigation presents data on the orientation dependence of the yield stress and the active slip plane of beta Ti-40 a/o V (42 w/o V) single crystals.

Experimental Procedure

The Ti-40 a/o V (Ti-42 w/o V) alloy was prepared by arc melting iodide titanium and zone refined vanadium. The arc cast ingots were homogenized at 1350°C for four hours at a vacuum of about 10^{-7} torr. Oriented single crystals were grown in an electron beam zone melting apparatus at a vacuum of 1×10^{-6} torr. The crystals were helium gas quenched from approximately 800°-1000°C immediately after growing. To avoid any possible hydrogen contamination, electro- or chemical polishing was not used on samples mechanically tested. Crystals of χ = +30° and -30° were submitted for analysis. Vacuum fusion analysis indicates that the tested crystals contained approximately: 190-250 pp O, 50 pp N, and 20-30 pp H by weight. The vanadium content was analyzed to be 40.5 ± 1 a/o (42 ± 1 w/o). Since only small strains (axial strain of 0.02 to 0.03) were required, the data was obtained by compression testing samples 4.3 mm in diameter by 10.2 mm long. All tests were performed in an Instron at an axial strain rate of 2×10^{-4} sec^{-1}.

Results

The effect of orientation can be conveniently summarized by the $\psi(\chi)$ curve (5,6). The angles ψ and χ are shown in Figure 1 and represent the angular rotations of the observed slip plane and the plane of maximum resolved shear stress, respectively, from the $(\bar{1}01)$ plane (assuming a [111] slip direction). If slip were determined exclusively by crystal symmetry, the slip geometry would be symmetric about $\chi = 0°$ and independent of whether the specimen is stressed in tension or compression.

The room temperature dependence of the active slip plane and of the yield stress is shown in Figure 2. The dotted line ($\psi = \chi$) indicates the behavior expected if the slip plane was the plane of maximum resolved shear stress. However asymmetry of slip occurs with {112} slip being dominant. Asymmetry of the yield stress is also observed at room temperature. Although slip is easy when the stress operates in the twinning sense (on the [111] $(\bar{2}11)$ system in compression), it is difficult when the stress operates in the anti-twinning sense (on the [111] $(\bar{1}\bar{1}2)$ system in compression). If crystals oriented at $\chi = +30°$ and $\chi = -30°$ represent nearly crystallographic $(\bar{2}11)$ and $(\bar{1}\bar{1}2)$ slip respectively, then the yield stress data in Figure 2 indicates that at 297°K: $\tau_{\bar{1}\bar{1}2}/\tau_{\bar{2}11} \approx 1.15$.

The asymmetry of slip increases at low temperatures. Figure 3 shows that at 77°K both the yield stress and slip geometry show considerable asymmetry. The $\psi(\chi)$ curve indicates a difficulty in activating $(\bar{1}01)$ slip, since at $\chi = 1°$ (in which the crystals are almost ideally oriented for $(\bar{1}01)$ slip), $(\bar{2}11)$ slip dominates, indicating that in compression: $\tau_{(\bar{1}01)} > \tau_{(\bar{2}11)}$.

The abrupt change from $(\bar{2}11)$ slip to primarily $(\bar{1}\bar{1}2)$ slip is not reflected in the orientation dependence of the yield stress at 77°K (see Figure 3). The yield stress decreases in a nearly linear manner from $\chi = -30°$ to $\chi = +30°$. A comparison of the $\psi(\chi)$ and the $\tau(\chi)$ data in Figure 3 shows that a large change in observed slip plane (for example, when χ varies from -15° to 0°) can have only a small effect on the yield stress. If the orientations at $\chi = -30°$ and $+30°$ are taken to indicate (112) and (211) slip respectively, then at 77°K: $\tau_{\bar{1}\bar{1}2}/\tau_{\bar{2}11} \approx 1.30$.

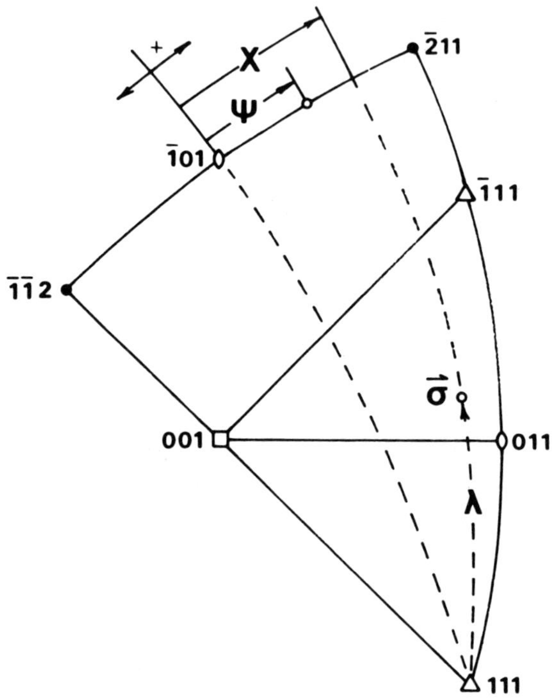

Figure 1. Standard stereographic projection illustrating ψ and χ ($\vec{\sigma}$ represents the compressive axis, ψ is the angle between the observed slip plane and the ($\bar{1}01$) plane, and χ is the plane of maximum resolved shear stress).

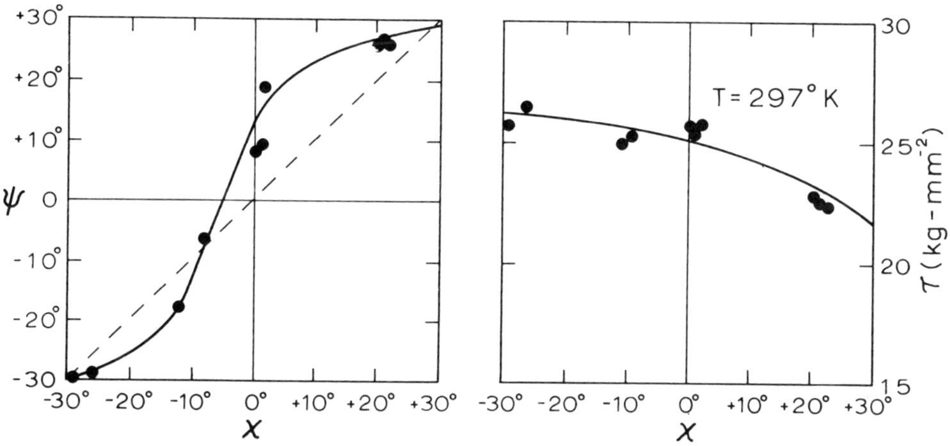

Figure 2. The dependence of $\psi = \psi(X)$ and the yield stress on X for Ti-40 a/o V tested in compression at 297°K.

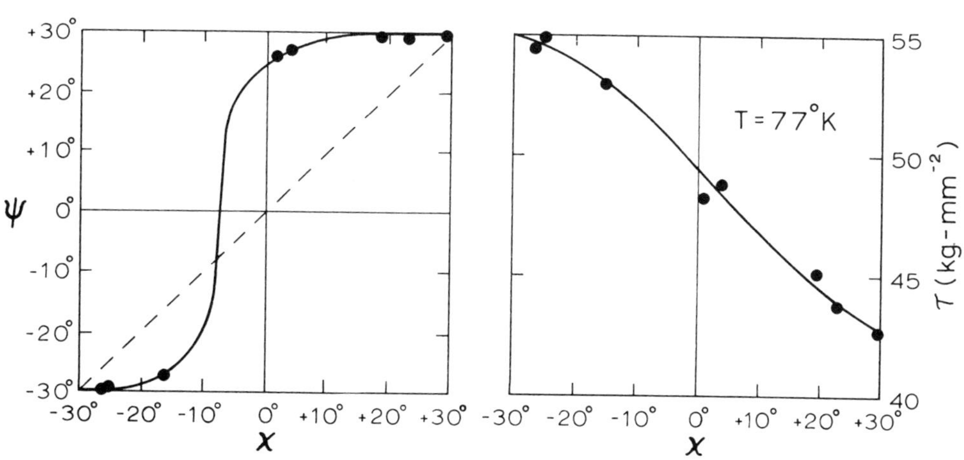

Figure 3. The dependence of $\psi = \psi(X)$ and the yield stress on X for Ti-40 a/o V tested in compression at 77°K.

Transmission electron microscopy has been performed on specimens deformed 1% to 2% axial strain. Many of the observations were similar to those of Paton and Williams who observed dislocation substructures in Ti-V alloys deformed in the temperature range 77° to 575°K (7). For example, dislocation substructures consisting of slip bands with jogged screw dislocations which protrude from the edges of the slip bands have been observed; a low density of jogged screw dislocations is found between the slip bands. Figure 4 shows a Ti-40 V crystal deformed at room temperature in which a region of relatively small strain is examined. A distribution of parallel, jogged screw dislocations is observed.

At 77°K, the dislocation structure becomes more heterogeneous. Slip is coarse and crystallographic, and as noted by Paton and Williams (7), the slip bands contain a high density of jogged screw dislocations. In regions showing relatively small strains, long screw dislocations loops are found (see Figure 5). The screw dislocations contain many jogs. In this particular micrograph the active slip plane was inclined 15-20° to the plane of the foil showing the very early stages of slip band formation. The reason for the mottled background in Figure 5 is not understood but may be an etching effect. A very small degree of diffuse streaking was observed, but no direct evidence of the omega phase was seen.

Discussion

Asymmetry of yielding has been observed in a number of bcc metals and alloys (1,2). There has been no detailed study of the asymmetry of slip and yielding in concentrated, near equatomic bcc solid solutions, but there is data on more dilute substitutional alloys. Asymmetry in the flow stress for slip on the {112} planes was observed in: Mo-Re (10), Fe-Si (6,8,9), Nb-6 W (11), and Nb-Mo alloys (12). These results indicate that slip is easy when the shear is favorable for twinning (on the [111]$(\bar{2}11)$ system in compression) and hard when slip occurs in the anti-twinning sense (the [111]$(\bar{1}\bar{1}2)$ system in compression). As shown in Figures 2 and 3, this is consistent with the present results where in compression: $\tau_{\bar{1}\bar{1}2} > \tau_{\bar{2}11}$.

Several studies have shown that asymmetry of slip increases with decreasing temperature for bcc alloys (4,8,11,12). Figures 2 and 3 in the present study show

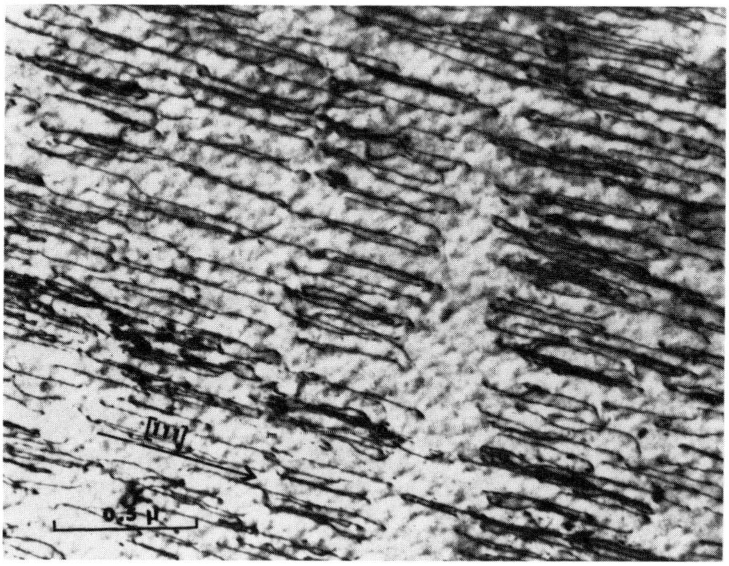

Figure 4. Dislocation arrangement in Ti-40 a/o V deformed in compression at 297°K (axial strain = 1.8%).

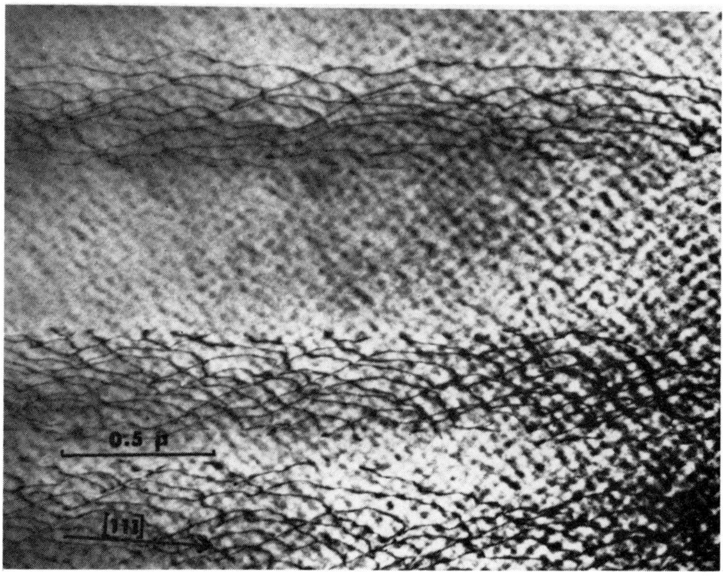

Figure 5. Dislocation arrangement in Ti-40 a/o V deformed in compression at 77°K (axial strain = 1.5%).

that asymmetry is greater at a low temperature. In addition, electron microscopy studies on Fe-Si (14) and Nb-Mo (15) alloys show that dislocation arrangements at small strains consist of predominantly jogged screw dislocations in the form of glide loops elongated parallel to the Burgers vector. In that respect the dislocation substructure of Ti-40 V is also similar to that of more dilute alloys. The presence of the screw dislocation substructure suggests that the mobility of screw dislocations may be controlling the macroscopic deformation rate.

The results cannot be explained solely by conventional solid solution models in which the yield stress of the alloy is determined by a solute of differing atomic size or modulus pinning a dislocation. The force of dislocation-solute interaction in such models does not depend on whether the dislocation approaches the solute on a {110} plane or on a {211} plane in the twinning or anti-twinning direction. Thus, one would expect no yield stress asymmetry; this is not observed (see Figures 2 and 3). The recent theory of H. Suzuki discusses the thermally activated yield stress of bcc solid solution alloys in the presence of a high Peierls stress (16). Suzuki's theory explains the asymmetry of slip and yielding in terms of double kink formation and reactions between kinks on differing {110} planes. Slip on a {112} plane is viewed in terms of composite motion on {110} planes, and hard and soft directions on a {112} plane are rationalized. However it would seem that {110} slip would be preferred, especially at low temperatures. The present results (Figures 2 and 3) clearly show the opposite; {112} slip is preferred and is dominant at 77°K even when the maximum resolved shear stress acts on the {110} plane ($\psi \approx 25°$ when $\chi = 0°$ at 77°K, see Figure 3). Thus, while solid solution hardening models may partly explain the rather high yield stress of Ti-40 V, these ideas cannot solely account for the strength observed.

The asymmetry of deformation in bcc alloys and metals is generally interpreted as indicating that the dislocation core structure influences the overall flow stress (17). The asymmetric deformation behavior of Ti-40 V can be rationalized on the basis that dislocation glide on the {211} planes is easier when shear is in the twinning sense than when in the opposite sense. This can be interpreted in terms of an asymmetry of lattice potential (18,19), but several investigators have proposed models in which the core of a screw dislocation is extended into two or more of the {211} and {110} planes containing the dislocation line (3,20-24). Recent computer calculations show that there is threefold symmetry about a core of a screw dislocation in a bcc lattice (25). The configuration in the core region is stress dependent, but the atom displacements are such that resultant configurations will lead to easy or hard motion on {112} planes, depending on direction of motion of the dislocation.

Theories of thermally activated motion of screw dislocations have been developed from the various dissociated core models (3,21-24). The theories have attempted to explain both the temperature dependence of the yield stress and the variation of slip geometry with orientation. In order for dislocation glide to occur, the dissociated screw must at least in part recombine. The recombination requires thermal activation or stress for formation since the dissociated configuration is more stable. Thus, the magnitude of resultant asymmetrical deformation would increase with decreasing temperature; this is consistent with the data in Figures 2 and 3. At a given stress and temperature, some models assume that dislocations move with differing velocities on all {110} and {112} planes of the slip zone. The macroscopic slip plane is a result of the vector addition of the dislocation velocities on the various planes. The ease of {112} slip in Ti-40 V suggests easy recombination (i.e., small energy required) for slip on {112} planes compared to slip on {110} planes. The data can be fitted to the theories (3,21-24) by choosing the stacking fault energy parameter γ such that $\gamma_{110} >> \gamma_{112}$. Thus the stacking fault energy could be large for {110} partial lying out of the primary {112} slip plane leading to easy motion on the {112} plane. There is no direct evidence to support this rationalization.

In summary, Ti-40 V single crystals exhibit asymmetry of slip such that (112) slip is predominant, especially at low temperatures. An anisotropy of the yield stress is also observed in compression with $\tau_{\overline{11}2} > \tau_{\overline{2}11}$; this inequality increases in magnitude at low temperatures. Transmission electron microscopy shows the presence of screw dislocations in lightly deformed samples. These results may be interpreted as indicating that the dislocation core structure influences the flow stress in a manner similar to that observed in bcc metals and dilute alloys.

References

1. Christian, J. W., Second International Conference on the Strength of Metals and Alloys, ASM, Metals Park, 1970, pp. 31-70.

2. Stein, D. F., Chen, C. P., and Ayres, R. A., to be published in Inhomogeneity of Plastic Deformation, ASM, Metals Park.

3. Duesbery, M. S., Phil. Mag., Vol. 19, 1969, pp. 501-519.

4. Guiu, F., Scripta Met., Vol. 3, 1969, pp. 449-454.

5. Sestak, B., and Libovicky, S., Czech. J. Phys., Vol. B12, 1962, p. 131.

6. Sestak, B., Zarubove, H., and Sladek, V., Can. J. Phys., Vol. 45, 1967, pp. 1031-1039.

7. Paton, N. E., and Williams, J. C., Second International Conference on the Strength of Metals and Alloys, ASM, Metals Park, 1970, pp. 108-112.

8. Sestak, B., and Zarubova, N., Phys. Stat. Solidi, Vol. 10, 1965, pp. 239-250.

9. Taoka, T., Takeuchi, S., and Furabayashi, E., J. Phys. Soc. Japan, Vol. 19, 1964, pp. 701-711.

10. Davidson, D. L., and Brotzen, F. R., Scripta Met., Vol. 5, 1971, pp. 303-307.

11. Koss, D. A., Phil. Mag., Vol. 22, 1970, pp. 493-499.

12. Statham, C. D., Veseley, D., and Christian, J. W., Acta Met., Vol 18, 1970, pp. 1243-1252.

13. Rogausch, K. D., and Mordike, B. L., Second International Conference on the Strength of Metals and Alloys, ASM, Metals Park, 1970, pp. 168-172.

14. Low, J. R., and Turkalo, A. M., Acta Met., Vol. 10, 1962, pp. 215-227.

15. Foxall, R. A., and Statham, C. D., Acta Met., Vol. 18, 1970, pp. 1147-1158.

16. Suzuki, H., to be published.

17. Christian, J. W., Comments Solid State Phys., Vol. 1, 1968, pp. 19-26.

18. Dorn, J. E., and Mukherjee, A. K., Trans. AIME, Vol. 245, 1969, pp. 1493-1500.

19. Kroupa, F., and Hull, D., Czech. J. Phys., Vol. 19, 1969, pp. 763-770.

20. Foxall, R. A., Duesbery, M. S., and Hirsch, P. B., Can. J. Phys., Vol. 45, 1967, pp. 605-629.

21. Bowen, D. K., Christian, J. W., and Taylor, G., Can. J. Phys., Vol. 45, 1967, pp. 903-937.

22. Vitek, V., and Kroupa, F., Phys. Stat. Solidi, Vol. 18, 1966, pp. 703-713.

23. Escaig, B., J. Phys., Vol. 28, 1967, pp. 171-186.

24. Duesbery, M. S., and Hirsch, P. B., Dislocation Dynamics, McGraw-Hill, New York, 1968, pp. 57-85.

25. Vitek, V., Perrin, R. C., and Bowen, D. K., Phil. Mag., Vol. 21, 1970, pp. 1049-1073.

A STUDY OF THE NATURE OF THE DUCTILE-BRITTLE TRANSITION IN BETA-TITANIUM ALLOYS

Igor V. Gorynin
Senior Engineer, Baltic Plant
Boris B. Chechulin, Senior Research Metallurgist
Stal S. Ushkov, Senior Engineer
Olga S. Belova, Senior Engineer
Institute of Shipbuilding Technology
USSR

The appearance of a crystalline component on the fracture surface of steel specimens with decreasing test temperature represents a convincing indication of the approaching brittleness in the material under study. The change in the fracture topology consisting in the appearance of facets with decreasing temperature is characteristic also for the bcc beta-titanium alloys exhibiting the ductile-brittle transition.

Despite a large number of works dealing with the fracture of beta-titanium, there are no publications devoted to the fractographic appearance of the beta-alloys as related to the ductile-brittle transition.

At the same time, an investigation of the changes occurring on the fracture surface of the beta-titanium alloys with decreasing temperature not only would be of practical value (by providing a method of evaluating metal quality by the fractographic appearance) but also permits to determine the nature of the low temperature brittle fracture of the beta-alloys.

Experimental Details

The subject of the study was a beta-titanium alloy similar to the American alloy B120VCA with an addition of small amounts of molybdenum and zirconium as stabilizers. The metal was prepared by arc melting in vacuum using titanium sponge and high purity stabilizers. The metal to be studied was subjected to solid solution treatment.

One studied the fracture of prismatic specimens subjected to static bend tests at low temperatures.

At a test temperature of -180°C the fracture surface revealed facets only with no fibres present.

The fractographic surface was studied by the following techniques.

1. By investigating a microsection normal to the fracture surface. The microsection was prepared on the specimen before the testing. A good quality of the section could be maintained due to the tests being conducted at -180°C.

2. By a direct investigation of the fractographic surface proper. The fracture topology was studied visually with a binocular microscope at magnifications of 4 to 20 times, with a Zeiss vertical metallographical microscope at a magnification of 100 times with the object stage in the lower position, as well as on a JEM-5 electron microscope using carbon replicas. The orientation of the facets was determined by studying the X-rays reflected from the surface of the facets set perpendicular to the X-ray beam. To set the facet surface perpendicular to the X-ray beam, a plane mirror was attached to this surface, and the specimen was moved in a position where the light beam passing through a collimator would again enter it after reflection from the mirror. After this, the mirror was removed carefully from the facet surface and the chamber housing the specimen was mounted in YPC-70 X-ray installation. A tungsten anode X-ray tube was used at 40 kV and 10 A. Exposure times were 6 to 20 hrs.

Experimental Results and Discussion

An investigation of the fractographic microstructure of the specimens indicates that the fracture is transgranular rather than occurring along the boundaries. It should be noted that just as in the case of steels, the fracture proceeds here along certain crystallographic directions (Fig. 1). The initial stages of fracture near the main crack are characterized by features indicating brittle fracture: the cracks are straight, with no plastic zone around them. Such a nature of the microstructure in the region of fracture gives grounds to conclude that the fracture proper and the facets on its surface are associated with the cleavage plane.

A study of the fractographic surface with a binocular microscope has shown the facets to be polyhedral in shape, their size varying proportional to the size of the beta-grain.

The pattern on the surface of the facets was investigated with a Zeiss microscope at a magnification 100X. Some facets reveal the "herringbone" pattern (Fig. 2a). It is characterized by two groups of lines oriented at right angles to one another and separated by a narrow smooth strip forming the bisectrix of the 90°-angle. This pattern representing a typical feature of cleavage fracture was observed with the electron microscope on the fractographic surfaces of tungsten, chromium, silicon iron and stainless steel 430.

In the majority of cases the facets represent parallel rectilinear step structures (Fig. 2b), sometimes with a fanlike pattern. That facets (Fig. 2c) are typical for low test temperatures.

A comparison of fractographic appearance studied at optical magnifications with the microstructure of sections perpendicular to the fracture gives grounds to consider the facets as cross sections of grains of the beta-phase along certain crystallographic planes, apparently the cleavage planes.

The micromechanism of fracture on the surface of brittle beta-alloys with facets was studied with the electron microscope using the technique of replicas.

In the fractographic structure containing only facets the cleavage fracture is indicated by cleavage steps (Fig.3a) arranged in parallel lines, sometimes with a river pattern. These formations typical for cleavage arise due to the fact that the cleavage cracks in an imperfect polycrystalline solid propagate not along

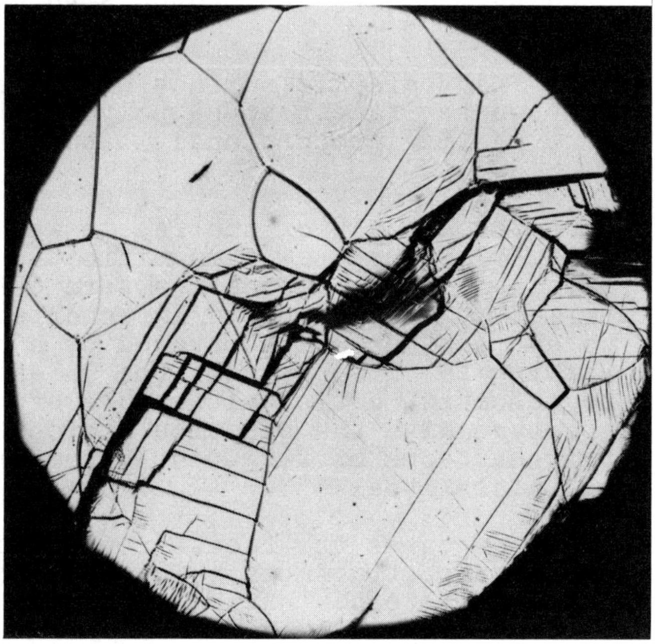

Fig. 1. Microstructure of fracture area in a specimen of beta-titanium alloys tested at -180°C (magnification 100X).

one plane but rather along a series of parallel cleavage planes thus forming steps on the parallel planes by slip or cleavage along secondary planes.

Another indication characteristic for cleavage fracture is the presence of tongues (Fig. 3c, A).

In addition, one observed on the brittle fractographic surfaces of beta-titanium perfectly flat facets with a clearly pronounced crystallographic orientation of very thin steps (Fig. 3b) and ridges (Fig. 3c, B). One should note the presence on the fracture surfaces of beta-titanium of closely intermingled areas with cleavage steps and other indications of brittle fracture and areas with transition and even ductile fracture (shown with arrows, Fig. 3b).

The beta-titanium alloys reveal the same electron micrographic patterns as are typical for the ductile and brittle fracture in steels.

To check the suggestion that the facets represent areas of brittle fracture propagating across the grain of the beta-phase along the cleavage planes, we determined the orientation of the facets by the X-ray technique. An X-ray micrograph typical for smooth facets without steps on their surface is shown in Fig. 4. X-ray micrographs of facets with steps on the surface could be obtained only after a preliminary etching which indicates that the formation of steps results in a distortion of the surface layer.

An analysis of the X-ray micrographs shows the facet plane to coincide, within the experimental error, with the (001) plane representing the plane of cleavage in the bcc metals. The X-ray data indicate that the facets are the cross sections of the beta-phase grains along the cleavage plane.

The formation on the facet surface of an array of rectilinear steps aligned in the same direction can be attributed to the intersection of the cleavage plane in which the crack propagates with one of the effective planes of slip, twinning or secondary cleavage. A comparison of the facet pattern and of the X-ray micrograph arranged parallel to one another showed the direction of parallel steps on the surface of the facets to coincide with the /010/ or /100/ directions. Of all

A

B

Fig. 2. Typical facets in beta-titanium alloy (magnification 100X).

C

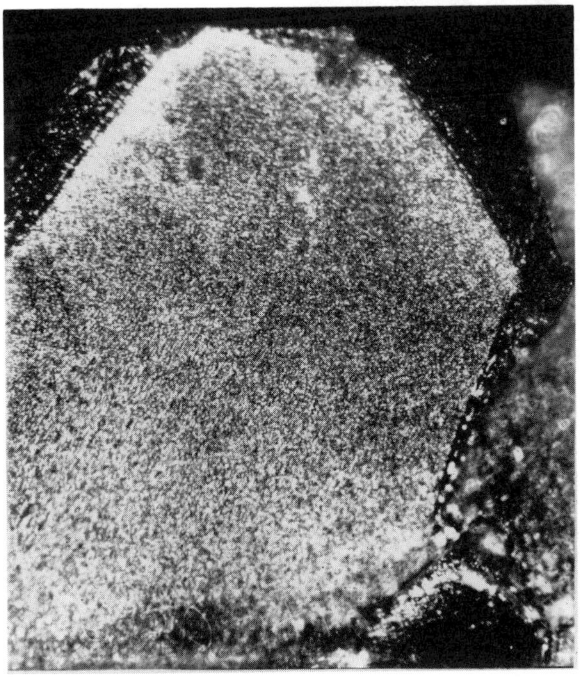

D

Fig. 2 (cont'd).

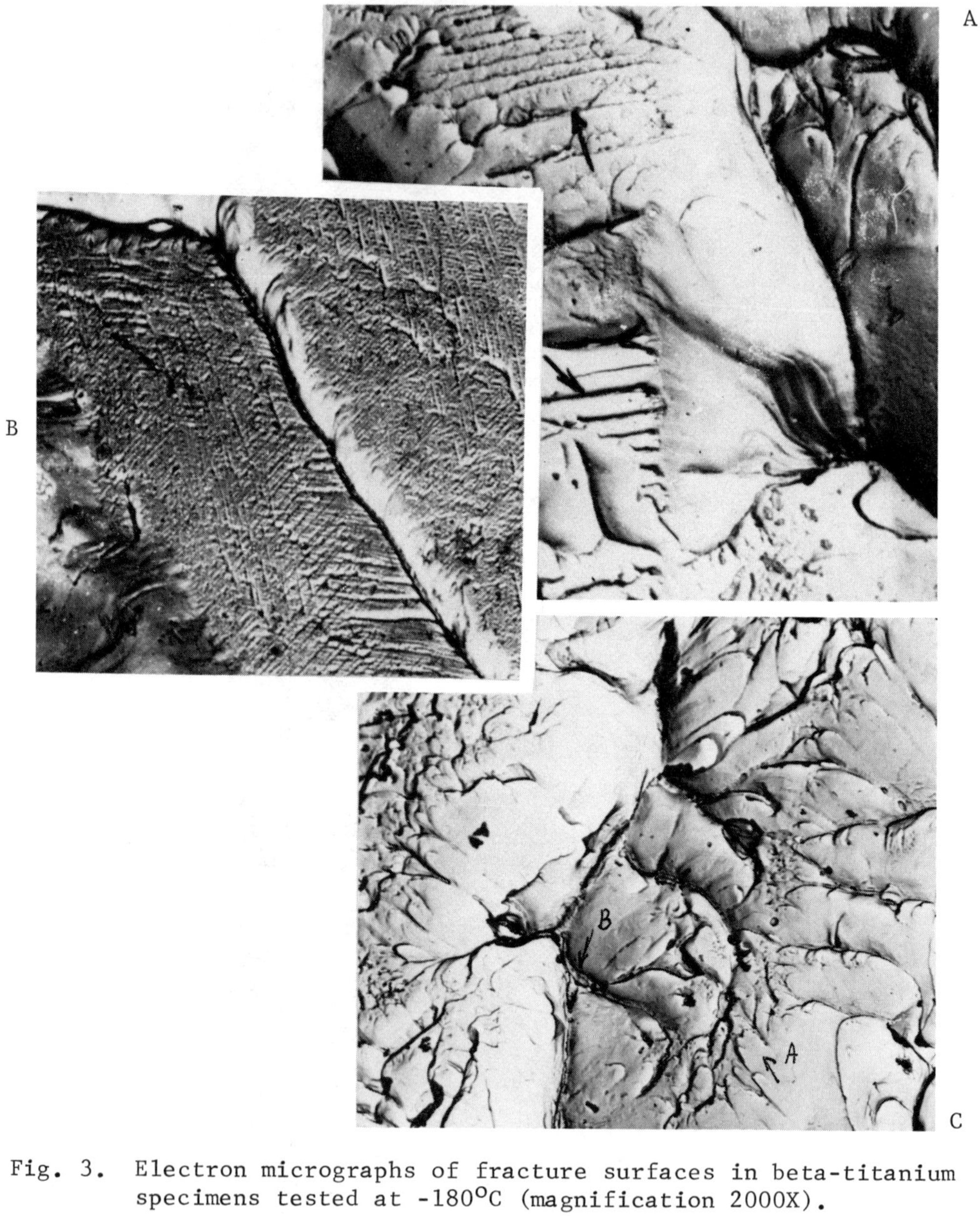

Fig. 3. Electron micrographs of fracture surfaces in beta-titanium specimens tested at -180°C (magnification 2000X).

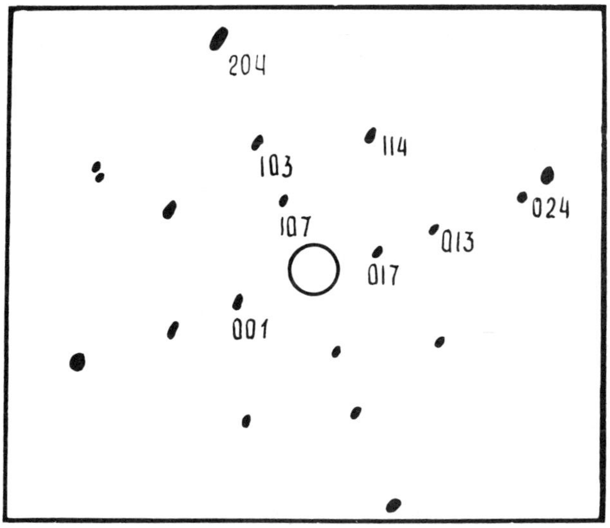

Fig. 4. X-Ray micrograph of facet in beta-titanium.

slip systems in a bcc lattice acting along these directions (the cube edges) determined by the X-ray technique, only the (101) and (011) planes intersect with the cube facets. This means that the steps result from the cleavage crack propagating into the parallel cleavage planes by slipping along the $\{110\}$ planes.

The established close similarity between the main features of low temperature fracture in beta-titanium and steel gives grounds to conclude that the nature of the low temperature brittleness in beta-titanium is the same as in steel. Within the framework of the modern physical concepts concerning the crystal structure of solids, the ductile-brittle transition in bcc metals is considered to be a result of increasing resistance to plastic deformation with decreasing temperature, due to specific dislocation movement in the lattice of this kind and to the presence of foreign atoms forming an interstitial solid solution.

Such facts associated with the ductile-brittle transition and experimentally established for beta-titanium as the cleavage mode of fracture, a strong temperature dependence of the yield strength and an impurity-induced shift of the ductile-brittle transition temperature indicate that also in this bcc metal the motion of dislocations is connected with a thermally activated process of overcoming some potential barriers and is brought about by the interaction with the impurity atoms.

Conclusion

The low temperature ductile-brittle transition in beta-titanium is accompanied by the appearance on the fractographic surface of facets indicating brittle fracture, similar to the appearance of a crystalline component on the fracture surface of steel specimens.

In beta-titanium undergoing the ductile-brittle transition, fracture occurs by cleavage.

On the basis of common features in the fracture mechanism of beta-titanium and iron-based alloys, a conclusion is drawn of the identical nature of the ductile-brittle transition in these materials associated with the blocking of dislocations in bcc metals.

TITANIUM ALLOYS' BEHAVIOUR UNDER VARIOUS LOADING CONDITIONS

Ivan N.Bogatchev, Professor, Doctor
Maria A.Dyakova, Assistant Professor
The Urals Polytechnical Institute
Sverdlovsk, USSR

The paper deals with Ti-alloys on the base of α, β - solid solutions, two-phase $\alpha + \beta$ -alloys (Table I). The subject of investigation was the mechanism of deformation as well as the changing of mechanical properties under various loading conditions: static tension, impact bending, micro-impact affect (cavitation) in a wide temperature range from -196 °C up to +500 °C.

Table I. Chemical Composition of Alloys Investigated

Nos.	Alloy	w/o of alloying components					
		Al	Mo	Cr	Mn	V	Sn
1	Technical titanium	-	-	-	-	-	-
2	Ti3Al	2.7	-	-	-	-	-
3	Ti4Al	4.41	-	-	-	-	-
4	Ti6, 5Al	6.5	-	-	-	-	-
5	Ti5Al,2,5Sn	5.2	-	-	-	-	2.5
6	Ti4Al 3Mo 1V	4.23	3.0	-	-	1.2	-
7	Ti5Al 2Mo 2Cr	5.05	2.05	1.56	-	-	-
8	Ti2,5Al 7Mo	2.6	7.0	-	-	-	-
9	Ti8Mo	0.30	8.1	-	-	-	-
10	Ti3Al 8Mo	3.47	8.1	-	-	-	-
11	Ti5Al 8Mo	5.2	8.4	-	-	-	-
12	Ti2,5Al 11Mo	2.5	11.0	-	-	-	-
13	Ti9Mn	-	-	-	9.06	-	-
14	Ti3Al 6Mo 10Cr	3.18	6.18	9.8	-	-	-

The results of the research showed that plastic deformation under static tension at 20°C and in the field of negative temperatures takes place both by gliding and by twinning. Temperature lowering in the range of negative temperatures causes an increase of plastic deformation by twinning (1).

The possibility of realizing various mechanisms of plastic deformation determines the nature of mechanical properties change.

In technical titanium parallel with the growth of yield point and strength under temperature lowering in the range of negative temperatures relative elongation grows as well from 20% at 20°C up to 30% at -196°C. The growth of plasticity at temperature lowering results from, firstly, the possibility of essential development of twinning parallel with plastic deformation by gliding. Though twinning as such very little contributes to deformation, it provides crystal lattice reorientation in separate volumes of metal and so creates conditions for consequent gliding, more favourably arranged planes for plastic deformation development by gliding (2). Secondly, the possibility of realising various mechanisms of plastic deformation at lower temperatures results in increased strengthening during deformation due to bigger defects density of crystalline structure, which in its turn gives rise to increased resistance to big plastic deformations and destruction, and indirectly promotes the growth of relative elongation.

At temperatures from 20° up to 500°C plastic deformation of technical titanium under static tension occurs principally by gliding, but faulting takes place as well. The latter reaches its greatest development at tension temperatures of 200--300 C.

The investigation of plastic deformation nature of α -alloys, Ti-Al and Ti-Al-Sn showed that aluminium and tin composing with titanium solid solution of substituation hamper the development of plastic deformation, particularly twinning and faulting processes. In Ti3Al, Ti4Al alloys, besides gliding, twinning is observed only in the negative temperatures range: at tension temperatures range of -120°C and -196°C. In higher alloyed α-alloys Ti6,5Al, Ti5Al 2,5Sn plastic deformation at static tension at room temperature as well as in the negative temperature range processes by sliding only. The fact that the development of plastic deformation in alloyed α-alloys is hampered causes the decrease of plastic properties in the negative temperature range. In alloyed α- alloys Ti6,5Al and Ti5Al 2,5Sn plastic deformation at tension temperatures of 20-500°C processes principally by gliding while in less alloyed α- alloy Ti3Al side by side with gliding faulting is observed at tension temperatures 200-300°C.

The investigation of the influence of notches (holes, grooves) on the nature of plastic deformation under static tension of technical titanium flat specimens and α-alloys showed that the change of stressed state under given loading conditions causes essential changes in the plastic deformation mechanism. With this plastic deformation begins at higher stress (the yield point is in the average 15 kg/mm^2 higher in comparison with the specimens without stress concentrators and the elongation decreases approximately twice). Specimens with stress concentrators (with holes) of technical titanium an essential decrease of gliding in plastic deformation is observed, and at -196°C deformation occurs principally by twinning. In α-alloy Ti4Al under similar loading conditions the ability of plastic deformation development by gliding diminishes more sharply and already at -60°C plastic deformation takes place principally by twinning. At lower temperatures the extent of plastic deformation developing by twinning decreases. The difficulty in development of plastic deformation by gliding and by twinning in specimens with notches in α-alloys Ti-Al, Ti-Al-Sn, brings about a lowering of plastic properties and brittle destruction in the negative temperatures range (at -60°C).

The study of plastic deformation mechanism of technical titanium and α-alloys in more severe loading conditions (impact bending of notched specimens (ГОСТ 9454-60 type I) (3) in the temperature range of 20°C to -196°C of technical titanium and α-alloys processes principally by twinning. The degree of development of plastic deformation by twinning with temperature lowering decreases. Impact viscosity of α-alloys lowers from 8-10 kgm/cm^2 at 20°C up to 3 kgm/cm^2 at -196°C (Fig. 1). At higher test temperatures 100-500°C in such loading conditions plastic deformation takes place principally by gliding, impact viscosity grows with temperature up to 16 kgm/cm^2.

The study of plastic deformation nature in high-alloyed Ti3Al6Mo10Cr alloy, hardened at β-solid solution showed that with static tension of specimens without notches and with holes and render impact bending of notched specimens as well plastic deformation in β-phase in a wide temperature range from -196°C up to 500°C processes by gliding. The increasing severity of loading (specimens with holes, impact bending of notched specimens, temperature lowering down to the range of negative ones) hampers plastic deformation by gliding causes reducing of plastic properties and brittle destruction in the negative temperature range already at -70°C.

Tension diagrams from the specimens of Ti3Al6Mo10Cr alloy hardened at β-solid solution showed that at tension temperatures of 200, 300, 400°C diagrams have toothlike character which results from the Portven-Le-Chatelle effect (Fig.2a). In this case in the process of plastic deformation of instable β-solid solution in the temperature range of 200-400 C occurs an accumulation of atoms of admissive elements forming clouds of these atoms around the dislocations with consequent tearing dislocations out of the clouds under plastic deformation. With annealed specimens of the same heat the structure of which consists of β-solid solution separating α-phase, such a toothlike nature of tension diagram was not observed (Fig.2c). It is found that the dissociation of instable β-solid solution of Ti2,5Al7Mo, Ti3Al6Mo10Cr alloys in the process of tension at 550°C with separation of disperse α-phase causes growth of yield points and strength with simultaneous growth of elongation (Fig.3). The growth of relative elongation results from diffusive plasticity displayed under transformation of $\beta \rightarrow \alpha$ in the tension process.

Plastic deformation of Ti3Al6Mo10Cr alloy in heat-treated state for maximum strength σ_b =150 kg/mm², δ =4% (hardening from 800°C + tempering at 450°C for 25 hours + holding at 560°C for 15 min) develops principally by faulting at tension temperatures of 20-300°C. In this case on tension curves jumps and derangements are observed (Fig.2b). At tension temperatures of 400°C and higher, when side by side with faulting plastic deformation by sliding gets its essential development jumps on the tension curves are not observed.

Availability of an instable β-phase in titanium alloys brings about substantial changes in the behaviour of alloys under various loading conditions (4). In the investigated alloys of Ti-Al-Mo at the temperature of thebeginning of martensitic transformation lower than 300°C under hardening from β-field side by side with martensite, instable remanent β-phase is fixed able to dissociate during the process of deformation forming martensite and under heating in the range of 400-500°C with ω-phase formation. It is stated that in these alloys non-stable remanent β-phase in the process of deformation under tension at 20°C and in the negative temperatures range desintegrates forming martensite which results in a low value of yield point ($\sigma_{0,2}$ =50-60 kg/mm² at 20°C, 80-100 kg/mm² at -120°C), higher alloys strengthening during deformation, high ultimate strength (95-110 kg/mm² at 20°C, 115-130 kg/mm² at -120°C), keeping sufficient level of plasticity in the negative temperatures range (relative elongation

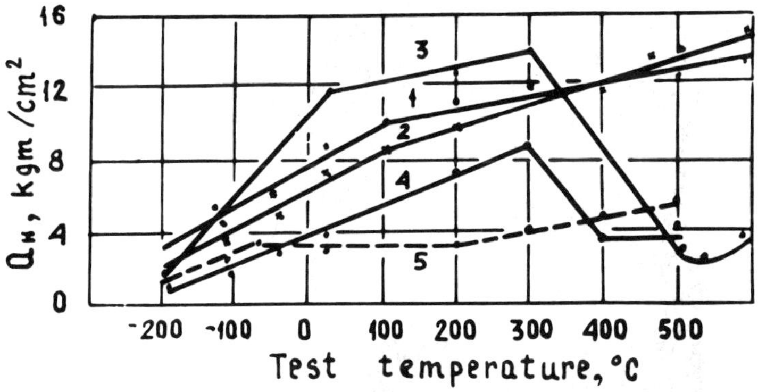

Fig. 1. The change of impact ductility of Ti-alloys depending on the testing temperature. 1) Ti5Al2,5Sn; 2) Ti2,5Al 7 Mo, annealed; 3) Ti2,5Al 7 Mo, hardening from 900°C; 4) Ti5Al 2 Mo 2 Cr, hardening from 860°C; 5) Ti5Al 2 Mo 2 Cr, hardening from 860°C, tempering 550°C.

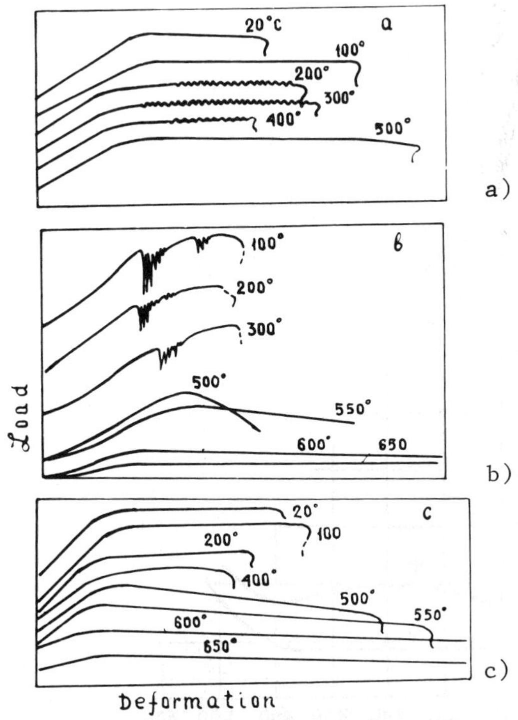

Fig. 2. Tension diagrams of Ti3Al 6 Mo 10 Cr after various heat treatment. a) Hardening from 800°C; b) hardening from 800°C, tempering 450°C 25 hours holding at 560°C, 15 min; c) annealed at 650°C.

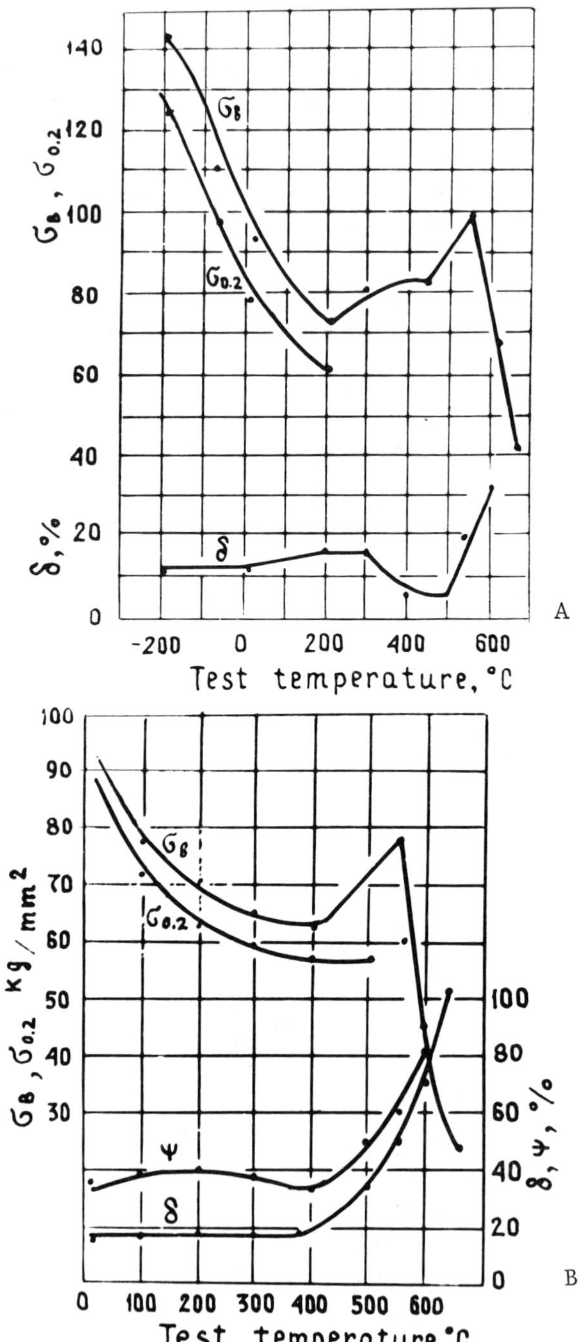

Fig. 3. Change of mechanical properties of Ti-alloys hardended from from β-field depending on testing temperature.
a) Ti2,5Al 7 Mo; b) Ti3Al 6 Mo 10 Cr.

10–14% at 20°C) (5). At lowering temperature from 20°C up to -196°C impact ductility decreases from 8 kg/cm^2 to 3 kg/cm^2. Such a change of properties results from the mechanism of plastic deformation. In this case plastic deformation mostly occurs on account of shearing process under transformation of β-phase into martensite along moving division boundaries between β-phase and martensite which results from low resistance to shearing at β-phase - martensite transformation. With the growth of the extent of deformation as the result of martensite transformation the defect density of cristalline structure increases. This provides an essential strengthening during the process of deformation and a high ultimate strength. Under tension test of alloys with an unstable β-phase in the temperature range of 400–500°C a sharp fall of plastic characteristics takes place. A relative elongation under tension in temperature range of 20–300°C is maintained at the level of 12–14%, and under tension is temperature range of 400–500°C it falls up to 4–3%. The value of impact viscosity at lower test temperatures grow from 8 kgm/cm^2 at 20°C up to 14 kgm/cm^2 at 300°C and fall sharply up to 4 kgm/cm^2 in the test temperature range of 400–500°C (Fig.1). Such fall of plasticity and viscosity of alloys in the temperature range of 400–500°C results from desintegration of β-solid solution forming ω-phase under the process of heating during 15 min. before the test.

The analysis of the metallographic picture of plastic deformation development and the distruction of Ti-alloys under microimpact loading in the process of cavitation affect*) showed that the weakest resistance to cavitation affect is offered by α-solid solution. Under cavitation affect lines of gliding and twins appear. The distraction of α-solid solution has a viscouse nature, develops from the grain boundaries, twins, and spreads over the whole grain (6). β-phase offers higher resistance to cavitation affect. Plastic deformation in β-phase under given loading conditions develops by gliding. At the initial period of cavitation an irregularity of plastic deformation in the volume of metal is observed: there are grains without any noticeable lines of gliding. The destruction of β-solid solution under micro-impact loading begins from the grains boundaries, lines of gliding, and

*) The cavitation affect was accomplished on an impact errosion stand at the circumferential speed of the disc with the specimens fastened on it of 78 m/sec., the distance between the nozzle and the disc - 14 mm., and the water head - 0.28 atm.

finishes in shear of metal between the planes of gliding. The destruction of Ti28Ni, Ti25Co alloys with titanides separation begins with α -solid solution. The destruction of alloys containing martensite parallel with β -solid solution processes from the martensite plates boundaries and spreads towards the latter. The destruction of alloys containing, besides β-solid solution, the ω -phase takes place by shearing of metal between the planes of sliding and ends in brittle destruction.

It is stated that α-alloys as their stableness against cavitation affect is concerned are at about the same level as steel IXI8H9T (Fig.5).

Alloys on the base of α-solid solution with titanides separation in annealed state at hardness of 25-30 HRC as to their stability against cavitation are twice as stable as α - alloys (7). In these alloys under hardening from β -field the ω -phase is formed which results in increased hardness (50 HRC) and in brittle destruction during the process of cavitation. Stableness against cavitation of $\alpha + \beta$ alloys phases is determined by relative correlation of structural components of α - and β -phases, as well as by the degree of disperseability of α-phase. The most stable against cavitation in annealed state is the high alloy Ti3Al6Mo10Cr, which can be explained by relatively high quantity of β -phase and by high dispersity of α-phase. As a result of studying the titanium alloys behaviour under cavitation affect, with the instable solid solutions structure, it is stated that hardening alloys from two-phase field to the structure consisting of $\alpha + \beta$ phases as compared with hardening from β - field to the martensite structure provides higher stability against cavitation.

Stability against cavitation of titanium alloys is greatly influenced by the ability of an instable β-solid solution to decompose under the micro-impact loading with martensite formation.

The formation of martensite during the process of cavitation was observed in Ti5Al8Mo, Ti3Al8Mo, Ti8Mo, Ti2,6Al7Mo alloys hardened from β -field to the structure of martensite with remanent instable β-phase. The stability against cavitation of such alloys is the higher the more of remanent β-phase is transformed into martensite during the process of microimpact affect. The latter is determined by the position of M_s. For the investigated alloys M_s was in the temperature range from 370°C to 30°C. The lower the position of M_s in this range of temperatures, the more intensively the instable β -

phase is transformed into martensite in the process of cavitation and the higher is the stability against cavitation of the alloys. The destruction of the instable β-phase with the formation of martensite at the initial stages of the process of deformation results in stress relaxation on the account of shear transformation and at the later stages of deformation provides an intensive strengthening due to additional increase of defect density as the result of shearing martensite transformation. This increase the resistance of the material to big plastic deformation and destruction.

There exists an analogy in the nature of strengthening and destruction of titanium alloys at tension and under cavitation effect. Technical titanium has a low yield point, shows a slight strengthening at room temperature both at tension and at micro-impact affect in the process of cavitation, and offers a small resistance to destruction at tension as well as under cavitation affect (Fig.5). Alloys containing a non-stable β-phase (Ti5Al8Mo, Ti3Al8Mo, Ti8Mo, Ti2,5Al7Mo) dissociating during the process of deformation with martensite formation, possess a low yield point, but a marked ability to strengthen under plastic deformation (Fig.4a). The higher is the alloys strengthening coefficient, the higher resistance these alloys offer to destruction both at tension and at micro-impact affect. β-phase alloys and two-phase $\alpha + \beta$ alloys, strengthened by heat treatment possess high values of yield points and strength with a sufficient plasticity which guarantees high resistance to destruction both at tension and under micro-impact affect, despite a slight strengthening during the process of deformation (Fig.5).

At a certain extent of alloyage with transient elements in titanium alloys hardened at β-solid solution under micro-impact loading in the process of cavitation affect the dissociation of a nonstable β-phase may take place forming ω-phase which results in brittle destruction of alloys during the test process. Alloys Ti9Mn, Ti3Mn1Fe1Cr1V pertain to such type of alloys. They have little losses under cavitation affect but undergo brittle destruction after 30 min of testing (Fig.5).

Applying strengthening heat treatment causing dissociation of β-solid solution with formation of a disperse α-phase eliminates the chance of ω-phase formation under micro-impact loading during cavitation affect and guarantees growth of stability against cavitation about twice as big as in annealed condition. To highten stability against cavitation it is reasonable to harden two-phase alloys from the two-phase $\alpha + \beta$ field and to temper them at temperatures of 480-500°C. The optimum schedule of hardening and tempering are defined by

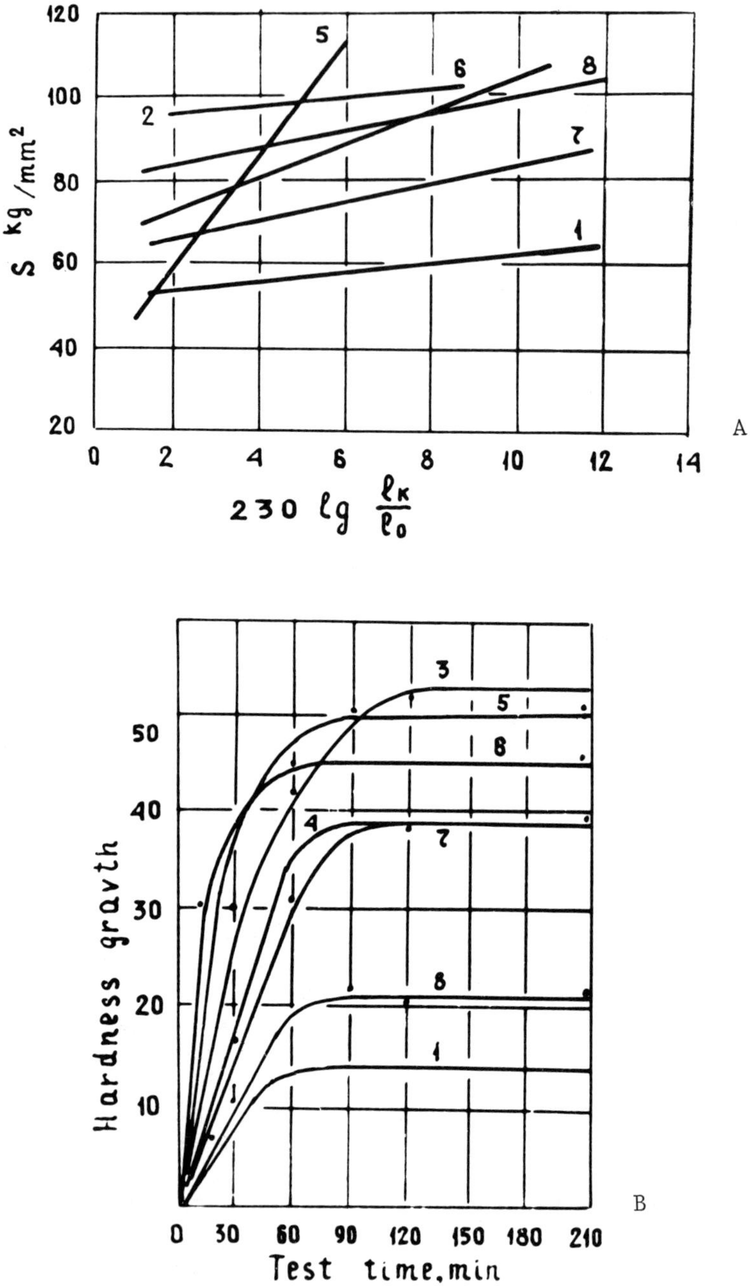

Fig. 4. Strengthening of Ti-alloys under static tension (a) and under cavitation affect (b).
α'-phase: 1) technical titanium; β-phase: 2) Ti9Mn; 3) Ti2,5Al 11 Mo; 4) Ti3Al 6 Mo 10 Cr; $\alpha''+\beta$: 5) TiAl8Mo; 6) Ti3Al 8 Mo; 7) Ti8Mo; 8) Ti2,5Al7Mo.

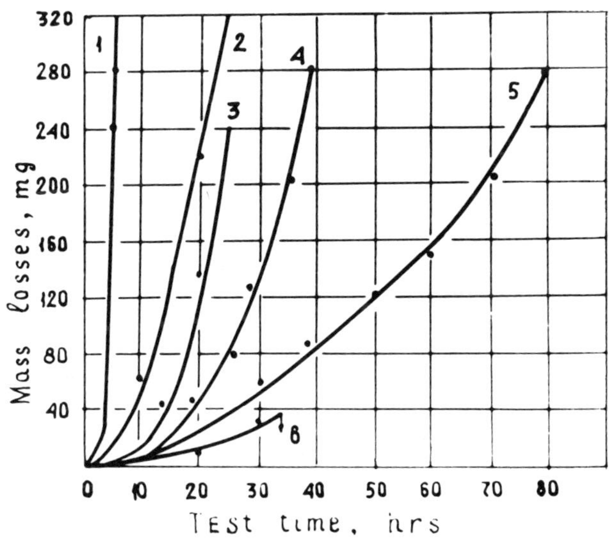

Fig. 5. Stability of Ti-alloys against cavitation depending on phase composition:
1) α-phase: technical titanium; 2) $\alpha + Ti_2Me$; 3) $\alpha+\beta$: Ti5Al2Mo2Cr, Ti4Al3Mo1V; 4) $\beta+\alpha$ high-dispersed; 5) β-phase: Ti2,5Al11 Mo, Ti3Al6Mo10Cr; 6) $\beta+\omega$: Ti3Mn1Fe1Cr1V, Ti9Mn.

the chemical composition of the alloy. Tempering at temperatures lower than the optimum ones does not eliminate the probability of ω-phase forming, and results in brittle destruction. Tempering at temperatures higher than 500°C leads to destrengthening the solid solution, and to enlarging structural components (α-phase) and lowers stability against cavitation.

Conclusion

The character of Ti-alloys behaviour render various loading conditions is defined by the stressed condition, by the possibility of plastic deformation processing under given loading conditions, by the phase composition and the dispersity degree of structural components, as well as by the possibility of dissociation of the instable β-phase with martensite formation of the ω-phase and α-phase during the process of loading.

References

1. И.Н.Богачев, Л.П.Одинокова, Физика металлов и металловедение, т. I9, в. 6, I965, стр. 908.

2. Е.Шмидт, В.Боас, Пластичность кристаллов, в особенности металлических, ГОНТИ, I938.

3. Металлы. Методы механических и технологических испытаний, сб. Государственные стандарты СССР, I970, стр. 45.

4. Masao Adachi, Toxuzo Tsujimoto, Journal of the Japan Institute of Metals, vol.25, 1961, p.568.

5. М.А.Дьякова, И.Н.Богачев, А.К.Безрукова, С.И.Селицкая, Металловедение и термическая обработка металлов,I970,№ I0, стр. 36.

6. И.Н.Богачев,Р.И.Минц, Т.М.Петухова, М.А.Дьякова, Физика металлов и металловедение, т.2, в.4, I96I, стр. 557.

7. И.Н.Богачев, М.А.Дьякова, Цветные металлы, I966, №3, стр. 80.

THERMALLY ACTIVATED DEFORMATION OF Ti-N ALLOYS

K. Okazaki, M. Momochi
Department of Metallurgy
Kyushu Institue of Technology
Tobata, Kitakyushu, Japan
and
H. Conrad
Department of Metallurgical Engineering
and Materials Science
University of Kentucky
Lexington, Kentucky, U. S. A.

The effects of temperature, strain rate and nitrogen in interstitial solid solution on the yield and flow stresses at low temperatures ($T<0.4T_m$) of alpha Ti-N alloys with 0.035 at.% to 0.613 at. %N were studied to ascertain the rate controlling dislocation mechanism. The effect of interstitial nitrogen atoms on the thermal component of the flow stress increases with decreasing temperature. The relationship between the thermal component of the flow stress, σ_0^* at 0°K and nitrogen concentration, C is well-ruled by $\sigma_0^*=0.34\mu_o\sqrt{C}$, where μ_o is the shear modulus C_{66}. A linear relation between $T(\partial\ln\dot{\varepsilon}/\partial\sigma)_T$ and $(-\partial T/\partial\sigma^*)_{\dot{\varepsilon}}$ observed for the present Ti-N alloys implies that an Arrhenius-type equation describes the plastic deformation over the temperature and nitrogen concentration ranges considered. The activation energy varies linearly with temperature, yielding 1.5 eV ($0.21\mu b^3$) at 625°K regardless of the nitrogen concentration. The experimental data are utilized to construct a force-activation distance curve, yielding $F_M \simeq 83\times10^{-6}$ dynes ($0.21\mu b^2$) and $X_o \simeq 1.5b$ which are in good accord with the thermally activated overcoming of interstitial nitrogen atoms as the rate controlling mechanism. These activation parameters obtained for the dislocation-nitrogen atom interaction are in accord with the experimental observation that nitrogen strengthens alpha titanium more effectively than oxygen, consistent with the larger size of the nitrogen atom relative to that of the oxygen atom.

Introduction

The interstitial impurities oxygen, nitrogen and carbon are known to have a marked effect on the strength of titanium, nitrogen being twice as effective as oxygen, which in turn is twice as effective as carbon in raising the flow stress (1,2). Conrad and coworkers (3-9) have studied in considerable detail the effects of temperature, strain rate, strain, grain size and total interstitial content on the flow stress of alpha titanium of various purities at temperatures below about four-tenths the melting temperature. They concluded that the dislocation mechanism associated with the thermal component of the flow stress τ^* was thermally activated overcoming of interstitial solute obstacles by dislocations moving on the first-order prism planes. By analyzing the experimental data, the activation energy for this thermally activated process was obtained as a function of τ^* yielding

$$H \simeq H_o(1 - \tau^*/\tau_o^*)^2 \qquad (1)$$

with H_o having a value of 1.2 eV and $\tau_o^* = 0.15\mu\sqrt{C}$[†], where C is the total interstitial content expressed by an oxygen equivalent (4). This square root dependence of τ_o^* on the concentration of interstitials led Conrad (4,5) to conclude that the strengthening due to interstitial solutes in titanium was essentially Fleischer's model of rapid solution hardening due to asymmetrical distortions (11). As all the materials previously used were not of the same base composition, it was necessary to estimate the hardening contribution in terms of an oxygen equivalent.

Kratochvil and Conrad (12) theoretically studied the interaction between dislocations and interstitial atoms in titanium and compared the theoretically calculated thermal component τ^* and activation volume V^* as functions of interstitial content and temperature with the experimental values (6,7). Agreement between the theoretical and experimental values was found to be satisfactory, however, there still remains the question whether it is adequate to describe the concentration of interstitial solute atoms in terms of an oxygen equivalent, since there is the possibility of second order effects reported by Gupta and Weinig (13).

The objective of this investigation was to measure systematically the effect of nitrogen, which is the most effective interstitial strengthener, on the mechanical behavior of titanium by starting with high purity material and adding various amounts of nitrogen and thereby determine in greater detail the strengthening mechanisms involved in the deformation of titanium at low temperatures.

† In the previous work (3-9) it was assumed that the average resolved shear stress in polycrystals $\tau = \sigma/2$, where σ is the tensile stress. However, recent work (10) indicates that $\tau \simeq \sigma/5$. On this basis $\tau_o^* \simeq 0.06\mu\sqrt{C}$.

Experimental

Five Ti-N alloys (A to E) were prepared from MARZ grade zone-refined Ti and 99.8% TiN powder from the Material Research Corp. by melting them in a vacuum of less than 10^{-6} Torr with an electron beam. The zone-melted ingots were first cold-rolled to 6x6x150 mm bars and then homogenized in vacuum for 48 hours at 800°C. No increase in oxygen was found, the analysis being within the range of as-received MARZ grade Ti. The N content was analyzed by the Kjehdahl technique (Table I). Homogenized bars were swaged at room temperature to 1.2 mm dia. wires without any intermediate annealing, representing a reduction in area of approximately 96%.

The swaged wires were chemically polished in a solution of HNO_3 and HF, and then annealed in vacuum at several temperatures and for certain periods, on referring to the grain growth data for Ti-N alloys obtained by the present authours (14), to obtain the grain sizes of 1, 2, 4 and 15 microns. Grain sizes were measured under an optical microscope using the conventional linear intercept method. The total length of the tensile specimens was 50 mm and the effective gauge length was about 25 mm.

Tensile tests were conducted with an Instron-type Shimadzu tensile testing machine, IS-500 over the temperature range of 77 to 700°K (0.04 to $0.4T_m$) and a tensile strain rate of $3.3\times10^{-4}sec^{-1}$. To ensure that the deformation partials $(\partial \ln\dot{\gamma}/\partial\tau)_T$ and $(\partial\tau^*/\partial T)_{\dot{\gamma}}$ were obtained at a constant structure, (5:1) incremental changes in strain rate from the base strain rate of $3.3\times10^{-4}sec^{-1}$ were performed during the stress-strain determinations. Shear stresses and shear strains were calculated from their tensile counterparts by assuming that $\tau=\sigma/5$ and $\gamma=5\varepsilon$ (10), where σ and ε are the tensile stress and strain.

Results and Discussion

The macroscopic yield or flow stresses in titanium can be separated into two components; a thermal component τ^* which depends sensitively on temperature T, strain rate $\dot{\gamma}$ and impurity content C, and an athermal component τ_μ which is dependent on T only through the shear modulus C_{66} (15) and strongly on strain γ and grain size d; hence

$$\tau = \tau^*(T,\dot{\gamma},C) + \tau_\mu(\gamma,d,C) \qquad (2)$$

Table I. Chemical analysis, σ_0^* and A of Ti-N alloys

As-receive Ti (ppm in wt)	O	C	N	H	Fe
	63	78	6	4	50
Alloy	A	B	C	D	E
N (wt.%)	0.0103	0.0276	0.0511	0.104	0.180
σ_0^* (Kg/mm^2)	31.6	51.2	63.8	88.2	123.0
A (Kg/mm^2/°K$^{1/2}$)	1.264	2.04	2.56	3.52	4.92

The athermal component $\tau_{\mu T}$ at any temperature T can be calculated by using the conventional extrapolation method, from the equation

$$\tau_{\mu_T} = (\mu_T/\mu_{T_o})\tau_{\mu_{T_o}} \quad (3)$$

where μ_T and μ_{T_o} are the shear moduli at T and T_o and $\tau_{\mu T_o}$ the athermal component at T_o. Here, T_o =625°K was determined by finding the temperature where the flow stress varies with temperature only as the modulus, i.e. where $d\tau/dT=(\tau/\mu)(d\mu/dT)$. Then, the thermal component τ^* can be obtained by subtracting the athermal component from the applied stress. The validity of this extrapolation method to the low temperature deformation of unalloyed titanium has been discussed (9), and hence it is used for analysis of the present experimental data.

The effect of temperature on the thermal component σ^* of the yield and flow stresses of a series of Ti-N alloys is given in Fig.1. The data points presented here represent the average for four grain sizes and all strains, since σ^* was found to be essentially independent of grain size and strain. Thus, the work hardening was reflected almost entirely in σ_μ. The value of T_o'(extrapolated to $\tau^*=0$) is equal to 625°K for all Ti-N alloys within the experimental errors. It is seen in Fig. 1 that σ^* decreases in a linear manner with the square root of temperature, giving

$$\sigma^* = \sigma_o^* - A\sqrt{T} \quad (4)$$

with respective values of σ_o^* and A for each N concentration in Table I.

The dependence of σ_o^* on the nitrogen concentration is depicted in Fig. 2 in which an approximately linear relation is observed between σ_o^* and the square root of the atomic fraction of interstitial nitrogen atoms. The slope of the best fitting line through the experimental points is $\simeq 0.34\mu_o$ where μ_o is the shear modulus C_{66} at absolute zero, giving

$$\sigma_o^* = 0.34\mu_o\sqrt{C}. \quad (5)$$

A similar relation was reported by Conrad (4,5) for materials of various interstitial contents. In terms of Fleischer's nomenclature (16), the Ti-N alloys represent the case of rapid hardening. The value of the slope for Ti-N alloys is larger than those reported for unalloyed titanium (0.26-$0.30\mu_o$) in which the total interstitial concentration was expressed in terms of an oxygen equivalent. This is understood from the fact that nitrogen is a more effective strengthener than oxygen. However, for a strict comparison of the effectiveness it is necessary to obtain the experimental data for Ti-O alloys with the same base composition.

A square root dependence on concentration also existed for the athermal component σ_μ. The slope of the relation between the N content and the athermal component of the yield stress at 625°K varies from $0.017\mu_{T_o}$ to $0.021\mu_{T_o}$ with the grain size. This small value implies a case of weak hardening.

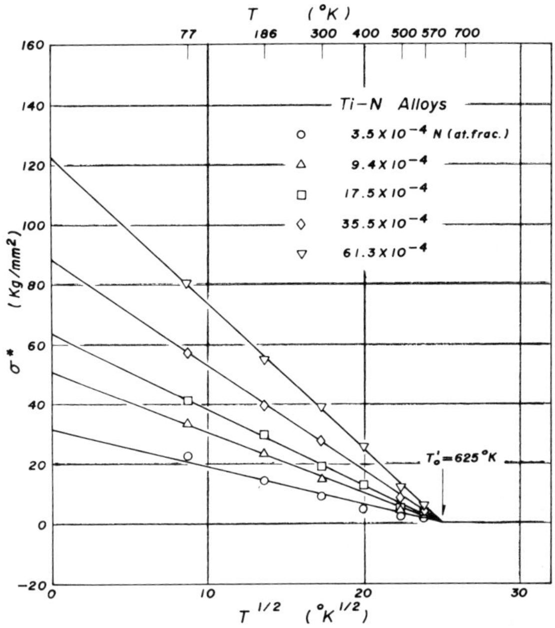

Fig. 1. The dependence of the thermal component of the flow stress on temperature and nitrogen content for Ti-N alloys.

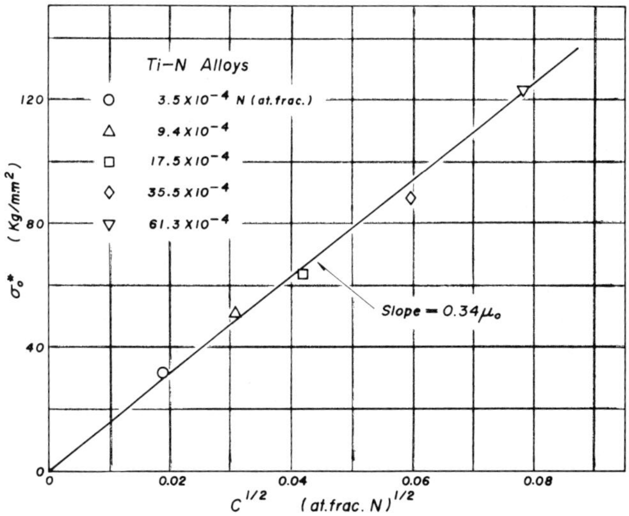

Fig. 2. The relation between nitrogen content and the extrapolated value of the thermal component of the flow stress at absolute zero for Ti-N alloys.

Thermal Activation Analysis

If a single thermally activated dislocation mechanism is rate controlling over a given temperature and strain rate range, then the shear strain rate, $\dot{\gamma}$ can be written by

$$\dot{\gamma}= \rho b\bar{v}=(\rho/\ell^*)Ab\nu^*\exp(-H(\tau^*)/kT) \quad (6)$$

where ρ is the mobile dislocation density, b the Burgers vector, $\bar{v}$ the average dislocation velocity, ℓ^* the dislocation segment length involved in the thermal activation process, A the area of the slip plane swept out per successful thermal fluctuation, ν^* the frequency of vibration of the segment ℓ^* and H the activation energy which is a decreasing function of τ^*. Relations following directly from Eq. (6) and used in the present analysis are as follows:

$$(\partial\tau^*/\partial T) = (1/T)(\partial\tau/\partial\ln\dot{\gamma})_T \ln(\dot{\gamma}/\nu), \quad (7)$$

where $\nu= (\rho/\ell^*)Ab\nu^*$,

$$-(\partial H/\partial\tau) = V^* = kT(\partial\ln\dot{\gamma}/\partial\tau)_T, \quad (8)$$

where V^* is the activation volume,

$$H = -kT^2 (\partial\ln\dot{\gamma}/\partial\tau)_T(\partial\tau^*/\partial T)_{\dot{\gamma}}, \quad (9)$$

and

$$H = kT \ln(\nu/\dot{\gamma}). \quad (10)$$

The deformation partials $\beta=(\partial\ln\dot{\varepsilon}/\partial\sigma)_T$ and $(\partial\sigma^*/\partial T)_{\dot{\varepsilon}}$ were evaluated from the strain rate cycling tests and the temperature dependence of σ^* over the temperature range of 77 to 625°K. The values of these deformation partials are given in Table II, the values being averaged over four grain sizes. If Eq.(6) applies to the present data, these deformation partials should be related according to Eq. (7), i.e., $(\partial\sigma^*/\partial T)_{\dot{\varepsilon}}$ should be proportional to $(1/T)(\partial\ln\dot{\varepsilon}/\partial\sigma)_T$. This is shown in Fig. 3, in which a reasonably straight line could be drawn through the data points for a series of Ti-N alloys. Using the values of these deformation partials, the activation energy was calculated as a function of temperature and nitrogen content (the values of H are given also in Table II). The values of H are plotted in Fig. 4 against temperature. It is seen in Fig. 4 that the data points are ruled to a good first approximation by a linear relation of temperature, independent of the N content, yielding H=1.5eV at T_o'. These results show that the experimental data obey Eqs.(7) and (10) and indicate that the deformation dynamics of Ti-N alloys at low temperatures is, in fact, described by Eq.(6) and that the assumptions implicit in this equation are correct. The slopes of Figs. 3 and 4 yield values of 27.7 and 27.6 for $\ln(\nu/\dot{\varepsilon})$. This represents reasonably good agreement, and gives additional support that Eq.(6) applies to the present data.

Values of the activation volume V^* derived by using Eq.(8) are listed in Table II. V^* in general increases with increasing temperature and decreases with increasing the N content. All the curves of the stress dependence of V^* rise to high values at τ^* equals zero, and above about τ^*=5 Kg/mm , the data can be represented by straight lines, depending on the N content, passing through V^* equals zero at τ^*=0. The linear portions have been extrapolated to τ^*=0 to obtain values of $V^*_{T_o'}$, which are given also in Table II. An approximately linear relation was found between the values so obtained and the inverse square root of the N content. The linear dependence of V^*

Table II. Activation parameters for plastic flow of Ti-N alloys

Temp. (^{o}K)	Parameter*	A	B	C	D	E
77	σ^*	22.5	33.5	41.0	57.2	80.5
	$1/\beta$	0.109	0.138	0.151	0.173	0.232
	$-(\partial\sigma^*/\partial T)_{\dot\varepsilon}\,x10^2$	8.92	12.7	14.5	20.2	28.0
	V^*/b^3	48.4	38.0	34.9	30.5	22.7
	H	0.105	0.117	0.123	0.150	0.155
	F $x10^6$	68.7	64.3	59.7	58.9	61.8
	d^*/b	0.407	0.508	0.615	0.760	0.759
186	σ^*	14.7	23.5	29.5	39.5	55.0
	$1/\beta$	0.148	0.168	0.185	0.210	0.277
	$-(\partial\sigma^*/\partial T)_{\dot\varepsilon}\,x10^2$	6.24	8.02	9.36	12.9	18.0
	V^*/b^3	86.1	75.7	68.7	60.6	45.9
	H	0.315	0.356	0.377	0.460	0.486
	F $x10^6$	49.9	48.8	46.3	44.2	46.2
	d^*/b	0.651	0.936	1.12	1.39	1.40
300	σ^*	8.9	14.8	19.0	27.6	39.0
	$1/\beta$	0.154	0.208	0.241	0.291	0.342
	$-(\partial\sigma^*/\partial T)_{\dot\varepsilon}\,x10^2$	5.20	7.04	7.36	10.2	14.2
	V^*/b^3	133	98.4	85.1	70.5	59.9
	H	0.656	0.656	0.533	0.680	0.806
	F $x10^6$	34.1	34.6	33.1	33.7	35.3
	d^*/b	0.891	1.08	1.26	1.48	1.70
400	σ^*	4.4		12.7		25.5
	$1/\beta$	0.156		0.306		0.575
	$-(\partial\sigma^*/\partial T)_{\dot\varepsilon}\,x10^2$	4.12		6.38		12.3
	V^*/b^3	176		89.3		47.7
	H	0.914		0.719		0.740
	F $x10^6$	20.5		24.3		25.5
	d^*/b	0.971		1.20		1.22
500	σ^*	2.0	3.9	5.0	8.2	12.0
	$1/\beta$	0.095	0.163	0.233	0.354	0.500
	$-(\partial\sigma^*/\partial T)_{\dot\varepsilon}\,x10^2$	2.30	3.82	5.70	7.88	11.0
	V^*/b^3	360	209	147	96.4	68.0
	H	1.31	1.26	1.32	1.20	1.18
	F $x10^6$	11.6	13.1	12.5	13.8	14.8
	d^*/b	1.59	1.60	1.51	1.47	1.41
570	σ^*	1.0	2.0	2.7	3.7	5.7
	$1/\beta$	0.053	0.109	0.140	0.204	0.288
	$-(\partial\sigma^*/\partial T)_{\dot\varepsilon}\,x10^2$	1.24	3.48	5.34	7.40	10.3
	V^*/b^3	737	358	278	191	135
	H	1.64	2.24	2.67	2.55	2.51
	F $x10^6$	7.08	8.13	8.02	7.82	8.74
	d^*/b	2.67	2.26	2.40	2.32	2.26
625	$V^*_{T_0'}/b^3$	225	150	135	105	80

* σ^*;(Kg/mm^2),$1/\beta$;(Kg/mm^2),$(\partial\sigma^*/\partial T)_{\varepsilon}$;(Kg/mm^2/^{o}K),H;(eV),F;(Dynes)

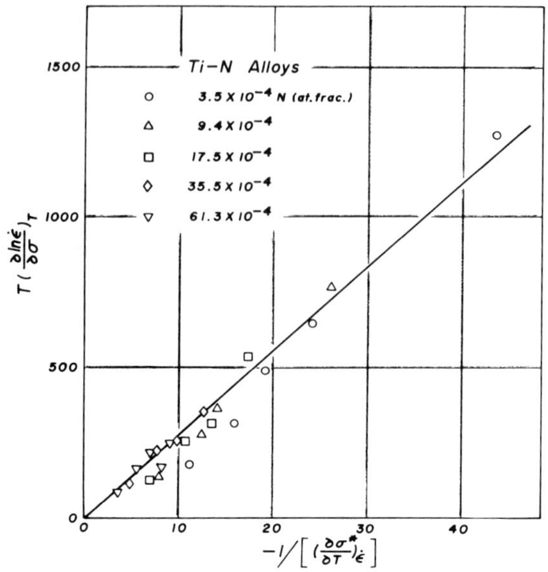

Fig. 3. $T(\partial \ln\dot{\varepsilon}/\partial\sigma)$ versus $(-\partial T/\partial\sigma^*)$ for plastic flow of alpha Ti-N alloys of various purities at low temperatures.

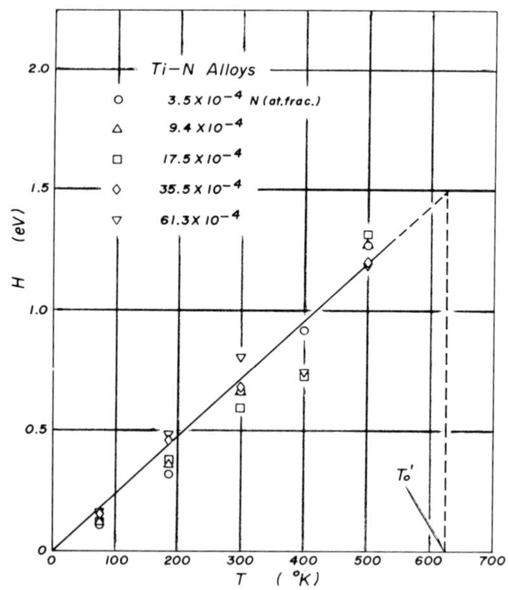

Fig. 4. Effect of temperature on the activation energy for plastic flow of alpha Ti-N alloys of various purities.

on τ^* can be written as

$$(V^*/V^*_{T_0'}) = 1 - (\tau^*/\tau_0^*) \qquad (11)$$

which is independent of the N content and Fig. 5 shows that the experimental data plotted in this way exhibit a single curve independent of the N content which satisfies Eq.(11) over most of the stress range. If there is a single invariant rate controlling process H itself should be independent of the N content, as is indicated by Fig.4. At T_0' we have H equals to $H_{T_0'}$, V^* equals to $V^*_{T_0'}$ and τ^* equals to zero. Consequently

$$H_{T_0'} = (1/2)\tau_0^* V^*_{T_0'}. \qquad (12)$$

The calculated values of $H_{T_0'}$ are 1.13eV, 1.29eV, 1.36eV, 1.46eV and 1.55eV for 3.5×10^{-4}, 9.4×10^{-4}, 17.5×10^{-4}, 35.5×10^{-4} and 61.3×10^{-4} N atomic fractions, respectively. These values compare reasonably well with the value of 1.5eV obtained directly from Fig.4, the main sources of error for the higher purity alloys being the few points available for the extrapolation to $V^*_{T_0'}$.

Assuming that overcoming interstitial solute atoms is the rate controlling mechanism for the low temperature deformation of the Ti-N alloys, a force activation distance curve can be constructed. If there is no interaction between the nitrogen atoms, their average spacing λ on the slip plane is

$$\lambda \simeq b/\sqrt{C}. \qquad (13)$$

Further, assuming a random distribution of the N atoms and taking into account the dislocation line tension the average distance ℓ^* between the obstacles along the dislocation is (17-19)

$$\ell^* = \left(\frac{2\mu}{\tau^*}\cdot\frac{b}{\lambda}\right)^{1/3}\lambda = b\left(\frac{2\mu}{\tau^* C}\right)^{1/3}. \qquad (14)$$

The force F opposing the dislocation motion at each obstacle is then

$$F = \tau^* b \ell^* = (\tau^*)^{2/3}\left(\frac{2\mu}{C}\right)^{1/3} b^2. \qquad (15)$$

Moreover, the activation volume is (20,18)

$$V^* = d^*(dF/d\tau^*) = \frac{2}{3} b^2 d^* \left(\frac{2\mu}{\tau^* C}\right)^{1/3} = \frac{2}{3} b d^* \ell^* \qquad (16)$$

where d^* is the activation distance. It follows that

$$d^*/b = \frac{3}{2}\left(\frac{V^*}{b^3}\right)\left(\frac{\tau^* C}{2\mu}\right)^{1/3} \qquad (17)$$

The values of F and d^*/b derived using Eqs. (15) and (17) are given in Table II. Depicted in Fig. 6 is the force activation distance plot constructed using the experimental data for the present series of Ti-N alloys. Also presented is the average curve based on the results by Tanaka and Conrad (10) for various alpha titanium materials with a total interstitial content ranging from 0.05 to 1.0 at.%O_{eq}. To be noted is that although the shape of the two curves is similar, the one for the Ti-N alloys is on the average slightly above that for the various Ti-O_{eq} materials, in accord with the slightly higher value of the activation energy $H_{T_0'}$ for Ti-N alloys(1.5eV versus 1.2eV) and the fact that the strengthening effect by nitrogen is higher than that for oxygen. This is all consistent with the larger size of the N atom compared to that of the O atom. Also to be noted is that the rapid rise in the force-acti-

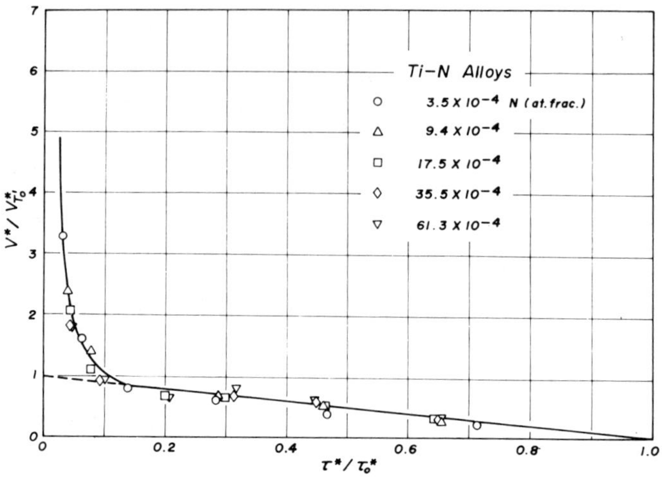

Fig. 5. The normalized stress dependence of the activation volume for plastic flow of alpha Ti-N alloys at low temperatures.

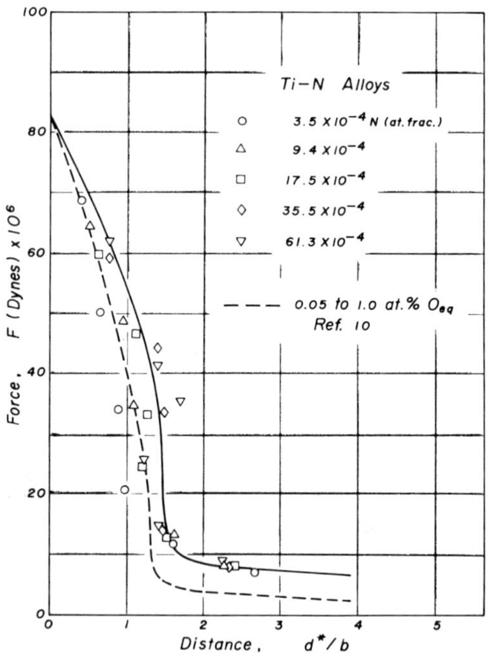

Fig. 6. Force-distance curve for the interaction between a dislocation and a barrier for alpha Ti-N alloys of various purities.

vation distance curve first occurs at $d^* \simeq b$, in accord with Tyson's (21) suggestion that the interaction between dislocations and interstitial solutes is a short-range interference by solute atoms with atomic motions at the core of gliding dislocations.

Finally, of significance is that the relative effects of O and N on the strength and thermal activation parameters for Ti is very similar to the effects of these interstitials in Zr (21).

References

1. Jaffee, R. I., Progress Metal Physics, vol.7, 1958, p.65
2. Jaffee, R. I., Ogden, H. R. and Maykuth, D. J., Transactions AIME, vol.188, 1950, p.1261
3. Orava, R., Stone, G. and Conrad, H., Transaction of ASM , vol.58, 1966, p.171
4. Conrad, H., Acta Metallurgica, vol.14, 1966, p.1631
5. Conrad, H., Canadian Journal of Physics, vol.45, 1967, p.581
6. Jones, R.L. and Conrad, H., Acta Metallurgica, vol.15, 1967, p.649
7. Jones, R. L. and Conrad, H., Transaction of TMS-AIME, vol.245, 1969, p.779
8. Conrad, H. and Jones, R. L., The Science, Technology and Application of Titanium, Pergamon Press, 1970, p.489
9. Okazaki, K. and Conrad, H., to be published Transaction of Japan Institute of Metals, vol.13, 1972
10. Tanaka, T. and Conrad, H., submitted to Acta Metallurgica
11. Fleischer, R. L., Acta Metallurgica, vol.10, 1962, p.835
12. Kratochvil, J. and Conrad, H., Scripta Metallurgica, vol.4, 1970, p.815
13. Gupta, G. and Weinig, S., Acta Metallurgica, vol.10, 1962, p.292
14. Okazaki, K., Momochi, M. and Conrad, H., Grain Growth Kinetics in Ti-N alloys, presented 2nd International Conference on Ti, May 3, 1972
15. Fisher, E. S. and Renken, C. J., Phsycal Review, vol.135, 1964, p.482

16. Fleischer, R. L. and Hibard, W. R., The Relation between the Structure and Mechanical Properties of Metals, HMSO, London, 1963, p.261

17. Friedel, J., Dislocations, Addison-Wesley, Reading, Mass., 1964, p.224

18. Kocks, U. F., Canadian Journal of Physics, vol.45, 1967, p.737

19. Foreman, A. J. E. and Makin, M. J., Canadian Journal of Physics, vol. 45, 1967, p.511

20. Conrad, H., AIME Journal of Metals, vol.16, 1964 July, p.582

21. Tyson, W. R., Canadian Metallurgical Quartery, vol.6, 1968, p.301

THERMALLY ACTIVATED DEFORMATION IN Ti 1% Si AND Ti 5%Zr 1%Si

H.M. Flower, P.R. Swann and D.R.F. West
Department of Metallurgy
Imperial College
London, S.W.7

Synopsis

An investigation has been made of the deformation characteristics of Ti-1wt%Si and Ti-5wt%Zr-1Wt%Si alloys in the range 77°-973°K; martensite structures and aged structures containing silicide particles were studied. Silicon and zirconium were found to produce substantial strengthening of martensite structures up to 850°K and the martensitic ternary alloy can be strengthened significantly by silicide precipitation. The binary alloy shows strain ageing at 750°K to an extent depending on the concentration of silicon in solution. From activation analysis the rate-controlling mechanism of thermally activated deformation in both alloys up to 400°K was found to be the overcoming of interstitial barriers.

Introduction

Deformation mechanisms in α-titanium have been studied by various workers, e.g. (1,2) and it has been reported that below ~0.4 Tm the step which controls the rate of deformation is the thermal activation of dislocations moving on $\{10\bar{1}0\}$ planes over interstital atom barriers. There appears however to have been little study of the operative mechanism of thermally activated deformation of martensitic α'titanium alloys.

The present results concerning Ti-Si and Ti-Zr-Si alloys form part of a wider study of alloys in the Ti-Al-Zr-Mo-Si system: in alloys of this complex system small amounts of silicon (typically 0.5% or less) contribute significantly to the mechanical properties at elevated temperatures. In previously reported work (3) it was shown that precipitation of $(TiZr)_5Si_3$ occurs on ageing martensitic titanium-silicon and titanium-zirconium-silicon alloys containing 1% silicon at $823^oK(550^oC)$. In the binary alloy the age hardening effect is very small because of the unfavourable precipitate dispersion and the loss of solid solution strength by depletion of the matrix in silicon. In the ternary Ti5%Zr1%Si alloy ageing at 823^oK gives a greater hardening, associated with the matrix nucleation of G.P. zones and their transformation to the equilibrium precipitate. The work reported here is aimed to correlate structure with strength over a wide temperature range and to determine the rate-controlling deformation mechanisms by using activation analysis.

Experimental methods

The alloys were prepared by arc melting using high purity sponge titanium and zirconium and semiconductor grade silicon. The oxygen and nitrogen content of the sponge titanium is typically $<\sim500$ ppm. The specimens, in the form of rod approximately 1 cm in diameter, were wrapped in molybdenum foil, sealed under argon in silica capsules and given a solution/homogenisation treatment of 24 hours at $1473^oK(1200^oC)$. The specimens were then water quenched to produce α'martensite, the capsules being broken on contact with the water. Compression specimens 2.5 mm square and 6 mm in length were cut from the rods using an oil cooled carborundum slitting wheel. The surfaces were lightly polished and some specimens were aged to peak hardness at 823^oK in argon filled pyrex capsules (3000 minutes for Ti1%Si, 4000 minutes for Ti5%Zr1%Si).

Compression testing was carried out over a range of temperatures between 77^oK and $973^oK(-196^oC$ and $700^oC)$ using an Instron testing machine with specimen rigs which are described elsewhere (4). Specimens tested above 800^oK were protected from oxidation by passing a slow, continuous stream of high purity argon through the specimen chamber. The initial crosshead speed was $\sim8x10^{-7}$ m/sec giving a strain rate of approximately 10^{-4} sec^{-1}. In the immediate post yield region the strain rate sensitivity was measured by increasing the strain rate by a factor of ten. At room temperature stress relaxation tests were also carried out. The activation volume and activation enthalpy were calculated employing the methods used by Conrad(1).

Results

The microstructures illustrated in Figure 1 are described in detail elsewhere (3,5). Essentially, the martensitic structures

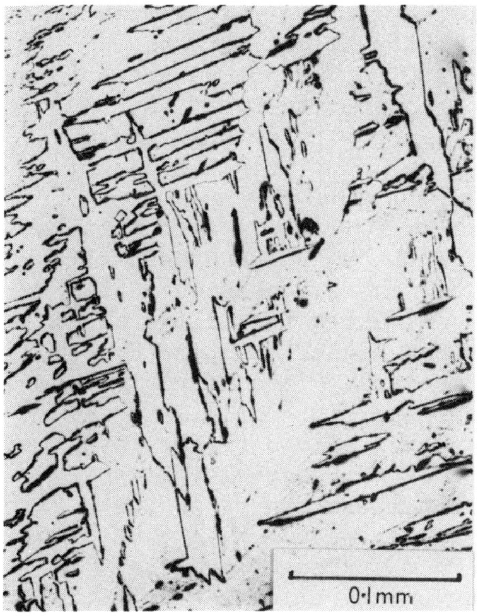

a) Light micrograph of Ti1%Si after water quenching from 1200°C.

b) Light micrograph of Ti5%Zr1%Si after water quenching from 1200°C.

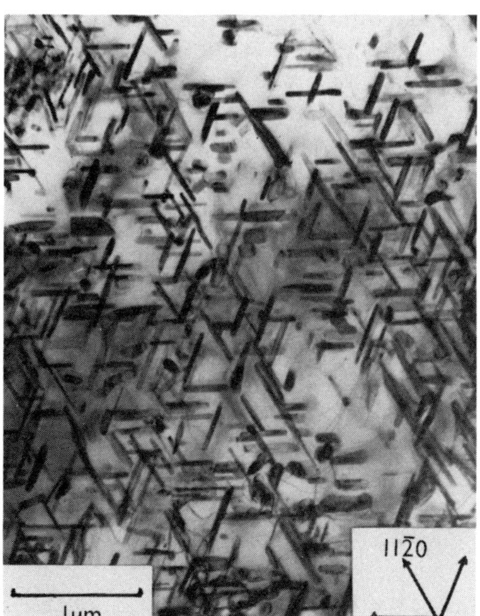

c) The precipitate dispersion in Ti1%Si aged to peak hardness at 550°C.

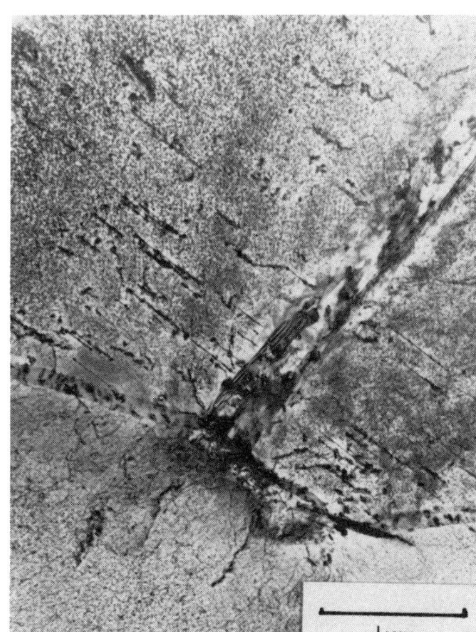

d) The precipitate dispersion in Ti5%Zr1%Si aged to peak hardness at 550°C.

Fig. 1.

consist of highly dislocated plates which exhibit a partly massive morphology in binary alloy; the aged binary alloy contains heterogeneously nucleated Ti_5Si_3 rods lying in the <11$\bar{2}$0> directions, whereas the aged ternary alloy contains a high density of matrix nucleated G.P. zones together with a small proportion of $(TiZr)_5Si_3$ precipitates.

The variation of the 0.1% proof stress (0.1%p.s.) with temperature is shown in figure 2(a) and 2(b) for the unaged and aged alloys respectively. Data obtained by Winstone (6) from quenched unalloyed titanium of similar purity as that used in making the alloys are also included. The specimens tested at the highest temperatures necessarily undergo some ageing during the test since approximately twenty minutes were required to achieve temperature stability and to carry out the test. In both of the unaged alloys (Figure 2(a)) the proof stress is a sensitive function of temperature below 500°K and above 850°K. Over almost the entire temperature range the 0.1%p.s. of the ternary alloy is ~150 MN m^{-2} greater than that of the binary alloy which itself is ~ 200 MN m^{-2} greater than that of the unalloyed martensitic titanium. The binary alloy has essentially the same form of 0.1%p.s./temperature relationship in both the aged and unaged states (cf figure 2(a) and 2(b). However the aged ternary alloy exhibits a significant increase in 0.1%p.s.(20%) over the as quenched material in the temperature range 150-700°K. Above 700°K the strength of the aged alloy drops rapidly and approaches that of the as quenched material at 850°K. Since testing at these temperatures inevitably involves some additional ageing, two specimens, aged to peak hardness, were further aged for 30 minutes at 870°K and were then tested at room temperature. The results are included in figure 2(b). The additional ageing produces a slight increase in 0.1%p.s. thus indicating that the loss in strength at 870°K is not due to overageing.

The stress-strain curves from unaged Ti 1%Si tested at 750°K exhibited a series of small (2-4 MN m^{-2}) intermittent serrations in the post yield region. Much larger serrations (10 MN m^{-2}) could be produced upon restraining after stopping the crosshead for some minutes during the test: these serrations died away to the smaller and intermittent serrations previously observed. Aged specimens of the binary alloy produced small serrations only after stopping the crosshead for prolonged periods (10 minutes). The ternary alloy did not exhibit serrated flow at any temperature.

Values of the activation for deformation (V*), calculated from the strain rate sensitivity, are presented in Figure 3; V* is given in dimensionless units obtained by dividing by b^3 where b is the Burgers vector of the 1/3<11$\bar{2}$0> dislocations. At 950°K and below 600°K identical results were obtained (within the limits of experimental accuracy) for both alloys, aged and unaged. In the intervening temperature range the activation volume of all except the aged ternary alloy rose to extremely high values indicating that

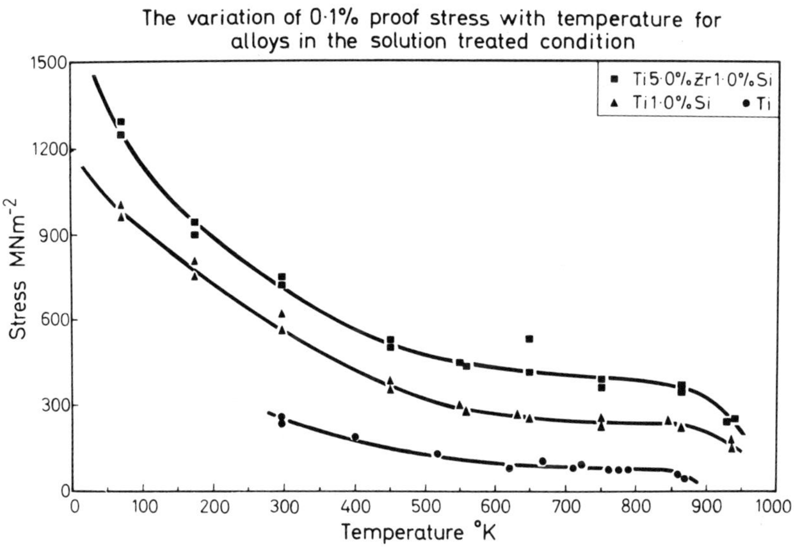

Fig. 2a. The variation of 0.1% proof stress with temperature for alloys in the solution treated condition.

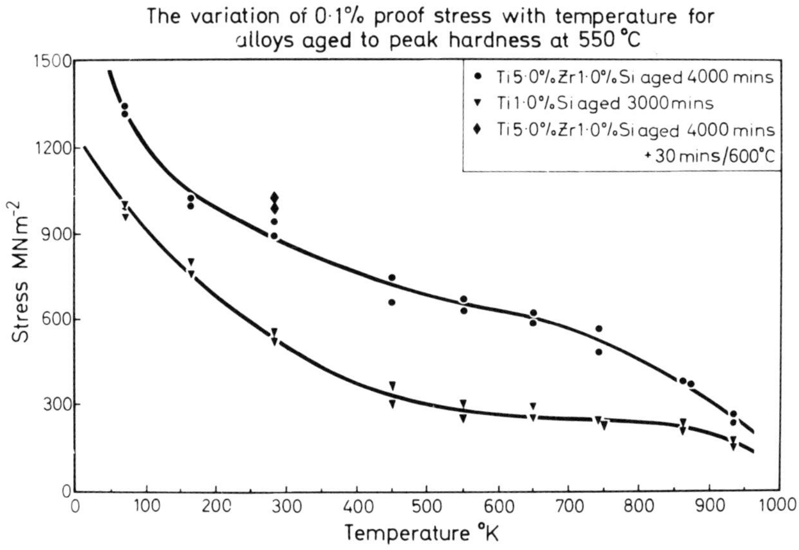

Fig. 2b. The variation of 0.1% proof stress with temperature for alloys aged to peak hardness at 550°C.

the deformation is not thermally activated. As a check, the activation volume was also calculated at room temperature from stress relaxation tests using the method described by Evans and Rawlings (7). These results, together with those obtained by the change of strain tests are presented in Table I and show excellent agreement.

Table I

Values of Activation Volume, V^*/b^3 at Room Temperature

	Stress relaxation	Strain rate sensitivity
Ti 1% Si		
Unaged Specimen A	21	27
" " B	19	21
Aged Specimen A	21	21
" " B	21	23
Ti5%Zr1%Si		
Unaged Specimen A	23	25
" " B	23	24
Aged Specimen A	26	24
" " B	25	24

The thermal component of the 0.1%p.s. (σ^*) was determined for both the aged and unaged alloys by extrapolation of the linear part of the stress/temperature curve (corresponding to the "athermal" deformation range) and subtracting the "athermal" component so obtained from the 0.1%p.s. In the temperature range 77 to 350-400°K the thermal component of the flow stress was found to be linearly related to the square root of the absolute temperature. The linear portion of the $\sigma^*/T^{\frac{1}{2}}$ graph extrapolates to $\sigma^* = 0$ at 425-450°K: this is approximately 100°K less than the temperature at which σ^* extrapolates to zero in Figure 2. However the latter value is of limited accuracy only as a result of the limited number of 0.1%p.s. measurements made in the temperature range over which σ^* approaches zero.

The graph of V^* as a function of $\tau^*(\frac{1}{2}\sigma^*)$(figure 4) shows the same relationship for both alloys and heat treatments. Extrapolation of the linear portion of the plot at high stresses back to $\tau^* = 0$ gives a value of 30 for V^*/b^3 which agrees closely with values reported for titanium containing only interstitial solutes (1,2,8).

Values of activation enthalpy and total activation enthalpy are presented in figure 5. The scatter above 400°K is due to the

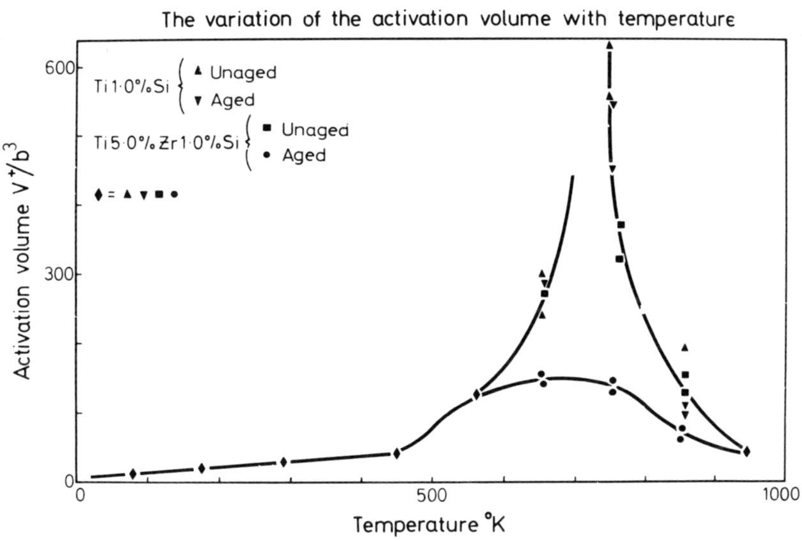

Fig. 3. The variation of the activation volume with temperature.

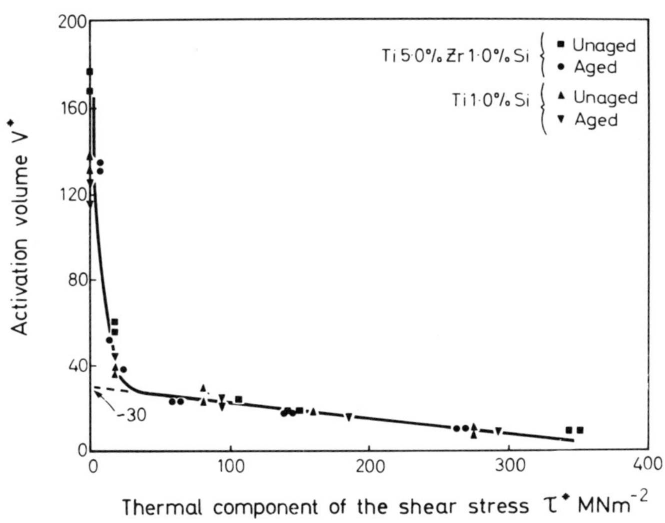

Fig. 4. The variation of activation volume with thermal component of the shear stress.

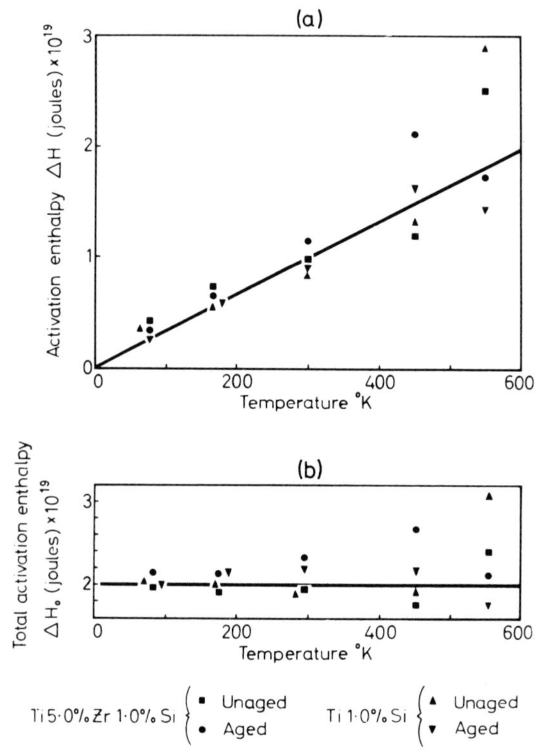

Fig. 5. a) The variation of activation enthalpy with temperature; b) The variation of the total activation enthalpy with temperature.

difficulty of evaluating accurately the gradient of the τ^*/temperature curve above this temperature where σ^* is very small. The total activation enthalpy is similar for both alloys and heat treatments and is equal to 2.10^{-19} joules atom^{-1} or 1.25 eV, which is identical to that reported previously for titanium containing interstitial solutes (2).

Discussion

The serrated yielding observed in the binary alloy is a manifestation of strain ageing. Since the aged binary alloy, whose matrix is solute depleted by precipitation, exhibited a much smaller tendency to serrated yielding than did the unaged alloy it appears that the amplitude of the strain ageing is a function of the concentration of silicon in solution at 750°K. Recent work by Winstone et al (9) on samples with oxygen and nitrogen levels in the range ~1000-1500 ppm has shown that the oxygen content also affects the strain ageing, so that some silicon-oxygen interaction may be involved. The absence of strain ageing in the ternary alloy may indicate that the silicon atoms are less mobile, possibly being trapped in zirconium silicon clusters: such clusters may act as the nuclei of the G.P. zones observed to form in this alloy, although there is insufficient evidence to substantiate this suggestion.

The insensitivity of the 0.1%p.s. to changes in strain rate (with consequent large values of V*) in the temperature range 550°K to 800°K indicates that the deformation is athermal at these temperatures; a possible exception is the aged ternary alloy for which the activation volume rises only to 150 b^3: if in this case the deformation is still thermally activated above 550°K a possible deformation mechanism would be the interaction of dislocations with the G.P. zones present in the aged alloy. In all cases the deformation is thermally activated above 800°K.

The observation that the activation volume and enthalpy results are similar for both alloys and heat treatments indicates that the same thermally activated deformation mechanism is rate controlling in all cases in the temperature range 77-400°K. In the range 400-550°K the deformation is still thermally activated but the limited number of results in this temperature range do not permit the identification of the mechanism although the breakdown of the linear relationship between σ^* and $\tau^{\frac{1}{2}}$ implies that the mechanism is different from the operating below 400°K.

The present results, including the linear $(\sigma^*)^{\frac{1}{2}}$/temperature relationship, are almost identical to those obtained by Conrad and co-workers (1,2,8) who employed similar methods to study titanium containing interstitial elements only. This suggests that the same rate controlling mechanism for deformation is operative for the range of microstructures shown by the alloys studied in this work (with the possible exception of the aged ternary alloy which exhibited a relatively small activation volume in the "athermal" deformation temperature range): according to Conrad the mechanism is one of the overcoming of interstitial atoms by mobile dislocations on the $\{10\bar{1}0\}$ planes.

It should however be noted that the methods of activation analysis used in this work and the quoted references of Conrad and his co-workers assume (i) that the temperature region of negligible strain rate sensitivity is a region of athermal deformation and (ii) that the athermal component of the flow stress varies linearly with temperature as a result of the temperature dependence of the shear modulus. Both these assumptions have been questioned recently (10,11).

Conclusions

1. Silicon and zirconium both produce substantial strengthening in martensitic titanium up to 850°K. The binary Ti1%Si alloy has a negligible ageing response. However the Ti5%Zr1%Si alloy shows a strength increase of 20% on ageing at 823°K and a substantial contribution from the precipitation is maintained throughout the range of testing temperatures from 77°-850°K.

2. Strain ageing has been shown to occur in the binary alloy at 750°K the strength of the effect being a function of the concentration of silicon in solution.

3. The rate controlling mechanism of thermally activated deformation in the temperature range up to 400°K is the same for both alloys, aged and unaged, and is the same as that previously determined for titanium containing only interstitial solutes, namely, the overcoming of interstitial barriers.

Acknowledgements

The authors wish to thank Imperial Metal Industries (Kynoch) Ltd. for the provision of the titanium used in this work. Thanks are also due to Professor J.G. Ball for the provision of laboratory facilities and to Dr. R.D. Rawlings for useful discussions concerning activation analysis.

References

1. Conrad, H. "Thermally Activated Deformation of α Titanium Below 0.4Tm", Canad.J.Physics, vol. 45, pp. 581-590.

2. Conrad, H. and Jones, R., "Effects of Interstitial Content and Grain Size on the Mechanical Behaviour of α Titanium Below 0.4Tm", The Science, Technology and Application of Titanium, 1970, Pergamon Press, pp. 489-501.

3. Flower, H.M., Swann, P.R. and West, D.R.F., "Silicide Precipitation in Ti-Zr-Al-Si Alloys", Metallurgical Transactions, vol. 2, 1971, pp. 3289-3297.

4. Pascoe, R.T., Radford, K.C., Rawlings, R.D. and Newey, C.W.A., "Instrumental Modifications for Compressive Testing at Temperatures up to 1800°C", Journal of Scientific Instruments, vol. 44, 1967, pp. 366-370.

5. Flower, H.M., Swann, P.R. and West, D.R.F., "The Effect of Si, Zr, Al and Mo on the Structure and Strength of Ti Martensite", Journal of Materials Science, To be published.

6. Winstone, M.R. Private Communication.

7. Evans, A.G. and Rawlings, R.D., "The Thermally Activated Deformation of Crystalline Materials", Phys.Stat.Solidi, vol. 34, 1969, pp. 11-31.

8. Sargent, G. and Conrad, H., "Stress Relaxation and Thermally Activated Deformation in Titanium", Scripta Met., vol. 3, 1969, pp. 43-50.

9. Winstone, M.R., Rawlings, R.D. and West, D.R.F., Unpublished work.

10. Monteiro, S.N., Santhanan, A.T. and Reed-Hill, R.E., "A thermal Plastic Deformation in Commerical Purity Titanium", The Science, Technology and Application of Titanium, 1970, Pergamon Press, pp. 503-516.

11. Tung, P.R. and Sommer, A.W., "Dislocation Energetics in α Titanium", Metallurgical Transactions, vol. 1, 1970, pp. 947-953.

SUPERPLASTICITY IN LARGE GRAINED

BETA TITANIUM ALLOYS

P. Griffiths and C. Hammond

Department of Metallurgy
University of Leeds
England

Introduction

Superplasticity is a deformation phenomenon which occurs at temperatures $\gtrsim 0.5T_m$ (where T_m = melting point in K) and which has hitherto been associated with very fine grained ($1\sim7\mu m$) materials. It is characterised (a) by very low flow stresses ($\sim10^7 Nm^{-2}$) at low strain rates ($10^{-4}s^{-1}$),(b) a very low (or zero) value of the work hardening index, hence the flow stress is almost independent of strain and (c) a very large dependence of the flow stress on the strain rate. In the absence of work hardening the empirical equation connecting stress (σ) and strain rate ($\dot{\varepsilon}$) is written

$$\sigma = K\dot{\varepsilon}^m \quad \ldots\ldots\ldots\ldots\ldots\ldots (1)$$

where K = constant and m = the strain rate sensitivity index of the deformation process. It is noted that $\frac{1}{m}$ = n, the stress sensitivity index of steady state creep. When m is large ($\gtrsim0.5$) a neck which forms in a tensile specimen is stabilized against further localised deformation and it is this factor which gives rise to the very large elongations which can be obtained.

A large m value is thus a necessary but is not a sufficient condition for a large elongation. Only if some fracture process (starting at cracks, notches or other heterogeneities) does not intervene are large elongations actually recorded. Hence superplasticity is distinguished from "extended plasticity" in that in the latter process m is relatively small ($\sim1/3$) and work hardening is not negligible. The largest possible value for m is unity, in which case the material behaves as a Newtonian fluid.

The associated single or two phase microstructures (of grain size 1∿7μm) are produced either by recrystallizing a heavily cold worked material or by working a eutectic or eutectoid microstructure to produce an equiaxed duplex structure. An important characteristic of superplastic deformation is that virtually no change in microstructure is observed throughout the whole of the deformation process.

Two theories are prominent in describing superplastic deformation. Grain boundary sliding satisfactorily explains the overall unchanged microstructure, but it requires in addition some accommodation mechanism for sliding at ledges or triple points and this may occur by diffusional flow of atoms or some "grain boundary" dislocation mechanism. The other mechanism is that of diffusion creep, mass transport occurring either through the grains (Herring-Nabarro Creep)[1] or along the grain boundaries (Coble Creep)[2].

$$\dot{\varepsilon} = \frac{5\sigma D_L \Omega}{L^2 KT} \quad \text{(Herring-Nabarro Formula)} \quad .. \; .. \; (2)$$

where D_L = lattice diffusivity, Ω = atomic volume, L = grain size, K = Boltzmann's constant and T = melting temperature.

$$\dot{\varepsilon} = \frac{75\sigma D_b \Omega w}{L^3 KT} \quad \text{(Coble Formula)} \quad .. \; .. \; (3)$$

where D_b = grain boundary diffusivity and w = grain boundary width.

Both equations predict m = 1 but have different grain size dependencies. Diffusion creep processes predict an elongation of the grains and again an accommodation mechanism of grain boundary migration is required in order to keep the grains equiaxed.

Diffusion creep mechanisms have been regarded as less satisfactory explanations of superplasticity than grain boundary sliding models (a) because the predicted strain rates are usually much too low and (b) because grain boundary migration in two phase materials requires large mass transport.

Experimental Methods

The compositions, transformation temperatures etc., of the three alloys studied are given in Table I. Tensile tests were carried out on thin (1.5mm) strip specimens of 30mm gauge length and on a number of cylindrical specimens 6.4mm in diameter and of 22mm gauge length. An Instron machine was used to test specimens at various temperatures and cross head speeds. A furnace with a 100mm uniform hot zone surrounded the specimen and was also used for solution treatment prior to testing. A pre-heat at 850°C for 15 mins. was used to achieve a large β grain size (∿500μm for Ti-Mn, ∿300μm for Ti-Mo and Ti-Cr-Al-V) and to minimise

Table I.

Alloy	Composition, wt.%	T_m(K)	Transformation Details
Ti-Mn	Ti-8%Mn-0.14%O_2 -0.04%C-0.03%N_2	1870	β→α+β at ∿670°C (0.5T_m) on quenching from β field a fine metastable ω precipitate occurs
Ti-Mo	Ti-15.5%Mo-0.18% O_2-0.019%C-0.01%N_2	2020	β→α+β at ∿700°C (0.48T_m) ω formation more sluggish
Ti-Cr-Al-V	Ti-13%Cr-11% V-3%Al-0.18% O_2-0.05%C-0.02%N_2	1850	β→α+β at ∿650°C (0.5T_m) transforms spontaneously to f.c.c. phase on electro-thinning

differential grain growth. m-values were determined by the strain rate cycling method which was repeated at increasing levels of strain in order to detect any work hardening or variation of m with strain. Specimens for microscopy were obtained by rapid cooling from the deformation temperature (in order to avoid α formation) and prepared in the usual way(3).

Results

(a) Tensile Tests

All three alloys behaved similarly when tested in the β field and a typical stress-strain curve for the Ti-Mn alloy is shown in Fig. 1. The true stress at a constant extension speed drops continually until fracture and a correction must be made for the decreasing strain rate as the specimen extends. This is carried out using Eqn. (1) and a knowledge of m from a strain rate cycling experiment at the temperature considered. The corrected flow stress is absolutely constant until fracture is imminent, apart from a small "lump" at 3∿4% strain. The constancy of the corrected flow stress indicates the absence of work hardening. The Ti-Mo and Ti-Cr-Al-V alloys showed the same behaviour but generally flowed at slightly higher stresses.

The maximum elongations obtained with the thin strip tensile specimens were ∿150% (Fig. 2), fracture appearing to take place by the propagation of a surface crack or imperfection rather than as a result of localised necking. It has been shown (4),(5) that maximum elongations may depend on specimen size and this was confirmed in the case of the cylindrical specimens where maximum

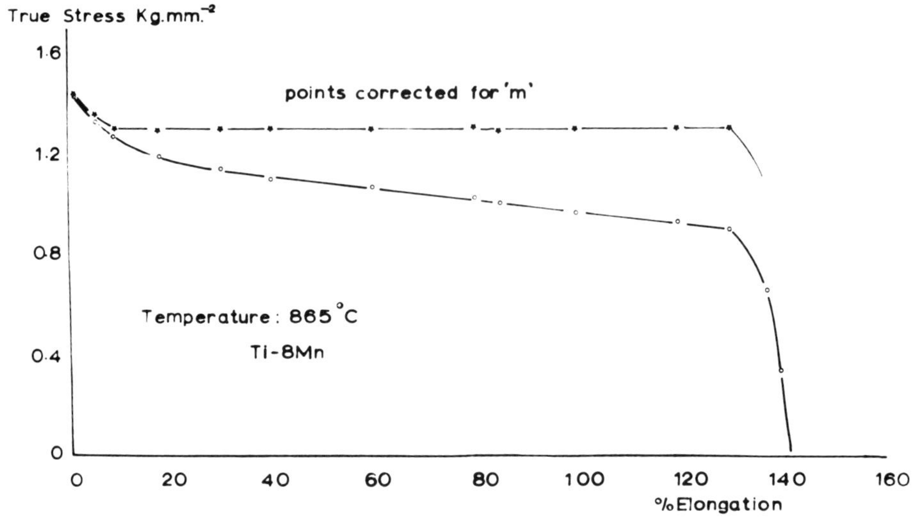

Fig. 1. True stress-elongation curve for Ti-Mn showing 'm' correction.

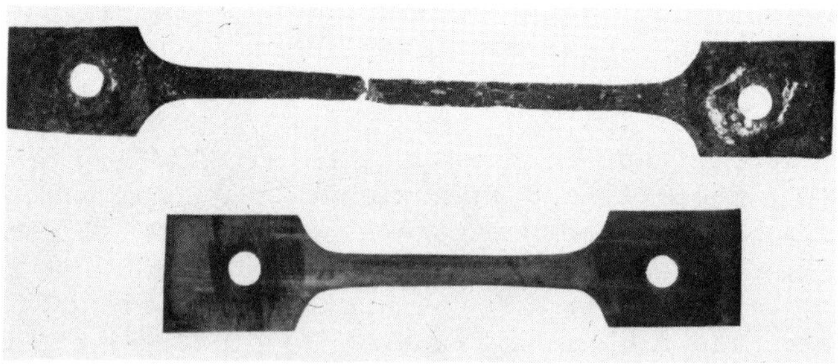

Fig. 2. Flate tensile specimens of Ti-Mn, before and after deformation at 800°C. Note uniform elongation up to ~100% strain.

elongations of ∿450% were only limited by the length of the uniform hot zone of the furnace. The values of flow stress and corresponding m-values were not affected by specimen dimensions.

The m-values for the deformation are given by the slope of a log σ vs. log $\dot{\varepsilon}$ plot. Fig. 3 shows the results for the Ti-Mn alloy at 760°C. It can be seen (a) that points plotted at increasing levels of strain all lie on a common curve, indicating again the absence of work hardening and the independence of m with strain and (b) that m is dependent on strain rate, increasing to a maximum value of unity at the lower strain rates. The variations in m and σ with temperature and strain rate for the alloys are given in Figs. 4 and 5. Whereas m approaches unity at the lower strain rates for all alloys tested in the β field, for alloys deformed in the $\alpha+\beta$ field m is always low (∿0.3) and independent of strain rate. The alloys do not therefore behave superplastically when α is precipitated in the large β grains.

The above results show that these large grained β alloys behave superplastically at the strain rates used. By increasing the annealing time beyond 15 mins. grain sizes up to 1mm were achieved and the results obtained were closely similar to those described above except that flow occurred at slightly higher stresses.

(b) Metallography

Prior to testing the β grains had the characteristic appearance of an annealed material. After superplastic deformation (above 3∿4% strain) the grains remain comparatively equiaxed; a small proportion are elongated but not in proportion to the total strain. (Figs. 6 and 7). The grain boundaries are now curved and the presence of a fine subgrain network can be seen. A proportion of the subgrain boundaries "cusp" the grain boundaries (Fig. 7), effectively pinning them and inhibiting grain growth. In addition the grain boundaries of elongated grains appear to migrate under the influence of the subgrain boundaries, the grain eventually splitting into two (Fig. 6). It is postulated that these two mechanisms balance the normal grain growth and that grain boundary migration results in an overall equiaxed grain structure. The subgrain boundaries were also revealed by electron microscopy and consisted mainly of dislocation walls of low misorientation (∿0.1°) (Fig. 8). However at higher strain rates, where m is decreasing, more complicated dislocation arrangements were seen (Fig. 9).

At a fixed temperature the subgrain size is a function of strain rate and hence stress. Fig. 10 shows typical results for the Ti-Mn alloy. At low stresses (where m $\approx$1) the subgrain size is independent of σ, but at the highest stresses (where m $\approx$0.3) the subgrain size tends towards an inverse relationship with the stress.

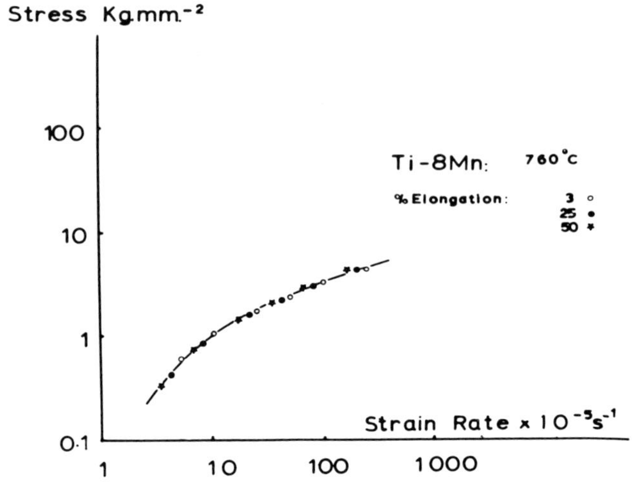

Fig. 3. Log σ vs log $\dot{\varepsilon}$ plot for Ti-Mn at 760°C.

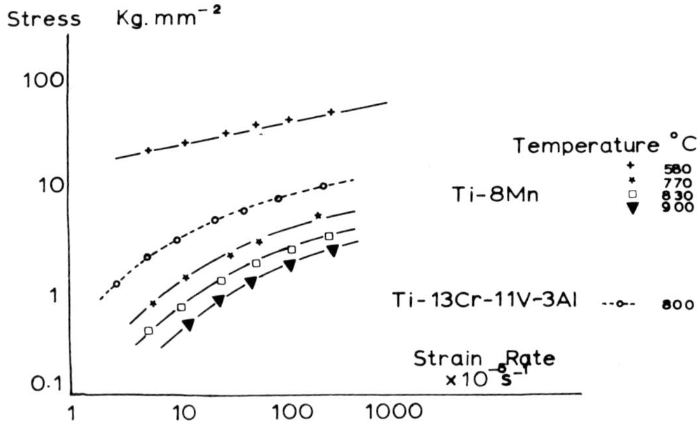

Fig. 4. Log σ vs log $\dot{\varepsilon}$ plots for Ti-Mn. A single plot for Ti-Cr-V-Al is also shown.

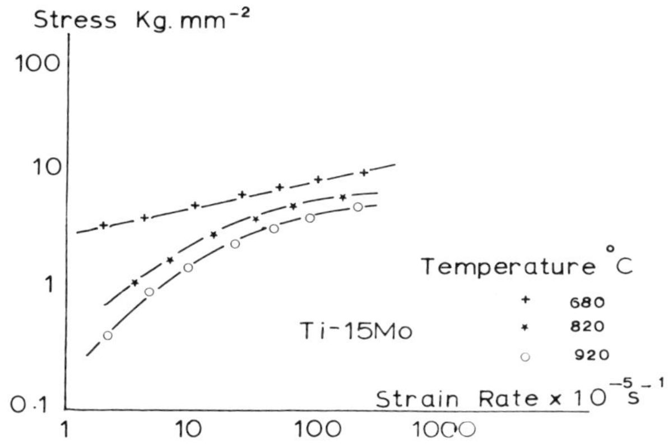

Fig. 5. Log σ vs log $\dot{\varepsilon}$ plots for Ti-Mo.

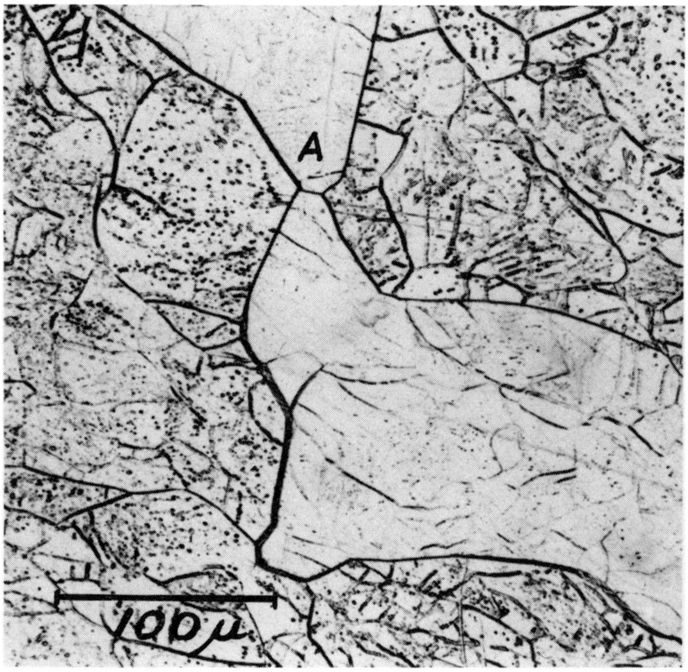

Fig. 6. Ti-Mn alloy after deformation at 865°C, showing subgrain network and division of grain (at A) into two. Optical micrograph.

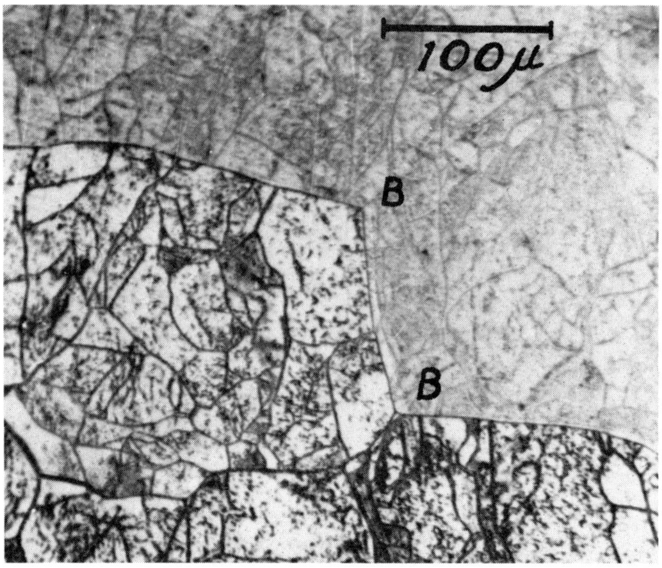

Fig. 7. Ti-Mn after deformation at 865°C, showing subgrain network and cusping (at B) of grain boundary. Optical micrograph.

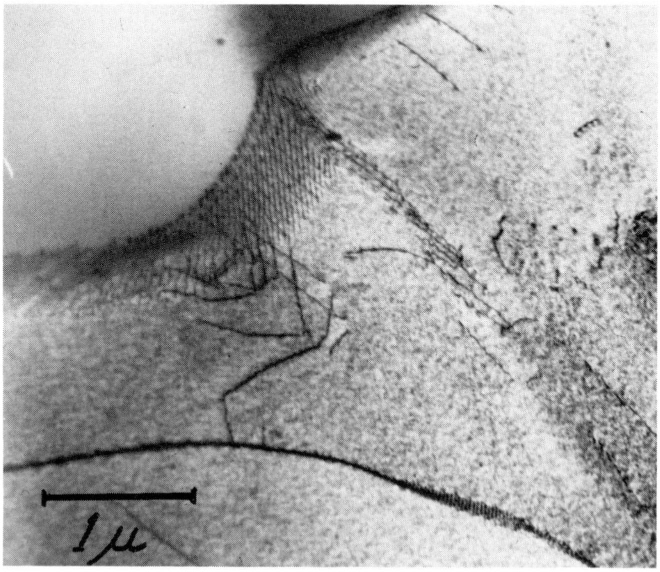

Fig. 8. Ti-Mo after deformation at 820°C, showing a migrating subgrain boundary. Electron micrograph.

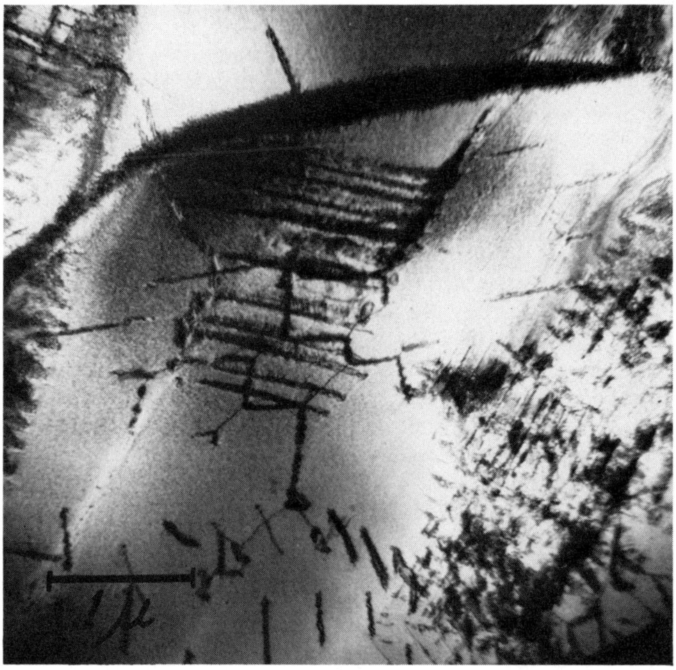

Fig. 9. Ti-Mn after deformation at 750°C, showing dislocation bands within a subgrain. Electron micrograph.

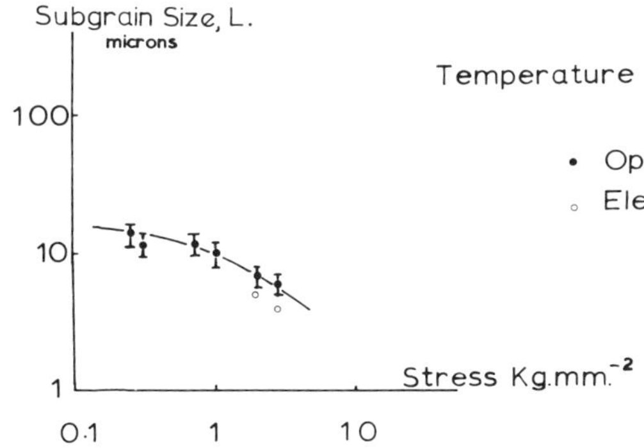

Fig. 10. Log L vs log σ for Ti-Mn at 850°C. L is nearly independent of the lowest stresses where m$\simeq$1.

Discussion

The results obtained conform to all the criteria characteristic of superplasticity except that the elongations do not compare favourably with the ∿1000% obtained on materials of similar m-values. However, it appears that elongations are dependent upon the number of grains in the specimen cross-section since an increase in specimen cross-section increased the elongations from 150% to 450%. A similar effect has been found by Backofen(4) on the Al-Zn eutectoid where elongations varied between 30% and 400%. A decrease in the grain size of superplastic alloys has an analogous effect. Nuttall(6) obtained >1000% elongations with the Al-Zn eutectoid of grain size 0.7μm (m = 0.55) but only 600% with a grain size of double this. Similarly, for the Ti-6%Al-4%V duplex alloy elongations of ∿240%(7) (17μm grain size) and ∿1500%(8) (∿3μm grain size) have been reported. These effects may arise from the presence of notches or surface defects which have an important effect on elongation, particularly at high m-values(5). In view of these and similar results the elongations of the present large grained specimens are not expected to be large.

Of the two possible deformation processes, grain boundary sliding would be expected to contribute little to the total deformation. Furthermore the diffusion creep equations predict strain rates orders of magnitude too small. However, Weertman has shown(9) that a diffusion creep process may operate between the subgrains, the condition being that subgrain boundaries act as perfect sources and sinks for vacancies. On this basis application of the diffusion creep formulae to the experimental results at lower strain rates where m∿1 gives good agreement between predicted and experimental strain rates. Typical calculations based on the Herring-Nabarro formula are presented in Table II.

Table II

	$\sigma(Nm^{-2})$	$D_L(m^2s^{-1})$	$\Omega(m^3)$	L(m)	T/T_m	T(K)	Predicted $\dot{\varepsilon}$	Observed $\dot{\varepsilon}$
*	0.5×10^{7}	1×10^{-13}	17×10^{-30}	1×10^{-5}	0.61	1140	2.7×10^{-5}	6×10^{-5}
**	0.6×10^{7}	5×10^{-14}	17×10^{-30}	8×10^{-6}	0.58	1180	2.4×10^{-5}	3×10^{-5}

* Alloy - Ti-Mn. ** Alloy - Ti-Mo.

L = subgrain size determined by the linear intercept method.

The diffusion creep process appears to be linked with a subgrain boundary migration process which maintains the subgrains equiaxed and keeps the diffusion path constant. Application of the Coble formula also gives good agreement but considerable numerical uncertainty is involved because of the difficulty in estimating w and D_b terms.

At higher strain rates m decreases and L is no longer independent of σ (Fig. 10). The deformation in these regions may either be explained by a dislocation creep model(9) or again by Herring-Nabarro creep by taking into account the dependence of L on σ (Eqn. 2) Eg. at the highest stress $L\alpha\sigma^{-1}$ (Fig. 10) and substitution into Eqn. 2 gives $\dot{\varepsilon}\alpha\sigma^3$, i.e. m = $^1/3$ - in good agreement with that found experimentally.

The Herring-Nabarro creep mechanism is normally only considered to be of importance at temperatures $\gtrsim 0.8T_m$. It accounts for the behaviour of the β Titanium alloys at $\gtrsim 0.5T_m$ only because of the anomolously high diffusivity of β-Ti. β-Ti, β-Zr and γ-U constitute the "anomolous" metals(10) with respect to their diffusion behaviour; the activation energy for self diffusion being both anomolously low and temperature dependent. This may arise from dislocation enhanced diffusion or from vacancy-impurity atom binding. However, the important point is that the anomolously high D values make diffusion creep an important deformation mechanism at substantially lower fractions of the absolute melting temperature. Table III shows that β-Ti alloys have equivalent diffusivities at $\sim 0.5T_m$ to Copper (a typical f.c.c. metal) at $\sim 0.8T_m$.

Table III

Metal	$D(m^2s^{-1})$	$Q(Kcals.mole^{-1})$		T/T_m
β Ti-Mn	$1x10^{-13}$	31	Temperature dependent	0.6
β Ti-Mo	$1x10^{-14}$	25		0.5
Copper	$1x10^{-14}$	51		0.8
Copper	$1x10^{-19}$	51		0.5
β Brass	$1x10^{-14}$	25		0.5

Also included is data for β brass which also has a well documented high diffusivity. On this basis β brass may also behave superplastically and this was found to be the case(3). At low strain rates ($\sim 10^{-4}s^{-1}$) m approached unity, work hardening was zero and a fine subgrain structure developed after a few percent strain.

Conclusions

The results described show that superplasticity is not necessarily confined to fine grained materials (excepting the special superplasticity arising in temperature cycling around a phase change) but may also occur in alloys having anomolously high diffusivities and which also develop a subgrain network on deformation above $0.5T_m$. This interpretation of the results on the basis of a diffusional creep process coupled with grain and subgrain boundary migration may not be directly applicable to conventional fine grained two phase alloys since it would require mass transport of material across a migrating two phase interface in order to maintain compositional differences. However, the importance of diffusivity in conventional superplasticity seems to have been neglected and recent work on steels has shown [5], [11] that alloy additions which raise diffusivity also raise strain rate sensitivities. In addition, the grain morphology of conventional materials resembles very closely that shown in Figs. 6 and 7, i.e. curved boundaries and the apparent division of grains into two.

The results are also of importance in connection with the β forming of titanium alloys. At controlled strain rates very low pressures would be required, dimensional accuracy would be achieved and the alloys would inherit the fine substructure on subsequent cooling and transformation at lower temperatures.

References

1. Herring, C., "Diffusional Viscosity of a Polycrystalline Solid" Journal of Applied Physics, Vol. 21, 1950, pp.437-445.

2. Coble, R.L., "A Model for Boundary Diffusion Controlled Creep in Polycrystalline Materials." ibid Vol. 34, 1963, pp.1679-1683.

3. Griffiths, P. and Hammond, C., "Superplasticity in Large Grained Materials." Acta.Metallurgica, (to be published).

4. Backofen, W.A., Turner, I.R., and Avery, D.H., "Superplasticity in an Al-Zn Alloy." Transactions Quarterly A.S.M. Vol. 57, 1964, pp.980-990.

5. Morrison, W.B.,"The Elongation of Superplastic Alloys". Transactions of the Metallurgical Society of A.I.M.E. Vol. 242, 1968. pp.2221-2227.

6. Nuttall, K., "Superplasticity in the Al-Zn Eutectoid Alloy." Ph.D. Thesis, Manchester University, 1969.

7. Grant, N.J., Ioup, W., and Kane, R.H. "The Hot Plasticity of Ti-6Al-4V." The Science, Technology and Application of Titanium. pp.607, Pergamon Press, 1970.

8. Lee, D., and Backofen, W.A., "Superplasticity in some Titanium and Zirconium Alloys." Transactions of the Metallurgical Society of A.I.M.E. Vol.239, 1967, pp.1034-1040.

9. Weertman, J., "Dislocation Climb Theory of Steady State Creep" Transactions Quarterly A.S.M. Vol.61, 1968, pp.681-694.

10. LeClaire, A.D., "Application of Diffusion Theory to Body Centred Cubic Metals." Diffusion in Body Centred Cubic Metals Gatlinburg Conference 1964, A.S.M. Publication 1965.

11. Schadler, H.W. "The Stress-Strain Rate Behaviour of Manganese Steel." Transactions of the Metallurgical Society of A.I.M.E. Vol. 242, 1968. pp.1281-1287.

TWINNING AND TEXTURE TRANSITIONS IN TITANIUM SOLID-SOLUTION ALLOYS

Frank R. Larson*, Anthone Zarkades* and Donald H. Avery†

The preferred orientation in cold-rolled sheet processed from laboratory heats of several titanium binary alloy series was studied. Titanium-based alloys of iron, maganese, chromium, cobalt, nickel, copper, molybdenum, vanadium, columbium, aluminum, tin, and zirconium were included in the investigation. It was found that additions of beta stabilizers caused a texture transition when the retained beta was in excess of about 16 volume percent. This texture transition, for the basal pole, may be roughly described as an unalloyed titanium to zinc-type transition. Copper was an exception and alloys of this element had a tendency to form an "ideal" (0001) [10$\bar{1}$0] type texture similar to that previously found for additions of the alpha stabilizer aluminum. Neutral alloying elements like tin and zirconium were found to have only a small effect upon the texture, merely sharpening the unalloyed type. Little effect of lattice parameters was noticed except for a trend of high c/a ratio alloys to produce a component of a stable end orientation of the (11$\bar{2}$0)[10$\bar{1}$0] type. A qualitative discussion of how these textures form is included; however, due to a lack of complete information about roll-gap stresses and deformation, along with absence of values for critical stresses for slip and twinning in these alloys, it was not possible to be quantitative.

*Army Materials and Mechanics Research Center
Watertown, Massachusetts

†Brown University
Providence, Rhode Island

Introduction

It is well known that polycrystalline materials are anisotropic (1). This anisotropy arises from the directional properties of single crystals and is dependent upon the degree of preferred orientation or texture of the polycrystalline aggregate. Many physical and mechanical properties exhibit this anisotropy and it forms the basis for improvement of processing techniques or materials for certain applications. One example of improved processing is that associated with increased drawability for sheet metals (2).

Probably the most successful practical use of anisotropy has been in the field of magnetic properties (3). Technologically, the control of preferred orientation, or texture control, in iron-silicon sheet is further advanced than in any other material. In iron single crystals the cube edge, or [100], direction is known as the easy direction of magnetization, and the [111], or cube diagonal, is the hardest direction of magnetization. Thus in a polycrystalline magnetic sheet material, control of the texture provides a means for improvement of magnetic properties. Recognition of this fact has provided the impetus for much research. Magnetic sheet materials and studies of texture have provided the necessary knowledge, and anisotropic magnetic sheet materials have been produced commercially for several years.

Another potential use of anisotropy is for the improvement of yield strength. Although this has received little commercial attention, it has been the subject of recent research. "Texture hardening"(4) has been demonstrated in the laboratory for plane stress loading of sheet materials. It has been shown that, particularly for tension-tension, the largest improvement in yield strength could probably occur in an hcp metal with an ideal texture (i.e., basal planes oriented parallel to the sheet surface). Obviously, beneficial effects can be achieved in other conditions. For example, a rod in which the c axis is parallel to the rod axis should be 50 to 80% stronger than one with a $[11\bar{2}0]$ fiber texture. Other metals should also show significant increases in strength. For example, a [111] fiber-textured wire should be 20% stronger than a randomly oriented material and 50% stronger than wire with a [100] fiber texture (5). Similar comments can be applied to textured face-centered cubic (fcc) sheet (6).

The potential improvements available through texture control seem to be adequately demonstrated. Why, then, are the practical applications so few? There appear to be many facets to the answer to this question; the most important is that, except for magnetic sheet material and a few cases of deep drawing sheet, industry is not in a position to supply commercial products controlled to a specified texture. The difficulty in obtaining quantitative preferred orientation information has probably impeded progress.

However, improved instrumentation for determining pole figures automatically will alleviate this bottleneck. The demonstration of yield strength improvements through the use of anisotropy has been successful. In most cases, prototypes or laboratory mockups of end items have not illustrated the predicted improvements, thus failing to convince management of the advantages of texture control. These failures can be directly attributed to the lack of specific intense textures in the material employed for the construction of these prototypes. The desired texture simply is not commercially available from the manufacturers. Thus it is imperative to develop substantial technical information so that the required texture control can be achieved, and it is part of the purpose of this program to add to our knowledge on this subject.

There are several mechanisms by which the texture or preferred orientation can be changed. The most important are deformation, deformation temperature, recrystallization, grain growth or secondary recrystallization, and phase transformation. Several complete reviews (7,8) have been published on this subject. Probably the most complete, at least on titanium, is that by Dillamore and Roberts (8). To summarize briefly, considerable study has been made of the deformation texture of heavily rolled sheet material of high purity, somewhat less work on the annealing texture of this material, and some, but very little reported, on the study of the effects of phase transformation of the same material. A limited amount of work has been reported on the effect of alloying elements on the cold-rolled texture of sheet materials in which large changes of preferred orientation were observed, utilizing both pole figure determinations and indirect measurement of mechanical properties. A detailed study of the limited prior work on the effect of alloy elements revealed that a more extensive study could provide a basic experimental program which might yield the information necessary for more fundamental understanding of how textures form. Thus, the main part of this program was to determine the effects of alloying elements on the deformation texture of titanium sheet metal, and to add further to our knowledge relating to texture control.

Experimental Procedure

Alloy Selection

The alloys upon which the program was carried out consisted of several elements of each of the four types (alpha stabilizers, beta isomorphous, beta eutectoid, and neutral). The range of compositions was chosen so as to cover (a) a large range of solubility in the alpha phase (alpha stabilizers and neutrals) and (b) beta stabilizer from pure titanium up to just enough to retain 100% beta (for the beta eutectoid and isomorphous alloys).

Melting, Casting, and Rolling

The alloys were melted in water-cooled crucibles under a positive pressure of welding-grade argon gas, employing thoriated tungsten electrode and a motor generator for a heat source. The castings were thoroughly melted once, then turned over and remelted several times to homogenize the buttons completely. The castings measured approximately 2-1/2" x 2-1/4" x 1-1/4" and normally weighed 450 grams.

These castings were heated in a furnace with argon flowing through. They were then upset to 3/4" thick at a variety of temperatures depending upon the alloy and prior experience with these compositions. The temperature was then lowered to place the alloy in the two-phase alpha-plus-beta phase field or, for the alpha alloys, in the alpha field, and then hot-rolled to a thickness of 1/4". These specimens were then cooled to room temperature and then cold-rolled in one direction, with intermediate anneals at a temperature which was chosen arbitrarily, until a final thickness of 0.040" was reached. No annealing was done after the final cold-rolling pass, which was about 10% in thickness in most cases. In some instances the sheets could not be reduced to the final size, particularly in the higher alloys. In some cases several attempts were made and in the aluminum series a second set of samples was made at 0.100" thick.

Texture Determination

The preferred orientation developed in each of the alloy sheets that were subjected to the above processing was determined by employing the method of Kula and Lopata (9). Such a method produces a first quadrant pole figure, and the other quadrants can be deduced from symmetry. Pole figures obtained by this method were obtained with isointensity contours 10, 20, 30, ... 80. Some of the data were obtained with an automatic pole figure computer* developed part way through the program. These pole figures are identified by isointensity contour levels 1, 2, 3, ... 8. The basal plane (0002) pole figure was determined for all sheets; the prism $(10\bar{1}0)$ pole figure was determined for most of the sheets. However since very little change occurred in the $(10\bar{1}0)$ pole figures they are not reported herein.

Chemical Analysis

A summary of chemical analysis is given in Table I.

* Martin, A., and Larson, F. R., to be published.

Table I. Summary of Chemical Analysis for Titanium Base Binary Alloy Series (WT %)

Alloying element		Melt a	b	c	d	e	f	g	h	i
Beta Eutectoid Stabilizer	Fe	0.12	0.31	0.60	1.18	2.32	4.26			
	Mn	.00	.46	0.73	0.91	1.42	5.88	7.09	7.68	
	Cr	.00	.09	0.27	0.51	0.97	1.71	3.43	6.81	
	Co	.00	.10	0.34	0.83	3.01	5.6	11.0		
	Ni	.00	.03	0.33	0.89	1.29	2.7*	5.6*	11.0*	
	Cu	.11	.55	1.61	2.0*	4.0*	8.0*	15.0*		
Beta Isomorphous Stabilizer	Mo	.00	.33	0.67	0.94	1.35	2.70	5.42	10.74	
	V	.00	.10	0.45	1.48	1.77	3.98	5.70	7.98	15.0*
	Cb	.225	.53	1.00	1.96	5.17	8.36	17.14	36.0*	
Alpha	Al	.01	.27	0.46	0.77	1.47	2.71	3.92	8.0	15.0*
Neutral	Sn	.01	.09	0.48	1.46	1.97	4.00	8.0*	15.0*	
	Zr	.03	.12	0.53	1.51	3.03	6.08	9.04	20.0*	

*These alloys were melted but could not be finished due to severe cracking during rolling.

Discussion of Experimental Textures

Effect of Beta Eutectoid Stabilizers

The textures obtained in this program are similar in many features to those reported by previous investigators (7,8). With the addition of the beta eutectoid stabilizer type elements, the changes in texture, with the exception of copper, formed a similar pattern. It was not possible to verify this pattern for cobalt and nickel because the higher composition alloys cracked during rolling and thus were not available for texture determination. However, manganese, iron, and chromium showed the same pattern. With zero to intermediate alloy additions the basal pole moves slowly from the unalloyed position toward the transverse direction. Sometimes two components in the texture are observed, one near the unalloyed position and one in the transverse direction. As will be shown later, the peak in the transverse direction, a $(11\bar{2}0)$ $[10\bar{1}0]$ texture, is a stable end orientation. With further increase in alloying addition, the beta phase becomes stable and at approximately 16 to 20% volume fracture beta an abrupt transition in the texture is observed. The new basal pole texture is similar to that of zinc or magnesium (7,8) sheet material. However, titanium still has the $[10\bar{1}0]$ parallel to the rolling direction, whereas in zinc and magnesium the $[11\bar{2}0]$ is parallel to the rolling direction. This behavior is best seen by examining the manganese series, which was one of the most complete, illustrated in Figure 1. The other series are shown in Figure 2 at alloy compositions above and below the transition.

An exception to the above general behavior was found in the copper alloy series which formed nearly an ideal (0001) [10$\bar{1}$0] type texture with the addition of 0.55 wt % (see Figure 2). One of the important factors in influencing twinning tendency is the grain size (10). The presence of a finely dispersed second phase within a grain has, in fact, the effect of reducing the "effective" grain size and probably would be influential in this suppression of {11$\bar{2}$2} twinning. Thus, the ideal texture produced in copper alloys above 0.55 wt % may be due to the finely dispersed second phase which is clearly visible in the microstructures. This finely dispersed second phase occurs because of the high solubility of copper in alpha titanium near the eutectoid temperature and the decreasing solubility with decreasing temperature.

Effect of Beta Isomorphous Stabilizers

For alloy additions of molybdenum, vanadium, and columbium, the beta isomorphous stabilizers, texture changes that occurred were similar to those of the beta-stabilized eutectoid alloys described in the previous section (see Figure 3).

Effect of Alpha Stabilizers

Additions of aluminum (Figure 4), an alpha stabilizer, caused tha basal pole to move toward the sheet normal, producing an ideal or (0001) [10$\bar{1}$0] type texture similar to that observed by Sparks et al (11). Comparison of the microstructures for the aluminum and the copper series, which also produced this type of texture, clearly showed a very fine structure which could probably suppress twinning.

Effect of Neutral Elements

Additions of tin and zirconium had only minor effects upon the unalloyed texture. The changes noted were primarily that of intensifying the unalloyed texture (Figure 4).

The Role of Lattice Parameters

Considerable effort was expended in determining lattice parameters, with very little evidence that they play a major role in texture formation. It may be that over the range of compositions and alloy types studied, there was too little change in lattice parameters to cause a pronounced effect on texture. There was slight evidence that the higher molybdenum (largest c/a ratio) and manganese (second largest c/a ratio) alloys had a greater tendency for the basal poles to lie in the transverse direction (i.e., to form a (11$\bar{2}$0) [10$\bar{1}$0]

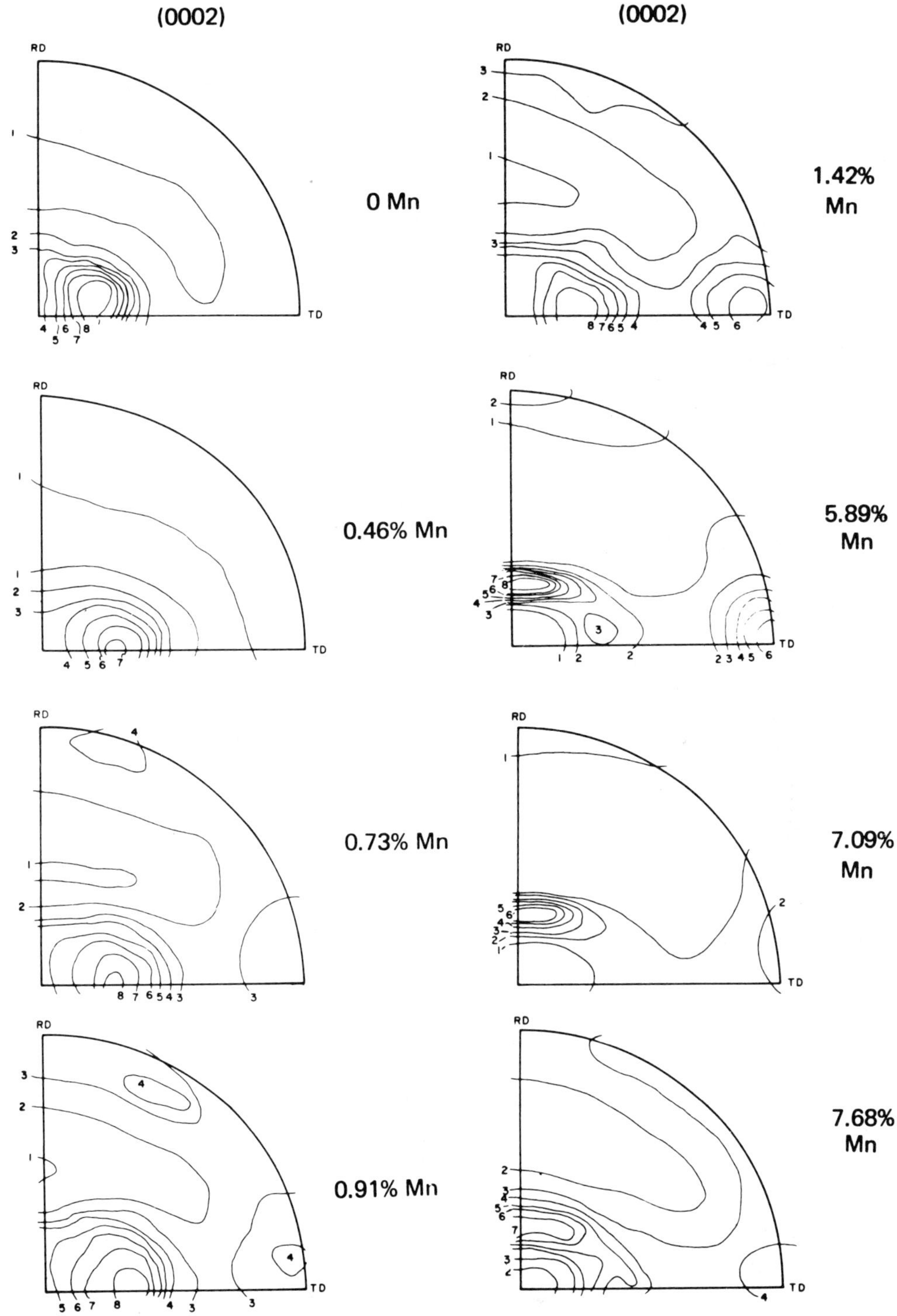

Fig. 1. Textures of Ti-Mn binaries.

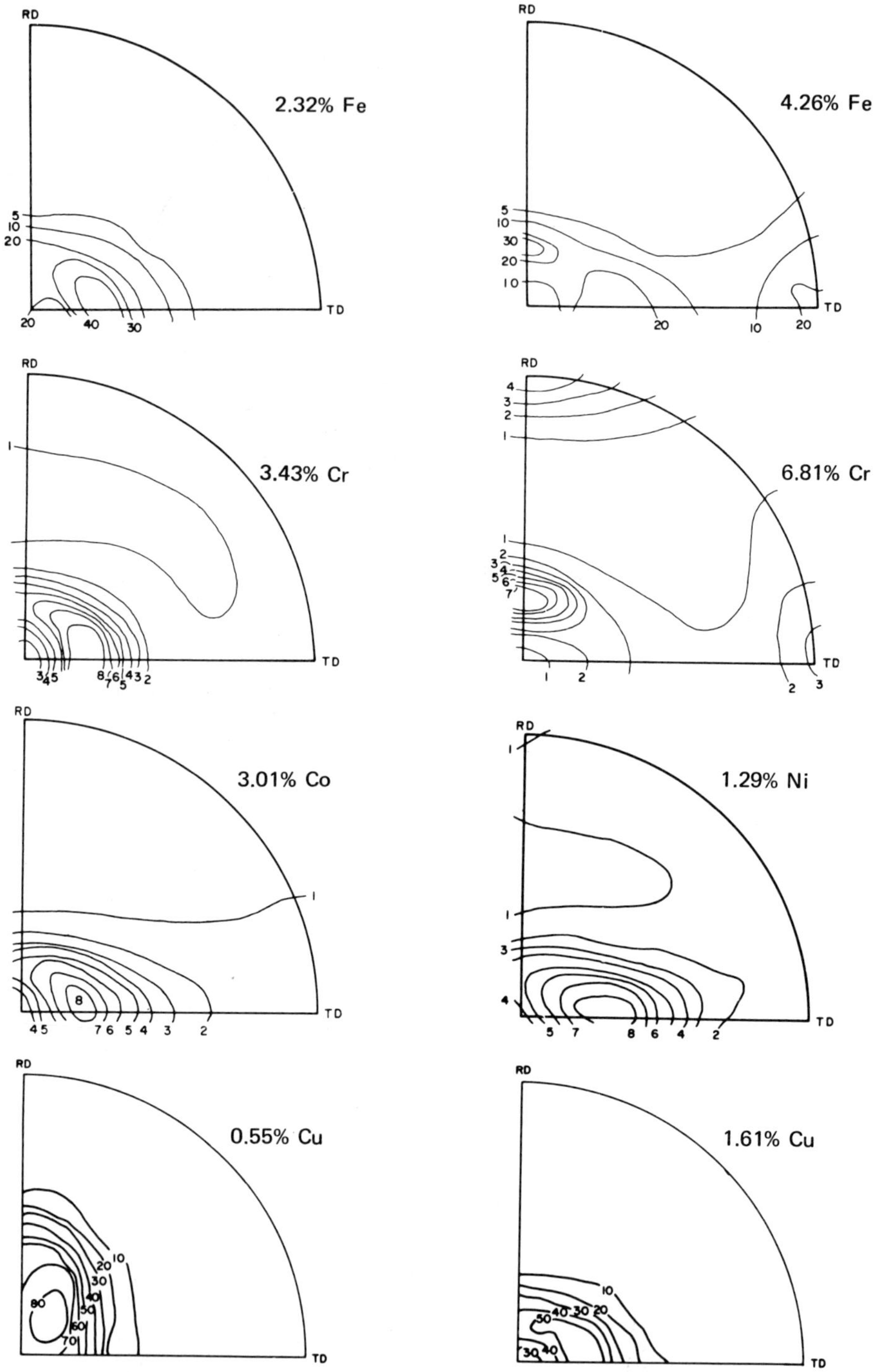

Fig. 2. (0002) textures of titanium binaries (beta eutectoid).

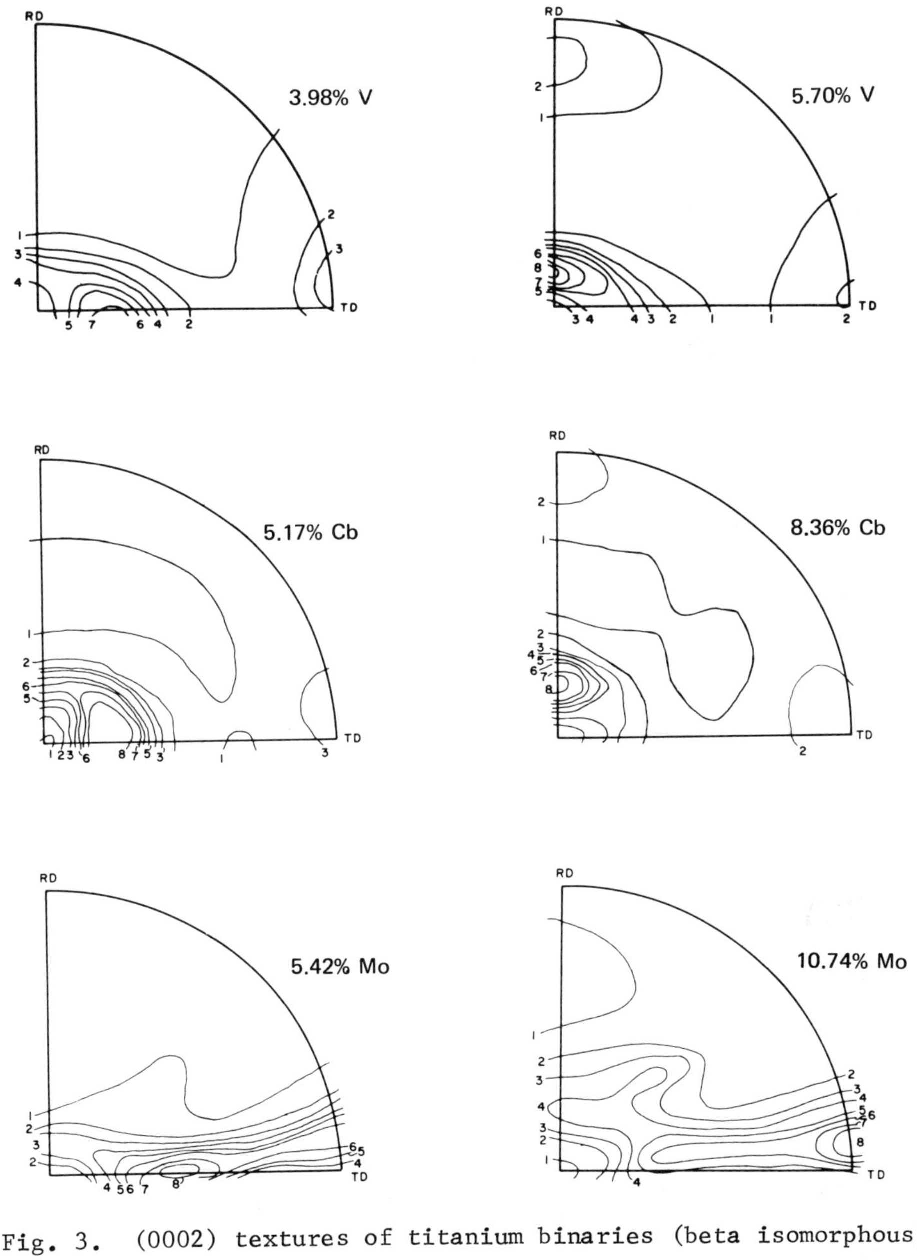

Fig. 3. (0002) textures of titanium binaries (beta isomorphous).

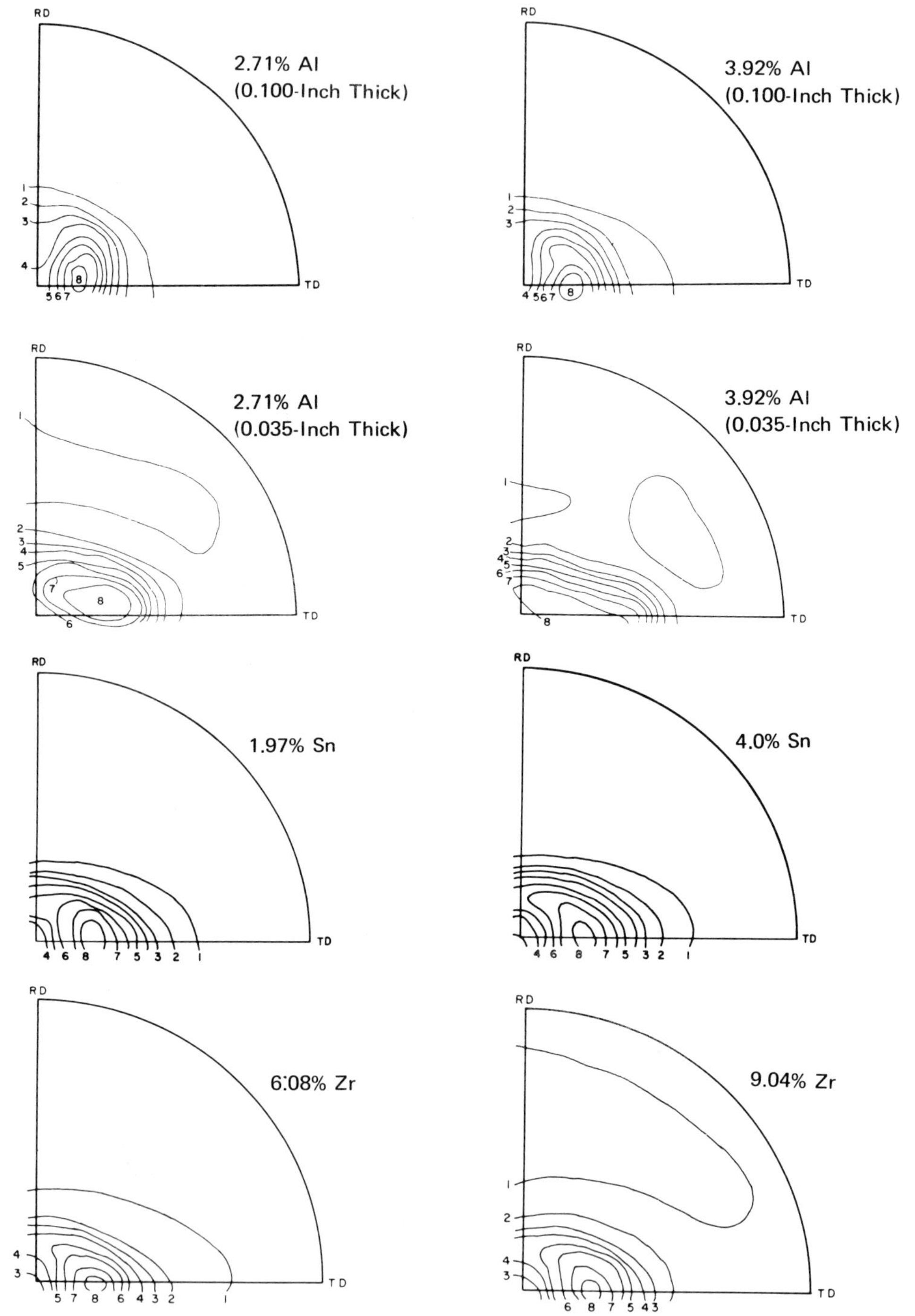

Fig. 4. (0002) textures of titanium binaries (alpha and neutral).

type texture). From a theoretical point of view, the increase in basal poles in the transverse direction could be attributed either to increased $\{11\bar{2}2\}$ twinning or to increased duplex $\{10\bar{1}0\}<11\bar{2}0>$ slip.

Theoretical Discussion of the Formation of Deformation Textures

Theories of the development of deformation textures are not too well advanced. This is probably due to the complex nature of polycrystalline deformation. The most complete treatment of the formation of deformation textures is that advanced by Calnan and Clews (12-15). Williams and Eppelsheimer have applied the Calnan-Clews method to titanium and conclude, after several assumptions, that by suitable choice of slip and twinning modes and their relative critical shear stresses, it is possible to develop a satisfactory explanation of deformation textures in tension, compression, and cold rolling. The weakness of this method is the lack of information about critical shear stresses for different slip planes and the assumption that twinning is achieved by a critical shear stress. In spite of all these difficulties, this method does have value because it allows qualitative discussion of how textures develop.

The Calnan-Clew Method

The main feature of the Calnan-Clew method is the determination of slip rotation and twinning reorientation caused by an applied stress system. To determine the general grain rotation behavior a resolved shear stress contour diagram for each slip and twinning system having the highest resolved shear stress ratio is plotted in a unit triangle. It is then assumed that only the slip or twinning system having the highest resolved shear stress ratio will be active. The resolved shear stress ratio is defined as the ratio of the resolved shear stress to the critical shear stress. The grain rotation tendencies for various slip and twinning systems are as follows:

1. Simple slip
   - (a) Tension - slip direction rotates toward stress axis
   - (b) Compression - slip plane normal rotates toward stress axis
2. Duplex Slip
   - (a) Tension - the great circle joining two active slip directions rotates toward the stress axis
   - (b) Compression - the great circle joining two slip plane normals rotates toward the stress axis

3. Multiple slip

   Rotation occurs until stress axis is located symmetrically

4. Twinning

   (a) Stress direction does not coincide with shear direction-no deformation

   (b) Stress direction coincides with shear direction - rotations obtained from analysis of twin mode

Thus, it can be seen that in order to determine the grain rotations, it is necessary to know the stress system which causes the deformation.

Roll-Gap Stresses and Deformation

One of the main problems in applying any method of determining rolling textures is that the deformation and stresses are known to vary tremendously through-out the roll gap (16,17). This makes computation of grain rotations an almost impossible task. This complexity can be overcome to some degree by making the gross assumption that sheet rolling can be described as a deformation caused by tension in the rolling direction and compression perpendicular to the rolling plane (16-18). This simple model of roll gap stresses has been used by Calnan and Clews (12-15), Williams and Eppelsheimer (19), and Hobson (20). From this model it is possible to calculate the resolved shear stresses on grains of any orientation within a sheet and to determine which slip or twinning mode has the highest stress. It is then assumed that when this stress exceeds a critical value, yielding will occur along with the grain rotation associated with that mode of deformation. Hobson (20), studying cold rolling of zirconium single crystals, reported a computerized program for carrying out the above calculations. Figure 5 shows the results of one such calculation, assuming all slip and twinning modes yield at the same critical stress and that the tensile stress is equal to the compressive stress.

Grain Restraints

Hobson's work (20) was performed on zirconium, which has similar slip and twinning modes to titanium, and it is implied that polycrystalline material behavior is similar to that of single crystals. It is known that this is not entirely true, for Taylor (21) pointed out that in order to maintain grain-to-grain continuity it was necessary that five independent deformation modes occur simultaneously This continuity requirement and other local grain-grain interactions will undoubtedly cause major effects upon the stresses and deformation modes. Thus it is convenient to accept from the work of

Calnan and Clews, of Williams and Eppelsheimer, and of Hobson that the main deformation mode is that which has the highest resolved shear stress ratio. Furthermore, this main deformation mode produces the principal changes which occur in the texture, even though other accommodation modes are also active.

Application to Unalloyed Titanium

Starting with a random initial orientation and considering that rolling textures are a combination of tension and compression textures, it is now possible to develop a description of how titanium textures form. First, it can clearly be seen that $\{10\bar{1}2\}$and $\{11\bar{2}1\}$ twinning reorient all the grains which have their basal poles in those areas labeled in Figure 5. This reorientation occurs rapidly with small reductions, because the shear associated with these twinning modes is small. The $\{10\bar{1}2\}$ twinning rotation (about 85°) is such that the new basal pole orientation would be in or near the area labeled $\{11\bar{2}2\}$. The $\{11\bar{2}1\}$ twinning rotation would be such that the basal poles move along a great circle toward the transverse direction, about 35°. This is illustrated in Figure 6. The absence of basal poles near the rolling direction in real pole figures supports the twinning theory. Since it is well established that titanium pole figures have a strong $[10\bar{1}0]$ component in the rolling direction, it is evident that tension in the rolling direction causes $\{11\bar{2}0\}$ duplex slip on either $\{10\bar{1}0\}$ or $\{10\bar{1}1\}$ planes or both. From the compression stresses, $\{11\bar{2}2\}$ twinning causes a rotation of the grains which have their basal poles near the sheet normal. This rotation, about 64°, occurs about the rolling direction toward the transverse direction. However, real textures of unalloyed titanium do not have basal pole peaks in the transverse direction, thus an additional rotation back toward the sheet normal is required. The most likely mechanism, considering basic principles, would be basal slip. There is a possibility that $\{11\bar{2}1\}$ twinning could also cause the rotation back toward the sheet normal, since the stress direction would be correct for second-order $\{11\bar{2}1\}$ twinning within the $\{11\bar{2}2\}$ twins. The rotation for this second-order twinning would be about 35° from the transverse direction toward the sheet normal; further rotation by slip would be required in order to orient the basal poles properly. In any event, the position of the basal poles along the transverse direction-sheet normal great circle may be considered as the end result of a dynamic process of $\{11\bar{2}2\}$ twinning, possibly $\{11\bar{2}1\}$ twinning and basal slip. Therefore the observed cold-rolled texture of unalloyed titanium of $(0001)[10\bar{1}0]$ rotated about 27° toward the transverse direction can be thought of as a combination of $[10\bar{1}0]$ tension textures and a rotated [0001] compression texture. Since rotation occurs about the $\langle 10\bar{1}0 \rangle$ axis, both are compatible. Some of the principal twinning reorientations are shown in Figure 6.

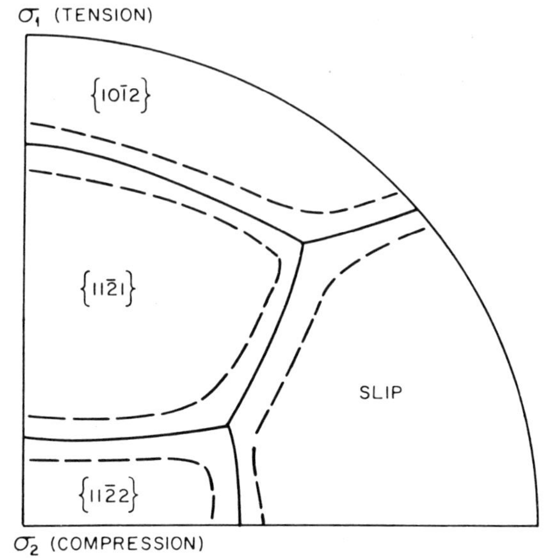

Fig. 5. Stereogram showing the deformation systems with the largest Schmid factor as a function of basal pole orientation. Calculations based on loading with $\sigma_1 = \sigma_2$ and the assumption that each mode has the same critical stress for activation. Between dashed boundaries more than one mode may operate depending on rotation of the basal pole.

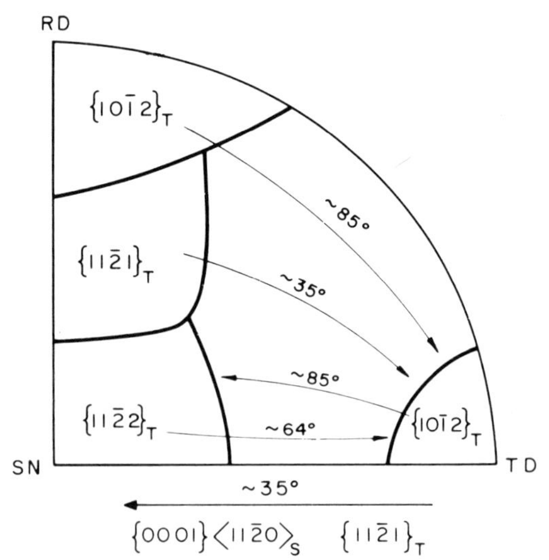

Fig. 6. Rotations of the (0001) pole due to various twin (T) and slip (S) modes.

Modification of Texture by Alloying

Changes in texture caused by alloy additions or variations in processing procedures can be explained by modification of the above mechanisms. For the aluminum and copper series the change is quite clear. The ideal (0001)$[10\bar{1}0]$ texture produced in these alloys seems to be a result of the suppression of $\{11\bar{2}2\}$ twinning. A decrease in the critical resolved shear stress for basal slip could also be effective in producing this texture. For the alloys where a strong $(11\bar{2}0)[10\bar{1}0]$ component shows in the texture, that is in high molybdenum and intermediate manganese alloys, it is similarly clear that this is caused by an increased $\{11\bar{2}2\}$ twinning tendency and/or a low critical resolved shear stress for $(10\bar{1}0)[11\bar{2}0]$ slip which produces a stable end texture of this type.

In order to explain the "texture transition" of titanium to magnesium type it is, however, necessary to introduce an additional second-order twinning mode. The texture transition is not a true one, for although the basal pole figures for highly beta alloyed titanium are similar to magnesium, zinc, and cobalt, the titanium has a strong $[10\bar{1}0]$ in the rolling direction whereas magnesium has a strong $[11\bar{2}0]$ in the rolling direction. Thus the main features of the tension texture in titanium still remain after the texture transition.

The texture formed by highly beta alloyed titanium can be produced by a second-order twinning of $\{10\bar{1}2\}$ type occurring within the $\{11\bar{2}2\}$ twin. The stress axis is in the favorable direction for this mode of twinning. It is interesting to note that the texture transition only occurs when there is a two-phase structure and the amount of the beta phase is greater than approximately 16 to 20 volume percent. It appears that the deformation of the beta phase, which is softer than the alpha, produces a strain field which is conducive to this twinning mode. The twinning rotation would be about 85° along a great circle, 30° from the sheet normal and passing through the transverse direction.

Conclusions

A program consisting of several series of titanium binary alloys was carried out in order to determine the effect of alloy additions on the cold-rolled sheet texture. Beta stabilizer-eutectoid type alloying elements included iron, manganese, chromium, cobalt, nickel, and copper. Beta stabilizer-isomorphous type elements included molybdenum, vanadium, and columbium. Aluminum represented an alpha stabilizer and tin and zirconium were included as neutral elements. A considerable range of composition was covered for each binary alloy.

It was found that most of the beta stabilizers caused a transition in the texture from a titanium to magnesium zinc type with increasing alloy additions. For some beta stabilizing elements which caused an increase in the c/a ratio there appeared an intermediate texture of the $(11\bar{2}0)[10\bar{1}0]$ type. This texture transition was not to true zinc type because after the transition, the titanium still had $[10\bar{1}0]$ in the rolling direction, not $[11\bar{2}0]$, which is characteristic of zinc. It is apparent that this transition is aided by the presence of a stable beta phase, since the transition in texture was observed only when the amount of beta exceeded 16 volume percent.

Additions of both aluminum and copper increased the tendency to form an ideal or $(0001)[10\bar{1}0]$ type texture. This was known for aluminum; however, the copper results are new. One possible suggestion for these results is that both copper and aluminum form a finely divided precipitate in the alpha phase, thus suppressing $\{11\bar{2}2\}$ twinning, which is thought to rotate the ideal titanium texture away from the "ideal" orientation.

Additions of tin and zirconium had little effect upon the texture of titanium, only serving to sharpen the unalloyed texture.

General analysis of these textures in the light of the Calnan and Clews method of texture development was carried out in a qualitative way. It was, however, concluded that a quantitative analysis was not possible now because of the complex state of stress and deformation in the roll gap and a lack of knowledge of the critical stresses for slip and twinning in these alloys.

References

1. McClintock, F. A., and Argon, A. S., eds. "Mechanical Behavior of Materials," Addison-Wesley, 1966.

2. Whiteley, R. L., "The Importance of Directionality in Drawing Quality Sheet Steel," Trans. ASM, Vol. 52, 1960, p. 154-169.

3. Graham, C. D., Jr., General Electric, TIS 64-RL-3752M.

4. Backofen, W. A., Hosford, W. F., and Burke, J. J., "Texture Hardening," Trans. ASM, Vol. 55, 1962, p. 246-267.

5. Chin, G. Y., and Mammel, W. L., "Computer Solutions of the Taylor Analysis for Axisymmetric Flow," Trans. AIME, Vol. 239, 1967, p. 1400-1405.

6. Hosford, W. F., and Backofen, W. A., "In Fundamentals of Deformation Processing," Syracuse University Press, 1964, p. 259.

7. Barrett, C. S., and Massalski, T. B., "Structure of Metals," McGraw-Hill, 1966.

8. Dillamore, I. L., and Roberts, W. T., "Preferred Orientation - Wrought and Annealed Metals," Met. Rev., Vol. 10, No. 39, 1965, p. 271-380.

9. Lopata, S. L., and Kula, E. B., "Reflection Method for Pole-Figure Determination," Army Materials and Mechanics Research Center, WAL TR 826.52/1, July 1961; also Trans. AIME, Vol. 224, 1962, p. 865-866.

10. Hall, E. O., "Twinning" Butterworths, Scientific Publication, 1954.

11. Sparks, C. J., Jr., McHargue, C. J., and Hammond, J. P., "Effects of Aluminum on the Cold-Rolled Textures of Titanium," Trans. AIME, Vol. 209, 1957, p. 49-50.

12. Calnan, E. A., and Clews, C. J. B., "Deformation Textures in Face-Centered Cubic Metals," Phil. Mag., Vol. 41, 1950[vii], p. 1085-10100.

13. Calnan, E. A., and Clews, C. J. B., "The Development of Deformation Textures in Metals. Part II - Body Centered Cubic Metals," Phil. Mag., Vol. 42, 1951[vii], p. 616-635.

14. Calnan, E. A., and Clews, C. J. B., "The Development of Deformation Features in Metals. Part III - Hexagonal Structures," Phil. Mag., Vol. 42, 1951[vii], p. 919-932.

15. Calnan, E. A., and Clews, C. J. B., "The Prediction of Uranium Deformation Textures," Phil. Mag., Vol. 43, 1952[vii], p. 93-104.

16. Cook, M., and Richards, T. L., "Fundamental Aspects of the Cold Working of Metals," J. Inst. Metals, Vol. 78, 1950-51, p. 463-482.

17. Tarnovskii, I. Y., et. al. "Deformation of Metals During Rolling" Pergamon Press, 1965.

18. Wever, F., "Texture of Metals after Cold Deformation," Trans. AIME Inst. of Metals Division, 1931, p. 51-53.

19. Williams. D. N., and Eppelsheimer, D. S., "A Theoretical Investigation of the Deformation Textures of Titanium," J. Inst. Met., Vol. 81, 1952-53, p. 553-562.

20. Hobson, O. O., "Textures in Deformed Zirconium Single Crystals," Trans. AIME, Vol. 242, 1968, p. 1105-1110.

21. Taylor, G. I., "Plastic Strain in Metals," J. Inst. Metals, Vol. 62, 1938, p. 307-324.

COLD-ROLLING TEXTURE DEVELOPMENT

IN TITANIUM AND TITANIUM-ALUMINUM ALLOYS

D. R. Thornburg† and H. R. Piehler

Carnegie-Mellon University

Pittsburgh, Pennsylvania 15213

Introduction

Most low c/a (<1.63) hcp metals and alloys have a cold rolling texture which has the (0002) or basal poles concentrated in a region $\pm$ 30-40° from the sheet normal (SN) in the transverse directions (TD) (1). This split TD texture occurs even when substantial amounts of alloying elements are present and severe cold reductions have been imposed. The two major exceptions to this trend are the alloys Ti-Al (2) and Ti-Cu (3), where the basal poles are aligned near the SN (basal texture). Earlier work on Ti-Al (4) seemed to indicate that there was a gradual transition from split TD to basal texture, i.e., a continuous rotation of the basal pole maximum toward the SN with increasing Al content. However, the manner in which the texture developed with various degrees of cold reduction and the micromechanical origins of the texture difference between Ti and Ti-Al alloys were as yet unexplained.

This manuscript is an experimental and theoretical investigation designed to study these questions. Experimentally, cold rolling texture development was studied in alloys containing 0 to 4% Al at six reductions to 95% RA. Theoretically, a micromechanical model of the isostrain type (5) was applied to the Ti-Al alloy system. This isostrain model is felt to be more realistic than the isostress model of Calnan and Clews (6), the only micromechanical model previously applied to hcp deformation.

†Present Address - Magnetics Department, Westinghouse Research Laboratories, Pittsburgh, Pa. 15235

Experimental Procedure

Alloys with 0, 0.25, 0.5, 1.0, 1.5, 2.0, 3.0, and 4.0% Al were melted in a laboratory non-consumable induction furnace. The raw materials were Osaka Sponge Ti (N + O = 0.07%) and 99.999% Al. The as-melted ingots were cut into slabs 2 in. x 1 in. x 0.65 in. and initially hot rolled to 0.325 in. in the β region at a temperature 50°C above the transus. All heating was done in an argon atmosphere. The samples were hot rolled to 0.2 in. in the α region 50° below the transus. Hot rolled samples were machined, pickled, and annealed for one hour at 50°C below the transus in a high-purity He atmosphere.

The samples were then cold rolled 95% RA to 0.010 in. Specimens for X-ray and metallographic analyses were taken from the hot rolled and annealed alloys and after reductions of 20, 40, 60, 80, 90, and 95% RA. Reflection (0002) pole figures were determined with CuK_α radiation using the Schulz technique (7). All samples were etched to remove any surface effects. The pole figures were plotted using a program developed by Westinghouse Electric Corp. The data were reduced in a Burroughs 5500 Computer and plotted on a Calcomp plotter. Textures were characterized by the computer plotted (0002) pole figures plus intensity histograms which tabulated the fraction of the total counts in 5° increments of the tilt angle.

Experimental Results

In general, the pole figures of the starting hot rolled and annealed samples tended to show a basal texture at the surface of the sample (Fig. 1a) with a nearly random texture at the mid-plane (Fig. 1b). The histograms also indicate the strong basal component near the SN at the sample surface (Fig. 1a) and the random nature of the texture at the mid-plane (Fig. 1b).

Following various degrees of cold reduction, three types of textures were observed: (1) split TD, (2) basal, and (3) split RD (basal poles concentrated at $\pm$ 20–30° from the SN in the rolling direction). The first and second are the classic textures observed in the Ti-Al system (2); the third texture type (Fig. 2a) was found only at low reductions (20–40% RA) in the higher Al alloys (2–4% Al). Above 40% RA, no split RD textures were noted in our alloys; similar observations of the appearance and disappearance of the split RD texture have been made by other investigators (8).

With the exception of the 2.0% Al alloy, stable terminal textures were developed by 60% RA (Table I). There was some evidence at lower reductions that split TD and basal texture developed, but this was masked by shear cracking which appeared at reductions up to 40% RA.

Table I. Position of the Maximum Intensity Peaks for Samples Cold Rolled 60 to 95%

% RA	% Al 0	0.25	0.5	1.0	1.5	2.0	3.0	4.0
60	30	30	30	30	25	20*	0	0
80	40	35	30	30	40	0	0	0
90	40	40	35	30	30	0	0	0
95	40	40	35	35	30	30	0	0

*Split RA texture

As shown in Table I, split TD textures are prevalent at 60% RA, even in the Ti-1.5% Al alloy. Both the histogram and pole figure in Fig. 2b indicate a split TD texture. As deformation proceeded, the stable terminal textures became more intense. At 90% RA (Fig. 3) samples with up to Ti-1.5% Al had a split TD texture. At 95% RA (Fig. 4) the Ti-2.0% Al alloy developed a split TD texture (Fig. 4c), while at 90% RA it had a strong basal orientation (Fig. 3c). The alloys with Al > 2% retained a basal texture at 95% RA (Fig. 4d), though there seemed to be more spreading in the TD at 95% than at 90% RA (Fig. 3d). The major conclusion from the pole figure analysis is that a terminal split TD or basal texture was obtained depending on the Al content of the alloy; no gradual rotation from one texture type to the other was noted in the ranges of reduction and Al content examined.

Metallographic examination of the various samples indicated that the cold-rolled alloys with 1.5% Al or less tended to have substantial amount of twinning and retwinning (Fig. 5a). Alloys with Al ≥ 2% had very inhomogeneous deformation patterns and regions containing almost no mechanical twins. In these alloys, some localized twinning was noted near the sample surfaces and at microcracks that formed during rolling (Fig. 5b). These cold-rolled higher Al alloys also tended to have large macroscopic shear cracks which usually were not associated with extensive twinning (Fig. 5c), suggesting that these cracks may have formed as a result of adiabatic shear. As rolling proceeded to higher deformations, both microscopic and macroscopic cracks tended to heal. The metallographic evidence correlated with the textures. Those alloys with a split TD texture showed substantial twinning, those with a basal texture almost no twinning, consistent with the work of Larson (3) among others.

Model and Discussion

A maximum-work isostrain model (5) was used to predict stable rolling texture components. Deformation by slip was assumed to occur only in the $\langle 11\bar{2}0\rangle$ direction on the $\{10\bar{1}0\}$, $\{10\bar{1}1\}$, and (0001) planes. Since these slip modes cannot provide for c-axis extension

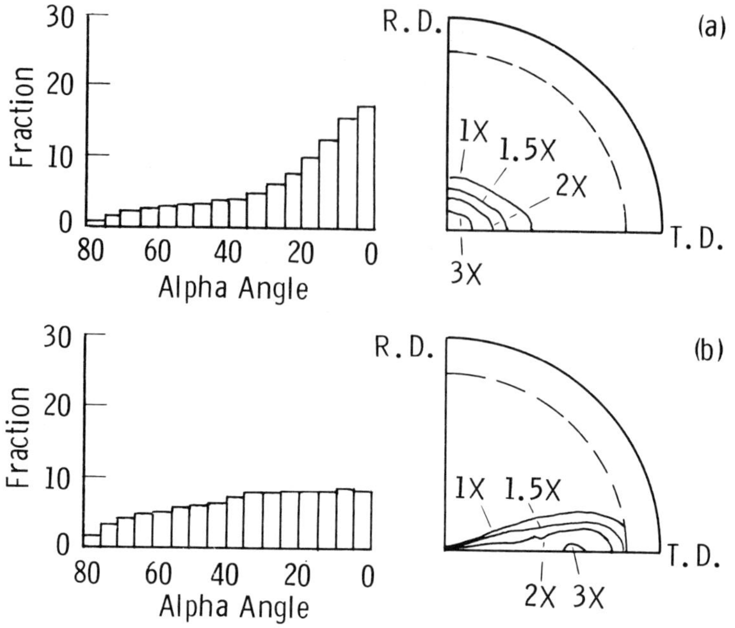

Fig. 1. (0002) pole figures and histograms for pure Ti hot-rolled and annealed (a) surface and (b) mid-plane. Contours show multiple of random intensity.

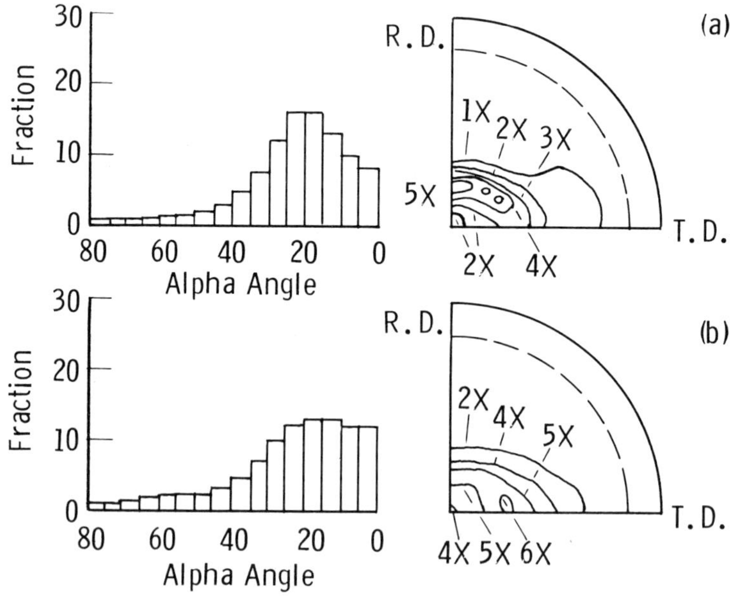

Fig. 2. (0002) pole figures and histograms of (a) Ti-20% Al cold rolled 20% (mid-plane) and (b) Ti-1.5% Al cold-rolled 60% (mid-plane).

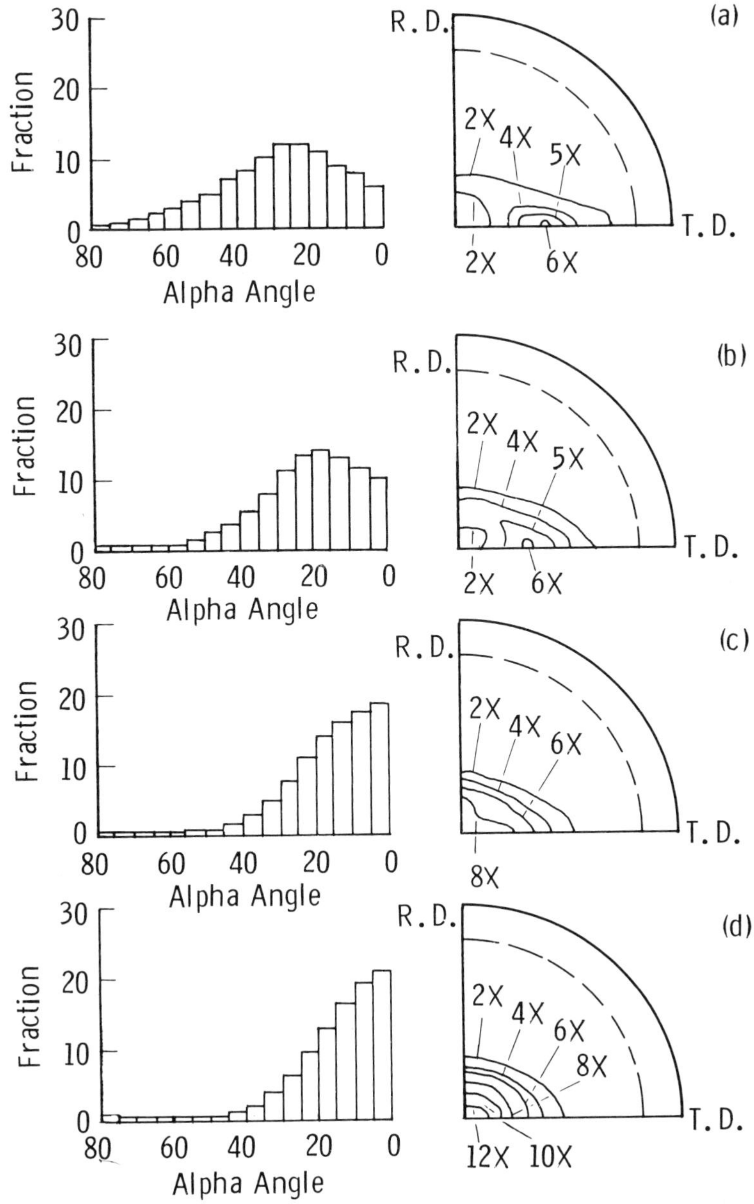

Fig. 3. (0002) pole figures and histograms of (a) pure Ti, (b) Ti-1.5% Al, (c) Ti-2.0% Al, and (d) Ti-3.0% Ti cold-rolled 90% (surface).

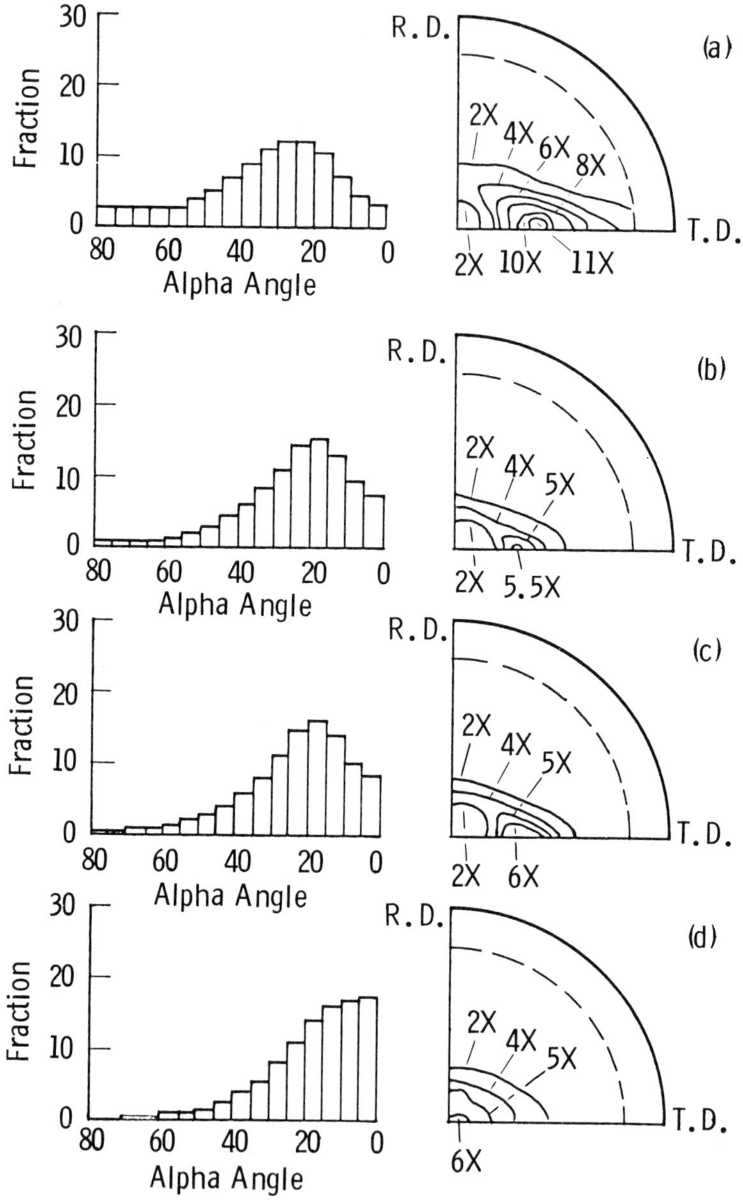

Fig. 4. (0002) pole figure and histogram of (a) pure Ti, (b) Ti-1.5% Al, (c) Ti-2.0%, and (d) Ti-3.0% Al cold-rolled 95% (surface).

or compression, deformation by twinning on the $\{10\bar{1}2\}\langle 10\bar{1}1\rangle$, $\{11\bar{2}1\}\langle 11\bar{2}6\rangle$, and $\{11\bar{2}2\}\langle 11\bar{2}\bar{3}\rangle$ systems was also included. The first two are tensile twins, the latter a compressive twin. Constrained deformation was assumed to occur by the simultaneous operation of four slip systems and one or more twinning systems. The localized twinning strain was assumed to be smeared out homogeneously throughout each grain, such that the imposed c-axis strain was accommodated uniformly in the entire volume. After subtracting the twinning strain, the remainder of the imposed strain state was accommodated by homogeneous deformation on four slip systems.

Using Churchman's data (9), the analysis predicted that 108 sets of four slip systems could operate simultaneously in impure Ti ($O + N > 0.01$). The specific operative slip as well as twinning systems were picked by computer using the maximum-work principle.

When the c-axis normal strain is not accommodated, i.e., no twinning occurs, there is a definite tendency for basal pole rotation toward the SN (Fig. 6). However, when twin accommodation is included, several interesting features are noted. First, there is a definite boundary for rolling (plane strain) which separates the regions of tensile and compressive twinning. This boundary is delineated by the dotted line in Fig. 7. Above the dotted line only tensile twins can operate, below the line only compressive twins, and on the dotted line the c-axis normal strain is zero. Secondly, the orientation of almost all (> 87 vol %) of the grains in all the materials studied lay in the zone of compression twinning, which results in grain rotations away from the SN. The rotations due to slip, which accommodate the remainder of the plane-strain deformation are shown in Fig. 7 for the c-axis accommodation by $\{11\bar{2}2\}\langle 11\bar{2}\bar{3}\rangle$ compressive twin and the more common $\{10\bar{1}2\}\langle 10\bar{1}1\rangle$ tensile twins. A stable region of near zero rotation due to slip is predicted at 25° in the TD. Since compressive twinning causes rotations away from the SN and tensile twinning causes rotations toward the SN, the formation of the stable split TD texture in the lower Al alloys is attributed to the rotations due to the slip which accompanies twinning.

In summary, contrary to previous results, the split TD texture was observed to develop suddenly at a given % RA and Al content, i.e., there was no continuous variation from basal to split TD texture. Terminal split TD textures at 95% RA were noted on alloys with Al ≤ 2.0%. Microstructurally the alloys with split TD textures contained large amounts of twinning, while alloys with basal textures showed little evidence of twinning, suggesting that twinning was required for the formation of split TD textures. The isostrain model successfully predicted a stable region of near zero rotation due to slip at 25° in the TD when twin accommodation was considered. The basal texture in the high Al alloys, which did not twin extensively, could not be predicted by the isostrain model, since only $\langle 11\bar{2}0\rangle$ slip was considered. However, if four of the five strain

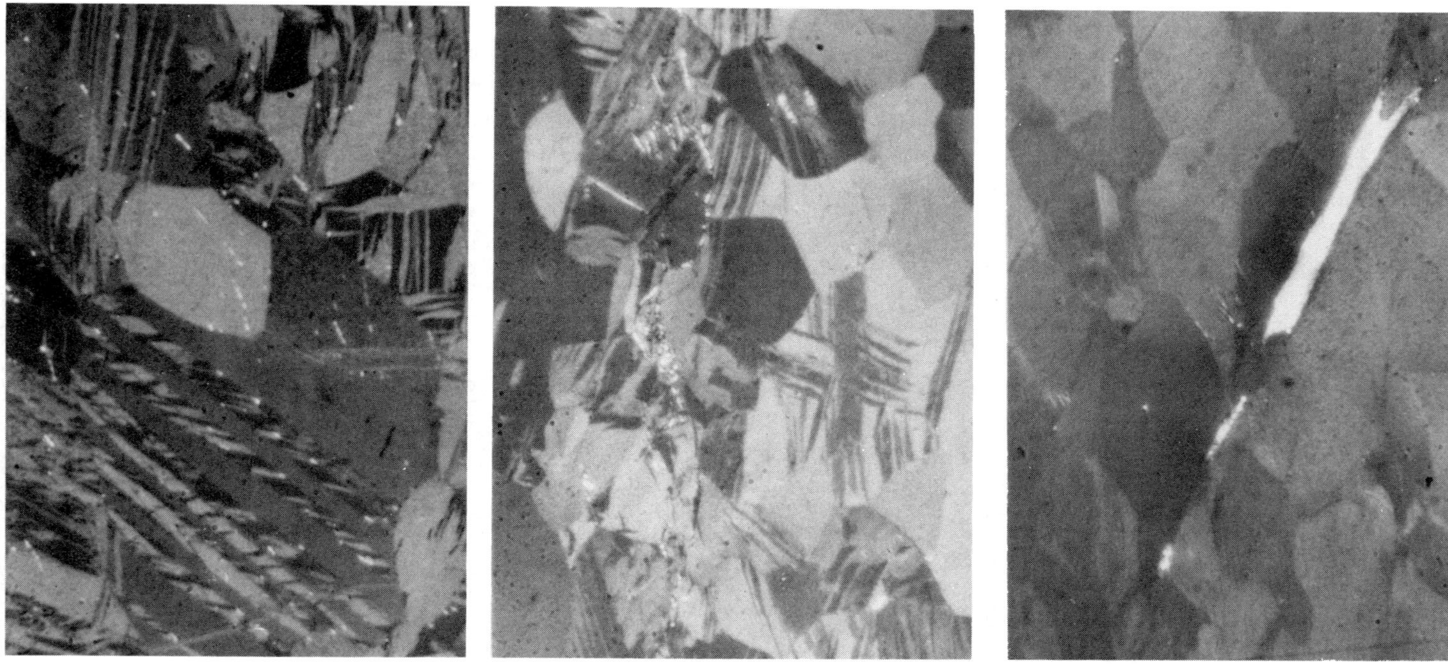

(a) (b) (c)

Fig. 5. Photomicrographs of (a) Ti-1.0% Al cold-rolled 20%, showing the presence of twins, (b) Ti-3% Al cold-rolled 20%, showing the presence of localized twinning in the region of microcracks, and (c) Ti-2% Al showing presence of macroscopic shear crack with no twinning. Etchant 1% HF, 2% HNO_3, 50% H_2O, Bal. H_2O. Polarized light, magnification 350X prior to reproduction.

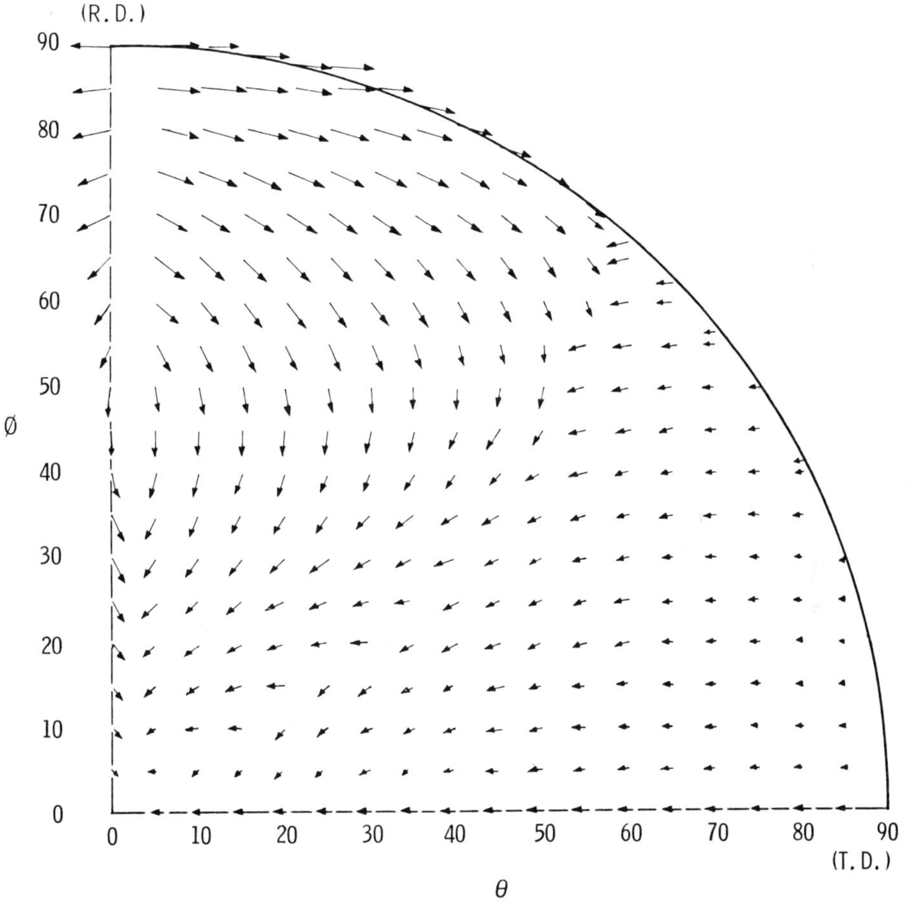

Fig. 6. Rotations due to only slip involved in the plane-strain deformation of crystals of various orientations for an applied strain of 0.05. The $\bar{2}110$ direction is aligned with the RD.

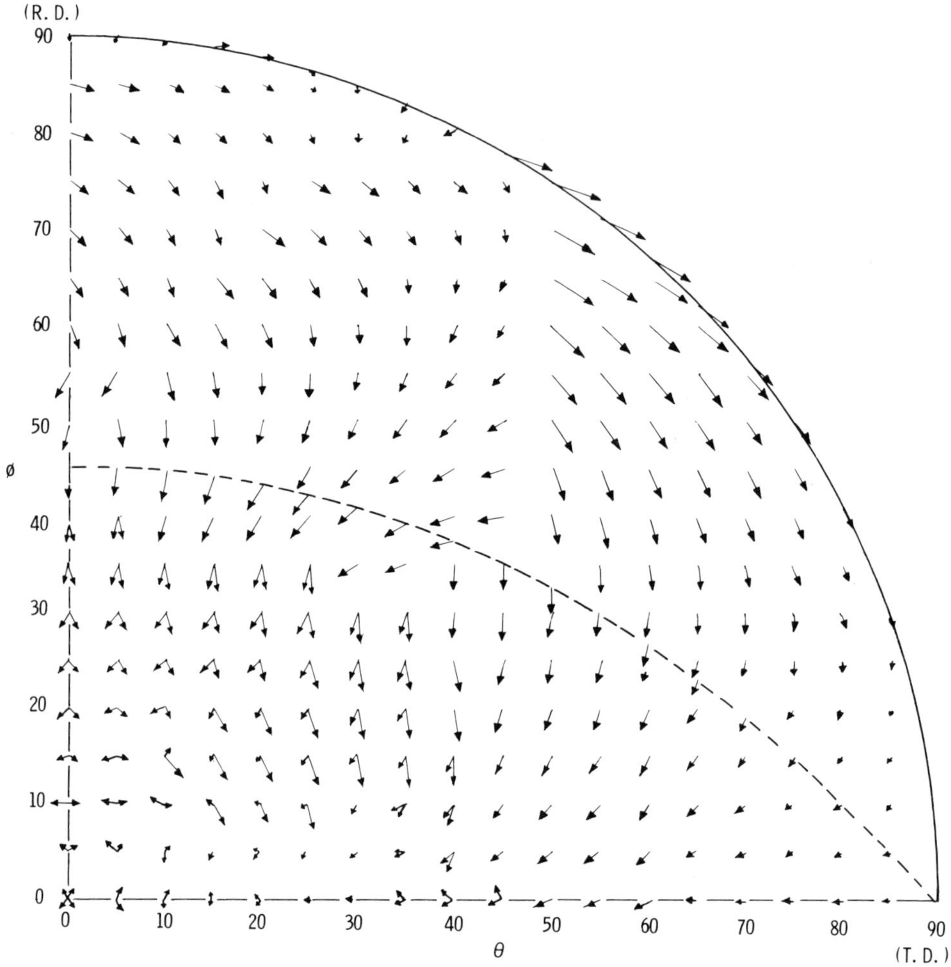

Fig. 7. Rotations due to slip involved in the plane-strain deformation of crystals of various orientations for an applied strain of 0.05. In addition to deformation by slip, $\{11\bar{2}2\}$ $\langle 11\bar{2}\bar{3}\rangle$ and $\{10\bar{1}2\}$ $\langle 10\bar{1}1\rangle$ twins also operate to help accommodate the imposed plane-strain deformation. The dotted line represents a boundary between compressive and tensile twinning; above the line only the $\{10\bar{1}2\}$ $\langle 10\bar{1}1\rangle$ tensile twins can operate; below the line only the $\{11\bar{2}2\}$ $\langle 11\bar{2}\bar{3}\rangle$ compressive twins. The $[\bar{1}010]$ direction is aligned with the RD.

components are accommodated by <11$\bar{2}$0> slip, there is a definite tendency to form a basal texture. The actual formation of the basal texture is probably due to the operation of both <11$\bar{2}$0> and <c + a> slip instead of twinning.

References

1. Dillamore, I. L. and Roberts, W. T., Met. Rev. 10, No. 39, p. 271, 1965.

2. McHargue, C. J., Adair, S. E., and Hammond, J. P., Trans. Met. Soc. AIME 197, p. 1199, 1953.

3. Larson, F. R., Ph.D. Thesis, Brown University, June, 1970.

4. Sparks, C. J., McHargue, C. J., and Hammond, J. P., Trans. Met. Soc. AIME 209, p. 49, 1957.

5. Thornburg, D. R., Ph.D. Thesis, Carnegie-Mellon University, November, 1971.

6. E. A. Calnan and C. J. B. Clews, Phil. Mag. 41, 1085 (1950, vii), 42, 616 (1951, vii), 42, 919 (1951, vii), 43, 93 (1952, vii).

7. Schultz, L. G., J. Appl. Phys. 20, p. 1030, 1949.

8. Mukhayev, V. V., Adamesku, R. A., and Geld, P. V., (Sverdlovsk) Rus. Met. (in English) V, p. 67, 1968.

9. Churchman, A. T., Proc. Roy. Soc. A226, p. 216, 1954.

PREFERRED ORIENTATION IN TITANIUM ALLOY THIN FOILS

P. K. Koh, Professor
Department of Mechanical Engineering
Texas Tech University
Lubbock, Texas 79409

According to experimental results obtained from thin Ti-75A and Ti-6Al-4V foils studied in recent investigations, titanium alloy thin foils possess multiple texture components. These texture components have a common axis in the c-axis of the hexagonal close-packed alpha titanium matrix phase. The measured integrated intensity under individual Bragg diffraction maximum was corrected for Lorentz-polarization, multiplicity, and structure factor differences to obtain the relative diffracting crystalline volume or crystalline population density index. In both Ti-75A and Ti-6Al-4V foils alpha titanium lines predominated. Beta titanium lines were revealed only after vacuum annealing of cold rolled foils. The end orientations of cold rolled foils and the preferred orientations in the vacuum annealed foils have the following multiple fiber axes: $\langle 1\bar{2}10\rangle$, $\langle 1\bar{1}00\rangle$ and $\langle 4\bar{5}10\rangle$. These multiple zone axes have the c-axis of alpha titanium phase as the common axis. The most prominently oriented planes in Ti-75A foils are (0002), $(01\bar{1}2)$, $(10\bar{1}3)$, $(01\bar{1}4)$, $(11\bar{2}4)$, and $(01\bar{1}5)$; while in Ti-6Al-4V foils, they are $(11\bar{2}0)$, $(20\bar{2}1)$, $(01\bar{1}4)$, and $(01\bar{1}5)$. Vacuum annealing depressed (0002), $(10\bar{1}3)$, and $(01\bar{1}4)$ texture components in Ti-75A foil.

Introduction

Objectives of the research reported herein were to demonstrate (a) that titanium alloy thin foils possess multiple texture components and (b) that the relative significance of the multiple texture components can be represented by atomic population index after the measured integrated intensity of an individual diffraction maximum is corrected for Lorentz-polarization factor, multiplicity factor and structure factor.

Slip, twinning, recrystallization, and grain growth cause reorientation in the close-packed metals. The reorientation tends to develop a preferred orientation. The preferred orientation might be of a simple type which can be adequately described by one or more "ideal poles." In many cases, however, the multiple texture components of the same materials such as in cold drawn wires (1) cannot be readily interpreted as "ideal poles." Furthermore, these texture components could have zonal axes corresponding to prominent crystalline directions.

The ideal pole concept for a polycrystalline pole figure which has been widely adopted for a single preferred orientation, becomes troublesome for multiple-component textures. An individual pole of a preferred orientation might be missing or underdeveloped due to incomplete grain growth or because of slight instrumentation misalignment. For a multiple-component texture, individual poles of different preferred orientations might also be superimposed upon each other. Another problem with the multiple-component texture concerns the true representation of diffracting crystalline volume for each preferred orientation. Classical studies (2) have shown that the measured intensity of a particular Bragg diffraction maximum is affected by the Lorentz-polarization factor, multiplicity factor, and structure factor (even if we discount the relatively minor and not easily calculable temperature and absorption factor) as well as by atomic volume and true crystalline population.

During cold rolling of a close-packed polycrystalline metal basal slip, as well as prism slip, commonly occurred. The terminal cold rolled orientation texture depends upon the type of metal and also upon the c/a axial ratio (1,2,3,4). The deviation from a simple (0001) $[2\bar{1}\bar{1}0]$, or (0001) $[11\bar{2}0]$ texture was found in pole figures of cold deformed titanium alloys as toward the transverse direction instead of toward the rolling direction as in zinc and cadmium (3,4). McHargue et al (5) showed that pole figures of various Ti-rich substitutional solid solutions are similar, with the exception of Ti-3.8 percent aluminum.

Thin metallic sheets furnish unique opportunities for texture studies in that their ultimate grain sizes remain less than sheet thicknesses and in the micron ranges, even after long time annealing at elevated temperatures. Therefore, thin metallic sheets or ultra-thin metallic tapes in any heat treated state can have grain structures suitable for quantitative texture evaluation using established diffractometer techniques, without the difficulties usually encountered with large annealed grains in thicker metallic sheets. Despite their relatively small grain sizes and restricted grain growths, complex textural changes did occur in ultra-thin molybdenum permalloy tapes (5,6).

Thin metallic foils are presently used to large extents in aerospace honeycombs and channel sections, as well as in applications requiring magnetic or mechanical anisotropy, through perfections in preferred orientations.

Materials

Ti-75A and Ti-6Al-4V strips, 0.0025-inches thick, were obtained from the Titanium Metals Corporation of America. The samples were both cold rolled as well as in various vacuum annealed states. No additional processing or handling was applied to these specimens before texture determinations were made.

Experimental Techniques

Flat foil specimens were used with a General Electric XRD-5 Diffractometer. Scanning was performed on each sample about the diffractometer axis which was parallel to the specimen surface. This was accomplished for all accountable Bragg maximum with the slowest possible scanning speed and the fastest possible chart speed (one degree two-theta angle measures five inches on recorder chart) in order to obtain the maximum possible area under each diffraction peak (integrated intensity). The integrated intensity was evaluated by Simpson's rule with 0.1-inch width as the smallest abscissa division.

The directly measured integrated intensity under each Bragg diffraction maximum (in square inches) was then corrected for Lorentz-polarization factor using procedures described elsewhere (1,7). The relative population density index, I_R, was tabulated as well as expressed in bargraphs as shown in Tables I and II and Figures 1 and 2.

In order to establish zonal relationship of preferred texture components in both cold rolled and annealed thin foils, and to locate a common axis to these texture components, each diffracting plane was plotted stereographically relative to (0001) as central pole. The interrelationship of the zone axes with the c-axis of the h.c.p. lattice was thus derived.

The splitting of the basal plane transverse to the rolling direction as reported by Clark and others (3) can be explained in

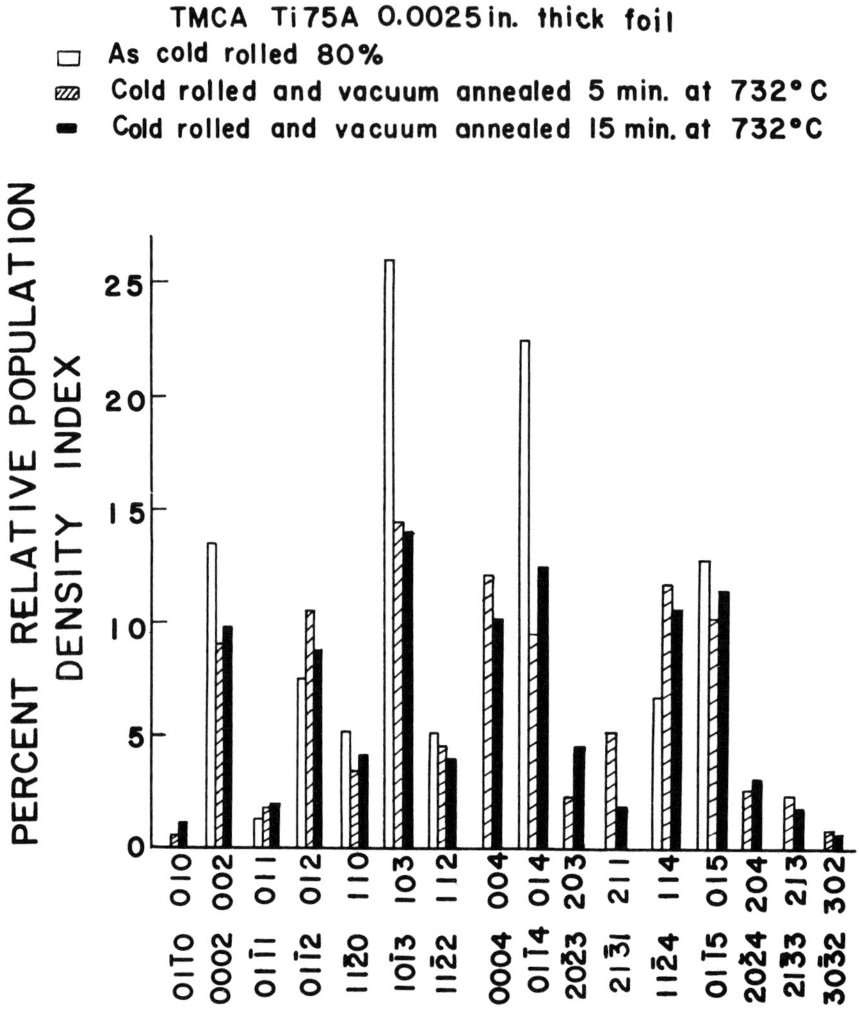

Fig. 1. Bargraph of relative population density index.

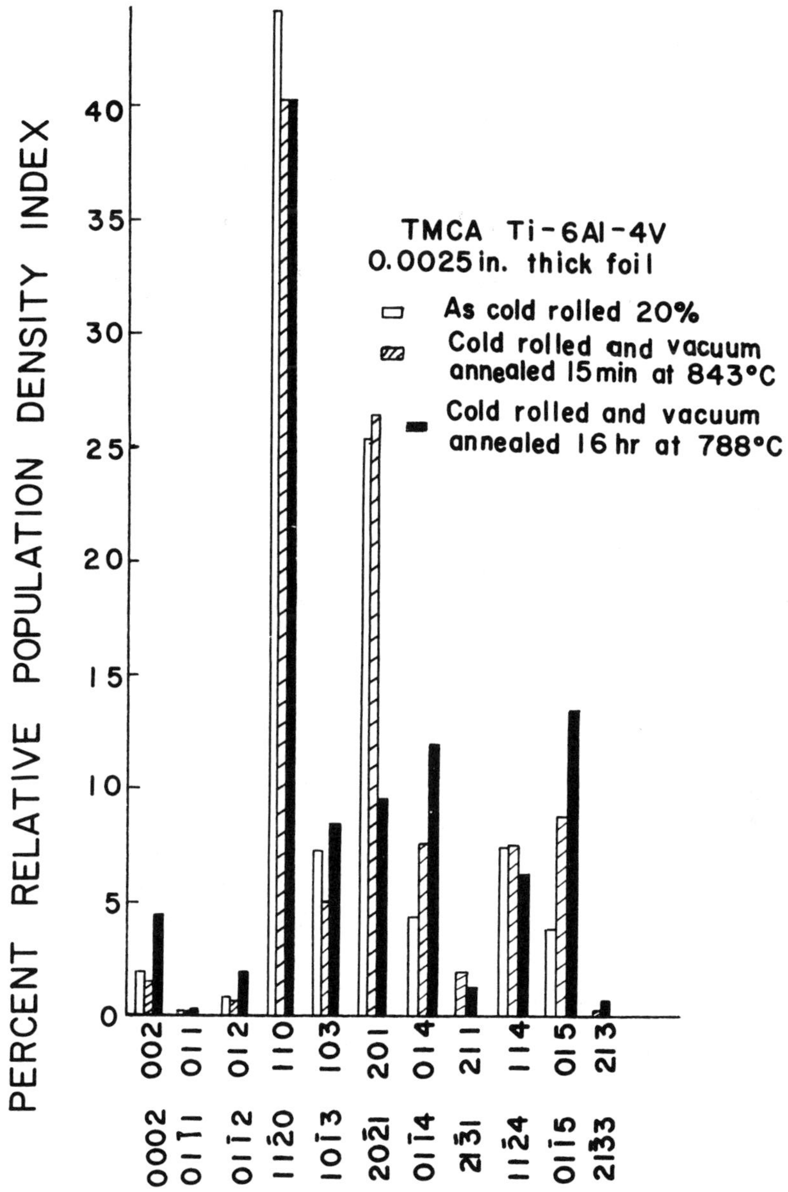

Fig. 2. Bargraph of relative population density index.

Table I. Ti-75A Foils

hkl α-Titanium	θ(degrees) Sample D	E	F	θ_{av}	f	m	L.P.F.[1]	I_c
$(01\bar{1}0)$		17.58	17.66	17°37'	15.7	6	19.12	28,397
(0002)	19.32	19.24	19.26	19°15'	15.4	2	15.71	29,805
$(01\bar{1}1)$	20.16	20.13	20.18	20°10'	15.1	12	14.18	116,389
$(01\bar{1}2)$	26.59	26.66	26.58	26°37'	13.1	12	7.567	15,582
$(11\bar{2}0)$	31.55	31.55	31.58	31°34'	12.0	6	5.156	17,819
$(10\bar{1}3)$	35.43	35.37	35.40	35°29'	11.2	12	4.034	18,221
$(11\bar{2}2)$	38.16	38.28	38.40	38°19'	10.9	6	3.497	9,972
$(20\bar{2}1)$			38.53	38°32'	10.9	12	3.459	14,793
(0004)		41.11	41.17	41°8'	10.5	2	3.124	2,756
$(01\bar{1}4)$	46.44	46.40	46.40	46°24'	9.9	12	2.771	3,259
$(20\bar{2}3)$		51.14	51.26	51°12'	8.4	12	2.749	6,984
$(21\bar{3}1)$		54.59	54.65	54°37'	8.3	24	2.864	4,736
$(11\bar{2}4)$	57.20	57.15	57.27	57°13'	8.2	12	3.060	9,878
$(01\bar{1}5)$	61.23	61.11	61.13	61°7'	8.0	12	3.468	7,990
$(20\bar{2}4)$		63.19	63.20	63°12'	8.0	12	3.764	2,891
$(21\bar{3}3)$		69.73	69.76	69°44'	7.80	24	5.176	22,683
$(30\bar{3}2)$		74.28		74°17'	7.75	12	6.873	19,811

Table I. Ti-75A Foils (continued)

hkl α-Titanium	Measured Integrated Intensity, I_m, (sq.in.) Sample			Relative Measured Integrated Intensity (%) Sample			Calculated Relative Intensity Factor[2] $(I_c/\Sigma I_c)$ Sample		
	D	E	F	D	E	F	D	E	F
$(01\bar{1}0)$		0.36	1.05		0.84	2.16		0.0895	0.0895
(0002)	8.45	8.22	9.95	26.75	19.12	20.45	0.1302	0.0940	0.0940
$(01\bar{1}1)$	2.82	6.56	7.92	8.93	15.26	16.28	0.5084	0.3670	0.3670
$(01\bar{1}2)$	2.42	5.02	4.62	7.66	11.68	9.49	0.0681	0.0491	0.0491
$(11\bar{2}0)$	1.93	1.80	2.52	6.11	4.19	5.18	0.0778	0.0562	0.0562
$(10\bar{1}3)$	9.89	8.01	8.70	31.31	18.63	17.88	0.0796	0.0574	0.0574
$(11\bar{2}2)$	1.07	1.34	1.34	3.39	3.12	2.75	0.0436	0.0314	0.0314
$(20\bar{2}1)$									
(0004)		1.02	0.95		2.37	1.95		0.0087	0.0087
$(01\bar{1}4)$	1.53	0.94	1.38	4.84	2.19	2.84	0.0142	0.0103	0.0103
$(20\bar{2}3)$		0.48	1.08		1.12	2.22		0.0220	0.0220
$(21\bar{3}1)$		0.74	0.30		1.72	0.62		0.0149	0.0149
$(11\bar{2}4)$	1.36	3.54	3.52	4.31	8.23	7.23	0.0432	0.0311	0.0311
$(01\bar{1}5)$	2.12	2.49	3.12	6.71	5.79	6.41	0.0349	0.0252	0.0252
$(20\bar{2}4)$		0.23	0.30		0.54	0.62		0.0091	0.0091
$(21\bar{3}3)$		1.67	1.41		3.88	2.90		0.0715	0.0715
$(30\bar{3}2)$		0.57	0.50		1.33	1.03		0.0625	0.0625

Table I. Ti-75A Foils (continued)

hkl α-Titanium	Population Density (∇_i)[3] Sample			Relative Population Density Index[4] I_R Sample			$\%I_R$ Sample		
	D	E	F	D	E	F	D	E	F
$(01\bar{1}0)$		4.02	11.73		0.0659	0.1734		0.41	1.08
(0002)	64.90	87.45	105.85	1.2181	1.4334	1.5647	13.53	8.96	9.78
$(01\bar{1}1)$	5.55	17.87	21.58	0.1042	0.2929	0.3190	1.16	1.83	1.99
$(01\bar{1}2)$	35.54	102.24	94.09	0.6670	1.6758	1.3908	7.41	10.47	8.69
$(11\bar{2}0)$	24.81	32.03	44.84	0.4657	0.5250	0.6628	5.17	3.28	4.14
$(10\bar{1}3)$	124.25	139.55	151.57	2.3320	2.2873	2.2405	25.91	14.29	14.00
$(11\bar{2}2)$	24.54	42.68	42.68	0.4606	0.6996	0.6309	5.12	4.37	3.94
$(20\bar{2}1)$									
(0004)		117.24	109.20		1.9217	1.6142		12.01	10.09
$(01\bar{1}4)$	107.75	91.26	133.98	2.0223	1.4958	1.9805	22.47	9.35	12.38
$(20\bar{2}3)$		21.82	49.09		0.3576	0.7256		2.23	4.53
$(21\bar{3}1)$		49.66	20.13		0.8140	0.2976		5.09	1.86
$(11\bar{2}4)$	31.48	113.83	113.18	0.5908	1.8658	1.6730	6.56	11.66	10.46
$(01\bar{1}5)$	60.74	98.81	123.81	1.1400	1.6196	1.8302	12.67	10.12	11.44
$(20\bar{2}4)$		25.27	32.97		0.4142	0.4874		2.59	3.05
$(21\bar{3}3)$		23.36	19.72		0.3829	0.2915		2.39	1.82
$(30\bar{3}2)$		9.12	8.00		0.1495	0.1183		0.93	0.74

Table II. Ti-6Al-4V Foils

hkl	θ(degrees) Sample							
α-Titanium	A	B	C	θ_{av}	f	m	L.P.F.[1]	I_c
(0002)	19.35	19.32	19.35	19°21'	15.4	2	15.75	29,805
$(01\bar{1}1)$	20.24	20.26	20.37	20°16'	15.1	12	14.03	116,389
$(01\bar{1}2)$	26.69	26.73	26.78	26°44'	13.1	12	7.946	15,582
$(11\bar{2}0)$	31.84	31.86	31.91	31°51'	12.0	6	5.054	17,819
$(10\bar{1}3)$	35.61	35.63	35.63	35°38'	11.2	12	3.999	18,221
$(20\bar{2}1)$	38.54	38.51	38.50	38°30'	10.9	12	3.463	14,793
$(01\bar{1}4)$	46.67	46.63	46.69	46°40'	9.9	12	2.766	3,259
$(21\bar{3}1)$		55.31	55.33	55°19'	8.3	24	2.922	4,736
$(11\bar{2}4)$	57.72	57.63	57.67	57°40'	8.2	12	3.098	9,878
$(01\bar{1}5)$	61.62	61.52	61.63	61°37'	8.0	12	3.541	7,990
$(21\bar{3}3)$		70.86	70.91	70°53'	7.80	24	5.530	22,683
β-Titanium								
(110)	19.10	(Broad & Superimposed)		19°6'	15.4	12	15.99	182,025
(200)	28.47	27.50	28.54	27°39'	12.8	6	6.941	27,174
(211)		34.22	34.22	34°36'	11.5	24	4.243	53,869

Table II. Ti-6Al-4V Foils (continued)

hkl	Measured Integrated Intensity, I_m, (sq.in.)			Relative Measured Integrated Intensity (%)			Calculated Relative Intensity Factor[2] $(I_c/\Sigma I_c)$		
	Sample			Sample			Sample		
α-Titanium	A	B	C	A	B	C	A	B	C
(0002)	1.21	1.15	3.87	3.56	2.61	8.83	0.1142	0.0871	0.0871
$(01\bar{1}1)$	0.17	0.11	0.56	0.50	0.25	1.28	0.4461	0.3401	0.3401
$(01\bar{1}2)$	0.29	0.28	0.91	0.85	0.64	2.08	0.0597	0.0455	0.0455
$(11\bar{2}0)$	16.44	21.65	21.03	48.38	49.22	48.00	0.0683	0.0521	0.0521
$(10\bar{1}3)$	2.76	2.78	4.55	8.12	6.32	10.39	0.0698	0.0532	0.0532
$(20\bar{2}1)$	7.75	11.78	4.17	22.81	26.78	9.52	0.0567	0.0432	0.0432
$(01\bar{1}4)$	0.30	0.75	1.14	0.88	1.70	2.60	0.0115	0.0095	0.0095
$(21\bar{3}1)$		0.28	0.16		0.64	0.37		0.0138	0.0138
$(11\bar{2}4)$	1.51	2.21	1.78	4.44	5.02	4.06	0.0579	0.0287	0.0287
$(01\bar{1}5)$	0.65	2.11	3.16	1.91	4.80	7.21	0.0249	0.0233	0.0233
$(21\bar{3}3)$		0.10	0.43		0.23	0.98		0.0663	0.0663
β-Titanium									
(110)									
(200)	2.90	0.64	1.93	8.53	1.45	4.41	0.1111	0.0794	0.0794
(211)		0.15	0.12		0.34	0.29		0.1574	0.1574

Table II. Ti-6Al-4V Foils (continued)

hkl	Population Density (∇_i)[3]			Relative Population Density Index[4]					
				I_R			$\%I_R$		
	Sample			Sample			Sample		
α-Titanium	A	B	C	A	B	C	A	B	C
(0002)	10.60	13.20	44.43	0.1942	0.1654	0.5704	1.94	1.27	4.39
$(01\bar{1}1)$	0.38	0.32	1.65	0.0070	0.0040	0.0212	0.07	0.03	0.16
$(01\bar{1}2)$	4.86	6.15	20.00	0.0891	0.0771	0.2568	0.89	0.59	1.98
$(11\bar{2}0)$	240.70	415.55	403.65	4.4108	5.2080	5.1823	44.11	40.06	40.06
$(10\bar{1}3)$	39.54	52.26	85.53	0.7246	0.6550	1.0981	7.25	5.04	8.45
$(20\bar{2}1)$	136.68	272.69	96.53	2.5047	3.4176	1.2393	25.05	26.29	9.53
$(01\bar{1}4)$	24.00	78.78	120.00	0.4398	0.9873	1.5406	4.40	7.59	11.85
$(21\bar{3}1)$		20.29	11.59		0.2543	0.1488		1.96	1.14
$(11\bar{2}4)$	39.84	77.00	62.02	0.7301	0.9650	0.7963	7.30	7.42	6.13
$(01\bar{1}5)$	21.24	90.56	135.62	0.3892	1.1350	1.7412	3.89	8.73	13.39
$(21\bar{3}3)$		1.51	6.49		0.0189	0.0833		0.15	0.64
β-Titanium									
(110)									
(200)	27.83	8.06	24.31	0.5100	0.1010	0.3121	5.10	0.78	2.40
(211)		0.95	0.75		0.0119	0.0096		0.09	0.07

Notes on Table I and Table II

Sample	Description
A	Ti-6Al-4V 0.0025-inch thick foil, cold rolled 20%
B	Ti-6Al-4V 0.0025-inch thick foil, cold rolled 20% and vacuum annealed 15 min. at 1550°F (843°C)
C	Ti-6Al-4V 0.0025-inch thick foil, cold rolled 20% and vacuum annealed 16 hr. at 1450°F (788°C)
D	Ti-75A 0.0025-inch thick foil, cold rolled 80%
E	Ti-75A 0.0025-inch thick foil, cold rolled 80% and vacuum annealed 5 min. at 1350°F (732°C)
F	Ti-75A 0.0025-inch thick foil, cold rolled 80% and vacuum annealed 15 min. at 1350°F (732°C)

Structure Factor

α-Titanium matrix, h.c.p. = (000; 1/2, 2/3, 1/2)

$$|F|^2 = 4f^2 \mathrm{Cos}^2\, \pi \left(\frac{h + 2k}{3} + \frac{1}{2}\right)$$

1. Lorentz-Polarization Factor (L.P.F.)

$$\mathrm{L.P.F.} = \frac{1 + \mathrm{Cos}^2 2\theta}{\mathrm{Sin}^2\theta \mathrm{Cos}\theta}$$

2. Calculated Relative Intensity Factor

$$= \frac{(I_c)_i}{\Sigma (I_c)_i}$$

Where $(I_c)_i$ = Calculated intensity per unit atomic volume (V = 1)

$$(I_c)_i = \frac{|F|^2 m(1 + \mathrm{Cos}^2 2\theta)}{\mathrm{Sin}^2\theta\ \mathrm{Cos}\theta}$$

3. Population Density, $V_i = (I_m)_i \,/\, \dfrac{(I_c)_i}{\Sigma (I_c)_i}$

4. Relative Population Density Index, I_R

$$I_R = \frac{V_i}{\left(\dfrac{\Sigma V_i}{n}\right)}$$

Where n = number of diffraction planes

Table III. Crystallographic Relationship Between Zonal Axes*

Ti-75A α-Ti	Ti-6Al-4V α-Ti	Zonal Axis $<1\bar{2}10>$ Zonal Plane with $(2\bar{1}\bar{1}0)$ Pole as Zonal Axis	Zonal Axis $<1\bar{2}10>$ Zonal Plane with $(1\bar{2}10)$ Pole as Zonal Axis	Zonal Axis $<1\bar{1}00>$ Zonal Plane with $(1\bar{1}00)$ Pole as Zonal Axis	Zonal Axis $<4\bar{5}10>$ Zonal Plane with $(4\bar{5}10)$ Pole as Zonal Axis
$(01\bar{1}0)$		$(01\bar{1}0)$			
(0002)	(0002)	(0002)	(0002)		
$(01\bar{1}1)$	$(01\bar{1}1)$	$(01\bar{1}1)$			
$(01\bar{1}2)$	$(01\bar{1}2)$	$(01\bar{1}2)$			
$(11\bar{2}0)$	$(11\bar{2}0)$			$(11\bar{2}0)$	
$(10\bar{1}3)$	$(10\bar{1}3)$		$(10\bar{1}3)$		
$(11\bar{2}2)$				$(11\bar{2}2)$	
	$(20\bar{2}1)$		$(20\bar{2}1)$		
(0004)		(0004)	(0004)		
$(01\bar{1}4)$	$(01\bar{1}4)$	$(01\bar{1}4)$			
$(20\bar{2}3)$			$(20\bar{2}3)$		
$(21\bar{3}1)$	$(21\bar{3}1)$				$(21\bar{3}1)$
$(11\bar{2}4)$	$(11\bar{2}4)$			$(11\bar{2}4)$	
$(01\bar{1}5)$	$(01\bar{1}5)$	$(01\bar{1}5)$			
$(20\bar{2}4)$			$(20\bar{2}4)$		
$(21\bar{3}3)$	$(21\bar{3}3)$				$(21\bar{3}3)$
$(30\bar{3}2)$			$(30\bar{3}2)$		

*All zonal axes lie perpendicular to the c-axis.

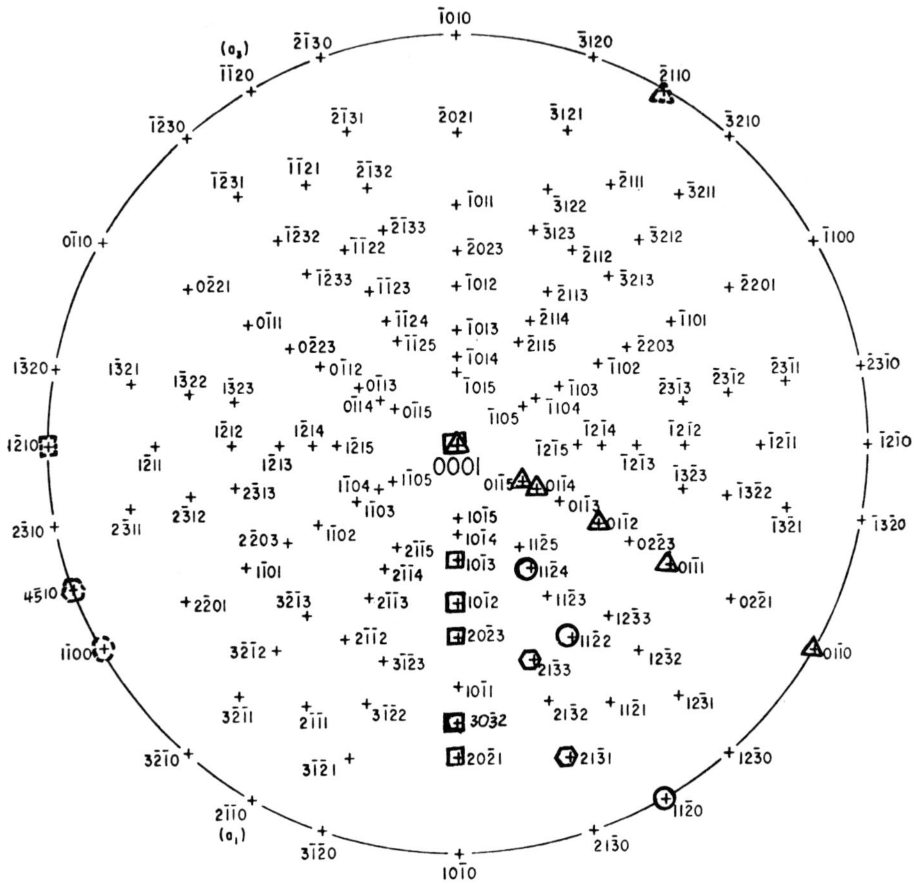

Fig. 3. (0001) standard projection for titanium (c/a = 1.5873).

Ti-75A and Ti-6Al-4V thin foils by multiple-component textures and their zonal relationships as described in this investigation.

Discussion

Table I lists the corrections for measured integrated intensities of various Bragg diffraction maxima of both cold rolled and vacuum annealed Ti-75A foils. Table II lists similar data for cold rolled and vacuum annealed Ti-6Al-4V foils. In both tables the population density index I_R which was plotted in Figures 1 and 2 was calculated from the measured integrated intensity I_m by the following correction factor:

$$I_R = \frac{V_i}{(\frac{\Sigma V}{n})} \quad \text{where } V_i = (I_m)_i \; / \; \frac{(I_c)_i}{\Sigma (I_c)_i}$$

Where $(I_c)_i = m|F|^2$ (L.P.) (for unit diffracting volume) and n = number of contributing diffracting planes.

Table III and Figure 3 show the zonal relationships between diffracting planes and zonal axes and also that of zonal axes and c-axis of the h.c.p. alpha titanium matrix phase (8).

Conclusions

The cold rolled and annealed textures in Ti-75 and Ti-6Al-4V 0.0025-inch thick foils were different from each other. In both cases alpha titanium lines predominated. Beta titanium lines were revealed only after vacuum annealing of the cold rolled Ti-6Al-4V foil.

In both Ti-75A and Ti-6Al-4V cold rolled and annealed thin foils, there are multiple texture components with zonal axes aligned with c-axis in alpha-titanium matrix.

Vacuum annealing depressed the (0002), $(10\bar{1}3)$, and $(01\bar{1}4)$ texture components in Ti-75A cold rolled foil.

Acknowledgments

The author would like to thank R. S. Nycum and A. B. Sauvageot of the Titanium Metals Corporation of America for providing cold rolled and vacuum annealed Ti-75A and Ti-6Al-4V 0.0025-inch thick foils.

References

1. Barrett, C. S. and Massalski, T. B., Structure of Metals, Third Edition, McGraw-Hill, 1966.

2. Wassermann, G. and Grewen, J., "Texturen Metallischer Werkstoffe," Springer, Berlin, 1962.

3. Williams, N. and Eppelscheimer, D. S., Transactions of The Metallurgical Society of AIME, Vol. 197, 1956, p. 1378. Keeler, J. H. and Geisler, A. H., Transactions of The Metallurgical Society of AIME, Vol. 206, 1956, p. 80. Clark, H. T., Transactions of The Metallurgical Society of AIME, Vol. 185, 1949, p. 588.

4. Keeler, J. H., Hibbard, W. R., and Decker, B. F., Transactions of The Metallurgical Society of AIME, Vol. 197, 1953, p. 932.

5. Koh, P. K., Journal of Applied Physics, Vol. 29, No. 4, 1958, p. 636.

6. Koh, P. K., Transactions of The Metallurgical Society of AIME, Vol. 221, 1961, p. 50.

7. Kasper, J. S. and Lonsdale, K. (Editors), "International Tables for X-Ray Crystallography - Vol. II. Mathematical Tables," International Union of Crystallography, Knock Press, Birmingham, England, 1967.

8. Hu, Hsun, Journal of Less Common Metals, 12, 1967, pp. 504-508.

DISCUSSION

SECTION VIIa. DEFORMATION

Co-Chairmen
Mr. I. Perlmutter
Air Force Materials Laboratory
Wright-Patterson AFB, Ohio

Professor U. Zwicker
Universität Erlangen-Nürnberg
Egerlandstr, Germany

Rapporteur
Dr. J.D. Boyd
Battelle Columbus Laboratories
Columbus, Ohio

Ramachandran (AMRL) stated that the true stress-true strain curves presented by Conrad showed a negative slop just before fracture, and requested an explanation for this. Also, he asked how Conrad selected the orientation factor for calculating the stress-strain curves? Conrad (University of Kentucky) replied that the apparent negative slope resulted from using elongation instead of reduction-in-area to calculate the true stress. The orientation factor was taken from earlier work by Armstrong on a variety of h.c.p. materials.

Fujishiro (AF Materials Laboratory) asked whether or not the ductility of titanium alloys should increase with aluminum additions if new slip systems are made available by the onset of $\langle c+a \rangle$ slip? Paton (North American Rockwell Science Center) replied not necessarily, since five independent deformation systems are provided by the operative slip and twinning systems in unalloyed titanium. The increase in critical resolved shear stress with increased aluminum content is probably responsible for the reduced ductility. Margolin (New York University) stated the work of Conrad and colleagues suggested that prism slip is the rate-controlling deformation mechanism in alpha titanium, and pointed out that other deformation mechanisms might be more important with respect to fracture. For example, it appeared that the fracture toughness of alpha-beta alloys was controlled by basal slip and void growth. Secondly, Margolin asked Conrad to comment on the observation that plots of H* against temperature for beta titanium fall on the same line as those for alpha titanium. Conrad replied, certainly other slip systems besides prism slip were operative, but prism slip was the rate-controlling deformation mechanism. This conclusion was based on comparisons of the data for polycrystalline specimens with those for oriented single crystals. The second point concerning the temperature dependences of H* for alpha and beta titanium illustrated that the interaction between dislocations and interstitials was not sensitive to crystal structure.

Margolin stated that the measurements of unit propagation energy (UPE) presented by Jones indicated that with increasing yield stress UPE initially increases, reaches a maximum, then decreases, and asked for an explanation for the decrease in UPE at high yield stresses. Jones (Franklin Institute Research Laboratory) replied that with decreasing grain size, the yield stress increased faster than the rate of work hardening, which eventually caused the UPE to decrease.

Weissmann (Rutgers University) said in the paper by Griffiths and Hammond that the authors emphasize that their material had a large grain size with a small subgrain size, and asked if the subgrains act as the effective grain size for a diffusion creep process. Hammond (University of Leeds) replied that the statement was correct and that the important point was that one can obtain superplastic behaviour without the usual prerequisite of obtaining a fine grain size. The subgrains are formed after 3-4 percent strain, and thereafter the materials behave superplastically. This has important consequences with respect to the beta forging of titanium alloys. In controlled beta forging the material flows superplastically, good die filling is ensured, and the subgrains refine the alpha precipitates which form upon cooling. Wilcox (Battelle Columbus Laboratories) asked whether the authors considered sub-boundary sliding as a possible mechanism of superplasticity? Hammond replied grain-boundary sliding is a complex process, presumably involving diffusion-controlled motion of defects in the grain boundaries. He found it difficult to envisage such a process occurring at well ordered sub-boundaries of low misorientation. Nevertheless, some form of sub-boundary diffusion, (Coble creep), may occur. Good agreement is found between the observed strain rates and those predicted by the Coble creep equation, although there is some uncertainty about the numerical values of subgrain boundary width and diffusivity.

Sommer (North American Rockwell Science Center) stated that with increasing temperature, titanium-oxygen alloys exhibit a well-defined transition in slip character from as-planar dislocation arrays to a tangled cell structure. He said it was surprising that this change in slip character did not appear in Conrad's thermal activation analysis. Conrad replied that he was not convinced that such a transition in slip character actually occurred, and that it was not obvious from transmission electron micrographs of deformed material Boyd (Battelle Columbus Laboratories) suggested that electron microscopy of surface replicas is a better technique than transmission electron microscopy for detecting marginal differences in slip character. This was illustrated by the observations on slip character in a titanium-aluminum alloy reported in the paper by R.G. Hoagland and himself.

PART B

FATIGUE AND FRACTURE

Fatigue Mechanisms in Titanium

D. E. MacDonald and W. A. Wood

School of Engineering and Applied Science

George Washington University

Washington, D.C.

Features of Titanium Fatigue

Features of special interest here are the following. First is the safe fatigue range, and the fact that the upper limit of this range may be no more than half the static elastic range (1). Both raise the question of how strain-amplitudes that are apparently elastic can cause an "elastic" fatigue only above a sharp limit.

Second is the observation that elastic amplitudes can cause some titanium grains to develop plasticity (1); they develop clear signs of slip, for example. The proportion of these grains is small for small amplitudes but may increase to some 20 or 30 per cent for amplitudes approaching yield. So it is tempting to ascribe the elastic fatigue to these plastic grains; and this previous workers have done (1). But evidence will be given below that it may occur by an independent mechanism at low cycle-frequencies (LF), frequencies of the order 100 c/sec. or less normally used for fatigue testing; and only by a special behavior of the plastic grains at ultrasonic or high frequencies (HF), frequencies of some 20,000 c/sec. or more.

A final feature, indicated by observation described below, is that titanium can tolerate surprisingly high internal stresses. It is to these that the peculiarities of its fatigue will be attributed.

To provide comparable illustrations of these features the observations will refer to the same commercially pure titanium annealed to a grain size ∿ 0.001 in. and having main impurities C 0.016, Fe 0.112, N 0.008, O 0.228, H 0.007 wt. per cent. The LF fatigue was produced by alternating torsion of given strain-amplitudes applied at 30 c/sec. to specimens with electropolished tubular test-portions 1 1/2 in. long, 1/4 in. o.d., and 3/16 in. i.d. The HF fatigue was produced in an apparatus (2) that applied alternating push-pull at 17.000 c/sec. to electropolished tubular test-portions 0.45 in. long, 0.20 in. i.d., and 0.18 in. i.d. The resulting microstructural changes were studied by scanning electron-microscope (SEM). All photographs reproduced here are mounted with their horizontal direction parallel to the specimen axis so as to permit comparison of slip and crack directions with directions of the applied cyclic stress.

The Plastic Grains

Study of these was facilitated by the SEM because their slip bands extruded debris from which the secondary emission under the incident beam is enhanced; so they stood out brightly, generally as bright ribbon or point extrusions, Figure 1. But a more significant aspect of this activity was revealed by a light etch in 3 per cent HF. This removed the debris and showed that it issued from holes and cavities in the slip bands, Figure 2a. The cavities, as was clear from stereomicrographs, were about as deep as long. Thus they indicated the early form that fatigue damage took in the plastic grains. It was in fact always possible to find too a still earlier form where the damage consisted merely of holes along potential slip bands, Figure 2b.

However the observation of special significance here was that the damage never progressed beyond these early forms. It never progressed to the extent that plastic fatigue normally progresses in a wholly plastic metal like copper (3). To emphasize this point it will be convenient to reproduce Figure 3 from similarly tested copper. This exemplifies the normal massive damage: a dense distribution of slip-band fissures traverses the grains; cross-linking fissures appear at grain and sub-grain boundaries. Therefore it was difficult to see how the sparse cavities in isolated titanium grains could provide a comparable mode of fatigue fracture. This conclusion was confirmed by observations on how the titanium did in fact fracture.

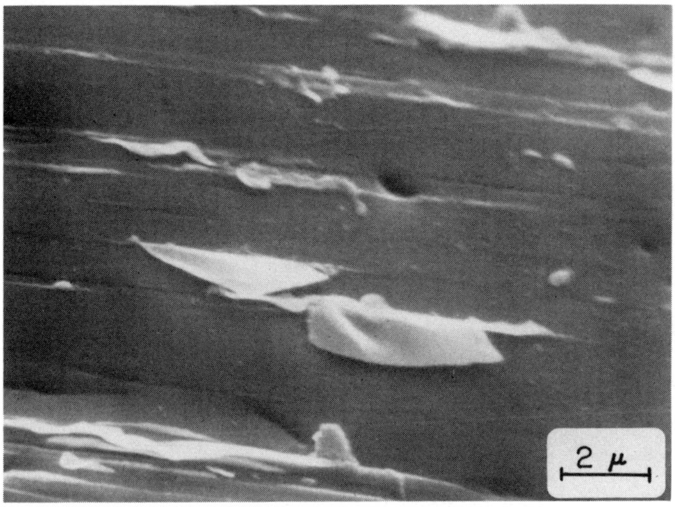

Fig. 1. Ribbon and point extrusions in slip zones in specimen with 5 x 10^3 cycles at $\pm$ 0.009 strain.

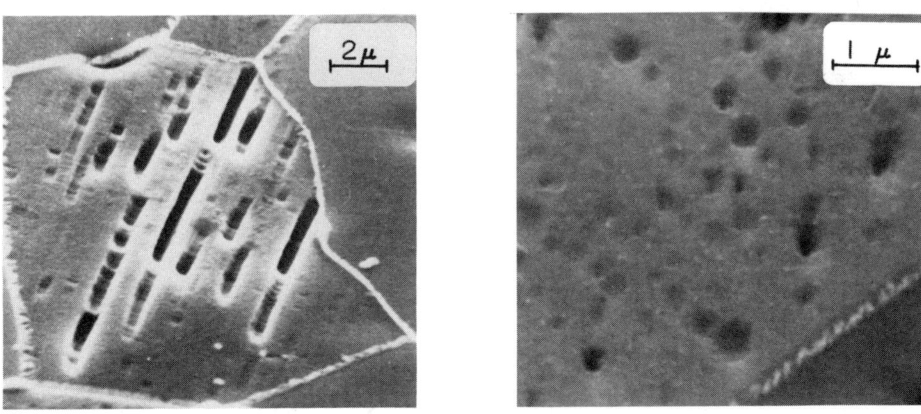

A B

Fig. 2. a) Etched up slip zone notches in specimen with 3 x 10^4 cycles at $\pm$ 0.0045 strain. b) Etched up holes along potential slip bands in a similar specimen.

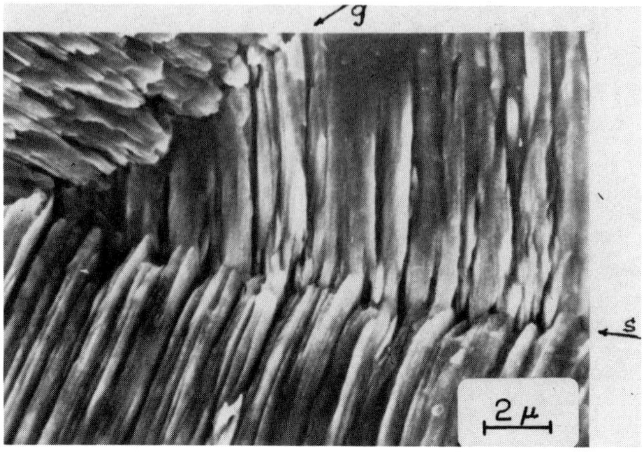

Fig. 3. Copper specimen with 10^6 cycles at $\pm$ 0.0045 strain etched to show slip band fissures in the grains and the cross-linking fissures at grain (g) and subgrain (s) boundaries.

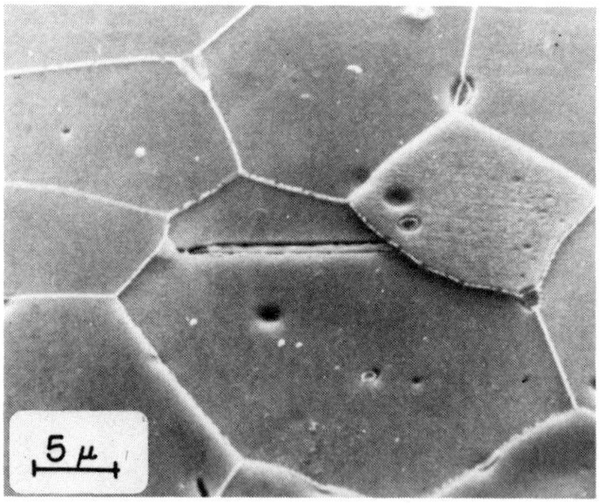

A

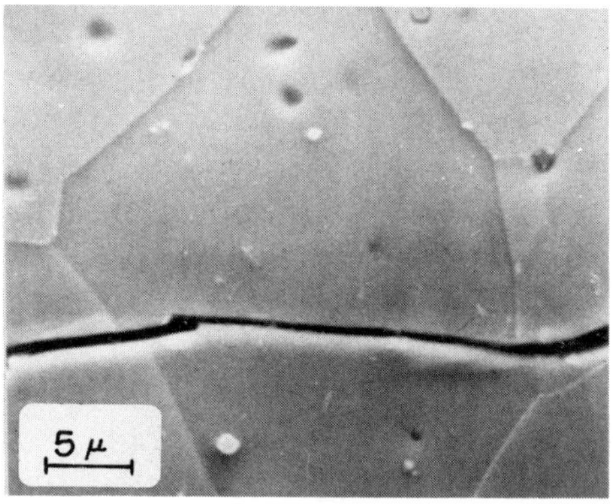

B

Fig. 4. a) Transcrystalline cleavage of a single grain in an etched specimen with 3 x 10^4 cycles at $\pm$ 0.0045 strain. b) Transcrystalline cleavage of several grains in the same specimen, unetched.

LF Fatigue by Cleavage

The difference in mode of fracture was especially clear in the LF fatigue. Here it was found that cracking in fact originated in the elastic grains, not the plastic ones. A grain developed a transcrystalline cleavage in a direction of maximum applied shear, and this crack then extended in the same shear direction by cleaving one grain after another, always without any accompaniment of noticeable plasticity or slip, Figures 4a, b. Further evidence that cracking was of this brittle type was that the crack continued in the same general direction despite the different crystallographic orientations of the grains that it traversed. Still further evidence was obtainable when a crack became wide enough for the SEM to penetrate between its walls. These then could be photographed in situ. They were typical cleavage surfaces, Figure 5a.

Other confirmation was made possible by similarly photographing in situ the internal walls of secondary cracks that the primary crack sometimes started when it encountered plastic grains, or triple points, or happened to coincide with a grain boundary. These fracture surfaces were of a rough ductile type; as shown by Figure 5b from the wall of a secondary grain-boundary crack they were quite different from the above cleavage type.

The HF Fatigue Cracking

This differed from the LF type in now utilizing a plastic grain, but the way it developed still differed basically from the way plastic fatigue normally developed. The HF deformation merely produced occasional slip-band fissures in a plastic grain. One of these it might propagate partially into the grain; but if this grain were elastic it did so with evident difficulty, for the crack would tail off in an irregular manner, Figure 6. Thus fracture in practice began in some area where enough of these small extended cracks might link into a large one.

That this HF fatigue have a safe range followed from observations on the internal friction Q^{-1} as the amplitude was slowly increased from a low strain of about 10^{-6}, Figure 7. The value of Q^{-1} at first remained roughly constant until the amplitude approached a critical value, here $\sim 10^{-3}$. Then it rose rapidly because it then started slip and cracking of the kind just illustrated. This strain amplitude was of the

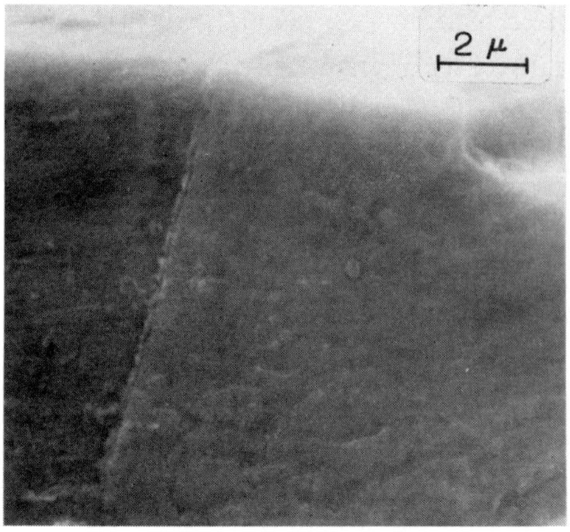

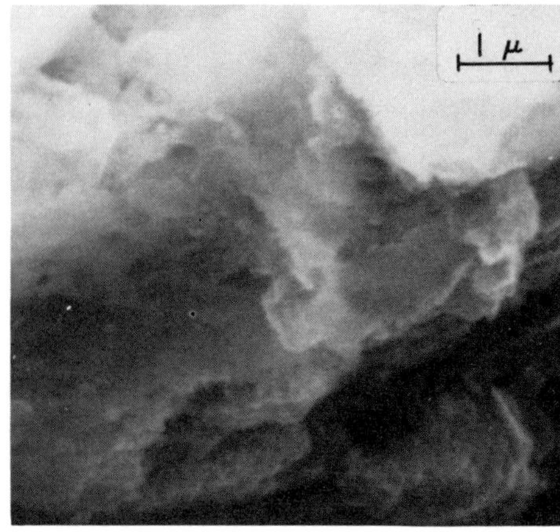

A B

Fig. 5. a) Fracture surface of cleaved grain in a specimen with 3 x 10^4 cycles at $\pm$ 0.0045 strain. b) Fracture surface of a secondary grain-boundary crack in the same specimen.

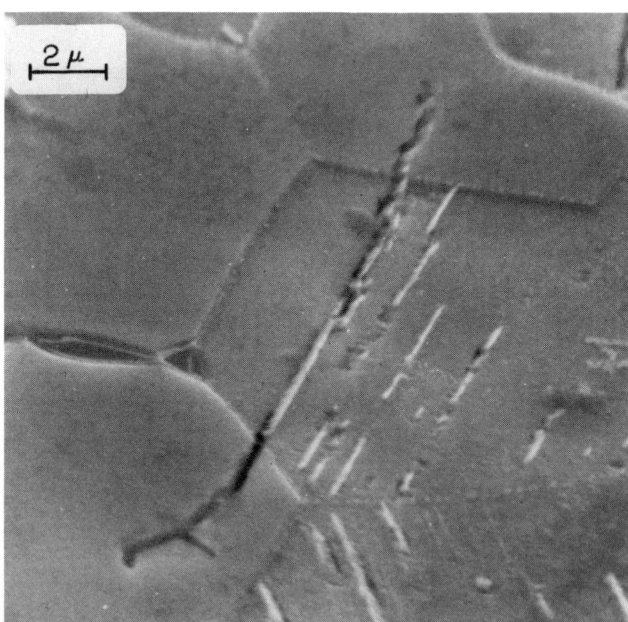

Fig. 6. Etched up slip band microcrack in an ultrasonically fatigued specimen.

same order as the safe limit at low frequency.

The High Internal Strain

This at low frequency made itself evident where one of the slipless cracks in its early stages crossed a grain boundary; there it would produce a striking shear in the boundary, Figure 8. From this it seemed reasonable to conclude that cracking of the grain had released a shear strain of an order given by the length of the step over the grain size. This strain in practice reached values of about 0.05, equivalent to a stress of 0.05 G, where G is the shear modulus, and of the same order as the theoretical strength of metal, generally put at about 0.1 G.

At high frequency, comparable internal stresses could make themselves evident where one of the HF slip bands crossed a surface marker. For such markers it was possible to utilize occasional fine straight lines that remained on the metal surface after fabrication, annealing, and electropolishing. An example of such a line and the steps produced in it by the HF slip appears in Figure 9a, b. The key diagram Figure 9b show that the slip has shifted point C to C' and D to D', but has not shifted the grain boundary at points F where the slip bands enter the boundary zone. Therefore the boundary is containing shear strains CC'/BC and DD'/CD; and these, as inspection at once shows, are $\sim$ 0.1, again indicating that internal streses of the order of the theoretical strength can develop locally in a titanium grain.

Discussion

A point of main interest is the last one, the high internal stress. In the material used here, bulk yield occurred as the elastic strain in torsion reached 0.009 shear. The elastic amplitudes employed were always less than this value and therefore much less than the local internal stresses $\sim$ 0.1 G to which they could give rise. Similarly the HF amplitudes were much less than the local internal stress they could produce. Thus in some way it is possible for a low cyclic stress to give rise at least momentarily to a high unidirectional one.

A full explanation must await further observations, but it could take the following form. We recall that to produce an elastic range of any significance the dislocations in a metal must be strongly pinned.

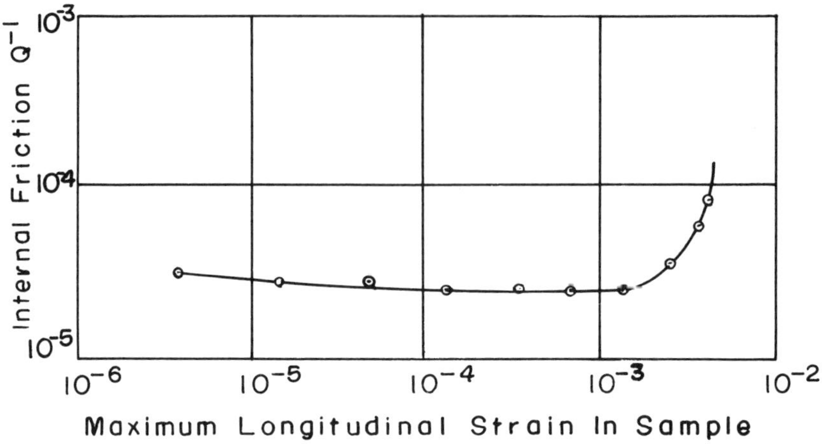

Fig. 7. Internal friction for ultrasonic specimen taken to fatigue.

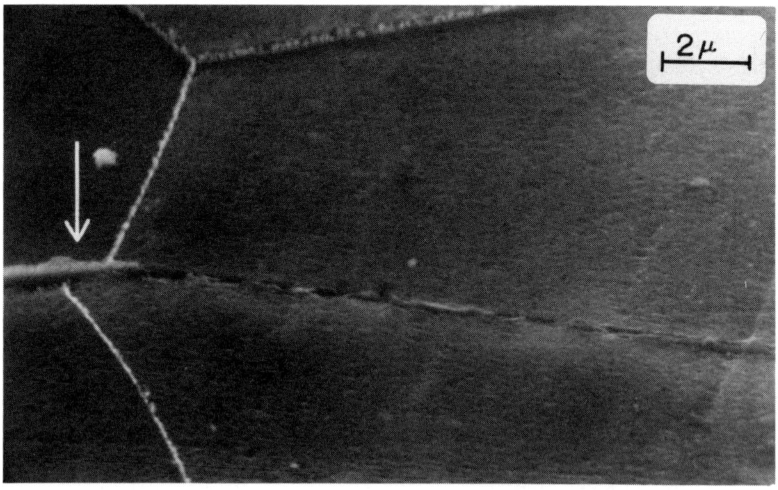

Fig. 8. Grain boundary displacement produced by a slipless crack in a specimen with 3 x 10^4 cycles at ± 0.0045 strain.

A

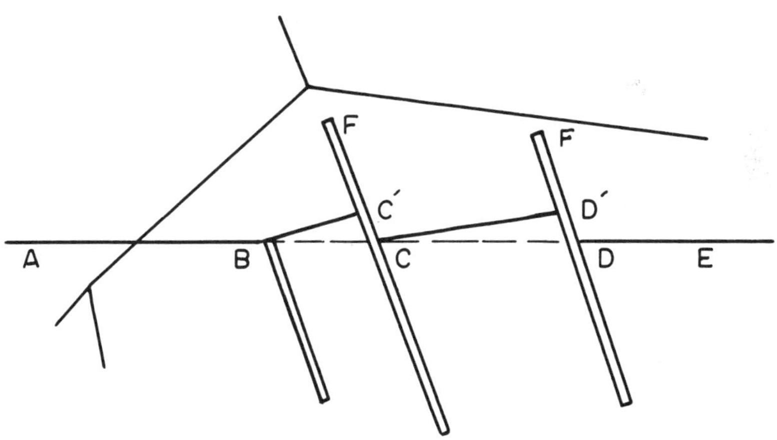

B

Fig. 9. a) Strain concentration at slip bands in ultrasonically fatigued specimen. b) Diagram shows that strains CC'/BC and DD'/CD are on the order of 0.1.

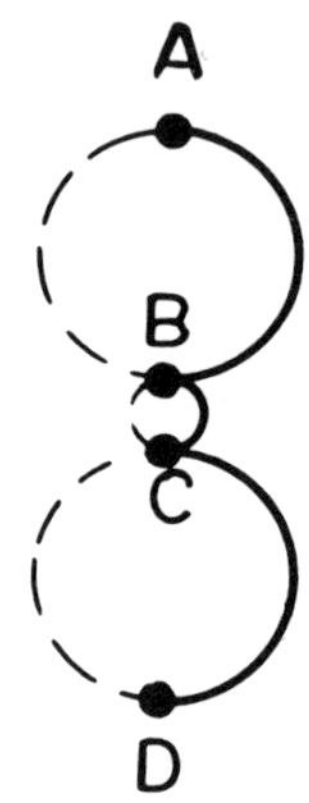

A

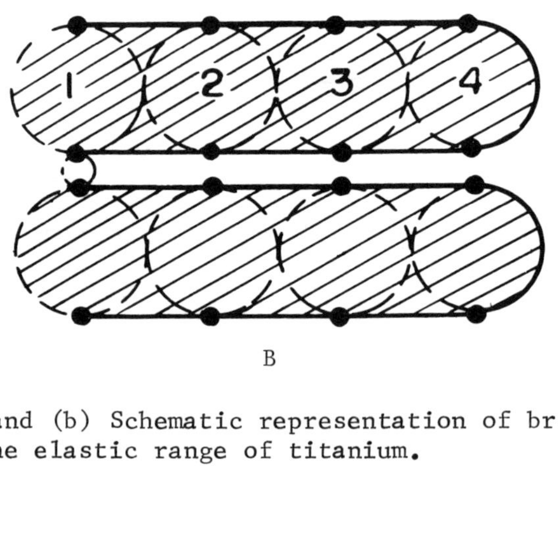

B

Fig. 10. (a) and (b) Schematic representation of brittle cracking in the elastic range of titanium.

Suppose AD in Figure 10a is pinned at intermediate centers represented here as points B and C (though pinning centers in practice might be of various kinds; sub-boundaries for example).

Up to a particular amplitude the dislocations simply oscillate about the pinning centers, and there is no reason for fatigue. The material has a safe range.

Above this amplitude, parts of the dislocation between more widely spaced pinning centers break away as in Figure 10b. But in an elastic matrix, with its many pinning centers, they cannot travel far before they are recaptured. Under cyclic strain they might wander to an fro between positions 1 to 4, for example.

But parts of the dislocation between nearer centers such as BC are less likely to break away; their amplitude is less; they are attracted to their own centers rather than to neighboring ones.

Therefore we may have a relatively stable layer between the unstable ones shaded in Figure 10b. And during the instants that a dislocation is moving between centers in the shaded layer, the externally applied load is borne mainly by the intervening layer. The layer is momentarily overloaded. Depending on the relative areas of the shaded and unshaded regions this overload could be large enough for fracture.

Further it should be noticed that the direction of the overload is simply the direction of the externally applied stress, whereas the direction of dislocation movement is likely to be a crystallographic slip direction. The two directions in titanium with its few slip directions may differ widely. Therefore the overload need not start slip.

Finally the wandering of individual dislocations does not constitute slip; for this a concerted avalanche of dislocations must break away. Therefore the fracture caused by the overload becomes slipless cracking in the direction of maximum applied stress.

Acknowledgement

This research has had the support of the Office of Naval Research and the Air Force Materials Laboratory.

References

1. e.g., Turner, N.G. and Roberts, W.T., "Fatigue Behavior of Titanium," Transactions of the Metallurgical Society of AIME, Vol. 242, July 1968, pp. 1223-1230.

2. Mason, W.P. and Wood, W.A., "Fatigue Mechanism in fcc Metals at Ultrasonic Frequencies," Journal of Applied Physics, Vol. 39, November 1968, pp. 5581-5584.

3. Wood, W.A., The Study of Metal Structures and their Mechanical Properties, Pergamon Press, New York, 1971, Ch. 10.

THE STRESS-RESPONSE OF CYCLICALLY-DEFORMED ALPHA-TITANIUM

J.I. Dickson[1], J.P. Owens[2,3] and A. Plumtree[2]

1 Département de Génie Métallurgique, Ecole Polytechnique, Montréal, Qué. Canada
2 Department of Mechanical Engineering, University of Waterloo, Waterloo, Ont., Canada
3 Now with Deloro Stellite Ltd., Belleville, Ont., Canada

Room-temperature fatigue tests were carried out on annealed, and on monotonically or cyclically prestrained titanium at controlled strain amplitudes between $\mp$ 0.25% to 1.5%, which gave fatigue lives in the 100 to 25,000 cycle range. At the lower cyclic strain amplitudes employed, the saturation stress depended on prior strain history but not at the higher amplitudes. For the same stress attained in prestraining, the saturation level was independent of the method of prestraining. The difference in saturation levels observed at low amplitudes is attributed to the higher twin densities present in the prestrained samples.

Prestraining significantly reduced the fatigue lives. Examination of the fatigue cracks showed that in non-prestrained samples, both stage I and stage II cracks had a stronger tendency to follow twins than in the monotonically prestrained samples. In the latter samples, stage I cracks had almost as strong a tendency to follow slip lines as to follow twins while stage II cracks made fewer deviations to follow twins, which suggests that the effect of a monotonic prestrain is to make crack paths along slip bands relatively easier. Cracks observed to run along or parallel to twins followed fragmented $\{10\bar{1}2\}$ twins more often than $\{11\bar{2}2\}$ or $\{11\bar{2}1\}$ twins.

Introduction

Low cycle strain-controlled tests on different metals have shown that the saturation stress depends on strain amplitude, temperature, stacking fault energy and cyclic deformation rate (1-3). Previous cyclic studies, such as the work of Feltner and Laird (4), have shown that the fatigue behaviour of metals can be determined by the slip mode character. At room temperature, the strain cycling of metals possessing a high stacking fault energy, i.e., having a wavy slip mode, will result in a unique saturation stress, which is independent of prior strain history. Metals having low stacking fault energy and deforming in a planar slip mode exhibit different saturation stresses for different strain histories. In these latter metals, recovery is difficult and prior deformation increases the saturation stress.

When mechanical twinning is important, its influence must also be considered and, indeed, has received considerable attention (5-10). Recently, Owens et al (11) found that prestraining titanium caused a reduction in the short to medium fatigue lives, which they associated with stress concentrations created at twins or at grain boundaries. The present reports a more detailed study of prestraining on the fatigue behaviour in the short to medium life range, where Stage II crack growth should be dominant.

Experimental Procedure

High purity titanium rods of 3/8 in. diameter prepared from electro-refined titanium sponge were machined into fatigue samples having threaded ends and gauge lengths of 0.3" over the minimum diameter of 0.2". On one sample, 0.1" flats were machined on the reduced section. Following a subsequent one-hour anneal at 750°C, the grain size was approximately 300 μ , the hardness R_B 10 and the oxygen and hydrogen contents determined by gas analysis were 300 p.p.m. and 17 p.p.m., respectively (0.03 and 0.0017 wt%).

Push-pull cyclic testing was carried out on annealed and on prestrained fatigue samples using a closed-loop electrohydraulic servo-controlled MTS machine. A clip-on extensometer permitted control of the total strain amplitude. A sinusoidal wave-form and constant strain amplitudes of $\mp$ 0.25, $\mp$ 0.50, $\mp$ 1.0 and $\mp$ 1.5% were employed. A frequency in a range of 0.33 to 2 cps, within which the material was frequency insensitive, was chosen to facilitate recording the load response.

Prestraining was produced either by a monotonic tensile strain of approximately 20% to a stress of 36,600 psi or by cyclic deformation at a total strain amplitude of $\mp$ 1.5% for approximately 10 cycles to the same stress. Both load and strain were monitored on a strip-chart recorder. An X-Y plotter recorded load-strain hysteresis loops of interest. The cyclically prestrained samples were fatigued at strain amplitudes of ∓ 0.25% or ∓ 0.50%. To prevent the introduction of residual stresses, the transition between the over-strain cycles and the lower strain level was closely controlled.

After a monotonic prestrain of 40%, the sample with machined flats was cycled for 50,000 cycles at a constant stress amplitude of 25,000 psi, followed by cycling at a constant strain amplitude of ∓ 0.25% to failure.

With the exception of a cyclically prestrained ∓ 1/2% sample, the surfaces and various sections of a full set of cycled and of uncycled prestrained samples were examined by optical microscopy. Two annealed ∓ 1/4% samples as well as two samples monotonically deformed to fracture were examined.

These experiments were supplemented by reverse three-point bending tests on rectangular parallelepiped samples, some of which were prestrained 10 to 20% in tension. Most cross-sections of these strain-annealed reverse-bending samples contained between 3-5 grains. The strain amplitude was controlled to give approximately the same amount of twinning in the central section as that observed in the fatigue samples cycled at amplitudes of ∓ 1 1/2%.

Results

Mechanical Testing of the Fatigue Samples

Graphs of average cyclic flow stress versus the logarithm of the number of cycles were plotted for each test (11). These graphs (e.g.,Fig. 1) indicated that cyclic and monotonic prestraining to the same stress level of 36,600 psi results in softening and in a common saturation stress on subsequent cycling at strain amplitudes of ∓ 1/4% and ∓ 1/2%. At these low strain amplitudes, the saturation level for the prestrained samples was higher than for the non-prestrained samples. This difference was greater for the smaller strain amplitude. For cyclic strain amplitudes of ∓ 1.0 and ∓ 1.5%, the saturation stress was larger than the stress reached on prestraining. A difference in saturation levels was not observed and the saturation stress may be regarded as unique. The cyclic stress-strain curve of saturation stress versus plastic strain amplitude (Fig. 2) was constructed from the stable load-strain hysteresis loops recorded. Fig. 2 also includes the sections of the monotonic deformation curves which cover the same plastic strain intervals for both non-prestrained and monotonically prestrained material. For a given plastic strain, the difference in stress levels between the monotonic and the cyclic curves indicates the amount of hardening or softening to saturation expected. For the ∓ 1.5% total strain amplitude samples, the saturation stress approached the ultimate tensile strength (59,000 psi).

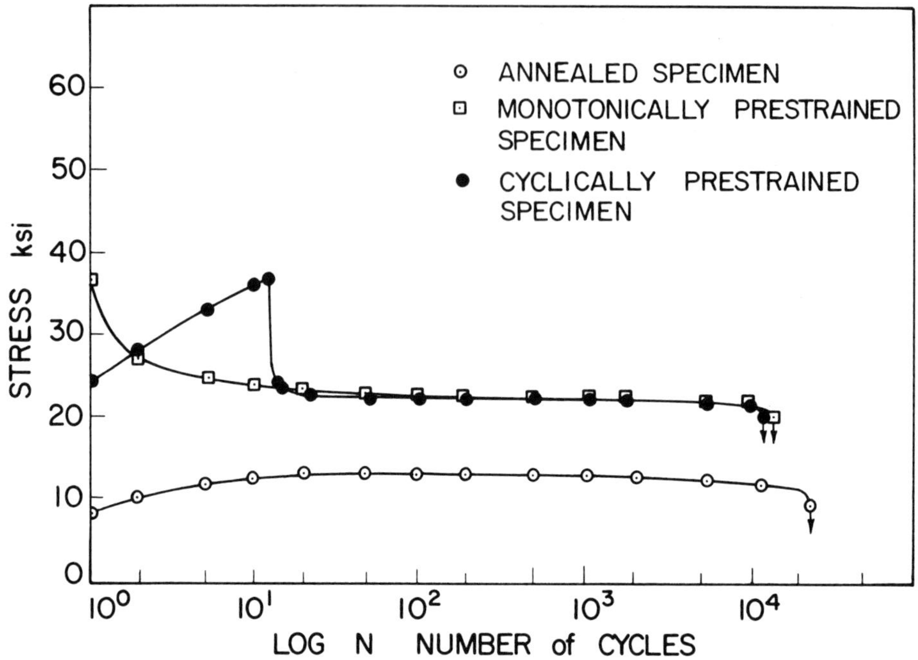

Fig. 1. Stress response of α-titanium cycled at $\mp$ 0.25% total strain.

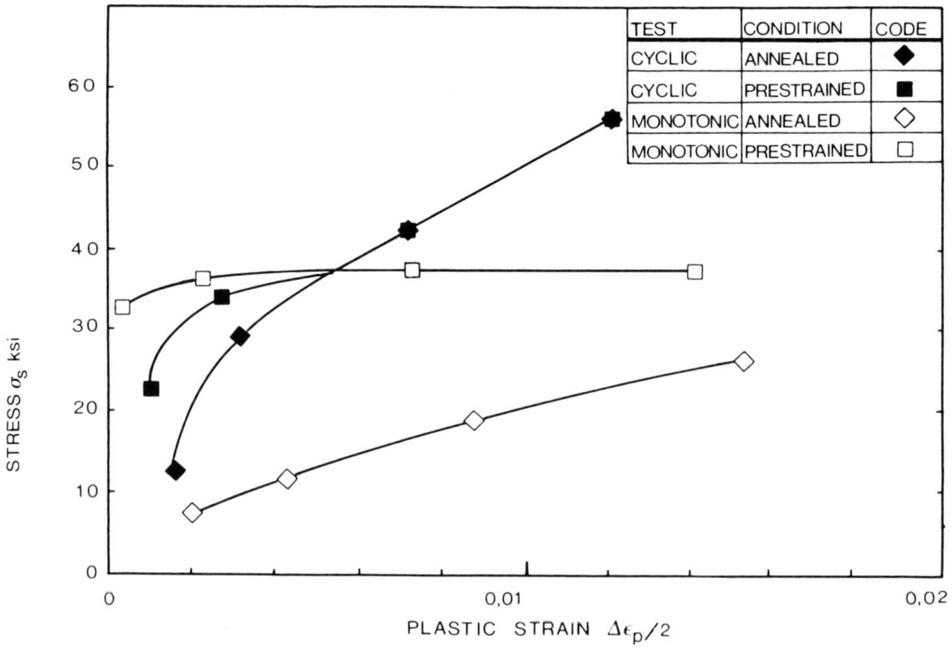

Fig. 2. Cyclic and monotonic stress-strain curves.

At the same strain amplitude, prestraining reduced the fatigue life by as much as 50%. This reduction was almost independent of the type of prestrain, but, if anything, was slightly larger for cyclic prestraining. The fatigue lives varied from 210 to 25,000 cycles in the non-prestrained samples and from 130 to 12,500 cycles in the prestrained samples.

Microscopical Observations

Reverse Bending Samples The twins present in the reverse bending samples were predominantly of the $\{10\bar{1}2\}$ and $\{11\bar{2}2\}$ types with $\{11\bar{2}1\}$ twinning occurring less frequently. When first formed, $\{10\bar{1}2\}$ twins were often wide and lenticular; $\{11\bar{2}1\}$ twins were thin with parallel interfaces and often occurred in zig-zag patterns; $\{11\bar{2}2\}$ twins had intermediate shapes. One or more types of secondary twins were often present in $\{11\bar{2}2\}$ twins, and, as the number of cycles increased, quite thick $\{10\bar{1}2\}$ secondary twins often formed in $\{10\bar{1}2\}$ twins, at times giving the impression of untwinned holes in the original twin since a misorientation of about 10° with the matrix can thus result.

Based on surface observations, untwinning completely removed some $\{10\bar{1}2\}$ and $\{11\bar{2}1\}$ twins within one cycle of their formation. In other cases, $\{10\bar{1}2\}$ and $\{11\bar{2}1\}$ twins detwinned to single long, thin twins. Frequently $\{10\bar{1}2\}$ twins and less frequently $\{11\bar{2}2\}$ twins untwinned to several parallel twin fragments (Fig.3). All marked untwinning of $\{10\bar{1}2\}$ or $\{11\bar{2}2\}$ twins introduced in tensile prestraining was of the fragmentation variety. In both types of twins, very irregular interfaces were at times produced by fragmentation and by reverse twin boundary motion. In this paper, the term "fragmentation" will be employed to denote the production of these irregular interfaces. In some cases, more or less cristallographic indentations, the number of which usually increased with increasing amounts of reverse twin boundary motion, were observed in twin interfaces (Fig. 3). The untwinning of $\{11\bar{2}1\}$ twins often left a number of widely separated twin sections behind but, otherwise, the $\{11\bar{2}1\}$ twin interfaces remained quite regular.

Similar to results described by Partridge (6, 7), extrusions were observed to form on slip bands as well as on some twin boundaries (Figs.3 and 4).

One reverse bending sample was cycled to fracture, which occurred after 90 cycles.

Microstructure of the Fatigue Samples The twins in the fatigue samples were identified from their shapes. Fragmented twins and large twins were either of the $\{10\bar{1}2\}$ or $\{11\bar{2}2\}$ type. With both large and thin twins, examination of the twin characteristics often permitted definite identifications. Identifications were verified by one-surface analysis on several grains whose orientations were

Fig. 3. Fragmentation of $\{11\bar{2}2\}$ twins having indented interfaces. An extruded line is present along one twin boundary. Mag. 700X.

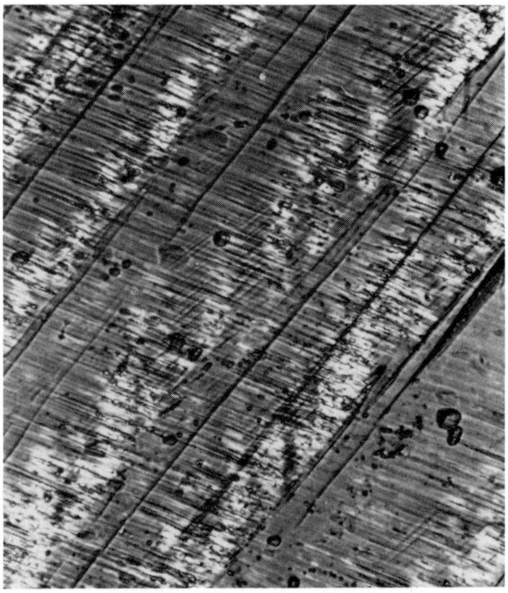

Fig. 4. Extrusions on prismatic slip lines in a region containing fragmented $\{10\bar{1}2\}$ twins. Mag. 300X.

obtained from slip traces produced by surface deformation and by utilisation of a technique to obtain basal plane traces (12).

As in the reverse bending samples, $\{10\bar{1}2\}$ twins were observed somewhat more frequently than $\{11\bar{2}2\}$ twins, with $\{11\bar{2}1\}$ twins occurring least frequently. Only the non-prestrained samples cycled at ∓ 1/4% contained a low twin density. One of these was almost free of twins. All other samples were moderately to heavily twinned. Prestrained samples were more heavily twinned than non-prestrained samples and the twin density increased with increasing cyclic strain amplitude. In the non-cycled monotonically prestrained sample, the twin boundaries were regular and some secondary twinning was observed. In all samples cycled at ∓ 1/4% only a small percentage of twins were fragmented. The cyclically prestrained ∓ 1/4% sample and the only cyclically prestrained sample had almost identical microstructures and contained twins which were slightly more fragmented. In the samples cycled at strain amplitudes equal to or greater than ∓ 1/2%, most of the twins were heavily fragmented. In the ∓ 1 1/2% samples and in the ∓ 1% prestrained sample, a large number of twins containing secondary twins were present. These twins containing twins also often had irregular interfaces but usually on a more microscopic scale (Fig. 5).

<u>Crack Observations</u> Besides the main crack that caused separation, many samples contained several large cracks which propagated at least 15% through the interior, as well as shallow cracks which could be 0.15 in long but which usually stopped at the first or second grain boundary below the surface. With the exception of the annealed sample cycled at ∓ 1/4% which contained almost no twins, the number of both large and shallow cracks present in the non-prestrained samples were several times higher than in the pre-deformed samples.

Observation of the cracks which had not completely propagated through the sample showed that shallow cracks and the sections near the surfaces of large cracks tended to follow irregular paths. From examination of the surface before and after polishing, it was found that, in the non-prestrained samples, the stage I cracks were parallel to twins along most of their lengths, while, in the monotonically prestrained samples, these cracks had almost as large a tendency to follow slip planes as to follow twins. Although the microscopic observations on the sample with machined flats has been previously described (11), on reexamination a strong tendency for the crack to follow slip traces, especially traces on which extrusions were discernible was observed. In the interior the crack had a stronger tendency to run perpendicular to the specimen axis; however, in all non-predeformed samples, sharp deviations in the stage II crack paths parallel to twins were observed (Fig. 7). In annealed samples cycled at ∓ 1/4% and ∓ 1/2%, some large cracks had stopped on reaching grains not containing twins. Especially in these more lightly twinned samples, it was evident that crack propagation often

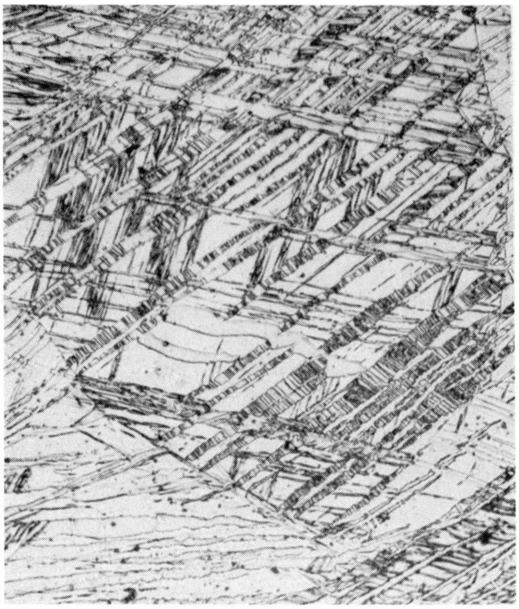

Fig. 5. The microstructure of the monotonically prestrained $\mp$ 1-1/2% sample consists mainly of fragmented $\{10\bar{1}2\}$ twins, many of which contain secondary twins. Mag. 300X.

Fig. 6. Portion of a crack on a monotonically prestrained sample, part of which is parallel to prismatic slip traces. Mag. 200X.

caused considerable twinning. Only slight, less frequent deviations parallel to twins were observed in the stage II cracks of monotonically prestrained samples. Some tendency for stage I and stage II cracks to follow grain boundaries was also observed, although at times such grain boundaries were parallel either to prism plane traces or to existing twins. The cyclically prestrained sample examined did not contain any noticeable cracks other than that which had caused fracture.

Both in surface and in interior grains, the twins which acted as preferential paths for crack propagation were predominantly quite large, fragmented twins including a few lightly-fragmented $\{10\bar{1}2\}$ and $\{11\bar{2}2\}$ twins containing secondary twins. The large majority of these appeared to be $\{10\bar{1}2\}$ twins. Only a small number of the crack deviations were parallel to thin twins where a $\{11\bar{2}1\}$ identification appeared reasonable.

On sections spark-cut very close to the surface of an uncycled sample monotonically strained to fracture, over a dozen very thin but distinct cracks, some of which were discontinuous, were observed. In several of the grains containing these cracks, chemical polishing had etched up the slip bands. The cracks had a strong tendency to follow these slip lines (Fig. 8) but only a weak tendency to follow twin boundaries.

Discussion

The cyclic stress-strain curve given in Fig.2 shows that at low plastic strains, the saturation stress levels for both annealed and prestrained titanium depend upon strain history; however, the saturation stresses are unique when they exceed the flow stresses of the monotonically predeformed metal. With the exception of annealed material cycled at $\mp$ 0.25%, a large number of twins were present in the cycled specimens, especially at high strain amplitudes. Fig. 2 shows that increasing the strain amplitude causes the saturation stress for the non-prestrained samples to eventually converge with that of the prestrained specimens. The difference in the saturation stresses at the lower amplitudes can then be explained by the different amounts of twins present. The hardening due to prestraining is partially associated with twinning and is not completely recoverable on strain cycling at lower strain levels. Since twin fragmentation is associated with twin boundary-dislocation interactions (5) and presumably with a high density of dislocations resulting from twin boundary motion, only "clean" removal of twins may be expected to contribute significantly to softening. The results suggested that clean untwinning only occurs when the twins are removed shortly after their formation; i.e., before significant amounts of dislocation-twin boundary interactions can pin sections of the twin interfaces. Since, in the reverse bending samples, all observations of untwinning of previously introduced twins was of the fragmentation

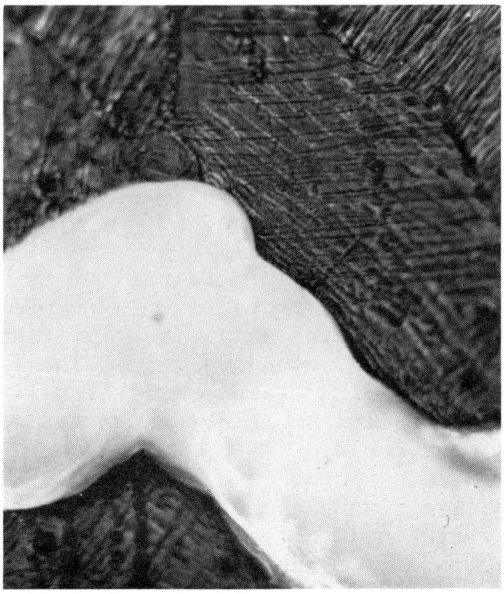

Fig. 7. A stage II crack in a non-prestrained sample changes direction to follow some fragmented twins. Mag. 90X.

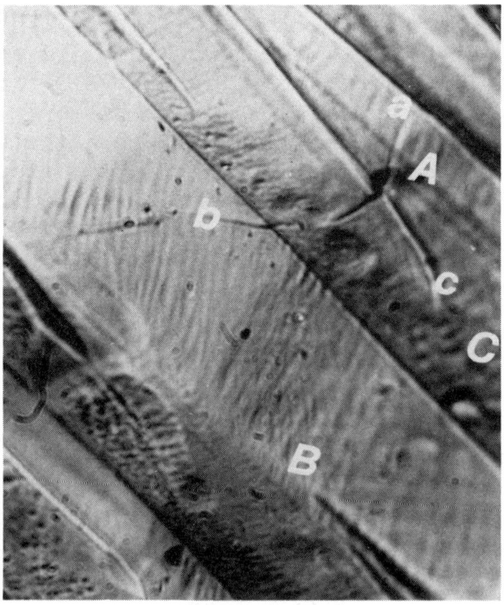

Fig. 8. One end of a thin crack (abc) observed on a section of a sample, monotonically strained to fracture. Branches of the crack are parallel to slip traces (A, B, and C) in the matrix and in the twin. Mag. 1500X.

variety, softening must be primarily associated with dislocation rearrangements. At high strain amplitudes, the stress required for subsequent strain cycling was higher than that attained in prestraining and no marked softening occurred. Prestraining, however, permitted saturation to be attained more quickly.

The results show that prior deformation causes a significant reduction in fatigue life which is almost independent of the type of prestrain. Although the saturation stress is not unique at low strain amplitudes, it is at high amplitudes and, thus, is not dominant in determining fatigue life. Prestraining reduces fatigue life whether hardening or softening occurs during subsequent cycling. As previously suggested (11), this reduction must be related to damage caused by the prestrain.

In non-predeformed samples, twin boundaries, especially fragmented $\{10\bar{1}2\}$ twins, were preferential crack paths in both stage I and stage II growth. The contrast between this result and that observed by Beevers and co-workers (8-10), who report that $\{11\bar{2}1\}$ twins are the predominant preferential paths in interior grains, may be due to the relatively small amounts of $\{11\bar{2}1\}$ twinning or to the shorter fatigue lives in the present case. While secondary twinning tended to limit the size of the twin boundary irregularities, some cracks also followed such twins.

In the monotonically prestrained samples, stage II cracks had only a slight tendency to follow twins, suggesting that easier propagation paths may now be more important. Stage I cracks had almost as large a tendency to follow slip planes as to follow twins. Interestingly enough, some thin cracks parallel to slip traces were observed near the original surface of a sample monotonically pulled to fracture. The results, therefore, suggest that, in the monotonically prestrained samples, stage II crack paths may have been determined by several different slip planes acting as easy propagation paths, while, in the non-prestrained samples, paths parallel to twin boundaries were usually easier.

The large number of long shallow cracks in some samples indicates some difficulty in the transition from stage I to stage II crack growth. The larger number of both large and small cracks in the annealed samples may have resulted from slower crack propagation permitting a comparatively longer time for the formation of cracks than that available in the prestrained samples.

Conclusions

In the low to medium life range, the strain cycling of titanium at low strain amplitudes results in a saturation stress which, because of the inability of the material to recover from twinning during prestraining, depends on the prior strain history. At high strain amplitudes, where the associated saturation level is higher than the prestrain flow stress, the saturation stress is unique.

In this life range, fragmented twins act as preferential crack propagation paths; however, monotonic prestraining decreases the tendency of cracks to follow twins, increases the tendency of stage I cracks to follow slip bands and also favors less deviations in the stage II crack paths. This behaviour appears to be associated with damage formed on slip planes during the prestrain.

Acknowledgements

The authors would like to thank the U.S.Bureau of Mines, Nevada, for supplying the electro-refined titanium sponge, the Department of Energy, Mines and Resources, Ottawa, for melting and swaging the material and Atlas Steels Company, Welland, Ontario, for carrying out the gas analysis. Also, the authors acknowledge Dr. T.H.Topper, Dr. P.Watson and Mr. H.Abdel-Raouf of the University of Waterloo for helpful discussions and Messrs. G.Guérin and J.Claudinon of Ecole Polytechnique for technical assistance.

Financial assistance from The National Research Council of Canada is gratefully acknowledged.

References

1. AVERY, D.E. and BACKOFEN, W.A., "Fatigue Hardening in Alloys of Low Stacking-Fault Energy", Acta Met., Vol.11,1963,pp.653-661

2. FELTNER, C.E. and LAIRD, C., "Cyclic Stress-Strain Response of F.C.C. Metals and Alloys-I and II", Acta Met., Vol.15, 1967, pp. 1621-1653

3. ABDEL-RAOUF, H. and PLUMTREE, A., "The Influence of Deformation Rate on the Cyclic Behaviour of Pure Iron", Metallurgical Transactions, Vol.2, 1971, pp. 1863-1867

4. FELTNER, C.E. and LAIRD, C., "The Role of Slip Character in Steady-State Cyclic Stress-Strain Behaviour", Trans. TMS-AIME, Vol. 245, 1969, pp. 1372-1373

5. PARTRIDGE, P.G., "Cyclic Twinning in Fatigued Close-Packed Hexagonal Metals, Phil.Mag., Vol.12,1965, pp.1043-1053

6. PARTRIDGE, P.G. and PEEL, C.J., "Effect of Cyclic Stresses on Unalloyed Polycristalline Titanium", Editors R. Jaffee and N.Promisel, Pergamon Press, New York, 1970, pp.517-534

7. PARTRIDGE, P.G., "Effect of Cyclic Stresses on the Microstructures of Hexagonal Close-Packed Metals", Czech. J.Phys., B19, 1969, pp. 323-332

8. BEEVERS, C.J. and ROBINSON, J.L.,"Some Observations on the Influence of Oxygen Content on the Fatigue Behaviour of α-Titanium", Journal of the Less-Common Metals, Vol.17, 1969, pp.345-352

9. BEEVERS, C.J. and HALLIDAY, M.D., "On the Formation of Internal Fatigue Damage in Association with Twins in α-Titanium", Metal Science Journal, Vol. 13, 1969, pp. 74-79

10. GOLLAND, D.I. and BEEVERS, C.J., "Some Effects of Prior Deformation and Annealing on the Fatigue Response of α-Titanium", J.of Less-Common Metals, Vol. 23, 1971, pp. 53-60

11. OWENS, J.P., WATSON, P. and PLUMTREE, A., "Effect of Prestrain on the Cyclic Behaviour of α-Titanium", International Conference on Mechanical Behaviour of Metals, Kyoto, August 1971

12. REED-HILL, R.E. and BALDWIN, D.H., "A Technique for Orienting Grains in Fine-Grained Polycristalline Hexagonal Close-Packed Metals Using the Polarized-Light Microscope" , Trans. TMS-AIME, Vol. 233, 1965, pp. 842-844

SOME OBSERVATIONS ON FATIGUE CRACK GROWTH IN ALPHA-TITANIUM

J.L. Robinson and C.J. Beevers

Department of Physical Metallurgy
and Science of Materials
University of Birmingham
Edgbaston, Birmingham, England

Fatigue crack propagation studies have been carried out on sheet specimens of three commercially pure α-titanium alloys, designated I.M.I. Ti 115, Ti 130 and Ti 155. These contained 0.060, 0.130 and 0.285 wt.% oxygen, respectively, as the principal alloying element. Fatigue tests were performed in air at room temperatures, at a frequency of ∿ 130 Hz. Tension-tension loading was used, with R = 0.35. A ΔK range of 4 to 16 MN $m^{-3/2}$ was covered, giving growth rates in the range 10^{-6} to 10^{-4} mm $cycle^{-1}$. Specimens of each material were annealed to give a grain size of ∿ 0.027 mm, and these specimens exhibited little influence of interstitial solid solution content on growth rate. Specimens of Ti 115 having a grain size of ∿ 0.080 mm were also tested, and in the ΔK range 5 to 10 MN $m^{-3/2}$ the crack growth rates were slower than in the finer grain size material by a factor of up to 10. Two modes of crack growth were identified, one where the crack path was strongly influenced by the orientation of the individual grains, and one where the grains appeared to exert little influence on crack propagation paths. It was suggested that grain orientation control of the crack path was strong when the extent of plasticity at the crack tip was less than the grain size.

Introduction

Considerable attention has been given to study of the fatigue crack growth characteristics of a wide range of titanium alloys, especially in corrosive environments (1-4). The influence of microstructure, however, has not been as extensively studied. The object of the present work was to examine the effects of two basic parameters, namely yield strength and grain size, on fatigue crack growth in commercially pure α-titanium. The test programmes were so arranged that the fatigue crack growth rates examined were in the range 10^{-6} mm cycle^{-1} to 10^{-4} mm cycle^{-1}, an experimental region which until recently has received little attention. The effects of yield strength were examined by using three grades of commercially pure α-titanium containing varying contents of interstitial alloying elements, principally oxygen. Higher oxygen contents lead to increased yield stresses and to more extensive cross-slip, through the occurrence of non-prismatic slip (5). The effect of grain size was examined in the lowest oxygen content alloy.

Experimental Details

The materials used in the investigation were three grades of commercially pure α-titanium, supplied by IMI Ltd., and designated Ti 115, Ti 130 and Ti 155. They were in the form of 2 mm thick cold rolled and annealed, descaled, pickled sheet. The suppliers analyses of the alloys for interstitial elements are shown in Table I.

The fatigue crack propagation tests were carried out on single-edge-notched sheet specimens, the dimensions of which are shown in Figure 1. Two specimens of each material were vacuum annealed for one hour at 973K, giving mean linear intercept grain sizes of about 0.027 mm. Two further specimens of Ti 115 were vacuum annealed for one hour at 1123K, giving a grain size of about 0.080 mm, and these specimens will be referred to subsequently as large grain sized Ti 115. The mean linear intercept grain sizes and tensile properties of the four pairs of specimens are given in Table II, these parameters having been measured on specimens cut from the fatigue specimens after testing.

Fatigue testing was performed on a 10,000 Kg Amsler Vibrophore

Table I. Interstitial contents of the alloys investigated.

Alloy Designation	Oxygen wt.%	Oxygen ppm at.	Carbon wt.%	Carbon ppm at.	Nitrogen wt.%	Nitrogen ppm at.
Ti 115	0.060	1800	0.020	800	0.0075	300
Ti 130	0.130	3900	0.020	800	0.0100	400
Ti 155	0.285	8500	0.025	1000	0.0060	240

Table II. The grain sizes and tensile properties of Ti 115, Ti 130 and Ti 155.

Specimen designation	Mean linear intercept grain size. mm	0.2% proof strength $MN\ m^{-2}$	U.T.S $MN\ m^{-2}$	% elongation to fracture.
Ti 115	0.023	247	380	24
Ti 115, large grain sized	0.080	250	385	42
Ti 130	0.030	390	525	25
Ti 155	0.031	435	585	26

at a frequency of ∿ 130 Hz. All tests were carried out at room temperature (293K) in laboratory air. Tension-tension loading was used with a stress ratio R (Minimum load/Maximum load) of 0.35. Fatigue crack lengths were monitored continuously by the potential drop method to an accuracy of approximately 0.05 mm. Crack growth rates were calculated by dividing each 0.5 mm increment of growth by the number of cycles taken to produce that growth increment. This rate was taken to be the fatigue crack growth rate at the centre of the growth increment. The stress intensity factor calibration used was that provided by Brown and Srawley (6).

At the low testing loads used commencement of crack growth direct from the notch root required a large number of cycles. Thus at the start of a test a load was used sufficiently high to give rapid initiation, this was then stepped down to the required test load in decrements such that fatigue crack growth was never completely arrested. No crack growth rate measurements were made until the crack had grown at least 1 mm at the intended test load. This procedure avoided the possibility of measuring growth rates affected by the plastic deformation associated with the initiation process. Fatigue tests were performed at a constant applied stress amplitude, with crack lengths in the range a/W = 0.1 to a/W = 0.6.

After fatigue testing, and macroscopic examination, one of the fatigue fracture surfaces was examined in a Mark I Cambridge Stereoscan scanning electron microscope. The other fracture surface was mounted in thermosetting bakelite resin to give a side view of the fracture path, that is with the viewing direction perpendicular to the plane of the sheet specimen. The mid-sections of these metallographic specimens were examined using an optical microscope.

Results

The crack propagation test results are presented as plots of macroscopic growth rate, da/dN, versus stress intensity range, ΔK, on log-log scales, in Figures 2 and 3. The estimated errors in da/dN and ΔK, resulting from errors in the measurement of crack length, load, and specimen dimensions, are shown on the plots as error bars, and were ± 14% and ± 7% respectively.

Figure 2 shows the results for the three pairs of specimens of Ti 115, Ti 130 and Ti 155 having grain sizes of approximately 0.025 mm. Above a growth rate of approximately 4×10^{-6} mm cycle^{-1} the rates in the three materials are the same within the estimated errors. Below this point growth rates in the Ti 115 are possibly slightly faster than in the other two materials. The straight line drawn through all the data points has a slope of 4.3, indicating that an expression of the form $da/dN = B\ \Delta K^{4.3}$, where B is a material constant, describes the fatigue crack growth in this region.

Figure 3 shows the results for the pair of large grain sized Ti 115 specimens compared with those for the Ti 115 specimens of smaller grain size. The points for each of the smaller grained specimens correspond with each other, within the estimated limits of error, except at low ΔK values, where they diverge slightly. The points for the two larger grain sized specimens also correspond with each other at high and low ΔK values, but at intermediate values they are significantly different. Despite this scatter it is evident that from ΔK values of between 5 and 10 MN m$^{-3/2}$ the growth rates at a given ΔK value in the larger grain sized specimens were significantly slower than those in the finer grained ones. This effect was also manifested in the a versus N curves for the specimen of each grain size that was tested at 29 ± 14 MN m^{-2}. Two and a half times as many stress cycles were taken for the crack to grow from an a/W of 0.09 to an a/W of 0.30 in the larger grain sized specimen as in the smaller grain sized one. On Figure 3 the data points for the fine grain sized specimens are fitted by a straight line having a slope of 3.5. The data for the coarse grain sized specimens can be better approximated by two lines having slopes of 6.6 and 1.5.

Examination of the fatigue fracture surfaces showed them all to be macroscopically perpendicular to the stress axis, with no significant shear mode growth. The fatigue fractures were similar in that they were rough and uneven near the notch, where fatigue crack growth rates had been low. At longer crack lengths the surface topography was macroscopically much smoother. This effect was particularly marked in the larger grain sized Ti 115 specimens.

Optical microscopic examination of the metallographically prepared specimens giving a side elevation of the fracture path revealed some features common to all specimens. The nature of the crack path at the lower growth rates is illustrated in Figure 4. This shows that crack growth on reasonably well defined planes occurred over distances of the order of the grain size, and that these growth planes were markedly misoriented with respect to the macroscopic crack growth plane. The subsidiary cracks associated

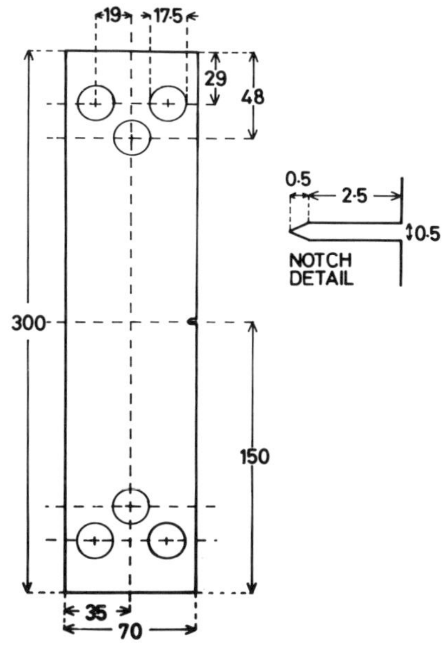

Fig. 1. Dimensions of the single-edge-notch sheet specimens, in mm. The notch root radius was approximately 0.07 mm.

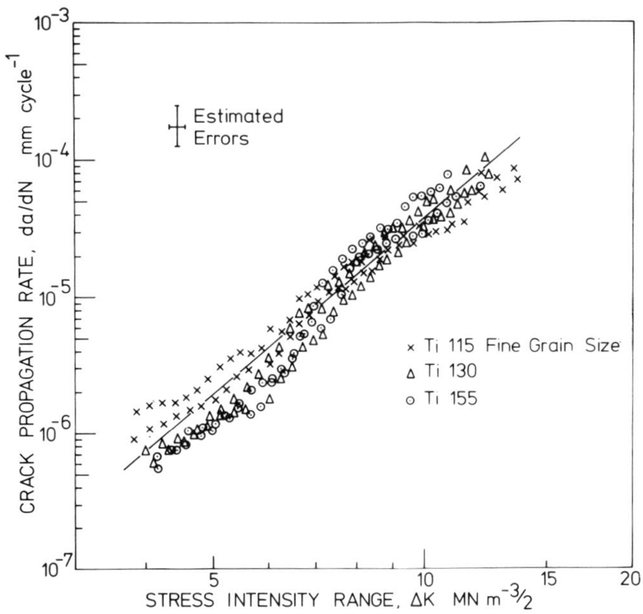

Fig. 2. A plot of stress intensity range against crack growth rate for the three pairs of specimens of Ti 115, Ti 130, and Ti 155 having grain sizes of about 0.027 mm.

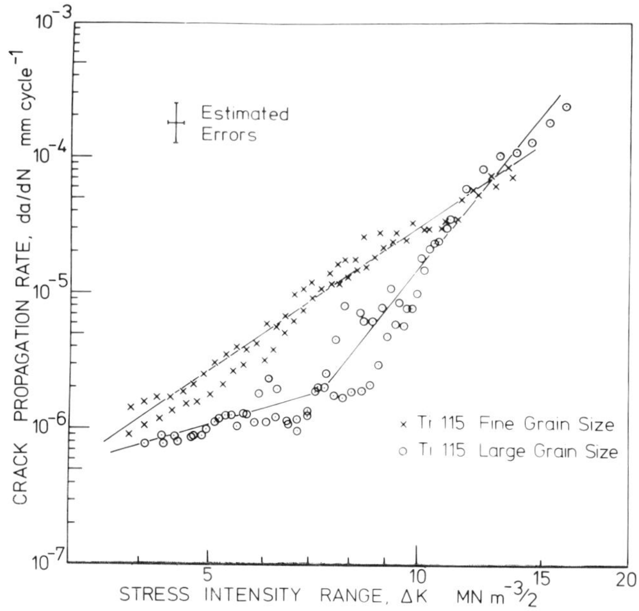

Fig. 3. A plot of stress intensity range against crack growth rate for the two pairs of specimens of Ti 115 having grain sizes of 0.023 mm and 0.080 mm.

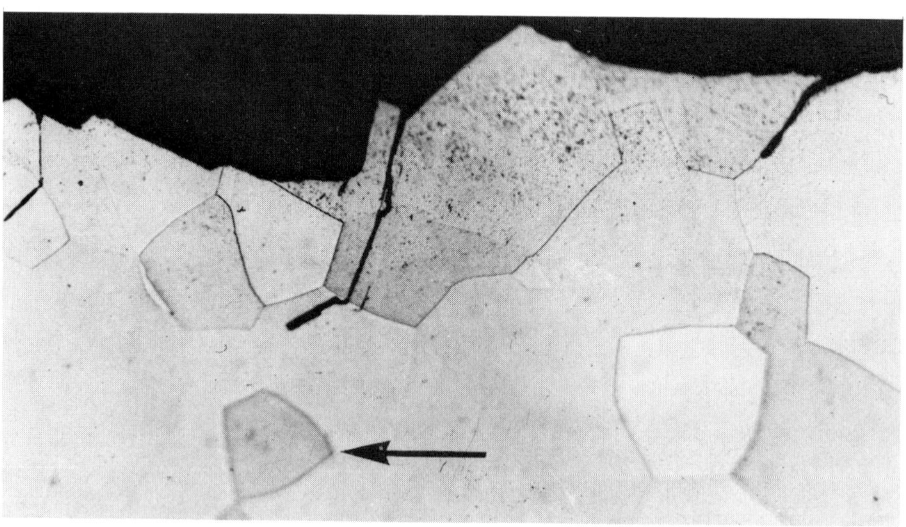

Fig. 4. An optical micrograph of part of the fracture path of a Ti 155 specimen, showing transgranular and intergranular branch cracking. X400.
$\Delta K \sim 6.7$ MN m$^{-3/2}$, da/dN $\sim 5.1 \times 10^{-6}$ mm cycle^{-1}.

with the main crack path were mainly transgranular, and this observation together with the planar character of the fracture surface, suggests that the main crack propagated in a predominantly transgranular mode. Examples of both transgranular and intergranular subsidiary cracks are shown in Figure 4. These subsidiary cracks were seen in all specimens, except those of the fine grain sized Ti 115, especially in the low growth rate regions. The crack profile became smoother with increasing crack length and growth rate.

Examination of the fatigue fracture surfaces in the scanning electron microscope showed that in all cases, especially at low growth rates, the fracture path and topography had been strongly influenced by the orientations of the grains through which the crack front had passed. This was indicated by the visibility of the grain structure on the specimen surface, an example being shown in Figure 5. The marker in this figure represents the mean linear intercept grain size, and the facets can be clearly identified as having dimensions of the order of the material's grain size. This effect will be referred to as "grain orientation control" of the fracture path. At higher ΔK values in all specimens there was much less evidence of the grain structure of the material, and an example of this is shown in Figure 6. The transition from strong to weak grain orientation control of the fracture path was always a gradual one, the proportion of the specimen surface showing evidence of the grain structure decreasing with increasing crack length, and thus the transition could not be said to occur at any particular crack length. Grain orientation control was strongest in the largest grain sized Ti 115 specimens, and was visible on the majority of the length of the fractures. In the smaller grain sized Ti 115 specimens it was only observed over about the first third of the fracture paths, and then relatively weakly. In the Ti 130 and Ti 155 specimens grain orientation control extended over the majority of the fracture paths, being strongest near the notch, and weakening with increasing crack length.

Discussion

Both the scanning electron microscopy and optical metallography showed that at low values of ΔK the fatigue crack path exhibited marked variations in growth direction from one grain to another, indicating that grain orientation could control local crack growth characteristics. (Figure 5). Further evidence of planar or crystallographic crack growth can be seen in the form of branch or subsidiary cracks, as in Figure 4. Microstructurally sensitive stage II fatigue crack growth similar to that in the present work has been reported by Hoeppner (7), Williams and Smith (8), and Birkbeck et al (9). Hoeppner, working with copper, found that during their early development cracks were markedly influenced by local crystallographic conditions, the effect being particularly strong in coarse grained material. Fracture paths were partly

Fig. 5. A scanning electron micrograph of the fracture surface of a coarse grained Ti 115 specimen. $\Delta K = 4.1$ MN m$^{-3/2}$, da/dN ~ 10^{-6} mm cycle^{-1}. X210

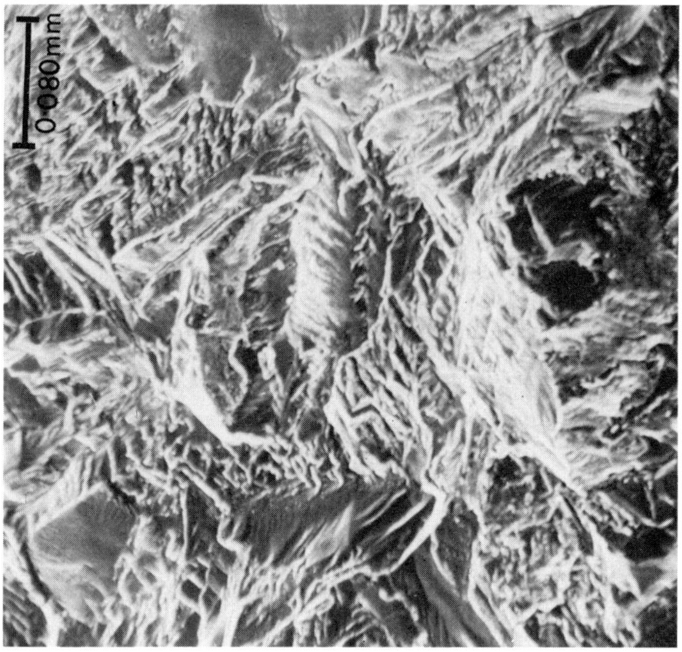

Fig. 6. A scanning electron micrograph of the fracture surface of a coarse grained Ti 115 specimen. $\Delta K = 14.2$ MN m$^{-3/2}$, da/dN ~ 10^{-4} mm cycle^{-1}. X210

intergranular and partly transgranular. Birkbeck et al found that in low and medium carbon steels stage II fatigue crack growth occurred by two distinct modes. The first mode, referred to as stage IIA, was microstructurally sensitive with the crack path changing direction at ferrite grain boundaries, revealing the grain structure in the fracture surface, the crack path being predominantly transgranular. The second mode, referred to as stage IIB, was structure insensitive, with a relatively featureless fracture surface, and no evidence of the underlying microstructure. The transition from stage IIA to stage IIB was fairly sharp, and was found to occur where the diameter of the plastic zone at the crack tip was equal to about four times the mean linear intercept grain size. The authors suggested that below this point the constraint on crack tip plasticity was high, with a restricted number of slip systems available, and specific planes in the ferrite were fractured. As the plastic zone size increased the degree of constraint was reduced, more slip occurred, the crack propagated in a larger pre-deformed zone, and the fracture became less sensitive to structure.

In the present work no sharp transition from structure sensitive to structure insensitive growth occurred, and thus relative degrees of grain orientation control of the fracture path are referred to. Because of the lack of a well marked transition point no definite correlation can be shown between the presence of grain orientation control of the fracture path and the extent of crack tip plasticity with respect to the grain size. However, some estimates of the extent of crack tip plasticity can be made to see if these were consistent with this factor being the main variable controlling the incidence of grain orientation control.

Estimates of the plasticity at the crack tip for plane strain conditions can be made using equation 1 and 2 (11,12). These define the maximum (W_{max}) and reverse plastic zone (W_{rev}) sizes in terms of the stress intensities at the crack tip, and the yield stress, σ_y, of the material.

$$W_{max} = \frac{K_{max}^2}{3 \pi \sigma_y^2} \quad \text{------------} \quad (1)$$

$$W_{rev} = \frac{\Delta K^2}{12 \pi \sigma_y^2} \quad \text{------------} \quad (2)$$

The maximum plastic zone size was considered important by Birkbeck et al (9). On the other hand reversed plasticity is an important factor in many models of fatigue crack propagation, and hence consideration of the reversed plastic zone size in the present context appears reasonable. The estimates of these plastic zone sizes, for the lowest and highest ΔK values used, are presented in Table III.

Table III. Plastic zone size estimates.

Material	Maximum plastic zone size at $\Delta K = 4.0$ MN $m^{-3/2}$ mm	Maximum plastic zone size at $\Delta K = 13.0$ MN $m^{-3/2}$ mm	Reversed plastic zone size at $\Delta K = 4.0$ MN $m^{-3/2}$ mm	Reversed plastic zone size at $\Delta K = 13.0$ MN $m^{-3/2}$ mm
Ti 115	0.065	0.68	0.0068	0.072
Ti 130	0.027	0.29	0.0028	0.030
Ti 155	0.021	0.23	0.0022	0.024

As the grain sizes were 0.023 to 0.031 mm for the finer grained Ti 115, Ti 130 and Ti 155 and 0.080 mm for the larger grained Ti 115 it is clear that the reversed plasticity at the crack tip could have been contained within individual grains throughout most of the ΔK range, especially in the Ti 130, Ti 155 and coarse grained Ti 115. On the other hand the maximum plastic zone size was larger than the grain size from close to the commencement of each test. The metallographic observations showed that grain orientation control was present over the majority of the ΔK ranges studied, but decreased with increasing ΔK. Thus a simple criterion that the plastic zone size had to be less than the grain size for grain orientation control to appear is not satisfied by consideration of either reversed or maximum plastic zone sizes. Neither does the criterion applied by Birkbeck et al (9) hold, namely that the maximum plastic zone size should be less than four times the grain size for grain orientation control to occur. Despite the inability at the present time to present a simple condition for the occurrence of grain orientation control, the general trend of the observations does indicate a correlation between this metallographic feature, and the influence of the restricted slip activity that would be expected if plasticity was confined to individual grains of a plastically anisotropic material, such as α-titanium.

Figure 3 shows that growth rates in the larger grain sized Ti 115 were markedly slower than those in the finer grained material over the middle range of ΔK values examined. Both Hoeppner (7) and Birkbeck et al (9) found grain size to have little effect on crack propagation rates in the materials they studied. The region where the growth rates in the Ti 115 differ is that where grain orientation control was strong in the larger grain sized specimens, and very weak in the others. This suggests that under certain circumstances growth where grain orientation control is strong may be slower than growth, under the same loading conditions, where growth occurs by more generalised plastic deformation. The reasons for this are not completely clear at the present time. However, one factor that may be involved is the nature of the crack tip plasticity when growth is on well defined crystallographic planes. McEvily and Boettner (13) and Ishii and Weertman (14) found that in aluminium and copper single crystals ease of cross-slip played a critical role in determining fatigue crack propagation rates.

Conclusions

In commercially pure α-titanium oxygen contents of from 0.060 wt.% to 0.285 wt.% had little effect on fatigue crack growth rates in the range 10^{-6} mm cycle^{-1} to 10^{-4} mm cycle^{-1}.

At these low propagation rates two modes of crack growth can be identified, one where the crack path is strongly influenced by the orientation of the individual grains, and one where the grains appear to exert little influence on crack propagation paths. There was no sharp transition between these two modes, but the results suggested that grain orientation control of crack growth was strongest when the extent of plasticity at the crack tip was less than the grain size. The extent of grain orientation control was enhanced by increased grain size in the Ti 115, and led to substantial decreases in growth rates over the ΔK range 5 MN m$^{-3/2}$ to 10 MN m$^{-3/2}$.

References

1. Frost, N.E. and Denton, K., "The Fatigue Crack Propagation Characteristics of Titanium and Two Titanium Alloys," Metallurgia, Vol. 70, 1968, pp. 113-120.

2. Hudson, C.M., "Fatigue crack propagation in several titanium and stainless steel alloys, and one superalloy," NASA TN D-2331, 1964.

3. Bates, R.C. and Clark, W.G., "Fractography and Fracture Mechanics," ASM Trans. Quart, Vol. 62, 1969, pp. 380-389.

4. Meyn, D.A., "An Analysis of Frequency and Amplitude Effects on Corrosion Fatigue Crack Propagation in Ti-8Al-1Mo-1V," Metallurgical Transactions, Vol. 2, March 1971, pp. 853-865.

5. Churchman, A.T., "The preparation of single crystals of titanium and their mode of deformation", Nature, Vol. 171, 1953, p. 706.

6. Brown, W.F. and Srawley, J.E., "Plane Strain Crack Toughness Testing of High Strength Metallic Materials," ASTM STP 410, Am. Soc. Testing Mats., 1966, p. 12.

7. Hoeppner, D.W., "The Effect of Grain Size on Fatigue Crack Propagation in Copper," Fatigue Crack Propagation, ASTM STP 415, Am. Soc. Testing Mats., 1967, pp. 486-504.

8. Williams, H.D., and Smith, G.C., "The Fatigue of Beta-brass," Phil. Mag., Vol. 13, 1966, pp. 835-854.

9. Birkbeck, G., Inckle, A.E. and Waldron, G.W.J., "Aspects of Stage II Fatigue Crack Propagation in Low Carbon Steel," Journal of Materials Science, Vol. 6, 1971, pp. 319-323.

10. Rice, J.R., "Mechanics of Crack Tip Deformation and Extension by Fatigue," Fatigue Crack Propagation, ASTM STP 415, Am-Soc Testing Mats., 1967, pp. 247-309.

11. Knott, J.F., "Mechanics and Mechanisms of Large-scale Brittle Fracture in Structural Metals," Mater. Sci. Eng., Vol. 7, 1971, pp. 1-36.

12. McEvily, A.J., and Boettner, R.C., "On Fatigue Crack Propagation in Face-centred Cubic Metals," Acta Met., Vol. 11, 1963, p. 725.

13. Ishii, H., and Weertman, J., "Fatigue Crack Propagation in Copper and Cu-Al Single Crystals," Metallurgical Transactions, Vol. 2, December 1971, pp. 3441-3452.

HIGH CYCLE FATIGUE PROPERTIES OF

TITANIUM IN AIRCRAFT APPLICATION

Walter J. Crichlow

Engineering Advisor
Aeronautical Systems Division
Wright-Patterson Air Force Base, Ohio 45433

Tjerand Lunde

Design Specialist
Structural Mechanics and Materials Dept
Lockheed-California Company
Burbank, California 91503

Introduction

Fatigue and corrosion are the twin plagues of the aircraft industry today. Military and commercial operators are demanding long-life trouble-free machines for use in hostile environments.

Titanium is an excellent corrosion resistant material so long as it is protected from halogen contamination at high stress and temperature. It has given excellent service in all sorts of environments from humid jungle, hot desert to sea coast exposures.

As with other materials, titanium is sensitive to the sequence of varying load amplitudes. For this reason, translation of constant amplitude S-N data into the response to variable amplitude stress spectra found in real life is most difficult. One fundamental flaw in this transformation, quite often overlooked, is the fact that in constant amplitude S-N testing, a different volume of material is involved in each different stress level data point. Information on the interactions and couplings of variable amplitudes of stress in a single volume of material simply does not exist in the data. In attempts to abstract non-existent information, transformation expedients, such as the Palgren-Miner cumulative damage hypothesis, are less than satisfactory. For these reasons, laboratory equipment and test techniques have been developed to apply the realistic flight-by-flight variable load amplitude spectra to material coupons, components and to full-scale airframes in fatigue tests. We will examine some titanium fatigue test results from the more realistic approach.

Included are some FAA-USAF Materials Laboratory (USAFML) sponsored experiments on supersonic transport fatigue spectra examining titanium response to various accelerated and fore-shortened test spectra compared with long term spectra tests at elevated temperatures cycled to represent real time and realistic flight-by-flight sequence. Another series of experiments examines an hypothesis of a threshold level of significance in variable load amplitude environment, analogous to the endurance limit stress level in constant amplitude test results.

In a program sponsored in part by the USAFML, application of titanium to actual production parts is reviewed with the results of a series of spectrum fatigue tests conducted on eight different alloy, heat-treat versions of a specially forged part.

Statistical variability of fatigue data is of major importance in establishing reliable safety factors for "Safe Life" parts with economically acceptable operating costs. The standard deviation of the results of a number of sets of variable amplitude spectrum fatigue tests conducted on titanium, aluminum and steel specimens are examined.

Titanium Constant Amplitude S-N Data

A large body of constant load amplitude high-cycle fatigue data on various titanium alloys and product forms has been generated in many laboratories for many purposes. Space restrictions prevent review of these data. However, two brief examples are included particularly to point out the potential for improved fatigue properties from dilligent process control of constituents such as oxygen content, and of grain size. The necessity for improved quality control is discussed in a later section of this paper.

Effect of Oxygen Content

For Ti-6Al-4V forging, the oxygen content appears to have no significant effect on the static yield and ultimate tensile strengths. It appears from Figure 1, however, that the fatigue strength for Ti-6Al-4V forging, for R = 0.1, increased when the oxygen content decreases, for K_t = 1 and 2.76 (Ref 1). Process control of oxygen and other interstitial elements may have an important contribution to the control of fatigue life.

Effect of Grain Size

For mill annealed, Ti-6Al-4V forging, for K_t = 1.0 and 3.0, R = 0.1, at room temperature, there appears to be a trend for the fine grained material to have a higher fatigue strength than the coarse grained material, as seen in Figure 2 compiled from Refs. 2-5. Process controls need determination to consistently reproduce the high side of the scatter band.

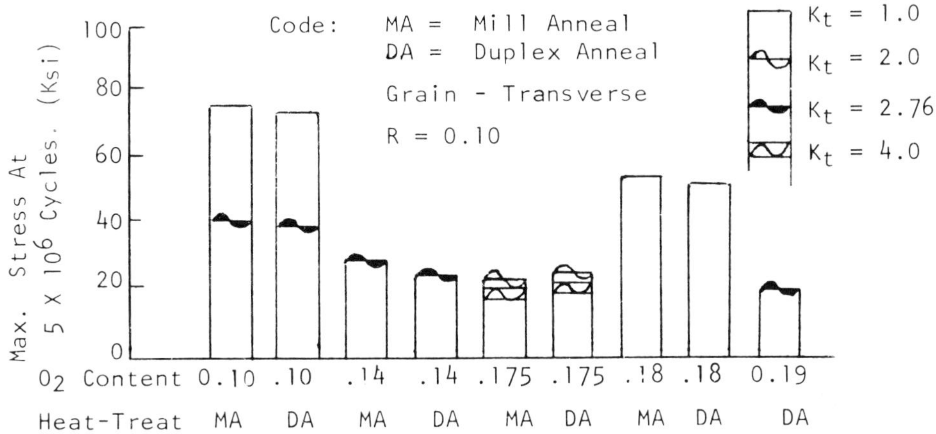

Fig. 1. Fatigue strength vs oxygen content for Ti-6Al-4V forged bar.

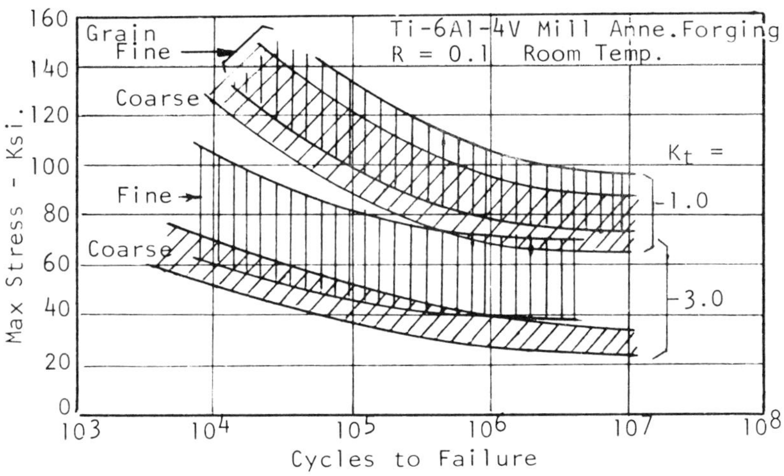

Fig. 2. Fatigue effect of grain size for titanium at K_t = 1.0, 3.0.

Titanium Spectrum Fatigue Test Data

It is well-known that the fatigue response of most structural materials is highly dependent on the sequence and magnitude of the load history. Some of the reasons for this response are well understood qualitatively, if not quantitatively. For example local yielding in a region of high stress concentration produces internal residual stress patterns which must be superimposed on the external stress field to fully describe the shifting stress history in the critical location. In addition, the repetitive local yielding under high peak stress cycles changes the material characteristics by work-hardening or-softening processes. The interaction and interdependence of these processes (and perhaps others not yet fully explored) during a random ordered flight-by-flight sequence of loads has as yet no viable analytical basis for prediction. Resort must be made to laboratory experiments to gather design data in the true loading environment, and to interpret foreshortened fatigue tests, necessarily modified for schedule economics. In this section we shall discuss a few cases of titanium response to variable load spectrum tests.

Effect of the Ground-Air-Ground Cycle

Conventional load spectra for laboratory tests, in the past, have been made up of block representations of the statistical load content of the expected or measured service history. Experimental evidence has shown the sensitivity of fatigue test results to the transition of a ground-to-air-to-ground load level each flight (GAG cycle) (Ref 6). For the supersonic transport program, FAA-USAFML sponsored research (Ref 7) explored various means of blocking a load spectrum for test purposes that would retain the closest approximation to the fatigue results of real time, real load sequence. The objective of this program was to determine the simplest and most practicable representations of accelerated test spectrum sequence for the conduct of the full-scale airframe fatigue tests. A mean-to-mean GAG cycle is defined by the transition from the minimum mean or steady stress level to the maximum mean or steady stress level during a flight. Peak-to-peak GAG cycle is defined by the transition from the minimum stress to the maximum stress on the average once per flight, Figure 3. The spectrum of GAG cycles is defined to include all (variable) load cycles occurring less than once per flight in the total life spectrum. The various spectrum loading sequences considered are illustrated on Figure 4. The program was conducted on titanium-8Al-1Mo-1V duplex annealed sheet (K_t = 2.76). The duration of flights in these tests were 0.09 minutes for constant temperature tests, 0.32 minutes for temperature cycling tests and 65 minutes for real time tests (Ref 8).

After more than seven years of continuous running, the real time flight-by-flight tests, including temperature cycling, are still going on (Feb 1972). Altogether 36 specimens were being tested. So far, nine of these have failed. Significant and consistent variations in fatigue life are introduced by the GAG cycle applied periodically

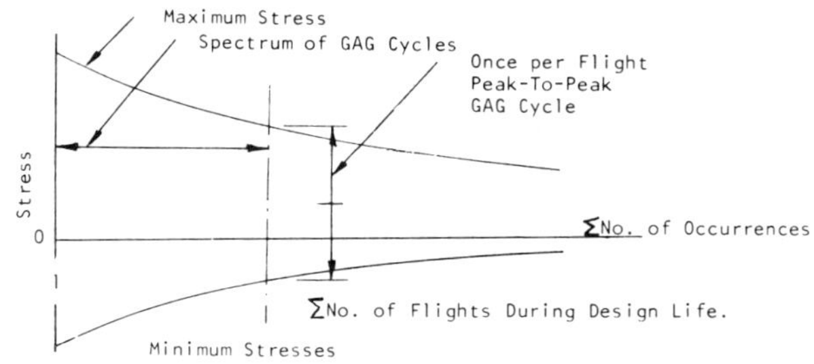

Fig. 3. Definition of once per flight peak-to-peak GAG cycle, and spectrum of GAG cycles.

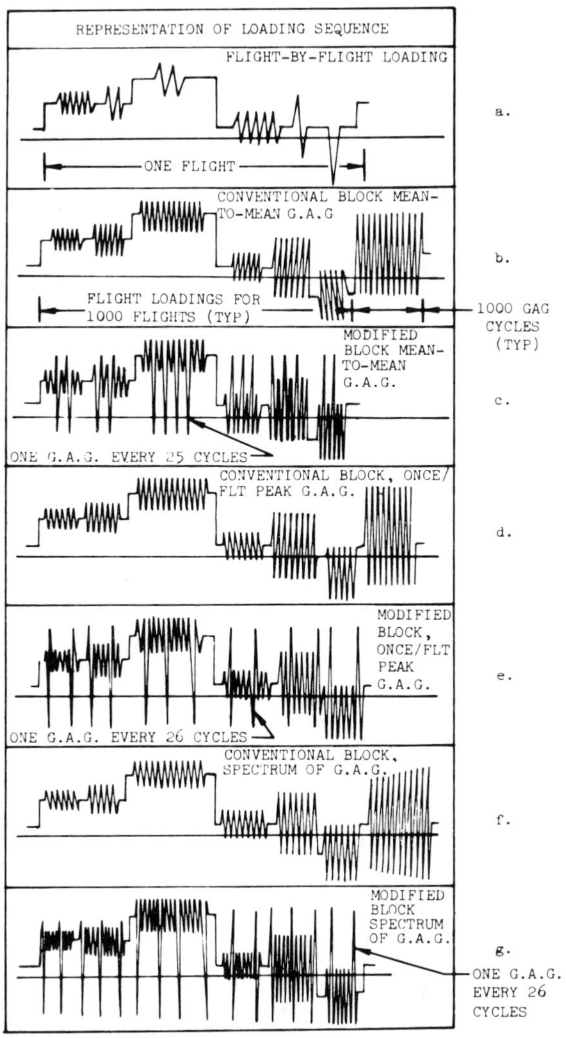

Fig. 4. Fatigue test loading spectra.

compared with the same loading spectra applied in a flight-by-flight loading sequence (in which the GAG cycle does not require specific definition). The use of groupings of GAG cycles may lead to unconservative test results, as shown in Figure 5 where a ratio as much as 6 to 1 exists between flight-by-flight loading and block loadings with the GAG cycles grouped.

Effect of Cycles Per Flight

Spectrum fatigue tests were conducted to determine the significance on life (in flights to failure) of the number of small amplitude stress cycles applied during each flight to notched Ti-6Al-4V mill annealed coupon specimens. Test loads followed the basic pattern on Figure 6, (Ref 9). The constants used in defining the various loading spectra for an intended helicopter application are as follows:

a = Normal mean stress in a given test.

b = This had two values (0.08a and 0.20a) and was the unit increment in normal mean stress due to overspeed loads.

k = 0.3 was the ratio of f_{vary}/f_{mean}.

As Figure 7 indicates, the number of flights to failure falls off rapidly and levels off between 40 and 400 cycles per flight. Therefore, at best for this type of spectrum, it is possible to truncate the load cycles down to about 50 to 100 cycles per flight without significantly affecting the test results.

Threshold Level

It is desirable to determine the extent to which a given fatigue spectrum can be truncated without significantly altering its representation of service loadings. Tests were conducted on Ti-6Al-4V coupons (K_t = 4.0) to investigate the relationship between truncation level and test life (Ref 10). Truncation was performed by eliminating entire load levels, and their corresponding number of cycles from a flight-by-flight fatigue test spectrum as shown on Figure 8. All load deletions were below the constant amplitude fatigue endurance limit of the material. The effect is similar to the reduction of number of cycles per flight discussed above except for the turbulence type of load spectrum.

The test results shown on Figure 9 indicate a threshold stress level below which there appears to be no appreciable effect on the fatigue test life. With omission of stress cycles above the threshold stress level, the number of flights to failure begins to increase. The threshold level may be used to determine which load levels may be deleted without compensation, while the upper portion of the curve provides trade-off information to further correct test results if additional truncation is deemed necessary.

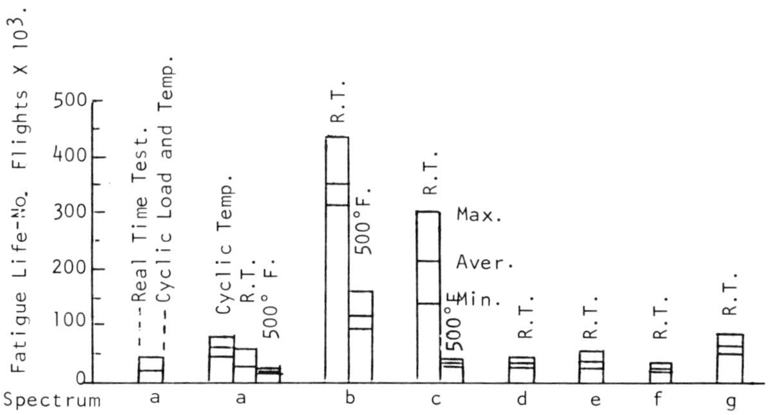

Fig. 5. Effect of spectrum testing Ti-8Al-1Mo-1V duplex annealed notched sheet specimens.

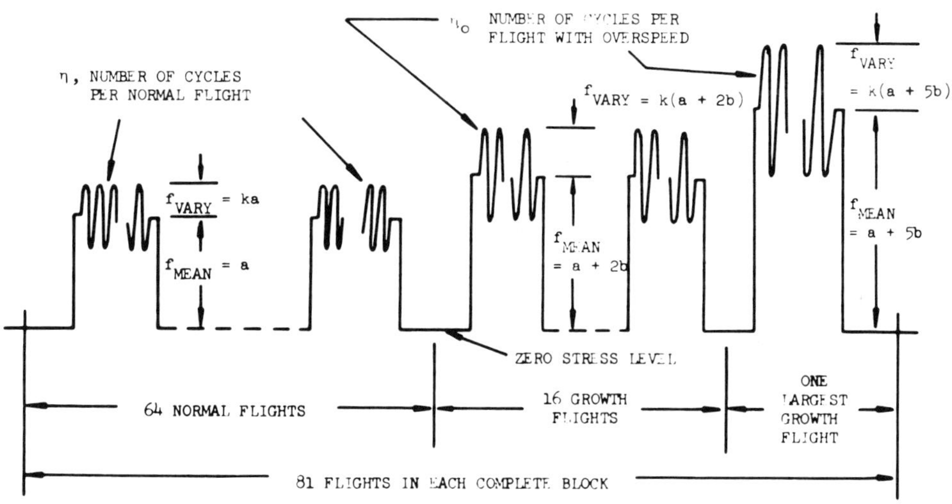

TEST	1	2	3	4
η	1000	100	10	1/2
η_o	90	90	10	1/2

Fig. 6. Loading schedules for tests varying the number of cycles per flight.

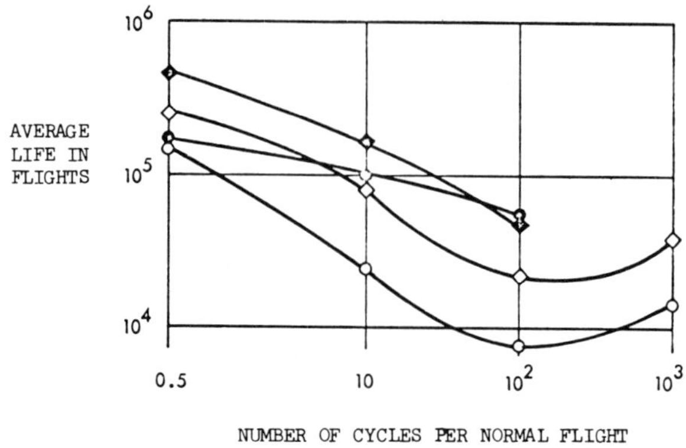

SYMBOL	GROWTH INCREMENT	VARYING STRESS RATIO	NET AREA STRESS FOR NORMAL FLIGHTS (KSI)	
	b	k	$a = f_{mean}$	$ka = f_{vary}$
◇	0.2a	0.3	22.0	6.6
◆	0.2a	0.3	18.5	5.55
○	0.08a	0.3	28.6	8.6
●	0.08a	0.3	24.3	7.3

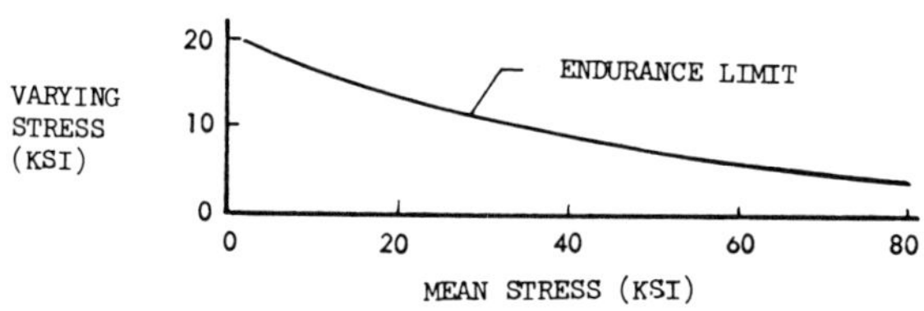

Fig. 7. Results of varying number of cycles per flight for Ti-6Al-4V mill annealed notched coupons, $K_t = 4.0$.

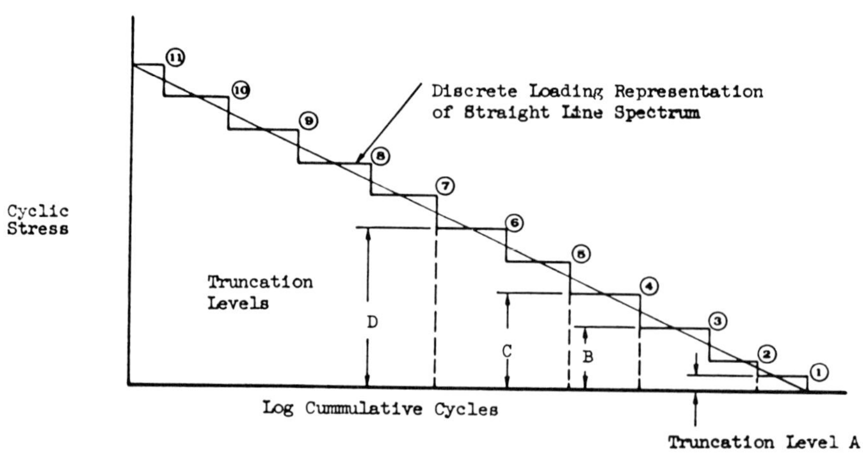

Fig. 8. Loading schedule with deletions for tests of threshold level of significance.

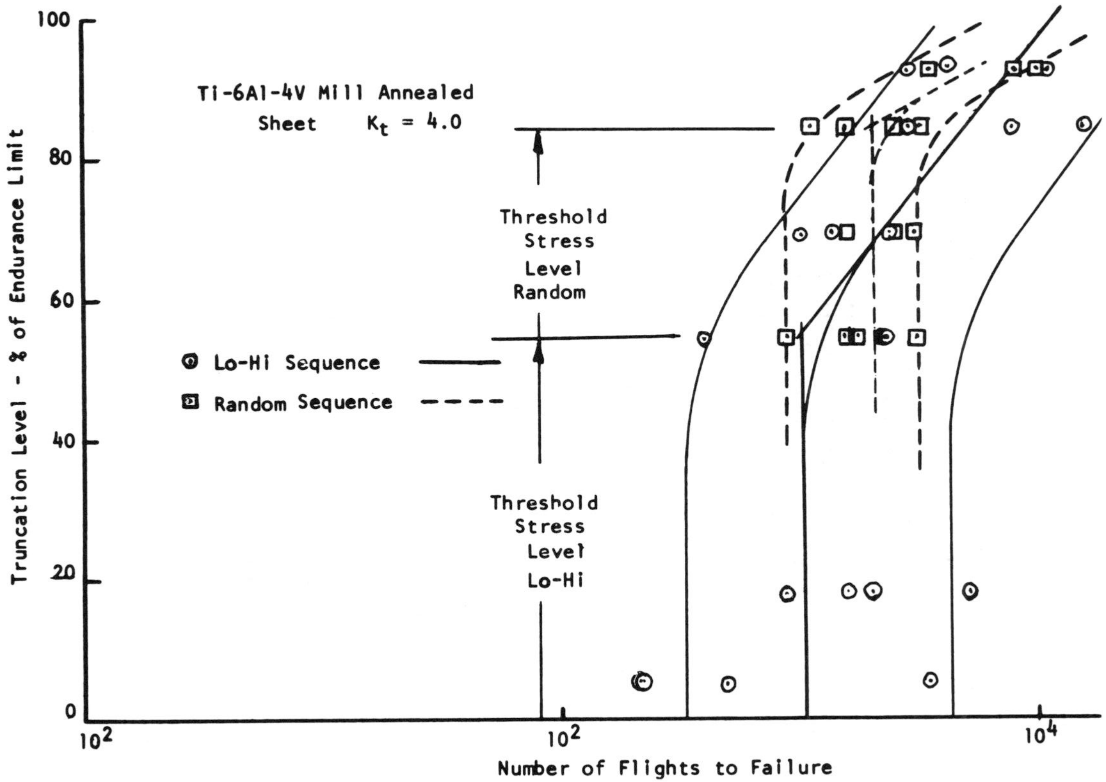

Fig. 9. Results of threshold level tests for Ti-6Al-4V mill annealed notched coupons, K_t = 4.0.

Titanium Forging Program

A program was initiated under USAFML sponsorship to evaluate the mechanical and fatigue properties of large titanium forgings (Ref 11). An F-104 fuselage ring forging was selected and the forging die was specially modified to provide a large central portion. The finished part supports the vertical stabilizer front spar and transmits horizontal and vertical tail forces into the fuselage. A production steel part of 4340 heat-treated to 180 ksi tensile strength was also tested for baseline data. Subsequent extensions (Ref 3), company funded work (Ref 2) and a joint USAFML program with North American Rockwell Corp (Ref 12), utilizing an equivalent diffusion bonded titanium fitting, produced in all seven titanium part tests. Each part was specifically redesigned and sized for minimum static margins of safety (minimum weight), but was not designed for a specific fatigue life. A diagram of the test arrangement is shown on Figure 10. The unit block fatigue test loading spectrum is defined in Table I.

The test results are summarized in Figure 11 in terms of the cumulative number of cycles of all load levels applied to crack or fail the part. Only the British alloy IMI 679 part, approximately sixty-nine percent of the weight of the original steel part, sustained equivalent number of cycles (11.2 vs 11.6 unit blocks) of the baseline steel part.

Test Scatter

For "Safe Life" parts on aircraft structure, the effects of scatter in fatigue test results must be properly accounted for, in part by reducing results of laboratory tests on components by a conservative life reduction factor. The variables that can affect scatter must be carefully assessed. A multivariate regression analysis on the results of 338 spectrum fatigue test groups (Ref 13) showed that material (aluminum and steel), the type of fatigue test, and the magnitude of the stress spectrum severity level were significant variables affecting scatter. The correlation between scatter on the log of test life and cyclic stress severity levels for both spectrum and constant amplitude S-N tests shows an increase in scatter with reduction in stress level. The scatter in S-N test results is much larger and increases markedly as the stress level approaches the endurance limit of the material.

A comparison of the standard deviation on log life of spectrum fatigue tests on aluminum, steel, and titanium is shown in Table II for a 90% confidence level. From these data, it appears that titanium materials exhibit more scatter than either aluminum or steel. A comparison was also made between the range of standard deviations for the various titanium alloys as shown in Table III. The scatter between the two titanium alloys is similar, but for both alloys there is a trend of decreasing scatter in going from single annealed to duplex annealed processing. For Ti-6Al-4V there is a further trend

Table I. Titanium Forging Fatigue Test Spectrum

% Py	Number of Cycles Per Repetition	Cumulative Number of Cycles Per Repetition
100	0.5	0.5
94	2.0	2.5
81	12.0	14.5
68	62.0	76.5
55	388.0	464.5
43	2,656.0	3,120.5
30	21,875.0	24,995.5

Table II. Comparative Scatter In Spectrum Fatigue Test Data Of Notched Coupons

Material	Standard Deviation On Log Life - 90% Confidence Level	No. of Test Groups
Aluminum	0.09	338*
Steel	0.06	
Titanium	0.17	64

* Aluminum and steel spectrum data were pooled and analyzed together using computerized regression techniques.

Table III. Comparative Scatter in Spectrum Fatigue Test Data For Notched Titanium Coupons

Material	Heat Treat	Range of Standard Deviation on Log Life
Ti-8Al-1Mo-1V	Single Annealed	.03 - .32
	Duplex Annealed	.07 - .20
Ti-6Al-4V	Single Annealed	.02 - .38
	Duplex Annealed	.09 - .21
	Solution Treated and Aged	.05 - .11

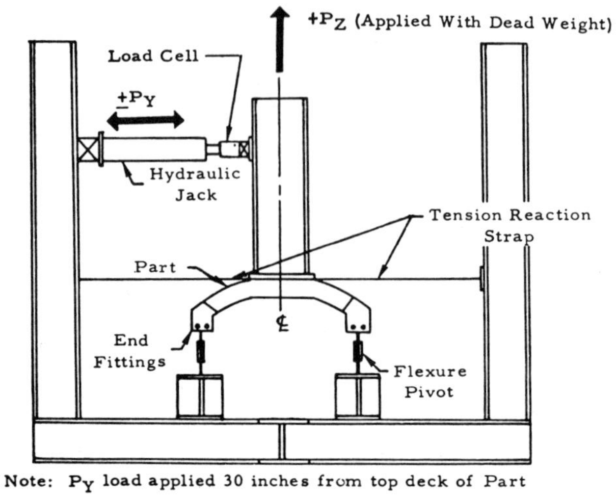

Fig. 10. Titanium forging spectrum fatigue test set-up.

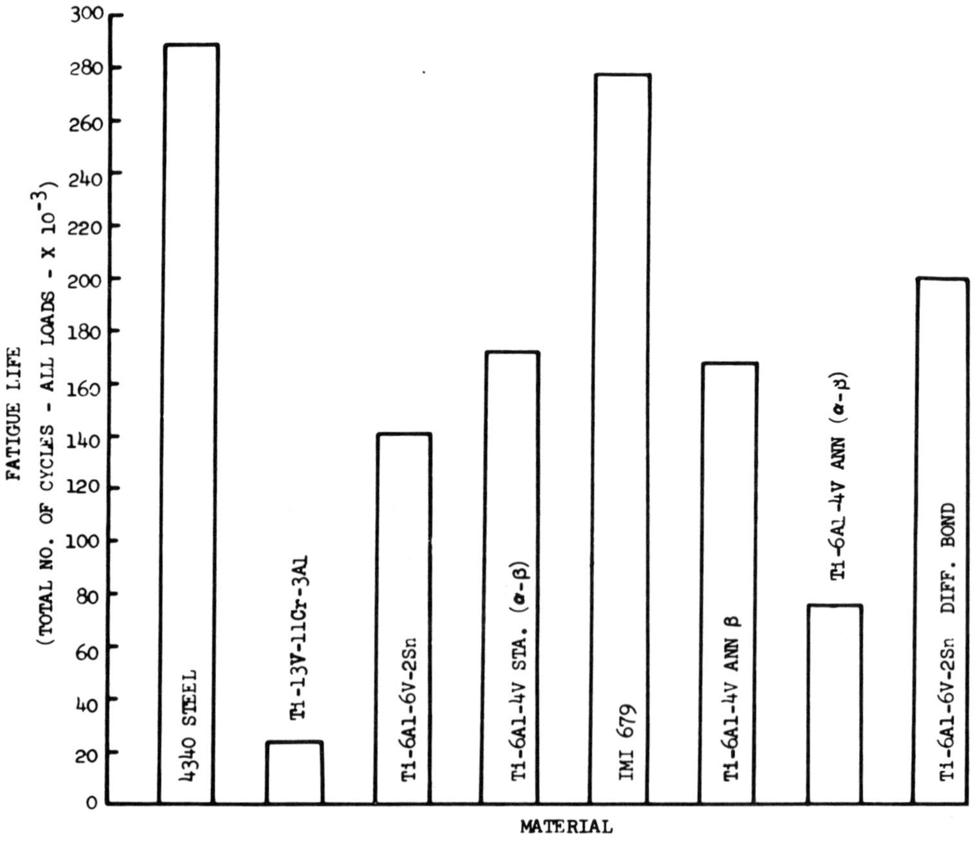

Fig. 11. Comparison of titanium forging spectrum fatigue test results.

to decreasing scatter when going from duplex annealed to solution treated and aged heat treatments. However, relatively few test results were available for duplex annealed and solution treated and aged specimens. When more test results become available, the scatter may increase. These data indicate that some measures must be taken in producing and processing of the material to reduce fatigue test scatter, if titanium is to remain competitive with steels and aluminums for fatigue sensitive "Safe Life" parts.

Summary and Conclusions

Titanium can be one of the best fatigue (and corrosion resistant) structural materials available for aircraft applications, provided costs can be justified. Indeed for some elevated temperature applications, titanium may be the only logical choice regardless of cost. The high-cycle spectrum fatigue properties of a number of alloys and product forms of titanium have been discussed primarily from the viewpoint of aircraft usage. Constant load amplitude S-N type data and the more realistic variable load amplitude spectra type data were reviewed. General conclusions drawn from these data include:

1. Titanium responds to variable load amplitude fatigue spectra in a manner similar to aluminum and steel, requiring very careful consideration of realistic spectrum type fatigue tests and special care in modifications of real spectra for laboratory test economics.

2. If titanium is to remain competitive in the future, major improvement in high-cycle fatigue properties and reduction of scatter must be achieved by more stringent control of all steps of production, processing and finishing, including tighter controls on heat-treatments, on grain size, oxygen, and other interstitial trace elements to which fatigue cracking is sensitive. There appears to be sufficient evidence of the potential, the task is now to achieve the improvement.

References

1. Kaplan, M.P., "Evaluation of Effects of Oxygen Content on Mechanical Properties of Ti-6Al-4V Forged Bar," LR 20511, June 2, 1967, Lockheed Aircraft Corp, Burbank, California.

2. Van Orden, J.M. and Soffa, L.L., "Evaluation of Beta Forged Ti-6Al-4V Forgings," LR 21363, February 20, 1968, Lockheed Aircraft Corp, Burbank, California.

3. Simenz, R.F., Macoritto, W.L., "Evaluation of Large Ti-6Al-4V and IMI 679 Forgings," AFML-TR-66-57, April 1966, US Air Force Materials Laboratory, Wright-Patterson Air Force Base, Ohio.

4. Van Order, J.M. and Soffa, L.L., "Below Beta Transus Press Working of Ti-6Al-4V Forgings," LR 21845, November 15, 1968, Lockheed Aircraft Corp, Burbank, California.

5. Beck, E., "Effect of Processing and Fabrication on Axial Loading Fatigue Behavior Of Titanium," MCR-68-257 (Issue 3), January 1969, Martin Marietta Corp.

6. McCulloch, A.J., Melcon, M.A., Crichlow, W.J., Foster, H.W., and Rebman, R., "Investigation of the Representation of Aircraft Service Loading in Fatigue Tests," ASD-TR-61-435, January 1962. US Air Force Flight Dynamics Laboratory, Wright-Patterson Air Force Base, Ohio.

7. "Supersonic Transport Development Program," LR 19839 Book 5, September 1966, Lockheed-California Company, Burbank, California.

8. "Fatigue Strength Evaluation of Titanium Materials for the Supersonic Transport Under Flight-by-Flight Loading Spectra, LR 19437, Technical Report SST-66-2, January 1966, Lockheed-California Company, Burbank, California.

9. Davis, S., "Use of Spectrum Fatigue Test Results to Calculate Damage Diagrams for Life Prediction," ASTM Symposium on Fatigue September 29 - October 4, 1968, Atlanta, Georgia.

10. Babcock, W., "Investigation of Fatigue Spectra Truncation Methods," LR 21881, March 10, 1969, Lockheed Aircraft Corp., Burbank, California.

11. Simenz, R.F., Macoritto, W.L., "Evaluation of Large Titanium Alloy Forgings," AFML-TR-65-206, July 1965, US Air Force Materials Laboratory, Wright-Patterson Air Force Base, Ohio.

12. Macoritto, W., "Evaluation of Ti-6Al-6V-2Sn Diffusion Bonded Fittings," LR 20921, 31 August 1967, Lockheed Aircraft Corp., Burbank, California.

13. Davis, C.S., and Young, L., "A Comparison of Fatigue Life and Reliability From Constant and Variable Amplitude Loading Tests," Journal of the American Helicopter Society, Vol. 12, No. 3, July 1967.

THE EFFECT OF TESTING DIRECTION ON THE FATIGUE AND TENSILE PROPERTIES OF A Ti-6Al-4V BAR

A W Bowen
Materials Department
Royal Aircraft Establishment
Farnborough, Hants, UK

Introduction

This paper reports the effect of testing direction on the fatigue and tensile properties of a Ti-6Al-4V forged bar. Directionality of mechanical properties can be of considerable significance in worked h.c.p. materials, firstly, due to their crystal structure, which is inherently anisotropic and secondly, due to any grain elongation introduced by the working operation, where there can be marked differences in the number of grains per unit area in different directions but, if little or no recrystallization occurs, there can be no change in the number of grains per unit volume (Fig 1). Any effects due to the mechanical fibering of inclusions etc., which markedly affects the mechanical properties of most cubic materials (eg refs 1-3), is absent in titanium alloys because of the purity of the starting material, their method of preparation (vacuum arc melting) and since titanium dissolves most of the common alloying or inclusion forming elements.

Material Composition and Processing History

Al	V	Fe	C	O_2	N_2	H_2	wt%
6.12	3.97	0.07	0.03	0.072	0.013	0.001	

21 in diameter ingot. $\xrightarrow{\beta \text{ forged}}$ 16in square bar $\xrightarrow{\alpha+\beta \text{ forged } (\sim 950^{\circ}C)}$ $9\frac{1}{4}$in wide x $2\frac{1}{4}$in thick bar (final dimensions) bar then annealed at $700^{\circ}C$ for $2\frac{1}{2}$h.

Testing

Test pieces were machined from the annealed bar in the longitudinal, transverse and short transverse directions. The longitudinal direction was parallel to the direction of maximum extension during forging (Fig 1). S-N curves were determined on longitudinally polished Rolls Royce rotating cantilever test pieces (2in long; 0.16in minimum dia) fatigued at a frequency of 6000 cycles/min. Tensile properties were determined on standard test pieces (1.3in gauge length; 0.16in dia) at a strain rate of 3×10^{-3} sec^{-1}.

Results

S-N curves are shown in Fig 2. The most noticeable features of this figure are the marked degree of scatter and the large variation in the fatigue strength with testing direction. Approximate fatigue strengths at 10^7 cycles are given in Table I.

Tensile properties for the three directions are also listed in Table I. Yield and tensile strengths in both the transverse and short transverse directions were considerably greater than in the longitudinal direction. A modulus change was observed only in the transverse direction. Ductility values remained almost constant except for a small decrease in the elongation value for the short transverse direction.

Table I - Mechanical Properties of the $9\frac{1}{4}$in wide x $2\frac{1}{4}$in thick forged and annealed Ti-6Al-4V Bar

Testing Direction	0.2% Proof Stress (ksi)	Tensile Strength (ksi)	E ($\times 10^{-3}$) (ksi)	Elong %	R of A %	Approx Fatigue Strength at 10^7 cycles (ksi)
Longitudinal	121.0	132.0	16.5	17.5	33	± 72
Transverse	135.5	143.0	18.7	17.0	30	± 62
Short transverse	129.5	142.0	16.5	12.5	30	± 82

Metallographic Observations

Microstructure

The grain elongation due to working could be clearly observed by examination of the etched bar at low magnifications (Fig 1). Although the bar showed small α grains surrounded by β at many of the grain boundaries, polarized light microscopy revealed clear evidence of a large prior β grain size, (ie large volumes of similar orientation) with each prior β grain containing many α grains (Fig 3). Only partial recrystallization had occurred, eg AA in Fig 3.

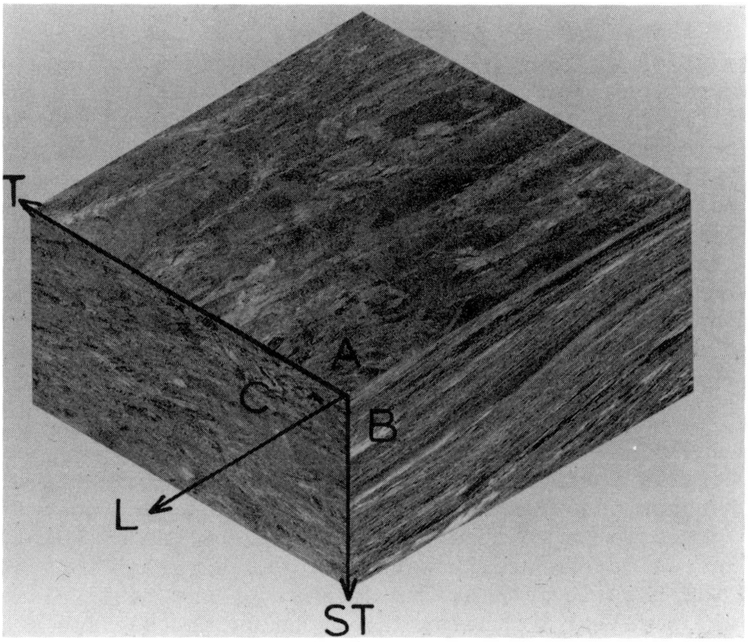

Fig. 1. Macrostructure in the three orthogonal sections of the 2¼ in thick forged and annealed Ti-6Al-4V bar showing the different number of grains per unit area in each plane. Etching solution 10% HF, 25% HNO_3, 65% H_2O. L, T, and ST - longitudinal, transverse and short transverse directions respectively. X2

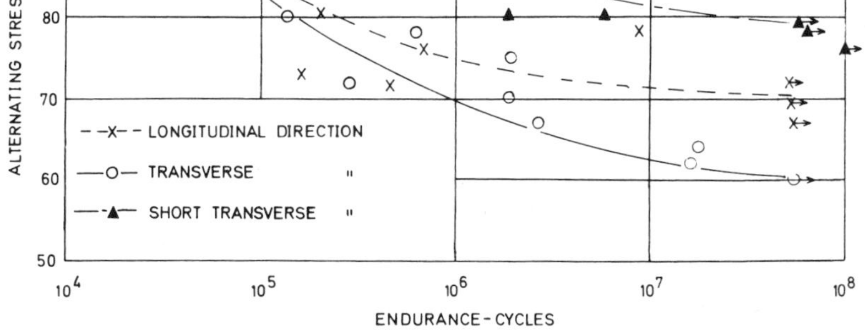

Fig. 2. The effect of testing direction on the fatigue life of the 2¼ in thick forged and annealed Ti-6Al-4V bar.

Many of the elongated regions observed on face B, Fig 1 were found to be up to 80 VPN harder than more equiaxed regions. These hardness variations can be attributed to crystallographic and/or chemical differences. Microprobe analyses revealed higher vanadium contents in the harder zones (by ~ 0.6 wt %) and also smaller aluminium variations (~0.2 wt %). However, in view of the texture analyses (see following section) which showed that a large proportion of basal plane poles are normal to face B, Fig 1, and hence would lead to hardness variations of the order of magnitude observed (4), it is most probable that crystallographic differences are primarily responsible for the hardness variations, with compositional differences enhancing this effect.

Texture

Considerable variations in preferred orientation were observed along and through the thickness of the bar and no one pole figure could be claimed to be representative of the bar as a whole. From the pole figures determined, however, it was apparent that at the surface of the bar (ie face A, Fig 1) most of the basal planes were approximately parallel to the surface; below the surface and through most of the thickness of the bar basal planes were mostly parallel to the bar edge (ie face B, Fig 1). In all cases the $\langle 10\bar{1}0 \rangle$ direction was in the longitudinal direction. Between these two textures it was possible to record a wide range of preferred orientations. The variations in texture are considered to be due to inhomogeneous deformation caused by temperature gradients or inhomogeneous working introduced into the bar during forging.

Fatigue Crack Initiation and Propagation

Fatigue cracks formed in α grains, usually in equiaxed regions such as AA in Fig 3. The extent of stage I or shear mode growth (5) seemed to be dictated by the applied stress amplitude for test pieces from all three directions. This is in agreement with other investigations (6,7).

Local variations in stage II growth rates were readily apparent on examination of fracture surfaces by optical and scanning electron microscopy. This was most noticeable on longitudinal and transverse test piece fracture surfaces, particularly when crack growth was in the short transverse direction. In these cases locally straight crack fronts could be observed frequently as faster growth in some prior β grains could be seen to terminate at prior β grain boundaries as the crack front entered a less favourably oriented prior β grain (Fig 4(a)). For growth in the longitudinal and transverse directions, serrated crack fronts were often observed (Fig 4(b)). Here faster growth in a favourably oriented prior β grain is held back by slower growth in less favourably oriented grains on either side.

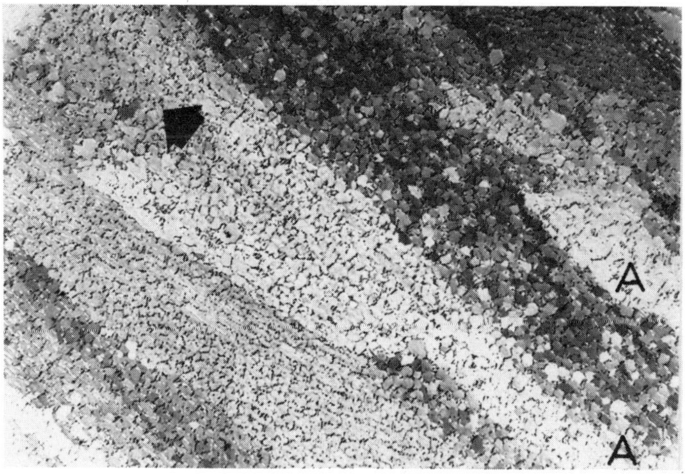

Fig. 3. Polarized light micrograph showing marked prior β grain size in the $2\frac{1}{4}$ in thick forged and annealed Ti-6Al-4V bar. Electropolished longitudinal section, longitudinal direction shown by arrow. X120

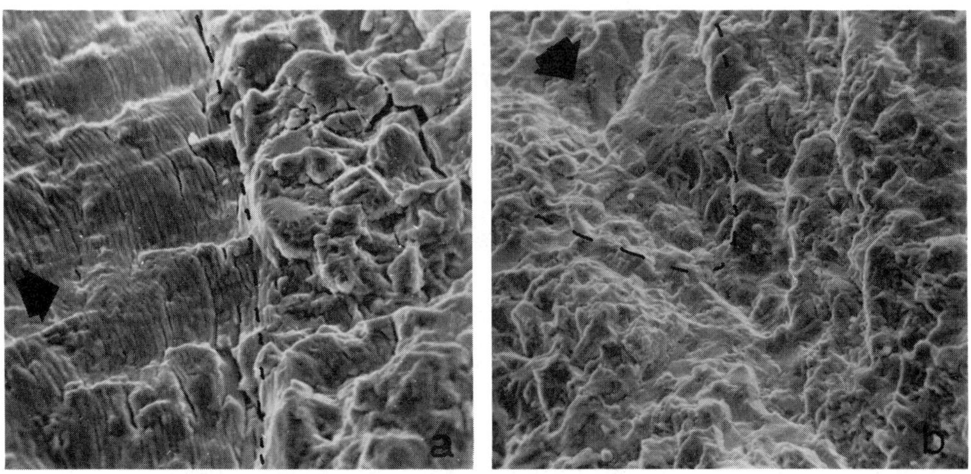

Fig. 4. Scanning electron micrographs showing fatigue crack growth in a longitudinal fatigue test piece cut from the $2\frac{1}{4}$ in thick forged and annealed Ti-6Al-4V bar (a) in the short transverse direction (X700) and (b) in the transverse direction (X340). Crack growth directions shown by arrows. Crack fronts shown by dotted lines.

Different growth rates, with changes in orientation, are not unexpected since growth is dependent, among other factors, on:-

(i) the plastic zone size, which is inversely related to yield strength (8), and changes in yield strength with orientation are common in h.c.p. materials (4, 9),

and (ii) Elastic modulus, which enters into many crack growth equations (eg 10, 11), usually as a measure of strain (ie $\frac{\text{stress term}}{E}$) and E has been shown to be very sensitive to crystal orientation (4, 12).

Crack propagation measurements on 0.5 in thick edge notched test pieces cut from $2\frac{1}{4}$in thick rolled and annealed Ti-6Al-4V bar have shown little difference in the macroscopic rate of stage II growth with testing direction, for ΔK values in excess of $\sim$ 14 ksi $\sqrt{\text{in}}$ (13). Since stage II growth rates in forged and rolled bars would not be expected to be very different, it can be assumed that the overall stage II growth rate in the forged bar used in this investigation is also independent of testing direction, and that the macroscopic growth rates average out the local variations described above.

Texture analysis of individual test pieces, in addition to the general studies described earlier, reaffirmed the texture variability, and revealed that test pieces oriented with a large proportion of prism planes approx parallel to the stress axis had shorter fatigue lives than those with a large spread of orientations or with a large proportion of basal planes approx parallel to the stress axis.

It may therefore be concluded that the differences in fatigue strength with testing direction arise predominantly from differences in the ease of crack initiation and stage I growth.

Tensile Deformation

Anisotropic deformation behaviour was common to tensile test pieces from all three directions in the bar. The effect of the prior β grain size could be observed to dictate the extent of slip (Fig 5). The effect of preferred orientation was apparent in the non-uniform necking of all test pieces. Minimum to maximum diameters of the neck were found to be in the range 0.75-0.85. This is due to the large proportion of basal planes lying approx parallel to the stress axis, slip occurs on prism planes and little contraction takes place in the c-direction. Examples of such behaviour are shown in Fig 6. Slip on intersecting prism planes, at 60° to one another, could be observed on many of the fracture surfaces.

There was evidence of a more uniform reduction of area in transverse test pieces where many of the basal planes were perpendicular to the stress axis. (C, Fig 6.) A modulus of 18 x 10^6 psi was recorded in this case (Table I). However, even then there must still be a considerable number of basal planes at angles < 90° to the stress axis, since a value of E = 21 x 10^6 psi would be expected if all basal planes were normal to the stress axis (4, 12). In these transverse test pieces, extensive $\{10\bar{1}2\}$ twinning occurred in favourably oriented prior β grains which, again, was limited by the size of those grains (Fig 7). It should be noted that it is elongated grains, such as that shown in Fig 7, which give rise to the hardness variations described earlier.

Discussion

This investigation has shown that considerable variations in fatigue and tensile strength can occur with changes of testing direction in large forgings of Ti-6Al-4V. In the following discussion explanations will be proposed to account qualitatively for this mechanical anisotropy in terms of texture and grain shape.

Fatigue Behaviour

Consideration of texture and stressing mode might lead to the conclusion that there should be no effect of texture during the early stages of fatigue, since a tension-compression cycle parallel to the basal plane is equivalent to a compression-tension cycle normal to the basal plane. However, this is not strictly the case since the elastic and plastic Poisson's ratios (μ_e, μ_p) are very sensitive to crystal orientation (12, 14). Thus a strain applied parallel to the basal plane will, on average, result in smaller strain changes normal to the stress axis (because of the higher values of μ_e and μ_p) than an equivalent strain applied perpendicular to the basal plane. It may therefore be proposed that on fatiguing test pieces with basal planes parallel to the stress axis (longitudinal and short transverse directions), the constraining influence of the higher Poisson's ratios enhances the fatigue life compared with test pieces with basal planes normal to the stress axis (transverse direction). Experimental results and texture analyses are in agreement with this suggestion.

It then remains to be explained why the longitudinal direction, while nominally having the same texture as the short transverse direction, has a lower fatigue strength. This is to be expected in view of its lower tensile strength (Table I) and is considered to be due to the changes in grain shape caused by working (Fig 1) and their orientation with respect to the stress axis. Longitudinal test pieces have the large volumes of similar orientation oriented along the length of test pieces and hence there is a greater probability that a smaller

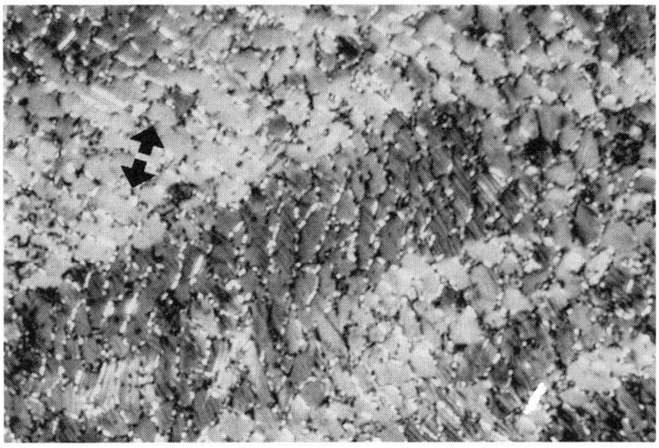

Fig. 5. Polarized light micrograph showing prior β grain size limiting extent of surface slip. Electropolished short transverse tensile test piece cut from the $2\frac{1}{4}$ in thick forged and annealed Ti-6Al-4V bar. Stress axis shown by arrow. X330

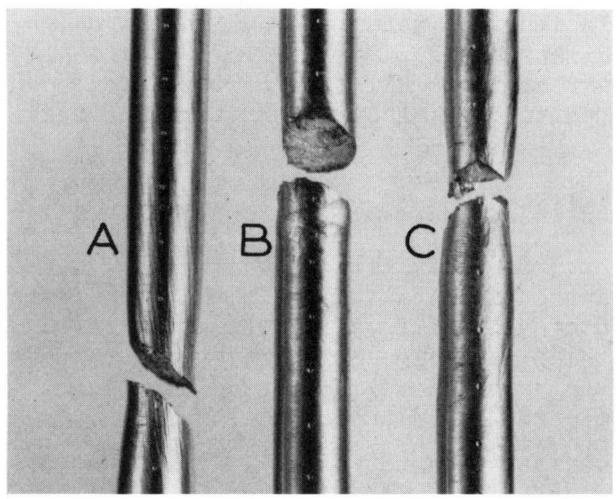

Fig. 6. Photograph showing deformation on longitudinal (A), short transverse (B) and transverse (C) tensile test pieces cut from the $2\frac{1}{4}$ in thick forged and annealed Ti-6Al-4V bar. X2

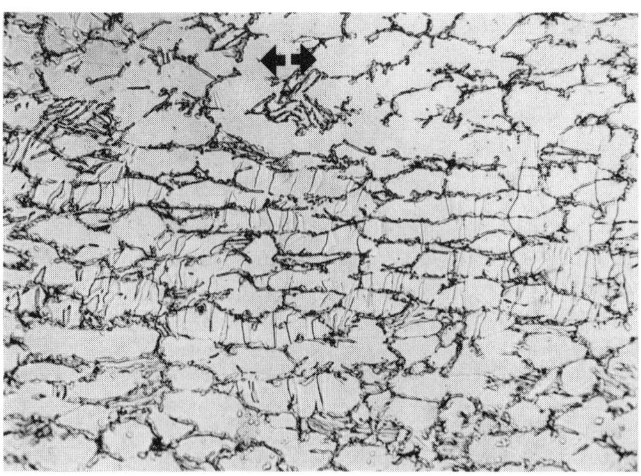

Fig. 7. Photomicrograph showing prior grain size limiting extent of $\{10\bar{1}2\}$ twinning. Electropolished and etched longitudinal section of a transverse tensile test piece cut from the $2\frac{1}{4}$ in thick forged and annealed Ti-6Al-4V bar. Stress axis shown by arrow. Etching solution 1% HF, 12% HNO_3, 87% H_2O. X500

number of these prior β grains per unit area are at the surface thus leading to easier crack initiation and stage I growth and resulting in a lower fatigue strength. This effect of grain shape is also probably a contributory factor to the low transverse fatigue strength, complementing the effect of texture.

The considerable scatter in Fig 2 can be attributed to the variable texture and coarse macrostructure common to all three directions. Variable surface finish is not the cause of scatter, firstly, since further working (7) and heat treatment (15) of this bar, using the same test piece preparation techniques and surface finish, produced very much less scatter, and secondly, electropolished test pieces cut from the as received bar showed approximately the same degree of scatter as mechanically polished ones. Small local variations in oxygen content are also a possible source of scatter.

Tensile Behaviour

Anisotropy in tensile properties was only noticeable in changes in strength (Table I). Little influence of crystallography on ductility was observed since, even when extensive twinning was favoured, as in the transverse direction, it can be shown that slip accounted for nearly all the deformation (9).

The effect of the bar textures can be correlated closely with the changes in yield and tensile strength values. In transverse test pieces (basal planes normal to the stress axis), a higher modulus and yield strength and a smaller increase in tensile strength, compared with the longitudinal and short transverse directions, would be expected from theoretical considerations (4, 9, 12, 14). Experimental results are in good agreement with such predictions (Table I). The lower tensile strength in the longitudinal direction is believed to reflect the smaller number of prior β grains per unit area in these test pieces compared with test pieces cut in the short transverse direction, which has the same nominal texture as the longitudinal direction. The large volumes of similar orientation extended over considerable distances along the gauge length in the longitudinal direction, with the result that lower strengths were observed in this direction.

Conclusions

1 The rotating cantilever fatigue strength at 10^7 cycles of test pieces cut in the longitudinal, transverse and short transverse directions of a $9\frac{1}{4}$in wide x $2\frac{1}{4}$in thick forged and annealed Ti-6Al-4V bar are approx ±72, 62 and 82 ksi respectively. Explanations are proposed to account for these differences in terms of the predominant texture in each direction and also the change in grain shape (ie grain elongation) due to the

working operation, which results in different numbers of prior β grains per unit area in the three testing directions.

2 Considerable scatter in the fatigue lives were observed at all stress levels for all three tested directions. The scatter is attributed to the variable texture and coarse macrostructure of the bar.

3 Yield strengths for the longitudinal, transverse and short transverse directions of this bar were 121.0, 135.0 and 129.5 ksi respectively, and the tensile strengths for these directions were 132.0, 144.0 and 142.0 ksi respectively. These trends are also explained in terms of the predominant texture in each direction and the different number of prior β grains per unit area in the three testing directions.

4 The modulus increased from 16.5×10^6 psi for the longitudinal and short transverse directions to 18.7×10^6 psi for the transverse direction. No large changes in ductility were observed with testing direction. These observations can be explained in terms of the predominant texture for each testing direction.

Acknowledgements

The author would like to thank N Gunn for carrying out the fatigue and tensile testing, D Clark for the texture determinations and J E Restall for the microprobe analysis. Thanks are also due to C A Stubbington for his continued advice and comments during the course of this work.

The paper is published by permission of the Controller, HMSO.

References

1 Zinkham, R E, "Longitudinal and Short Transverse Fatigue and Fracture Properties of Heavy Aluminium Alloy Plates, Produced by Forging and Rolling." Trans. Met Soc AIME Vol 245 July 1969 pp 1519-1525.

2 Gazza, G E and Larson, F R, "Impact Properties of Air and Vacuum Melted AISI 4340 Steel" ASM Trans Quart Vol 58 June 1965 pp 183-194.

3 Lauchener, E A "The Microstructure and Ductility of 17-4PH and 15-5PH Steels" J of Materials Vol 5 March 1970 pp 129-139.

4 Roberts, W T "Preferred Orientation and Anisotropy in Titanium" J Less Comm Metals Vol 4 Aug 1962 pp 345-361.

5 Forsyth, P J E "A Two Stage Fatigue Fracture Mechanism" Proc Crack Propagation Symposium Cranfield 1962 Vol 1 pp 76-94

6 Stubbington, C A and Bowen, A W "The Effect of Section Size on the Fatigue Properties of Ti-6Al-4V Bars" to be presented at the Second International Conference on Titanium, Boston, May, 1972.

7 Bowen A W and Stubbington C A "The Effect of $\alpha + \beta$ Working on the Fatigue and Tensile Properties of Ti-6Al-4V Bars" to be presented at the Second International Conference on Titanium, Boston, May, 1972.

8 Rice J R "Mechanics of Crack Tip Deformation and Extension by Fatigue" American Society for Testing and Materials STP 415 (1967) pp 247-309.

9 Reed-Hill R E "Role of Deformation Twinning in the Plastic Deformation of a Polycrystalline Anisotropic Metal" Deformation Twinning Met Soc AIME Conf Proc. Vol 25 (eds R E Reed-Hill, J P Hirth and H C Rogers) Gordon and Breach N.Y (1964) pp 295-320.

10 McEvily A J and Johnston T L "The Role of Cross-Slip in Brittle Fracture and Fatigue" Proc First Int Conf on Fracture Sendai Japan Vol 2 1965 pp 515-546.

11 Pearson S "Fatigue Crack Propagation in Metals" Nature Vol 211 (5053) Sept 3 1966 pp 1077-1078.

12 Zarkades A and Larson F R "Elasticity of Titanium Sheet Alloys" The Science, Technology and Application of Titanium (eds R I Jaffee and N E Promisel) Pergamon Oxford 1970 pp 933-941.

13 Bowen A W, to be published.

14 Larson F R "Anisotropy of Titanium Sheet in Uniaxial Tension" ASM Trans Vol 57 1964 620-631.

15 Bowen A W and Stubbington C A "The Effect of Heat Treatment on the Fatigue and Tensile Properties of a Ti-6Al-4V Bar" to be published.

THE EFFECT OF SECTION SIZE ON THE FATIGUE PROPERTIES OF Ti-6Al-4V BARS

C A Stubbington and A W Bowen
Materials Department
Royal Aircraft Establishment
Farnborough, Hants, UK

The fatigue strengths and tensile properties of a number of Ti-6Al-4V bars, ranging in size from $\frac{5}{8}$in diameter rolled bar to $2\frac{1}{4}$in thick rectangular section forged bar, have been determined, and their fatigue strengths, fracture modes and tensile deformation related to their microstructures. The investigation was carried out to determine the metallurgical features responsible for the reported differences between the fatigue strengths of large titanium forgings and those of small diameter bars and small forgings. The direct stress fatigue strengths of the $2\frac{1}{4}$in thick bars were in the range ± 50 to ± 60 ksi at 10^7 cycles, while those of the small diameter rolled bars were normally in the range ± 80 to ± 90 ksi. These differences were due to macro and microstructural differences between the bars. In the small diameter rolled bars, which had received large $\alpha\ \beta$ reductions, complete recrystallisation of the billet structure had occurred and the grain size was uniformly small. In contrast, the $2\frac{1}{4}$in thick bars were only partially recrystallised, so that large volumes of α of similar orientation were present, revealing the prior β grain size of the billet stock. There were microstructural differences between the small diameter bars and the $2\frac{1}{4}$in thick bars in the distribution of α and β. These macro and microstructural differences influenced fatigue crack initiation, fatigue fracture, and tensile deformation and fracture. The large volumes of similarly oriented α in the $2\frac{1}{4}$in thick forged bars are thought to be the chief reason for the low fatigue strengths of these bars as they favoured Stage I fatigue crack initiation and growth. Stage I crack initiation and early growth also occurred on test pieces cut from small diameter bars, but was very limited in extent. The change from Stage I to Stage II crack growth occurred at the same stress intensity for the different microstructures. Stage II fatigue crack growth occupied a very small percentage of the total life of smooth test pieces.

Introduction

In listing the problems that are still a real handicap to the full exploitation of titanium's potential, Jahnke[1] stated that the fatigue strengths of large titanium forgings were only 60% those of small bars and forgings. The causes of this difference were said to be unresolved. During an investigation whose primary aim was to determine the fatigue strength of electron beam welded joints in 2in thick rectangular section forged Ti-6Al-4V bars, and the effect of a stress relieving treatment on the fatigue strength of the alloy, differences in fatigue strength of the magnitude stated by Jahnke were observed between thick forged sections and small diameter rolled bars. The fatigue strengths of $2\frac{1}{4}$in thick forged bars at 10^7 cycles were in the range ±50 to ±60 ksi, while the comparable figures for small diameter bars were in the range ±80 to ±90 ksi. The work now reported compares the fatigue strengths of a number of section sizes, and, using metallographic and fractographic techniques, attempts to explain the differences observed.

Materials

Table I. Sizes and analyses of materials used in the investigation

Material	Composition wt%						
	Al	V	Fe	C	H	O_2	N_2
$9\frac{1}{4}$in wide x $2\frac{1}{4}$in thick forged bar	6.12	3.97	0.07	0.03	0.001	0.072	0.013
13in wide x $2\frac{1}{4}$in thick forged bar	6.11	4.01	0.03	-	0.0012	-	-
1 3/16in diameter rolled bar	5.89	3.99	0.19	-	-	-	-
1.2in diameter rolled bar	6.03	3.72	0.16	0.04	0.0013	0.16	0.023
$\frac{5}{8}$in diameter rolled bar	6.09	3.84	0.09	0.02	0.0024	0.14	0.080

Fatigue Test Pieces and Tests

Direct stress, 0 ± P loading, and rotating cantilever fatigue tests were carried out on test pieces machined from annealed, and annealed and stress relieved blanks. The stress relieving treatment, where applied, was $4\frac{1}{2}$h at 625°C. The direct stress Avery-Schenck Pulsator and Vibrophore test pieces had test diameters of 0.226in and 0.16in respectively, and parallel test sections 0.50in and 0.35in in length. The minimum diameter of the Rolls Royce rotating cantilever test pieces used in the investigation was 0.16in. The tests listed in Table II were carried out.

Table II. Fatigue tests and results

Material	Condition	Type of test	Fatigue Strength at 10^7 cycles ksi	Fatigue/ Ultimate Ratio
$9\frac{1}{4}$in wide x $2\frac{1}{4}$in thick forged bar	A	Rotating cantilever (R)	± 70	0.53
	A and SR	Direct stress (P)	± 53	0.40
	A and SR	Rotating cantilever (R)	± 75	0.75
13in wide x $2\frac{1}{4}$in thick forged bar	A	Direct stress (P)	± 59	0.43
	A and SR	Direct stress (P)	± 57	0.42
1.2in diameter rolled bar	SR	Direct stress (P)	± 85	0.55
	SR	Direct stress (V)	± 88	0.57
	SR	Rotating cantilever (R)	± 87	0.57
1 3/16in diameter rolled bar	A	Direct stress (V)	± 60*	0.43
$\frac{5}{8}$in diameter rolled bar	A	Direct stress (V)	± 87*	0.57
	A and SR	Direct stress (V)	± 87*	0.57

A - Annealed
SR - Stress relieved
R - Rolls Royce test piece
P - Avery Schenck Pulsator test piece
V - Vibrophore test piece

*Estimated

The results of the tests on the 13in wide x $2\frac{1}{4}$in thick forged bar, and those on the 1 3/16in and $\frac{5}{8}$in diameter rolled bars are given in Fig 1. The bars tested showed a section size effect, the magnitude of which is similar to that indicated by Jahnke. This can most readily be seen by comparing the fatigue strengths of the $2\frac{1}{4}$in thick bars with that of the $\frac{5}{8}$in diameter bar. An exception to the normally high fatigue strengths of small diameter bars was provided by the 1 3/16in diameter bar. The limited number of tests on this bar indicated a low fatigue strength and the lives obtained were within

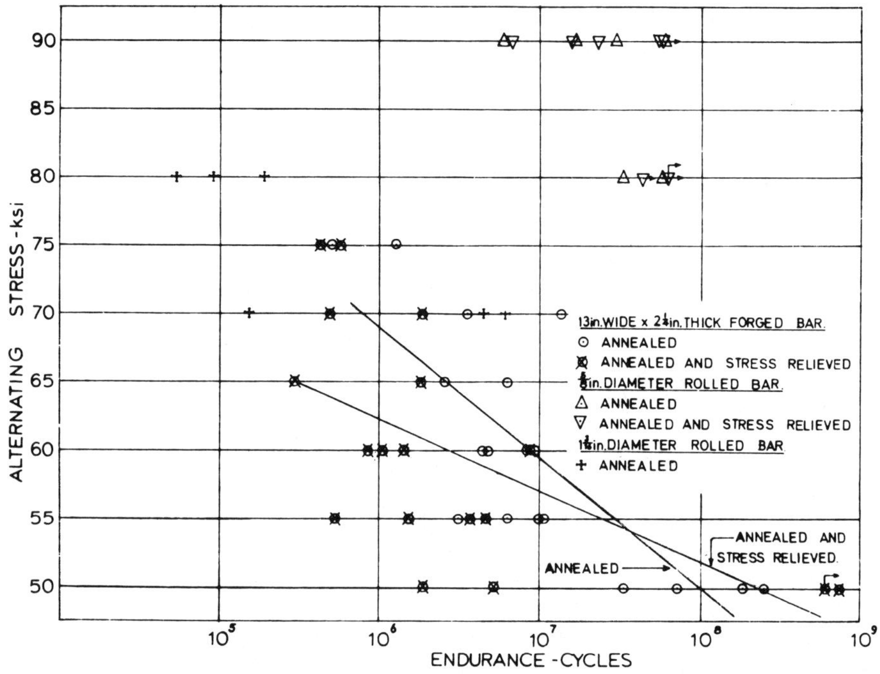

Fig. 1. Results of fatigue tests on 13 in wide x 2¼ in thick forged bar, and on 1-3/16 in and 5/8 in diameter rolled bars.

the scatter band of the results for the 13in x $2\frac{1}{4}$in forged bar. An explanation of this apparent anomoly was forthcoming when the microstructures of the bars were examined.

Examination of Fatigue Test Pieces

The fractures of test pieces machined from the $2\frac{1}{4}$in thick forged and annealed bars were examined, and the position of crack initiation determined. The test piece surface was then examined in this region and an example of the fracture path observed is shown in Fig 2. The fatigue cracks were at an angle of between 30 and 60° to the test piece axis, and it was apparent that Stage I[2] or shear mode crack initiation and early growth had occurred. Examination of the fractures of test pieces machined from $2\frac{1}{4}$in thick forged bar revealed a step, indicated A Fig 3, where Stage I crack initiation and growth had occurred. Bands of high reflectivity, B, on the Stage II or tensile mode fracture surface could also be seen.

Examination of fatigue test pieces machined from small diameter bars revealed some similarities and some differences in the initiation and growth of fatigue cracks to those discussed above. An example of the fracture of a fatigue test piece machined from $\frac{5}{8}$in diameter bar is shown in Fig 4. Stage I crack initiation also occurred on these test pieces, indicated A, but the bands of high reflectivity on the Stage II fracture surfaces were absent. A test piece surface where crack initiation occurred is shown in Fig 5, with the region of Stage I initiation indicated A.

Analysis of Fatigue Fractures

The observation that the amount of Stage I fracture on test pieces fatigued at low stresses was considerably greater than that on test pieces fatigued at high stresses suggested that there was a critical stress intensity at which Stage I growth changed to Stage II growth. The fractures shown in Figs 3 and 4 were therefore analysed and the stress intensity at which the change occurred calculated. A number of a and 2c values were used for the test piece machined from the $2\frac{1}{4}$in thick bar, as the exact limits of the area of Stage I fracture were rather indeterminate. The substitutions and the resulting K values are given in Table III. It can be seen that the stress intensity at which the change from Stage I to Stage II fracture occurs is between 5.6 ksi$\sqrt{\text{in}}$ and 6.3 ksi$\sqrt{\text{in}}$ for both materials. It is suggested that a stress intensity in this range is required for Stage II or tensile mode fatigue crack growth in Ti-6Al-4V alloy test pieces, and this critical stress intensity will be designated K_{TM}. Thus ΔK_{TM} ~12 ksi$\sqrt{\text{in}}$.

Measurements[3] of Stage II crack growth rates in 13in x $2\frac{1}{4}$in forged bar and in 1.2in diameter rolled bar, using $\frac{1}{2}$in thick edge notched test pieces, have shown that the growth rates for the two section sizes are the same when ΔK exceeds ~14 ksi$\sqrt{\text{in}}$.

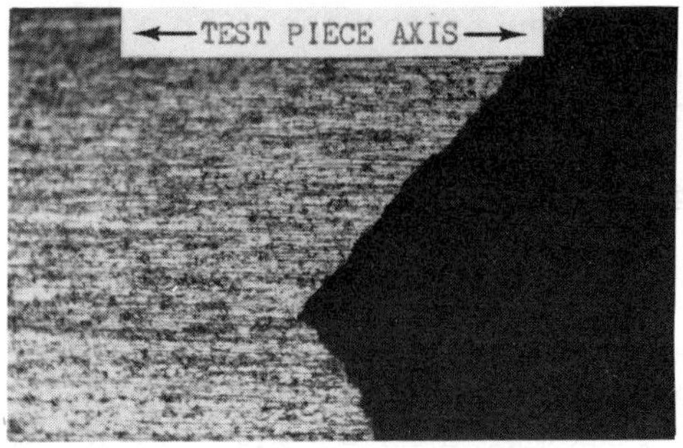

Fig. 2. Stage I fatigue crack initiation on the surface of a test piece machined from 2¼ in thick forged and annealed bar. X60

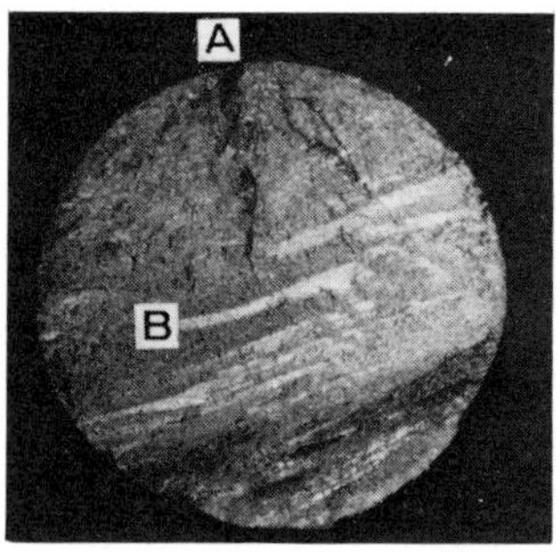

Fig. 3. Fracture surface of fatigue test piece machined from 2¼ in thick forged and annealed bar. X6

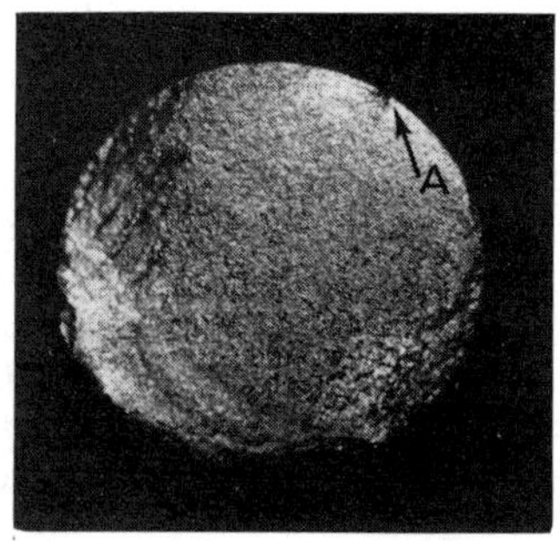

Fig. 4. Fracture surface of fatigue test piece machined from 5/8 in diameter rolled and annealed bar. X7

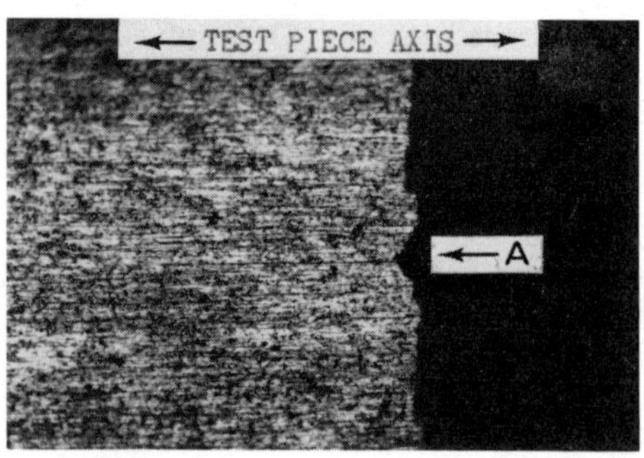

Fig. 5. Stage I fatigue crack initiation, A, on the surface of a test piece machined from 5/8 in diameter rolled and annealed bar. X60

Striation spacings on the Stage II fracture surface immediately adjacent to the periphery of the area of Stage I fracture have been measured on a test piece whose total life was 8.688 x 10^6 cycles, and calculation showed that $<$100,000 cycles, ~1% total life, of Stage II growth were required to cause final failure. Thus the ease of initiation and the rate of Stage I growth were of primary importance in determining the test piece life.

Table III. K_{TM} values for $2\frac{1}{4}$in thick forged bar and for $\frac{5}{8}$in diameter rolled bar, from equation derived by Tiffany and Masters[4]

Material and Condition	Test Piece Number	Applied Stress σ ksi	0.1% Yield Stress σ_y ksi	σ/σ_y	a in	2c in	$a/2c$	Q	K_{TM} ksi$\sqrt{in}$
13in wide x $2\frac{1}{4}$in thick forged and annealed bar	C8	60	124.5	0.48	0.0058	0.0117	0.5	2.45	5.7
					0.007	0.0117	0.6	2.85	5.8
					0.007	0.0140	0.5	2.45	6.3
$\frac{5}{8}$in diameter rolled and annealed bar	5	90	141.3	0.64	0.0025	0.005	0.5	2.45	5.6

Tensile Tests

Tensile Properties

With one exception the tensile properties of the small diameter bars were higher than those of the $2\frac{1}{4}$in thick forged bars. The exception was the 1 3/16in diameter rolled bar, whose properties approximated to those of the $2\frac{1}{4}$in thick forged bars. Again the reason for this became apparent when the microstructures were examined.

Tensile Deformation

The deformation of tensile test pieces was examined. Deformation typical of a test piece machined from $2\frac{1}{4}$in thick bar is shown in Fig 6. Secondary cracks and deformation near the fracture are indicated, A. It can be seen that the deformation bands extend over considerable distances suggesting relatively large volumes of material of similar orientation. In contrast, the deformation and fracture typical of test pieces machined from small diameter bar is shown in Fig 7. The length of the deformation bands is very much reduced compared with those shown in Fig 6, indicating very small volumes of similar orientation.

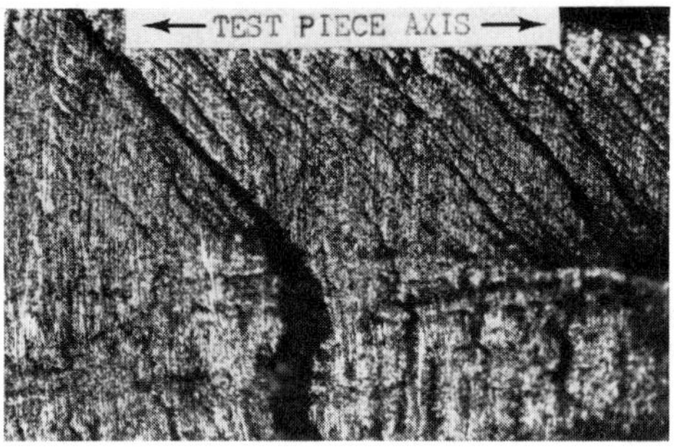

Fig. 6. Deformation on surface of tensile test piece machined from 2 ¼ in thick forged and annealed bar. X60

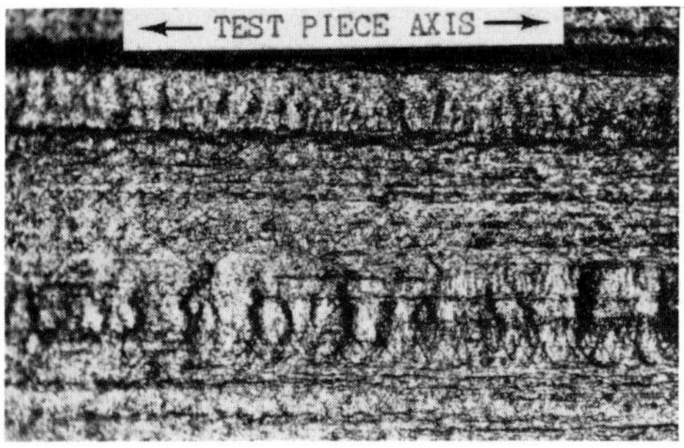

Fig. 7. Deformation on surface of tensile test piece machined from 5/8 in diameter rolled and annealed bar. X60

Microstructural Examination

Microstructure of $2\frac{1}{4}$in thick Forged Bars

When viewed with polarised light large areas of similar orientation, indicative of the macro (prior β) grain size, could be seen in these bars, Fig 8. Bands where recrystallisation had occurred, were also observed, P, and small recrystallised zones, Q, in the large areas of similar orientation. Microstructural inhomogeneities were also observed in the form of regions of elongated α between regions of equiaxed α. Hardness measurements showed that the regions of elongated α could be up to 70 HV 10 points harder than the equiaxed regions. Similar microstructural and hardness differences have been observed in Ti-6Al-4V alloy bars by Scanlon and Chambers[5]. They referred to these regions as regions of α stabilisation and concluded that aluminium segregation was responsible.

A longitudinal section through a fatigue test piece with bands of high and low reflectivity on the Stage II fracture surface, revealed that the high reflectivity bands were associated with the relatively hard zones of elongated α.

Microstructure of the $\frac{5}{8}$in diameter Rolled Bar

A transverse section across the $\frac{5}{8}$in diameter bar viewed with polarised light is shown in Fig 9. It can be seen that the grain size is very small and uniform, in marked contrast to the structure of the $2\frac{1}{4}$in thick bar shown in Fig 8. At higher magnification longitudinal sections revealed that the microstructure was fibrous and the small grain size could be clearly seen. The microstructure of the 1.2in diameter rolled bar was similar to that of $\frac{5}{8}$in diameter bar.

Microstructure of the 1 3/16in diameter Rolled Bar

The microstructure of the 1 3/16in diameter rolled and annealed bar, Fig 10, was found to be very different from the other small diameter bars. Such a microstructure would be produced by heating above the β transus, either inadvertently or deliberately, and cooling at a relatively slow rate. Subsequent annealing at 700°C to conform to specification would have very little effect on the overheated microstructure.

Discussion

The investigation has shown a clear relationship between the macro and microstructure of Ti-6Al-4V alloy bars and their fatigue strengths. High fatigue strengths are obtained with fine grained fibrous αβ structures found normally in small diameter rolled bars. Such bars receive very large reductions that result in complete breakdown and recrystallisation of the original billet structure. When the working reduction is insufficient to cause adequate

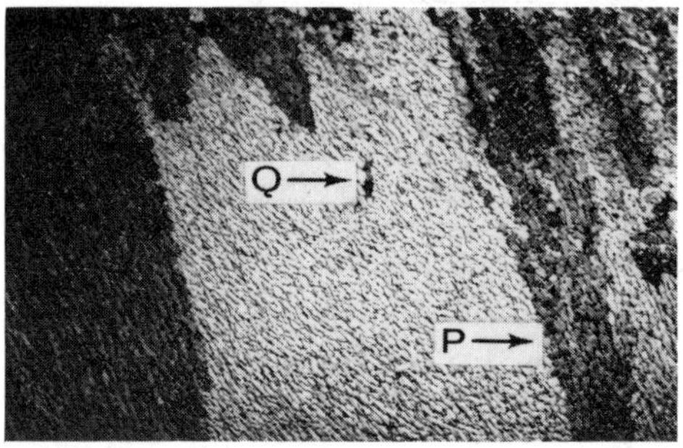

Fig. 8. Transverse section of 2¼ in thick forged and annealed bar, viewed with polarised light. X60

Fig. 9. Transverse section of 5/8 in diameter rolled and annealed bar, viewed with polarised light. X60

Fig. 10. Microstructure of 1-3/16 in diameter rolled and annealed bar. X150

breakdown, eg in the $2\frac{1}{4}$in thick forged bars, then large volumes of material of similar orientation - the prior β grains of the original billet - remain in the forged bar. These large volumes affect the tensile deformation and fracture, and Stage II fatigue crack growth, but in particular, Stage I fatigue crack initiation and growth.

Quantitative analysis of the fatigue fractures has shown that the change from Stage I to Stage II growth occurs at the same stress intensity for the two widely differing microstructures, fatigued at different stresses. This stress intensity factor is designated K_{TM}. It is suggested that in fine grained, fibrous structures, eg $\frac{5}{8}$in diameter rolled bar, fatigued at stresses above ± 85 to ± 90 ksi, Stage I fatigue cracks initiate and grow, and the stress intensity, K_{TM}, is eventually attained. Because the stress is high the amount of Stage I crack growth before K_{TM} is reached is small, and hence only a limited amount of Stage I fracture is observed on test pieces cut from such material. Fatigue at stresses below ± 85 to ± 90 ksi either does not initiate Stage I cracks, or they initiate and become non-propagating, so that the stress intensity K_{TM} is not reached.

In test pieces where there are large volumes of material of similar orientation, eg those cut from $2\frac{1}{4}$in thick forged bar, Stage I cracks can initiate and are able to grow with relative ease. K_{TM} is therefore ultimately attained, even with a relatively low applied stress, and Stage II growth then occurs to final fracture. These suggestions are summarised in Fig 11.

Thus if high fatigue strengths are to be obtained in the Ti-6Al-4V alloy - and probably in other $\alpha\beta$ titanium alloys - the thermomechanical fabrication sequence must completely break down the prior β grains of the original billet. This is accomplished in the small diameter bars by very heavy reductions, but in less heavily worked material only partial recrystallisation occurs and some of the prior β grains of the original billet remain. Sparks[6] states that in conventional $\alpha + \beta$ field forging of titanium alloys the macro (prior β) grain size as established by the billet supplier is not changed, although the grains are distorted. He also points out that the prior β grain size of forging stock can show extreme variations, and that for this reason the maximum acceptable macro (prior β) grain size for titanium billet stock for forging is now specified in the US for billet up to 14in diameter.

Another reason for large macrograins in finished components is inhomogeneous working. Scanlon and Chambers[5], and Turnbull and Gurganus[7] have shown dead areas of unworked metal adjacent to forging dies to have the large macro grain size of the billet stock. Such a macrostructure, with the low fatigue strength at the surface would be particularly undesirable in a component operating in a fatigue environment.

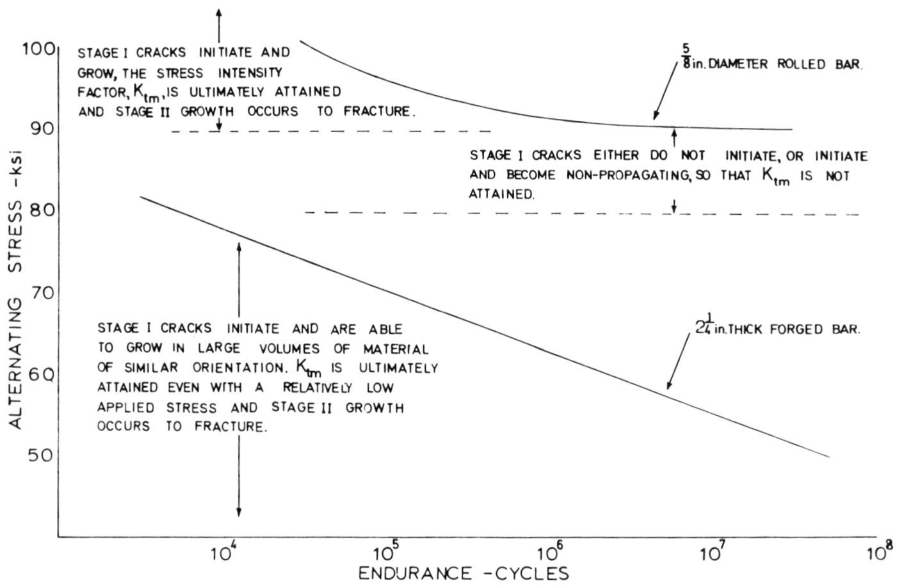

Fig. 11. Graphical summary of fatigue observations.

The investigation lends considerable support to the view that the prior β grain size of billet stock for forging should be specified and controlled. This will be particularly important when the billets receive only a limited amount of working to produce the finished component. In the case of billets that will be used for the production of small diameter bars, the macro (prior β) grain size of the billet stock is probably less important, as the total reductions are very large and adequate refinement of the structure then occurs. The results obtained on the 1 3/16in diameter bar, however, indicate that care must be taken in working and heat treating small diameter bars so that the β transus is not exceeded, either inadvertently or deliberately, resulting in the re-formation of a coarse macro-structure and a marked loss in fatigue strength.

Conclusions

1 The investigation has shown that the fatigue strengths of thick, forged Ti-6Al-4V bars can be approximately 60% those of small diameter rolled bars. Small diameter rolled and annealed bars had fatigue strengths at 10^7 cycles in the range ± 80 to ± 90 ksi, while the comparable figures for the $2\frac{1}{4}$in thick forged and annealed bars were in the range ± 50 to ± 60 ksi.

2 The fatigue strengths of Ti-6Al-4V bars are determined by their macro and microstructures. When the α β reduction is large, eg for small diameter rolled bars, complete recrystallisation occurs and the resulting fine grained, fibrous structures have high fatigue strengths. If the α β reduction is relatively small, eg in thick rectangular section forged bars, incomplete recrystallisation occurs and relatively large volumes of similar orientation, corresponding to the macro (prior β) grain size of the forging billet stock, remain in the final component. Such microstructures have low fatigue strengths.

3 Stage I fatigue crack initiation and early growth occurred in test pieces cut from small diameter rolled bars and on those from $2\frac{1}{4}$in thick forged bars. It was most marked on those cut from the forged bars, fatigued at stresses below those used for the rolled bar test pieces, and was favoured by the large volumes of similar orientation in the thick bars. The differences between the fatigue strengths of the rolled bars and those of the forged bars may be due (i) to greater ease of Stage I crack initiation in large grained material (ii) to differences in Stage I growth rates in the different microstructures, or (iii) a combination of these factors. Final fracture on all test pieces was by Stage II fatigue crack growth on a plane approximately normal to the direction of applied stress, but this occupied a very small percentage of the total fatigue life.

4 The change from Stage I to Stage II crack growth occurred at the same stress intensity in test pieces with different microstructures fatigued at different stresses. This stress intensity has been designated K_{TM} and was ~ 6.0 ksi$\sqrt{in}$, giving a stress intensity factor range, ΔK_{TM}, ~ 12.0 ksi$\sqrt{in}$.

5 If high fatigue strengths are required in forged components that receive only small reductions during fabrication, then the macro (prior β) grain size of the forging stock should be small and controlled. Grain size control on billets to be fabricated into small diameter bars by rolling is probably less important as the large working reductions ensure adequate refinement and distortion of the prior β grains. In the latter case however it is important to ensure that the bars are not overheated during forging or subsequent heat treatment, otherwise β grain growth will occur and the high fatigue strength of the refined grain structure will be lost.

6 The tensile deformation of Ti-6Al-4V alloy bars can give a qualitative indication of their probable fatigue strength. The tensile deformation of material containing large volumes of similarly oriented α, with low fatigue strength, is markedly different from that of fine grained, recrystallised material with high fatigue strength.

Acknowledgement

The authors are indebted to Mr N J Gunn for the tensile and fatigue testing. The paper is Crown copyright and is published by permission of the Controller, HM Stationery Office.

References

1 Jahnke L P Titanium in Jet Engines. The Science, Technology and Application of Titanium 1970, p1099. Pergamon.

2 Forsyth P J E Cranfield Symposium on Crack Propagation September 1961.

3 Bowen A W Unpublished.

4 Tiffany C F and Masters J N Fracture Toughness Testing ASTM STP 381 1965 p249.

5 Scanlon J V and Chambers G J G Forgings in Titanium Alloys. The Science, Technology and Application of Titanium 1970, p79. Pergamon.

6 Sparks R B The effect of macro (prior β) grain size on the fatigue of Ti-6Al-4V forgings TMS (AIME) Spring Meeting May 1969.

7 Turnbull G K and Gurganus T B Microstructural control of Ti-6Al-4V forgings. Materials Engineering Exposition and Congress. Philadelphia Pennsylvania October 1969.

THE EFFECT OF ROLLING TEXTURE ON THE FRACTURE MECHANICS PROPERTIES OF Ti-6Al-2Sn-4Zr-6Mo ALLOY

M. J. Harrigan, A. W. Sommer, P. G. Reimers
and G. A. Alers

ABSTRACT

An experimental research program was undertaken to assess the plane strain fracture mechanics properties of a highly textured titanium alloy. Ti-6Al-2Sn-4Zr-6Mo 1/2" thick plate was unidirectionally rolled to sheet to produce a strong basal texture with the major basal pole concentration in the transverse direction. The texture rolled sheets were diffusion bonded to produce a textured billet large enough to measure plane strain properties in all orientations. Fatigue crack growth rate data were obtained as well as K_{Ic} measurements. The results of this study show that the preferred orientation of basal planes controls modulus, strength, fracture toughness and fatigue crack growth rate in Ti-6Al-2Sn-4Zr-6Mo. Mechanisms by which each of these properties is controlled are presented.

M. J. Harrigan, A. W. Sommer, and G. A. Alers are members of the Technical Staff at North American Rockwell and P. G. Reimers is a member of the United States Air Force.

INTRODUCTION

Texture in metals is defined as the preferred orientation of individual grains in a poly-crystalline aggregate relative to a given set of reference axes. Textures arise as a result of extensive plastic deformation such as rolling and may be modified or strengthened by subsequent recrystallization annealing. In α titanium, or titanium alloys in which the hexagonal alpha phase predominates, the rolling textures which develop are usually described in terms of the orientation of the basal (0001) plane with respect to a set of reference axes (References 1-4). The basal plane textures which develop in titanium alloys can be broadly classified as those in which the basal planes are parallel with the rolling plane (Figure 1a) and those in which the basal planes are normal to the rolling plane (Figure 1b). The former textures have been found to result in improved biaxial strength properties (References 5 through 7) while the latter can produce anisotropy in mechanical properties in the rolling plane (References 3, 4, 8, and 9).

The textures which are produced by various rolling practices and the strength anisotropy which results are related to the available slip systems in the hexagonal alpha phase. These systems which are illustrated in Figure 2 all have slip vectors which lie parallel to the basal plane (0001). Applying a stress along the

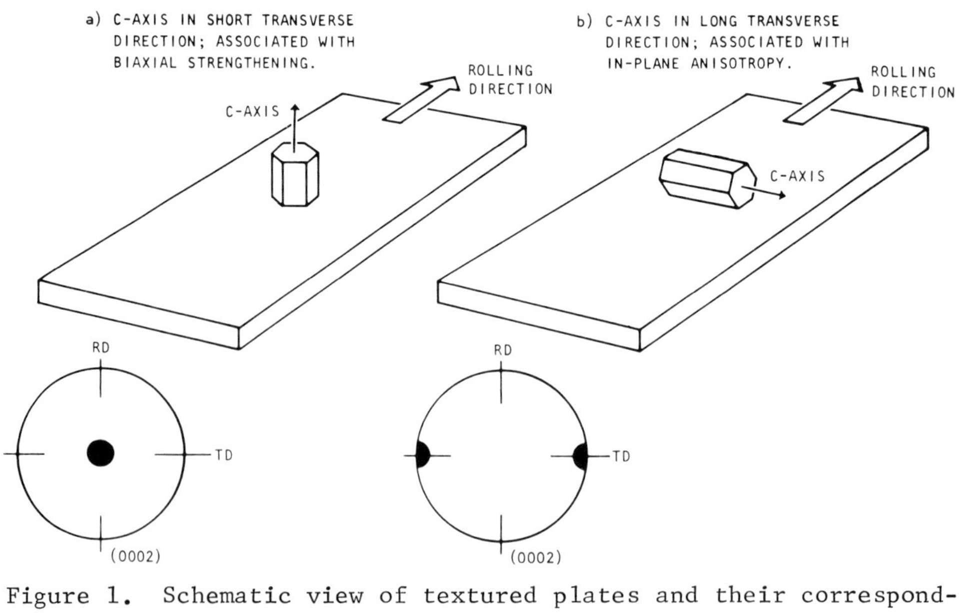

Figure 1. Schematic view of textured plates and their corresponding basal plane pole figures.

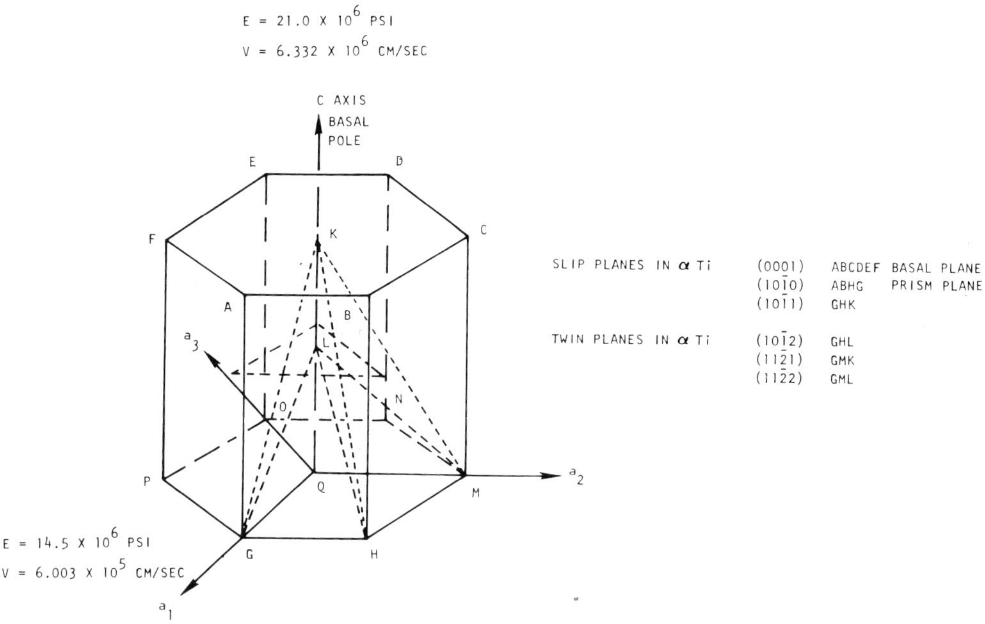

Figure 2. Slip and twin deformation planes in alpha titanium.

"C" axis of the hexagonal lattice will apply no critical resolved shear stress in the (0001) plane, or along the easy slip directions which lie in the basal plane, thus, producing the strength anisotropy. It must be noted that there do exist non-basal slip directions in alpha titanium of the type $\langle 11\bar{2}3 \rangle$ which are active on (10$\bar{1}$0), (10$\bar{1}$1), and (11$\bar{2}$2) planes (Reference 10). This type of slip, however, is felt to involve a quite high critical resolved shear stress, thus, the strength anisotropy described due to basal slip would still be observed. Furthermore, because basal or basal related slip predominates, extensive unidirectional deformation will result in a rotation of the hexagonal lattice into an orientation where further slip is difficult. It is this mechanism which is responsible for the various basal plane textures which arise (see Reference 2).

Another result of texture in alpha titanium is the anisotropy of the elastic modulus properties. This anisotropy is basic to alpha titanium and its hexagonal lattice structure. Figure 2 also illustrates the Young's modulus anisotropy of a pure titanium single crystal (References 11 and 12). A useful consequence of modulus anisotropy is that the basal plane texture of titanium can be at least qualitatively assessed using standard dynamic modulus measurement techniques (Reference 13) or through the measurement of the materials elastic constants using ultrasonic shear and longitudinal waves and the subsequent calculation of all of the elastic properties (References 14, 15, and 16).

Both the strength and modulus anisotropy have been observed in commercial titanium alloys (References 4 and 17). In one case, a strong basal plane texture was induced in Ti-6Al-4V by rolling (Reference 17). In this study, it was found that the strength was over 20 ksi higher in the directional parallel with the major basal pole concentration. This same product orientation also exhibited a Young's modulus (E) of 20.6×10^6 psi. The alloy Ti-6Al-2Sn-4Zr-6Mo has also been found to develop textures which produce anisotropic properties (Reference 17).

It was the purpose of this research to produce a strong texture in a commercial material in order to study the effect of textures on fracture mechanics properties, as well as on strength and modulus. Prior studies of the fracture mechanics properties (References 17 through 21) were inconclusive because either the texture which was studied produced isotropic properties in the plane of the sheet (Figure 1a) or the dimensions of the specimens used were inadequate to produce quantitative results. The alloy selected for this study was Ti-6Al-2Sn-4Zr-6Mo because this alloy has been shown to texture extensively and to produce highly anisotropic properties in the rolling plane (Reference 17, Figure 1b). In addition, the diffusion bonding process was utilized to produce highly textured specimens of sufficient thickness to allow characterization of the plane strain properties of the material.

Table I

TEST MATERIAL CHEMICAL COMPOSITION

Form	Composition %									
	Al	Sn	Zr	V	Mo	Fe	C	N	H	O
0.5-in. plate	5.9	1.9	4.0	-	6.0	0.07	0.021	0.009	0.008	0.09

EXPERIMENTAL PROCEDURE

For this study, a commercial plate of Ti-6Al-2Sn-4Zr-6Mo (1/2 x 36 x 48 inches) was obtained from TMCA. The producer supplied chemical analysis is shown in Table I. The plate was cut into pieces approximately 4 x 4 inches for texture rolling. Rolling was accomplished at TMCA's Henderson, Nevada Research facility. The desired texture was induced by unidirectionally rolling the material from 1/2 inch to 0.065 inch thick at a temperature of 1375 $\pm$ 25°F. Each piece was individually rolled requiring approximately 7 reheat cycles of 5 to 15 minutes each. At heavy gages, 4 to 5 passes could be made between reheating while below 1/8 inch thick only 1 to 2 passes could be made.

After texture rolling, two billets 7-1/4 inches long by 3-5/8 inches wide by 3 inches thick were fabricated by diffusion bonding. Prior to bonding, the textured sheets were grit blasted, chem milled and cleaned to produce surface conditions suitable for bonding. Diffusion bonding was accomplished at 1700°F for 8 hours under a vacuum of 10^{-4} TORR. The microstructure of the material is shown in Figure 3 in the as rolled condition and in Figure 4 in the as bonded condition.

The type and degree of texture induced by rolling was assessed using standard X-ray diffraction pole figures and by measuring the Young's modulus at several orientations normal to the plane of the sheet (i.e., longitudinal plane). Transmission pole figures were made using the technique of Bragg and Packer (Reference 22). A foil sample 0.002 inches thick was made by mechanically and chemically thinning from both sides of the as rolled sheet. A similar foil was made from the as bonded billet except that the foil was made in the longitudinal plane (the plane normal to the longitudinal or rolling direction) thus allowing the average texture in several full thickness laminates be measured. The as rolled pole figure was plotted via a Wolf net rotation so that it could be compared directly to the as bonded pole figure.

The second method of texture mapping was accomplished by observing the Young's modulus as a function of orientation. The effectiveness of this technique derives from the single crystal elastic anisotropy of the alpha phase in titanium (Figure 2) discussed previously. Using the resonant beam dynamic modulus technique of Liu and Alers (Reference 13) the Young's modulus can be determined using the following equation:

$$E = 1.45 \times 10^{-5} \rho (2L\ f)^2 \quad \text{(psi)} \qquad (1)$$

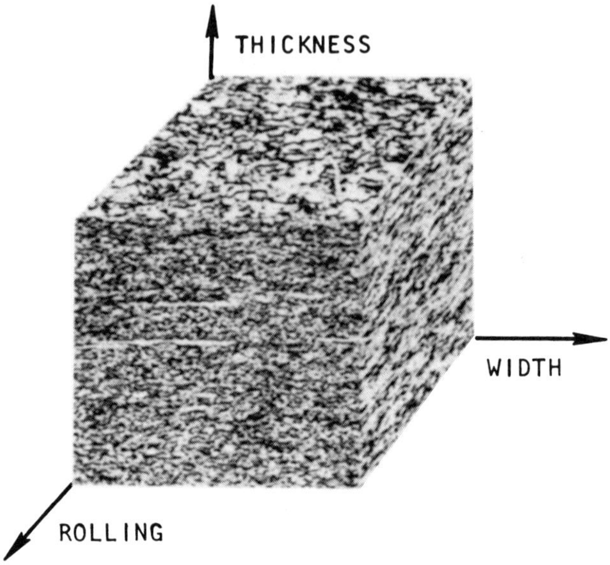

Figure 3. Microstructure of Ti-6-2-4-6 sheet in the as-rolled condition (500 X magnification).

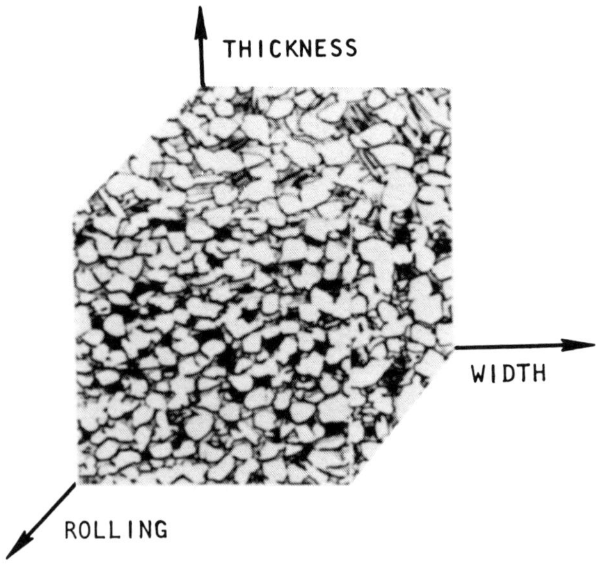

Figure 4. Microstructure of Ti-6-2-4-6 sheet following diffusion bonding (500 X magnification).

where

ρ is the density of the material (grams)
L is the length of the sample (centimeters)
f is the resonant frequency of the sample

In this study, Young's modulus measurements were made using rectangular beam specimens 2 x 0.125 x 0.065 inches. These beams were cut out of the longitudinal plane of the as bonded billet at 0, 22.5, 45, 67.5, and 90 degrees to the transverse direction.

Prior to use in this test program, the as bonded material was given a 1200°F anneal followed by air cooling. This was done in order to dissolve α_2 which has been found to precipitate in this alloy during slow cooling in the temperature range 900°F - 1200°F (Reference 22). In order to assure that the texture was not significantly altered by this treatment, the texture of the material was characterized using the Young's modulus dynamic measurement technique after this anneal cycle as well as several others at successively higher temperatures. The results shown in Figure 5 indicate that the 1200°F anneal produced only a slight change

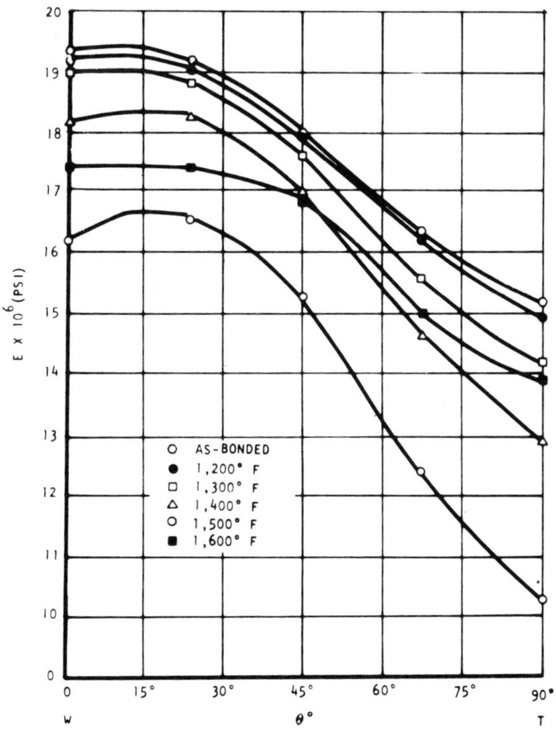

Figure 5. Variation of modulus with orientation and annealing temperature of Ti-6-2-4-6 obtained from ultrasonic resonance method.

in the modulus, thus, inferring that the texture of the material was not significantly altered.

After the 1200°F, standard round bar specimens were made oriented in the longitudinal (rolling) transverse (width) and short transverse (thickness) directions to determine tensile properties of the material. In order to evaluate the fracture toughness and fatigue crack growth rate properties of this material, compact tension specimens which fall within the limitations of the specimen designs specified in ASTM E399-70T were used. The basic configurations of the compact tension specimens used are shown in Figure 6 and Table II. The samples used were sized individually in order to conserve material. The sizing criteria were those developed by the ASTM (Reference 24) and NR/LAD (Reference 25). Fracture toughness testing was performed in accordance with ASTM E399.

Fatigue crack growth rate testing was performed in room temperature, low humidity air at a cyclic rate of 360 cycles per minute and a load range ratio (R factor) of 0.08. The crack growth loads were selected to produce initial crack growth rates of 10^{-7} inches per cycle. During testing, the loads were periodically lowered to produce marks on the fracture face. The test consisted of repeated sequences of growth for 0.1 inch and marking for 0.05 inch.

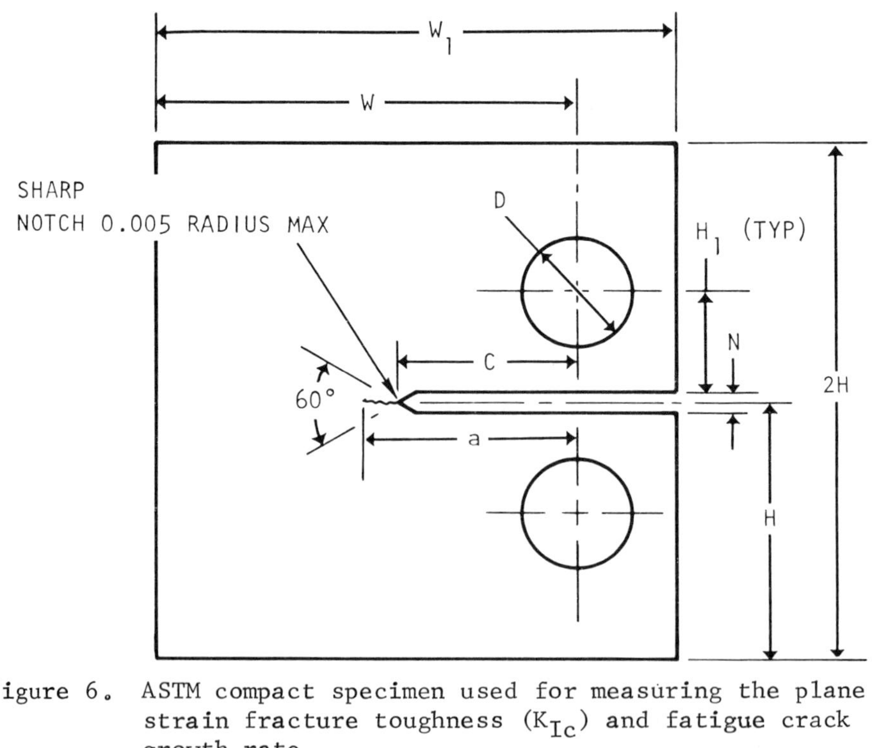

Figure 6. ASTM compact specimen used for measuring the plane strain fracture toughness (K_{Ic}) and fatigue crack growth rate.

Table II

COMPACT SPECIMEN DIMENSIONS

Thickness		W (±0.005)	W_1 (±0.010)	H (±0.005)	2H (±0.010)	H_1 (±0.005)	D (+0.005, -0.000)	N	C Fracture (±0.010)	C Fatigue (±0.010)
da/dN	K_{I_c}									
0.100	0.750	1.500	1.875	0.900	1.800	0.145	0.375	$\frac{1}{16}$	0.625	0.325
0.125	1.000	2.000	2.500	1.200	2.400	0.550	0.500	$\frac{1}{8}$	0.875	0.475
.1875	1.250	2.500	3.125	1.500	3.000	0.6875	0.625	$\frac{1}{8}$	1.125	0.625

EXPERIMENTAL RESULTS

The pole figures obtained from the as rolled sheet and the diffusion bonded billet are shown in Figure 7a and 7b, respectively. They reveal intense basal plane concentrations in the transverse direction as well as some lesser concentrations about the longitudinal (rolling) axis. The diffusion bonding cycle is seen to have caused a noticable increase in the intensity of the (0002) plane concentrations in the long transverse direction as well as a splitting of the secondary basal pole concentration about the longitudinal plane. Both of these effects are most likely due to the complete recrystallization which takes place during bonding (see Figures 3 and 4). Of less import would be the small amount of plastic flow which occurs during bonding (less than 5% compressive deformation) although the orientation of these secondary peaks is consistent with this type of deformation (Reference 26).

The tensile properties shown in Table III display anisotropy which correlates well with the basal plane orientations which appear in Figure 7. The magnitude of the tensile yield and ultimate strength are directly proportional to the basal plane concentrations. It should be noted in this case that the texture generated a tensile strength in excess of 190 ksi in this alloy in the annealed condition.

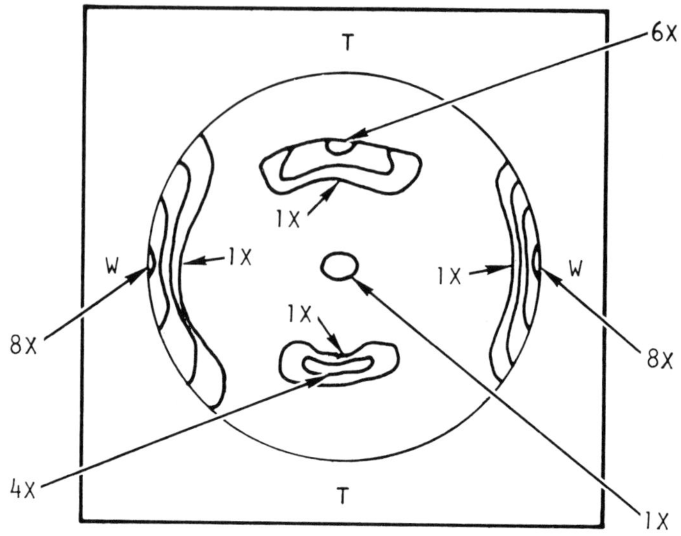

A. AS ROLLED

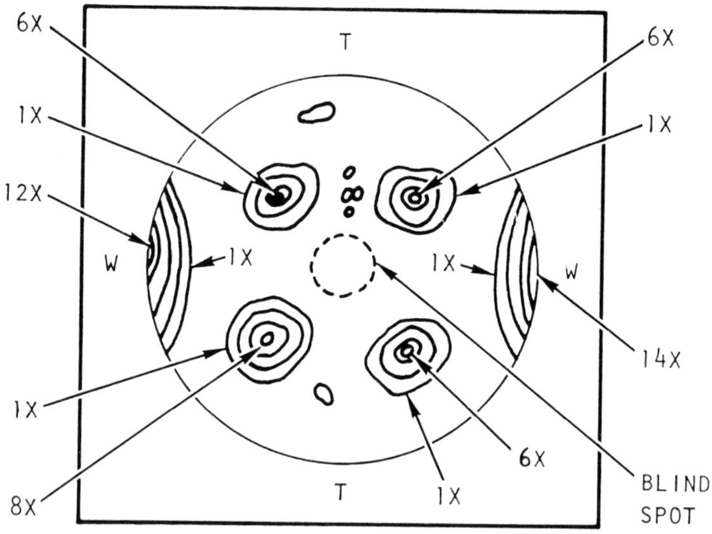

B. AFTER DIFFUSION BONDING

Figure 7. X-ray diffraction pole figures for textured rolled Ti-6Al-2Sn-4Zr-6Mo.

Table III

SUMMARY OF MECHANICAL PROPERTIES FOR TEXTURED Ti-6Al-2Sn-4Zr-6Mo

Orientation	F_{TU} (KSI)	F_{TY} (KSI)	% El	% R.A.	E ($X10^6$ PSI)	K_{Ic} (KSI$\sqrt{IN.}$)	K_{Ic}^2/F_{TY}^2 (IN.)	G_{Ic} IN. LB/IN.^{2}
L OR RW	148.7	138.2	11.5	18.0	15.5*	67.8	0.241	274
T OR WR	196.9	173.8	11.3	13.5	19.4**	83.0	0.212	284
ST OR TW	136.1	134.3	6.5	26.0	15.1**	44.5	0.110	118

*Measured dynamically after simulated diffusion-bonding cycle

**Measured using dynamic modulus technique

The modulus follows the pattern set by the tensile strength as would be predicted based upon the single crystal modulus anisotropy (compare Figures 2, 5, and 7). The ideal maximum modulus of 21×10^6 psi was not achieved in the transverse direction because other basal plane concentrations had developed and because the microstructure is approximately 15% BCC beta phase which has a significantly lower modulus (estimated to be 12×10^6 psi).

Table IV presents the fracture toughness test data and the K_{Ic} results which exhibit an anisotropy which is similar to that observed for the tensile strength and modulus properties. It should be noted that the transverse direction exhibits the maximum K_{Ic} which is somewhat anomolous because strength (also highest in the transverse direction) and fracture toughness are usually considered to be inversely proportional.

Additional evidence for the influence of texture on the fracture characteristics of this alloy was observed in the gross mode of failure of the toughness specimens as can be seen in

Table IV

FRACTURE TOUGHNESS DATA FOR TEXTURED Ti-6Al-2Sn-4Zr-6Mo

Orientation	Test Temp.	a (in.)	B (in.)	W (in.)	P_{max} (lb)	P_Q (lb)	K_{Ic} (ksi $\sqrt{in.}$)	ASTM Validity Tests Failed	$\left(\frac{K_{Ic}}{F_{ty}}\right)^2$
RW	RT	1.275	0.804	2.50	10,438.0	9,112.0	70.915	-	0.263
RW	RT	1.258	0.834	2.50	9,542.0	8,820.5	64.85	-	0.221
WR	RT	1.308	0.858	2.50	10,560.0	10,440.0	79.35	-	0.199
WR	RT	1.276	0.364	2.50	4,230.0	2,975.0	87.66	Thickness	0.254
TW	RT	1.258	0.778	2.50	6,123.0	6,123.0	48.26	-	0.130
TW	RT	1.020	0.566	2.001	5,287.0	5,091.5	41.085	-	0.094

Figure 8. Each specimen in the RW and WR orientations fractured at a specific reproducible angle to the plane of the notch. The RW specimens failed at 32.5 degrees to the plane of the notch while the WR specimens failed at 28.5 degrees to the plane of the notch. The TW specimens failed in the expected plane of maximum normal stress.

Finally, the fatigue crack growth rate data are plotted in Figure 9. Again, the data for the transverse orientation stands out from the other two orientations. Since the (WR) orientation has the highest K_{Ic}, it exhibits the lowest growth rate at high values of ΔK. The unusual feature of this data, however, is the fact that the WR orientation changes characteristics and exhibits the fastest growth rate at low ΔK values. These data emphasize the considerable influence which texture can have on the properties of titanium alloys. In this regard, it should be pointed out that the data for the RW orientation only describe a portion of the curve since the fracture path, beyond the last reported data point deviated from the plane of the notch to the same angle followed during rapid fracture in the K_{Ic} tests of this orientation. Therefore, ΔK could not be correctly calculated beyond this point.

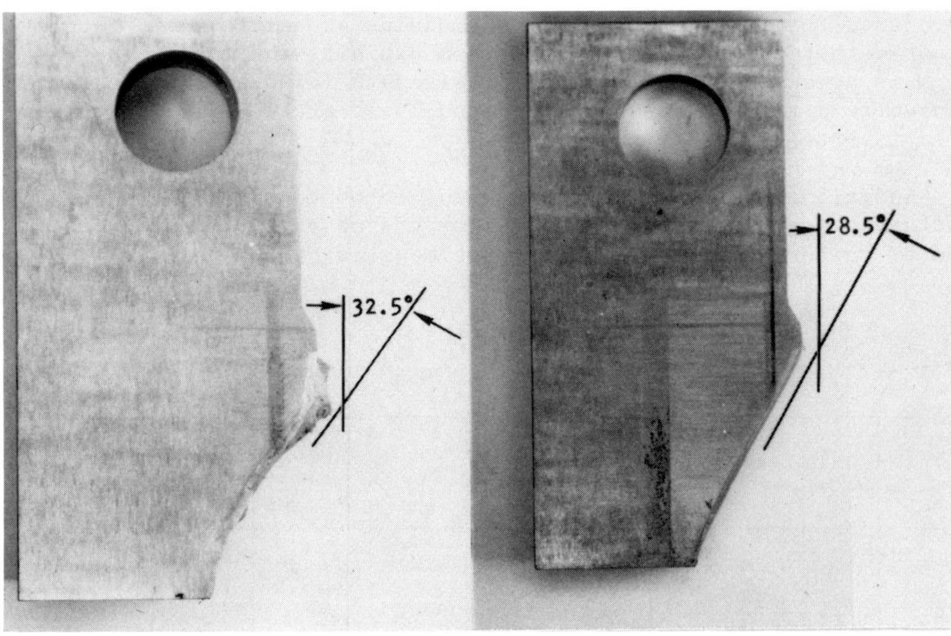

Figure 8. Fracture appearance of the RW and WR orientations.

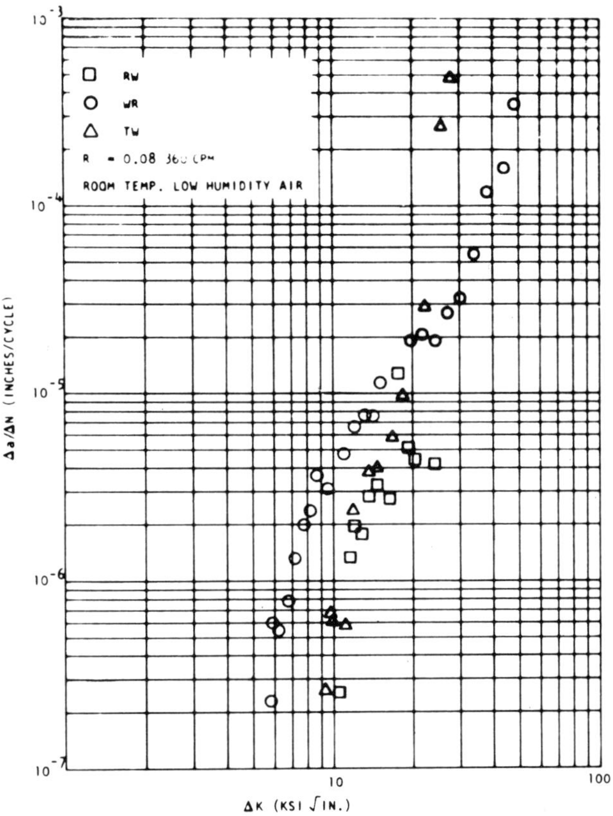

Figure 9. Fatigue crack growth rate. Characteristics as a function of orientation for textured Ti-6Al-2Sn-4Zr-6Mo.

DISCUSSION

The results of testing have shown that textures have a strong influence on the properties of Ti-6Al-2Sn-4Zr-6Mo and have brought to light the particular effect of basal plane textures on the fracture mechanics properties. Table III presents an overview of the results of this program. This summary shows that the transverse direction exhibited the highest strength, modulus and K_{I_C}. This combination of properties correlates directly to the highest concentration of basal planes in Figure 7. Thus, the basal plane orientation is seen to be a primary feature of textures which control property anisotropy. How this anisotropy arises is easily explained for modulus because of the known anisotropy of the hexagonal lattice (Figure 2) and for strength by considering the critical resolved shear stress for slip in various crystallographic directions in and out of the basal plane as discussed in the introduction. It is more difficult to explain the effect this same texture has on fracture mechanics properties. Therefore, this discussion will be concerned primarily with the fracture toughness and fatigue crack growth rate characteristics exhibited by this textured material.

The fracture toughness of this material is seen in Table III to be directly proportional to strength rather than the more typical inverse relationship. This may be due to the fact that strength in textured titanium is controlled by the orientation of available slip planes and slip directions rather than by the dislocation density (as in work hardened materials) or volume percent second phase (as in precipitation hardened materials). However, starting from first principles of linear elastic fracture mechanics, it should be recalled that toughness is directly related to elastic properties (modulus in Griffith's original formulation) rather than strength. The fracture toughness properties can be transformed into energy release rate by transforming K to G using the relationship

$$G_{Ic} = \frac{K_{Ic}\ \nu(1-\ 2}{E}$$

The results of this transformation are shown in Table III. As can be seen, the critical energy release rate G_{Ic} is the same for the RW and WR orientations. This indicates that the modulus anisotropy is primarily responsible for the "in plane" toughness anisotropy. The distinctly different results obtained in the TW orientation will be discussed later in the Section. The G_{Ic} equivalence finding is consistent with the newly emerging crack opening displacement theories of fracture (Reference 27) in which the crack tip displacements control failure. In the case at hand, the higher modulus orientation would require a higher load in order to achieve the same COD as a lower modulus orientation. Presuming that the microscopic critical strain to failure is identical for both orientations, each would exhibit a K_{Ic} which was directly proportional to the modulus. This is precisely the observation in this work.

Further evidence for the equivalence of toughness between the "in plane" orientations can also be obtained by normalizing the fracture toughness with the yield strength using the quantity $(K_{Ic}/F_{ty})^2$. This factor is proportional to the critical flaw size which a material can endure and is a useful means of comparison between materials. The results of this type of comparison are also shown in Table III. Again, the RW and WR orientations are seen to be equivalent.

One of the unusual aspects of the fracture toughness test results was the fracture mode behavior. As stated previously, the

RW and WR specimens fracture at an angle to the plane of the notch as can be seen in Figure 8. The angles were different for each orientation, however, both specimens from the same orientation failed at angles that were within 1 degree of being identical. This effect is felt to be a further consequence of the texture present in this alloy.

By considering that group of grains which give rise to the major basal plane concentration (that peak which lies in the transverse orientation of the pole figure) as a hexagonal single crystal, it can be seen that loads applied either parallel with or perpendicular to the C axis will result in the maximum shear stress being applied on the pyramidal planes, the twin planes, and certain of the prism planes (Figure 2). Specifically, it is found that the twin planes of the $(11\bar{2}2)$ type lie at an angle of 32.3 degrees to the longitudinal plane. The fracture toughness specimens which were loaded in the longitudinal direction (i.e., the RW orientation) fracture at 32.5 degrees. Therefore, preferential deformation by twinning in this plane may have led to fracture at the angle noted.

The (WR) orientation specimens failed at 28.5 degrees to the transverse plane. Loading in these specimens corresponds to a load applied parallel to the "C" axis of the hexagonal lattice. In this case, however, the closest possible deformation system is the $(10\bar{1}2)$ twin plane which lies at 42.4 degrees to the transverse plane. This does not seem to be a likely cause for the fractures which occurred at 28.5 degrees to the transverse plane, in the WR orientation specimens.

An alternate explanation for angle of the fracture plane in the WR specimens may be due to basal slip. This is possible since the basal planes are distributed in a cone about the transverse direction. This peak spreads to 20 degrees from the transverse direction so that applied stress components of sufficient magnitude could arise and cause basal slip.

Another possibility is that the deformation occurs in two steps, first, by twinning, then slip. Twinning on the (1012) plane could orient the normal slip systems in the twinned material for deformation.

Finally, the angle of failure may be due to slip on the prism planes $(10\bar{1}1)$ in a non-basal slip direction $\langle 1123 \rangle$. Because of the strong texture and the loading orientation, it is possible that this more difficult slip mode would be activated.

It is difficult to ascertain the precise cause for these fractures to any better extent at this time for several reasons. First, twin planes are forbidden reflections in X-ray diffraction so that detection of twinned material in the fractured specimens using diffraction technqiues is impossible. Furthermore, twins were not observed by light microscopy in this material and a limited transmission electron microscopic study yielded only inconclusive results. Further work is required in order to develop a totally satisfactory explanation for the failure modes observed. It is felt to be almost certain, however, that this phenomenon is a consequence of the texture induced in this material.

From the results and the discussion thus far, it is obvious that the properties measured in the TW orientation do not follow the pattern set by the other two orientations. The differences begin with the tensile properties which, upon first observation, show evidence of embrittlement. These signs of embrittlement include an atypical 2 ksi spread between the yield strength and ultimate strength and a much reduced total elongation to fracture. In addition, the fracture toughness in the TW orientation is comparatively low. On the other hand, the reduction in area is higher than in the other two orientations. Furthermore, fractographic examination of the fracture toughness specimens revealed that all three orientations exhibit almost identical fracture modes. Even the TW orientation specimen failed by ductile rupture. Finally, the fatigue crack growth rate results show that the RW and TW orientations have the same crack growth rate behavior.

One clue to the underlying cause of these divergent properties is seen in the fracture faces of the RW and WR specimens. This is shown in Figure 10. In this fracture face the material at the surface and midplanes of the laminates exhibit different appearances so that a striped pattern results. This pattern which appears on all RW or WR specimens suggests that there is a difference between the properties at the center and surface of the laminates. Confirmation that the properties differ can be seen in that the fatigue crack fronts in the WR orientation specimen in Figure 10 drape between the laminate interfaces. In this figure, it appears that the growth was slower in the surface material than at the midplane.

A surprising result is that similar examination of the RW specimen revealed the opposite behavior. In the RW orientation, the crack growth rate was highest in the surface material and lowest in the material at the midplane. This evidence suggests that there may be a difference in the texture present between the surface and center of the rolled sheets. This is reasonable since the fatigue crack growth rate behavior is much different for the RW and WR orientations as shown in Figure 9. Further evidence that the texture is different between the surface and center of the laminates is shown in Figure 11. In this figure, photomicrographs taken under polarized light are shown which reveal that the microstructure at the surface and center of the laminates respond

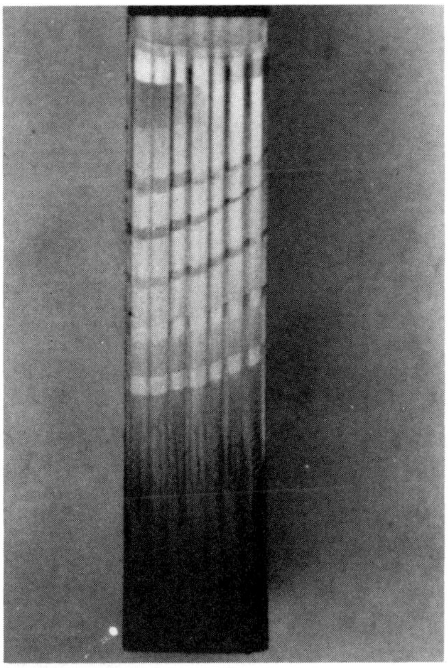

Figure 10. Fracture appearance for fatigue crack growth rate for specimen of Ti-6-2-4-6, WR orientation.

differently to polarized light. It is also shown that the response is reversed in changing from the longitudinal to the transverse view.

The presence of two different textures, one in the surface material and one in the center material, could be responsible for the poor tensile and fracture toughness results in the following manner. Textured material, regardless of its actual grain size, could be expected to exhibit a large effective grain size since each grain bears a close orientational resemblance to each adjacent grain. In this textured Ti-6-2-4-6 where bands of different textures appear to be present, deformation in these bands of large effective grain size could impinge at the interface between bands and cause high local stresses. The high local stress could result in early failure of tensile specimens and could cause the measurement of low fracture toughness properties. The fatigue crack growth rate behavior on the other hand would not be affected by this cooperative deformation in bands since the amount of material involved in fatigue crack growth is limited to a very small zone at the crack tip. At low ΔK levels, the fatigue plastic zone is very much less than the band width so that band interactions could not take place. It is felt, therefore, that differences

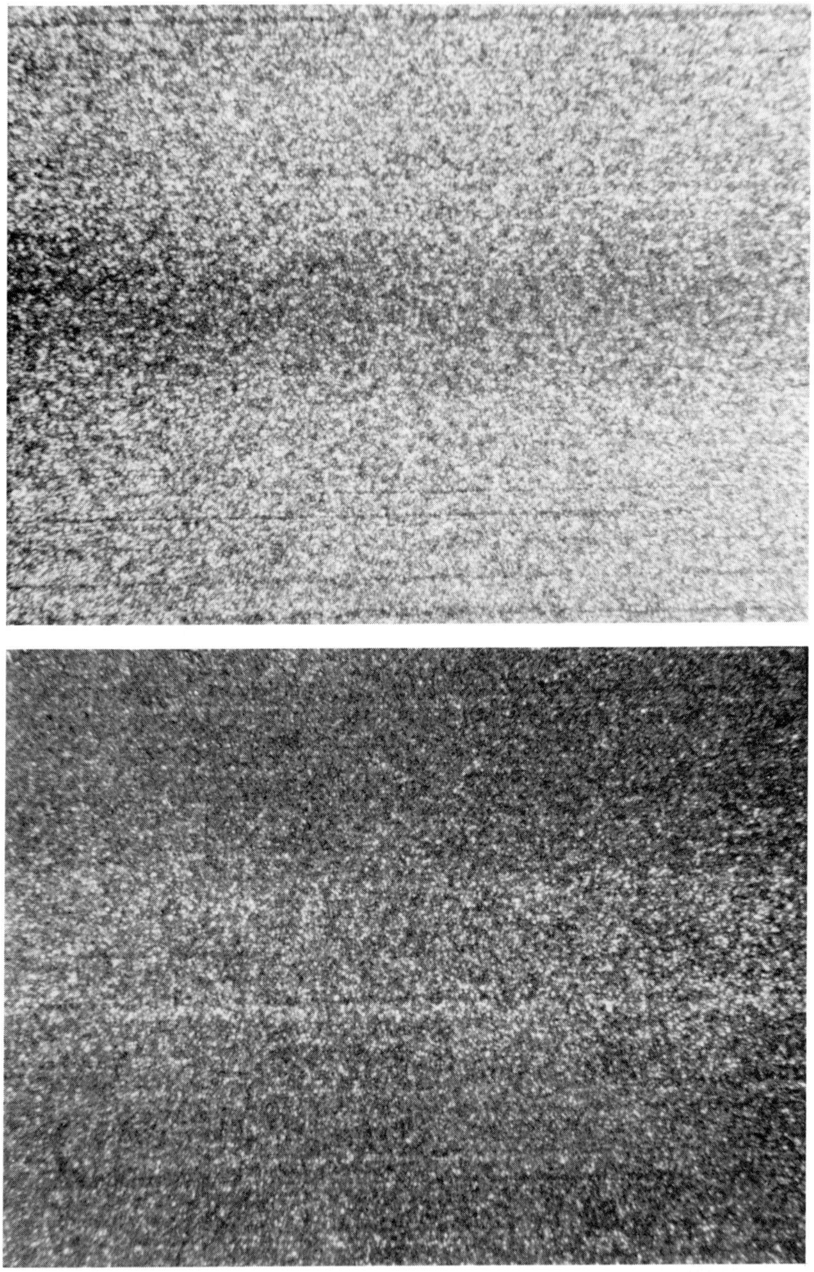

A

B

Figure 11. Polarized light micrographs of the bond zone of diffusion bonded Ti-6Al-2Sn-4Zr-6Mo illustrating the difference in texture between the edge and contour of each laminate. (a) Viewed in the longitudinal direction the surface material is dark; (b) Viewed in the transverse direction the surface material is light. (50 X magnification) in the texture which developed at the surface and in the center of the rolled sheets is responsible for the striped fracture appearance of the RW and WR specimens and for the poor properties of the TW orientation specimens. Evidence for this is seen in Figure 12 where it appears that the static fracture path in the TW orientation specimen did indeed follow a path near the interface between regions with different textures.

Finally, it is necessary to comment on the fatigue crack growth rate behavior of this material. As was mentioned previously, the difference in behavior between the WR orientation specimen and that exhibited by the specimens from the other two orientations is quite striking. At first glance, it appears obvious that the difference must be explainable in terms of elastic modulus, strength or K_{Ic}, all of which show similar variations. Many approaches have been developed to attempt to predict fatigue crack growth rate. Several of these (References 28 through 31) contain terms which are made up of these same material properties. In general, these equations are such that the growth rate is proportional to $\Delta K/E$ or $\Delta K/F_{ty}$. Based upon such approaches, it should be possible to plot data from various kinds of material which exhibit different crack propagation rates, as $\Delta K/E$, or $\Delta K/F_{ty}$ versus da/dn and compress these data onto a single curve. In the case at hand, this type of approach

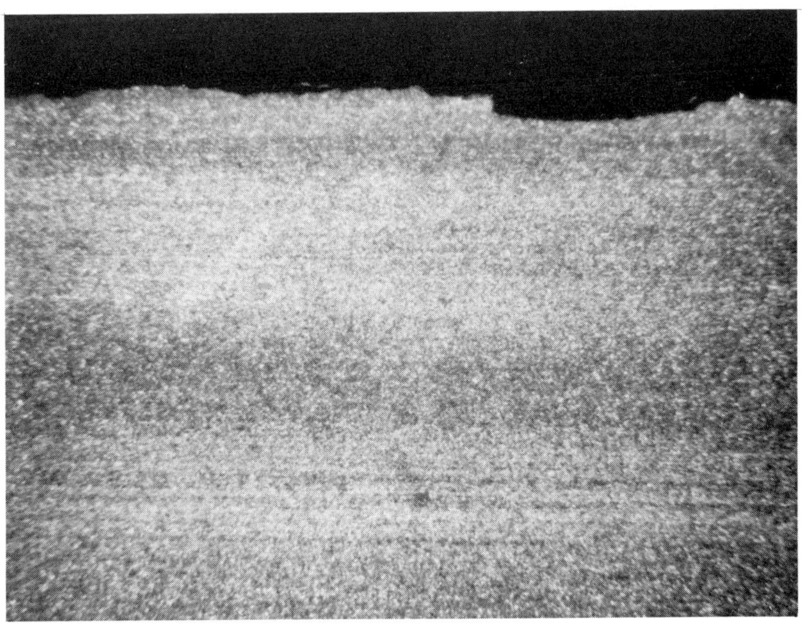

Figure 12. Polarized light micrograph of the longitudinal plane at the fracture surface of a TW orientation K_{Ic} specimen (25 X magnification).

leads to a worse misfit between the WR data and the RW and TW orientation data. Only in the upper portion of the curves is there a correlation. This suggests that the current attempt to predict fatigue crack propagation behavior using bulk material properties are only adequate where large scale plastic flow processes are involved such as at the high growth rates where K_{max} (in ΔK) approaches K_{Ic}. In this regard, very recently Hickerson and Hertzberg (Reference 32) also have demonstrated convincingly that there is no fundamental relationship expressed by current predictive equations between crack growth behavior and conventional properties such as cyclic stress: strain parameters, etc.

Despite the inability of existing concepts of fatigue crack propagation to shed light as to why the textured Ti-6-2-4-6 behaves as it does, the data may offer hints as to what factors are influencing fatigue crack propagation. For example, the microstructure in Figure 4 suggests that grain size is not important because the grains in this material are equiaxed. This is not surprising since the growth per cycle at low ΔK levels is at least 2 orders of magnitude smaller than the average grain diamter (grain size - 2×10^{-4} inches). Therefore, while existing theoretical constructs have not aided in interpreting these results, it is hoped that the data will serve to help in improving the search for better analytical tools.

CONCLUSIONS

1. It has been shown that the basal plane textures control modulus, strength, fracture toughness and fatigue crack growth rate in Ti-6Al-2Sn-4Zr-6Mo. The modulus was found to vary according to the known elastic anisotropy of the titanium hexagonal crystal structure, and the strength variations are explainable in terms of the available deformation systems as a function of orientation in this textured material. The fracture toughness properties are shown to be a function of the microscopic strain to failure which, in turn, is controlled by the modulus. Finally, the fatigue crack growth rate behavior could not be explained within the limits of current theories of crack growth mechanisms.

2. A difference in texture was found to exist between the surface and center of the laminates. This difference in texture is felt to have influenced the tensile and fracture toughness properties measured for those specimens loaded in the short transverse direction.

3. During fracture mechanics testing, specimens taken from the various directions in the rolling plane exhibit a tendency

to fail at an angle to the plane of the notch. This behavior is felt to be due to the limited number of deformation modes available, notably twinning or non-basal slip which could have occurred in these specimens.

4. Finally, the control of textures in titanium has been shown to be a powerful tool with which to control the properties of alpha-beta titanium alloys. In this alloy, it has been demonstrated that it is possible to produce fully annealed Ti-6Al-2Sn-4Zr-6Mo which will exhibit a Young's modulus of 19×10^6 psi, an ultimate strength of 190 ksi and a K_{Ic} of 80 ksi $\sqrt{in.}$ in one orientation in the plane of the sheet laminate product.

ACKNOWLEDGEMENT

This work was sponsored by North American Rockwell's corporate internal research and development funds. This paper is based in part upon work performed by Mr. P. G. Reimers in partial fulfillment of requirements for his Master of Science degree in Materials Science from the University of California at Los Angeles. Mr. Reimers participated in a cooperative program between North American Rockwell and UCLA under the supervision of Dr. A. W. Sommer of North American Rockwell and Dr. A. H. Tetelman of UCLA. The authors wish to express their thanks to North American Rockwell for funding this effort. In addition, the authors are sincerely appreciative of the patient and thoughtful counseling received from Dr. J. C. Williams, Mr. M. P. Kaplan, and Dr. L. J. Kashar during the preparation of this paper.

REFERENCES

1. Keeler, J. H. and Giesler, A. H., "Preferred Orientation in Rolled and Annealed Titanium," Journal of Metals, Vol. 8, 1956, pp. 80-90.

2. Holden, F. C., Williams, D. N., Riley, W. E., and Jaffee, R. I., "Flow Properties Deformation Textures and Slip Systems of Titanium and Titanium Alloys," Titanium Metallurgical Laboratory Report No. 30, 31 Jan 1956.

3. Roberts, W. T., "Preferred Orientation and Anisotropy in Titanium," Journal of the Less Common Metals, Vol. 4, 1962, pp. 345-361.

4. Zarkades, R. and Larson, F. R., "Sheet Tensile Properties of Titanium Alloys as Affected by Texture," AMMRC TR-68-03, Jan 1968, DDC No. AD666765.

5. Hatch, A. J., "Texture Strengthening of Titanium Alloys," Trans Met Soc AIME, Vol 233, Jan 1965, pp. 44-50.

6. Lockheed Missile and Space Company, "Development of Improved Biaxial Strength in Titanium Alloy Rocket Motor Cases Through Texture Hardening," AFRPL-TR-302, Dec 1967.

7. Babel, H. W. and Frederick, S. F., "A Review of Texture Strengthening of Titanium Alloys," Douglas Paper 4664, Jan 1968.

8. Larson, F. R., "Anisotropy of Titanium Sheet in Uniaxial Tension," Trans ASM, Vol. 57, No. 3, Sept 1964, pp. 620-631.

9. Larson, F. R., "Textures in Titanium Sheet and Its Effect on Plastic Flow Properties," Report No. AMRA-TR-65-24, Oct 1965, DDC No. AD 626410.

10. Williams, J. C. and Blackburn, M. J., "Identification of a Non-basal Slip Vector in Titanium and Titanium Alloys," Phys Stat Sol Vol. 25, p K-1, 1968.

11. Fisher, E. S. and Renken, C. J., "Single Crystal Elastic Moduli and the HCP-BCC Transformation in Ti, Zr, and Hf," The Physical Review, Vol. 135, No. 2A, July 1964, pp. A482-A494.

12. Flowers, Jr., J. W., O'Brien, K. C., and McEleny, P. C., "Elastic Constants of Alpha Titanium Single Crystals at 25°C," Journal of the Less Common Metals, Vol. 7, No. 5, Nov 1964, pp. 393-395.

13. Liu, Y. C. and Alers, G. A., "The Anisotropy of Young's Modulus in Cold-Rolled Sheets of Binary Cu-Zn Alloys," Trans AIME, Vol. 236, 1966, p. 489.

14. Borik, F. and Alers, G. A., "Measurement of the Elastic Properties of Rolled Sheet," Trans AIME, Vol. 233, 1965, p. 7.

15. Alers, G. A. and Liu, Y., "Calculation of Elastic Anisotropy Rolled Sheet," Trans AIME, Vol. 236, 1966, p. 482.

16. Morris, R., "Elastic Constants of Polycrystals," Int J of Engr Sci, Vol. 8, 1970, p. 49.

17. Harrigan, M. J., Sommer, A. W., Alers, G. A., "The Effect of Textures on the Mechanical Properties of Titanium Alloys," NA-69-909, Dec 1969.

18. Carman, C. M., Forney, J. W., and Katlin, J. M., "Plain Strain Fracture Toughness of 5Al-2.5Sn Titanium at Room and Cryogenic Temperatures," NASA-CR-54296, April 1966.

19. Sullivan, T. L., "Texture Strengthening and Fracture Toughness of 4Al-0.2 Oxygen and 5Al-2.5Sn ELI Titanium Sheet in Biaxial Fields at Room and Cryogenic Temperatures," NASA TM-X-52272, 1967.

20. Sullivan, T. L., "Texture Strengthening and Fracture Toughness of Titanium Alloy Sheet at Room and Cryogenic Temperatures," NASA TN-D-444, May 1968.

21. Hatch, A. J., "Mechanical Metallurgy and Fracture Toughness of Titanium Alloys," Presented at the Titanium Technical Conference, University of Dayton, Dayton, Ohio, 13 Nov 1967.

22. Bragg, R. H. and Packer, C. M., "Quantitative Determination of Preferred Orientation," J of Appl Physics, Vol. 35, p. 1322.

23. Williams, J. C. and Rhodes, C. G., "Phase Transformations and Mechanical Properties of Ti-6Al-2Sn-4Zr-6Mo," to be published.

24. ASTM E399-70T, Tentative Method of Test for Plane Strain Fracture Toughness of Metallic Materials, 1970.

25. Harrigan, M. J., "Fracture Toughness Testing for Qualification of Metallic Materials," ST0169LB0001, 12 August 1971.

26. Williams, D. N. and Eppelsheimer, D. S., "Compressive Texture of Iodide Titanium," Trans Met Soc AIME, Vol. 192, June 1952. pp 615-618.

27. Irwin, G. R., Lingaraju, B., and Tada, H., Lehigh University Fritz Engineering Laboratory Report No. 358.2, July 1969.

28. Liu, H. W., "Fatigue: An Interdisciplinary Approach," Syracuse University Press, 1964.

29. McClintock, F. A., "Fatigue Crack Propagation," ASTM STP 415, 1967.

30. Laird, C., "Fatigue Crack Propagation," ASTM STP 415, 1967.

31. Donahue, R. M., et al., "Crack Opening Displacement and the Rate of Fatigue Crack Growth,"

32. Hickerson, J. P. and Hertzberg, R. W., "The Role of Mechanical Properties in Low-Stress Fatigue Crack Propagation," Met Trans Vol. 3, Jan 1972, p. 179.

EFFECT OF TEXTURE ON THE CHARPY IMPACT ENERGY OF SOME TITANIUM ALLOY PLATE

Anthone Zarkades and Frank R. Larson
Army Materials and Mechanics Research Center
Watertown, Massachusetts

An investigation to determine the anisotropic nature of toughness due to texture in titanium and titanium alloy plate was conducted. Standard Charpy V-notch specimens were machined at ten-degree increments from the rolling to the transverse direction with notch orientation either parallel or perpendicular to the plate surface. Basal pole figures were determined along with impact energy. The study revealed that significant toughness variations can be found in titanium plate as a function of specimen and notch orientation. This impact energy anisotropy can be related to texturing with normal fibering effect being small or absent. An explanation of the toughness variation on a simplified crystallographic slip basis is given. Titanium plate studied includes commercially pure, Ti-6Al-4V, Ti-4Al-4V, Ti-4Al-4Mn and Ti-8Mn.

Introduction

The anisotropic nature of the mechanical properties of metals can arise from two sources. The first is related to crystallographic preferred orientation or texture, and the second is due to fibering or elongation of nonmetallic inclusions as it is commonly seen in steel. All mechanical properties are affected to some degree by both types of anisotropy. In the past, when the quality of steel was poor and more nonmetallic stringers were present, a large amount of effort was devoted to studying the anisotropic characteristics of fracturing. More recently, interest has shifted to the effects of preferred orientation upon yielding. Most of this work has centered around the increases in strength which can be expected in biaxial loading. This work on preferred orientation has dealt mainly with yield and flow with very little attention being directed toward fracture or toughness. The basic problem of predicting the effects of texture upon the toughness is extremely difficult and complicated because it includes yield and flow effects. The toughness property can be viewed as the integral or the area under the stress-strain curve. From this it can be seen that the total work or toughness is a function of both the strength level and the fracture ductility. The effects of texture on yielding and flow are fairly well understood from the work of Hill (1) and Backofen et al (2), yet the quantitative definition of the effects of texture on fracture are illusive and still ill defined.

In these situations an experimental program is useful in defining the parameters. It was the purpose of this investigation to determine the anisotropic nature of toughness in titanium and titanium alloy plate where a known texture was present. The Charpy impact specimen was utilized for convenience and because at the level of toughness found in some of the plates, a specimen larger than the plate thickness would be required to determine a valid K_{IC}.

Testing Procedure

Examination of ten titanium plates as listed in Table I was carried out in this investigation. Basal (0002) pole figures were determined along with impact energy. All tests were conducted at room temperature with the exception of Ti-75A-M230 in which impact energies were also determined at 100, 200, and 300 C. All material was examined in the as-received condition. Plate thickness ranged from 0.50 to 1.00 inch.

Single-quadrant basal pole figures were determined to describe the crystallographic preferred orientation for each material. The reflection technique as described by Lopata and Kula (3) was utilized.

Table I - Titanium Plates Examined

Titanium Alloy	Heat No.
Ti-75A	M230, L925
Ti-4Al-4V	H8839, G8821, G8839
Ti-6Al-4V	2804
Ti-RC130B	3210M
Ti-RC130A	A3249B

Standard 0.394-inch-square Charpy V-notch impact specimens were machined at ten-degree increments from the rolling direction to the transverse direction. Two notch orientations were examined. One orientation has the specimens notched parallel to the plate surface, and the second orientation has the notch normal to the plate surface or in a through-thickness direction. The ASTM designations for longitudinal and transverse directions specimens with notches as indicated would be RT, WT and RW, WR. In this form, the first letter indicates the direction the notch is normal to, and the second letter the direction of crack propagation. For the thickest plate examined which was the Ti-4Al-4V alloy, Charpy specimens were taken from the top and bottom of the plate. Impact results plotted are an average of these specimens.

Discussion of Test Results

Prior to a detailed discussion of the Charpy impact test results, a brief review of the textures in titanium and its alloys along with a discussion of the anisotropic nature of the mechanical properties would be helpful. An extensive review of preferred orientation in wrought products can be found in an article by Dillamore and Roberts (4). A number of other reports have been published concerning textures found in titanium (5-9). The results of these investigations indicate that the titanium textures can be schematically subdivided into five main types as illustrated in Figure 1.

If it were desired to demonstrate the variation of a specific property as influenced by texture, not all of the textures shown in Figure 1 would be suitable. Since the properties of hexagonal materials are nearly axisymmetric with respect to the c axis then test samples must be cut at various angles from the c axis. Because of difficulty in getting through-thickness test samples for sheet material, this is most easily done with a type texture illustrated in Figure 1c (i.e. where the basal poles lie at or near the transverse direction). In this case longitudinal sample would be parallel to the crystallographic "a" direction and a transverse sample parallel to the c direction. Prior work (8) has demonstrated clearly that the elastic and plastic properties do vary

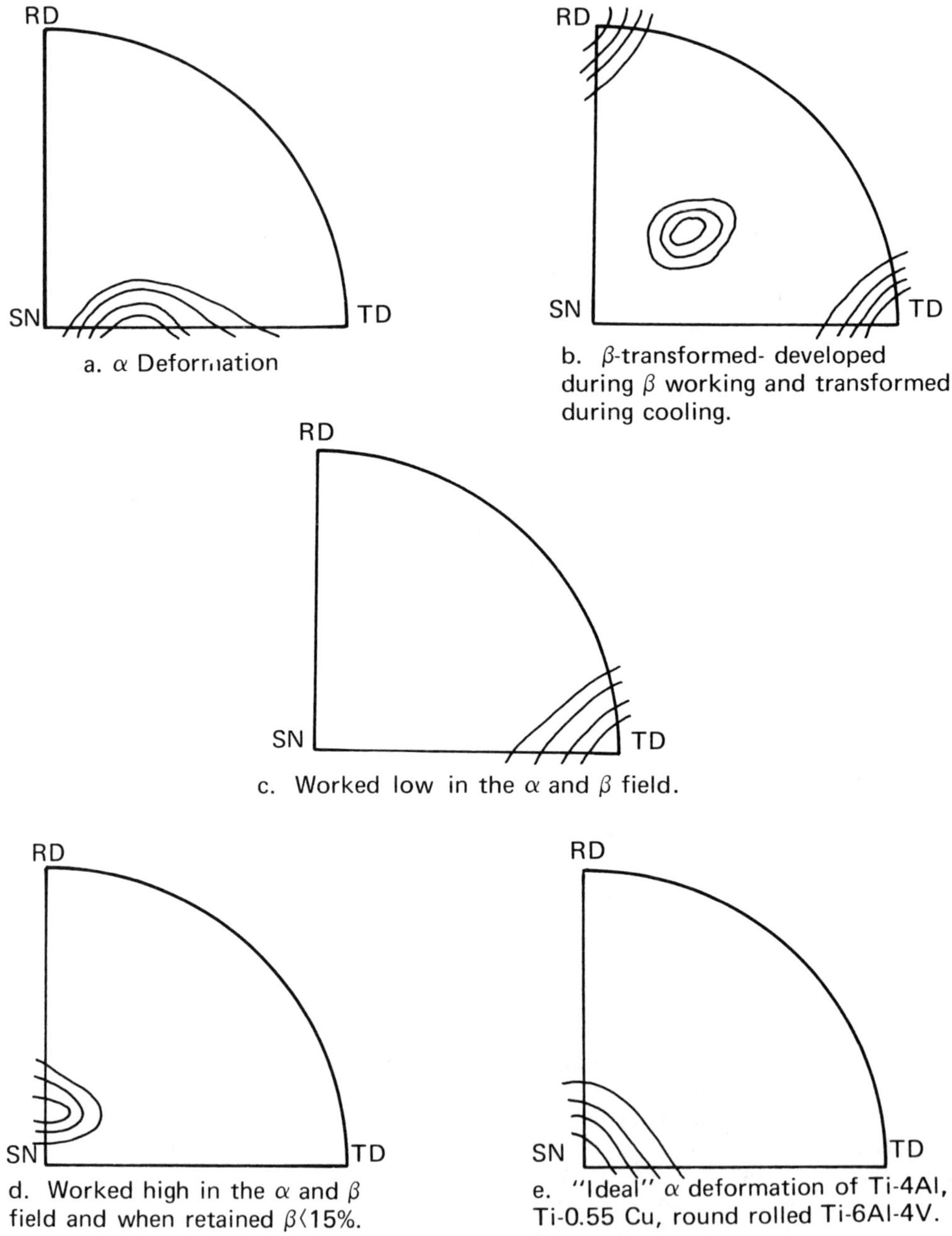

a. α Deformation

b. β-transformed- developed during β working and transformed during cooling.

c. Worked low in the α and β field.

d. Worked high in the α and β field and when retained β⟨15%.

e. "Ideal" α deformation of Ti-4Al, Ti-0.55 Cu, round rolled Ti-6Al-4V.

Fig. 1. (0002) texture types evident in various titanium sheet and plate.

in a predictable way and the average tilt of the basal poles from the sheet normal can be calculated from the measurements of elastic and plastic properties.

Since we are primarily concerned about the effects of texture it is necessary to rule out the effect of fibering. Although not completely conclusive, the variation of toughness caused by fibering can be detected by the metallographic observations of delaminations associated with the fractured surfaces (10). To a lesser degree the effect of an elongated second phase can also be evaluated in this manner. A large number of fracture samples were studied, and delaminations of the type associated with mechanical fibering were not found in the magnitude as described by English (10). There also did not seem to be much preferential fracture path associated with the elongated alpha-beta duplex structures found in some of the alloys. In addition, the Charpy impact energies did not show the anisotropy for the WT and WR orientation normally attributed to mechanical fibering or elongated grain structure. This will be discussed later in the report. Analysis of the data indicate that Charpy impact fractures can be catorgorized into two general types, brittle and ductile. In the case of brittle fracture (less than 15 foot pounds) very little anisotropy of toughness was found. Figures 2-5.

Thus it appears from this preliminary observation that the anisotropy of Charpy impact properties is related to plastic flow anisotropy. One of the most dramatic cases found was for the Ti-4Al-4V alloy, heat H8839, Figure 6, and other examples are shown in Figures 7-9. For heat H8839 the RW and WR specimen orientations gave about 25 foot-pounds. The WT was also low at about 35 foot-pounds. However the RT results were three times higher or about 75 foot-pounds. These results can be explained if the texture is examined in relation to specimen constraints and its ability to deform plastically. Considering the fact that slip occurs in the $[11\bar{2}0]$ direction on the prism planes: it can be seen that the tough orientation would be the "softest" and hence the greatest plastic deformation by slip would occur. For the other orientations, either $(11\bar{2}2)$ or $(10\bar{1}2)$ twinning would be required for deformation. Subsequent to the twinning reorientations, the lattice would not be favorably orientated for slip.

The high toughness orientation unfortunately was not demonstrated for "ideal" texture in this program for they were generally brittle. However, they were clearly shown in the work of Hatch (11). See Table II. For these ideal textures, the tough orientation would occur for all edge-notched specimens (RW and WR). The high R values are characteristic of a nearly ideal texture.

The effect of texture upon the toughness transition caused by varying temperature can be seen in Figure 10 (12,13). The data is for Zircalloy 2 which has deformation modes similar to titanium

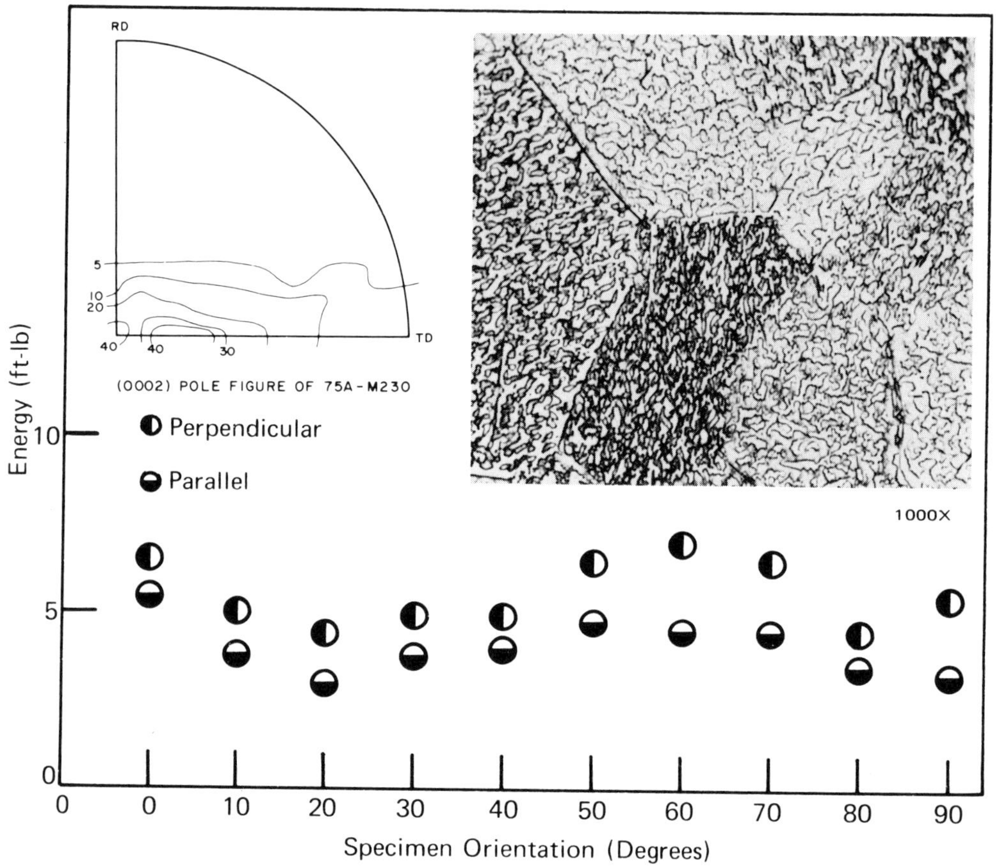

Fig. 2. Typical test data and (0002) pole figure for Ti-75A-M230.

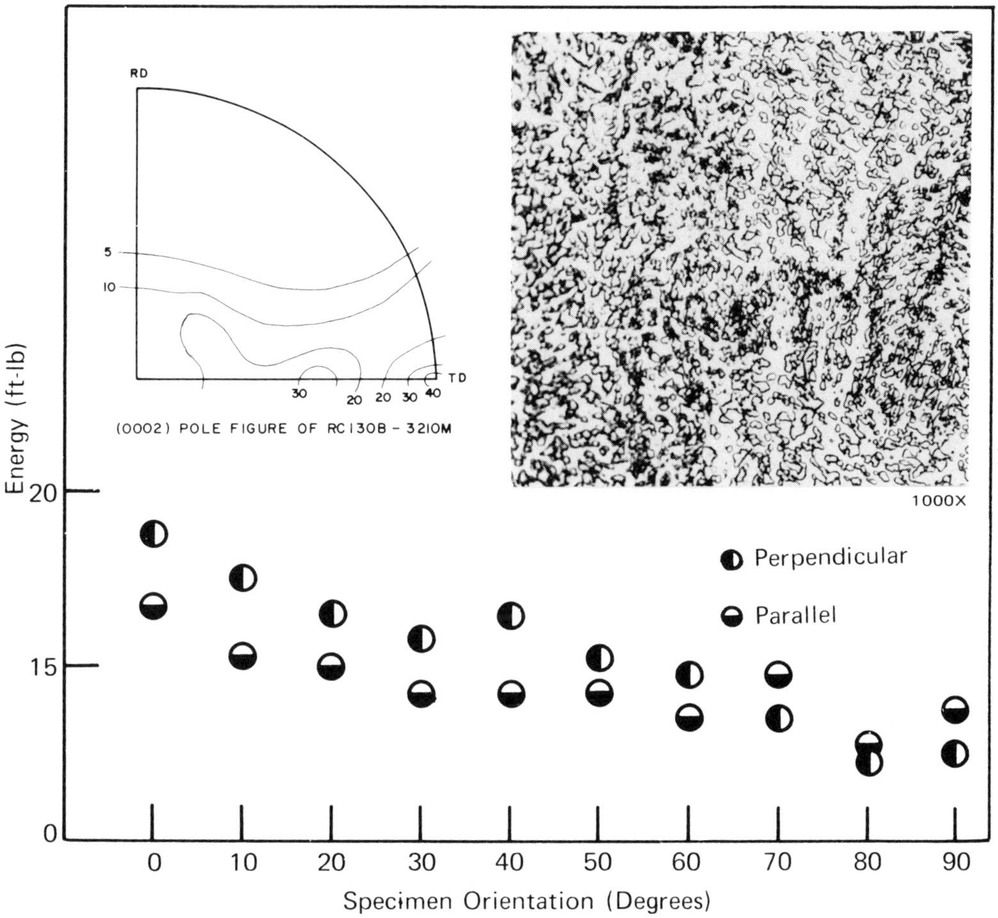

Fig. 3. Typical test data and (0002) pole figure for Ti-4Al-4Mn-3210M.

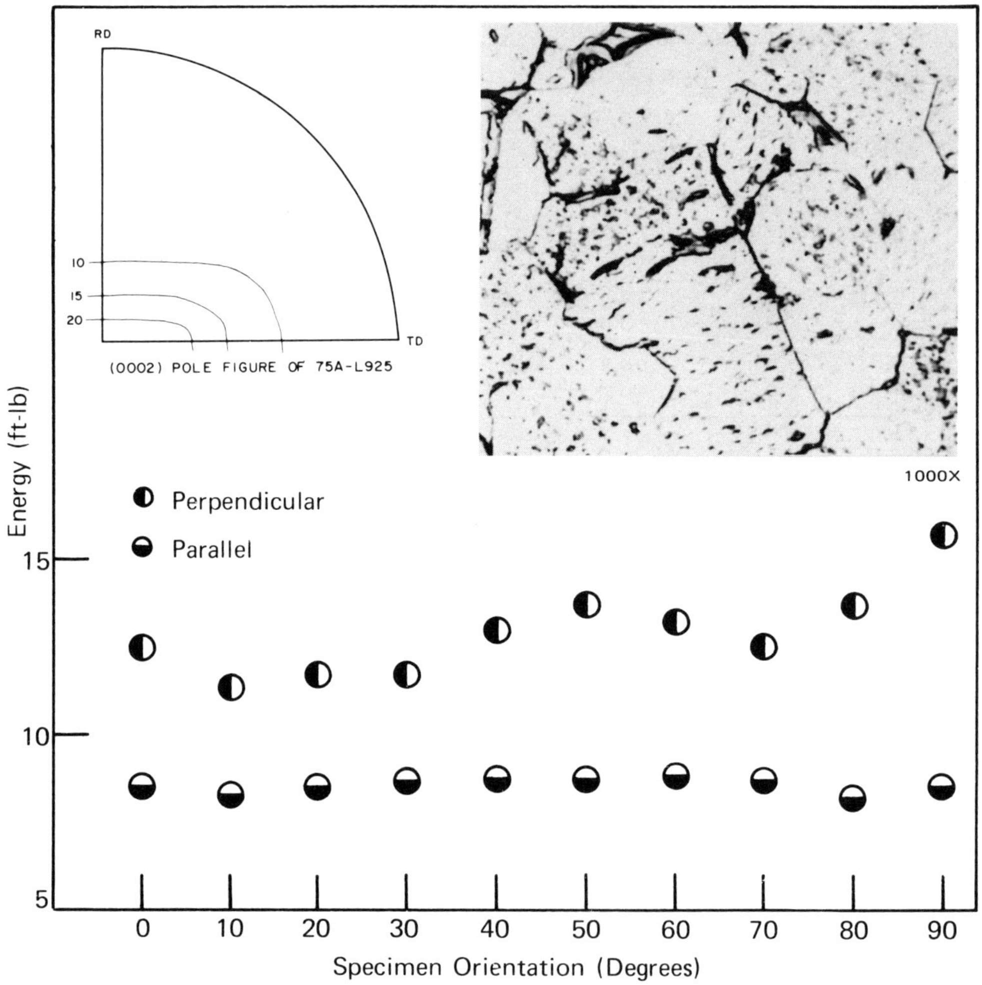

Fig. 4. Typical test data and (0002) pole figure for Ti-75A-L925.

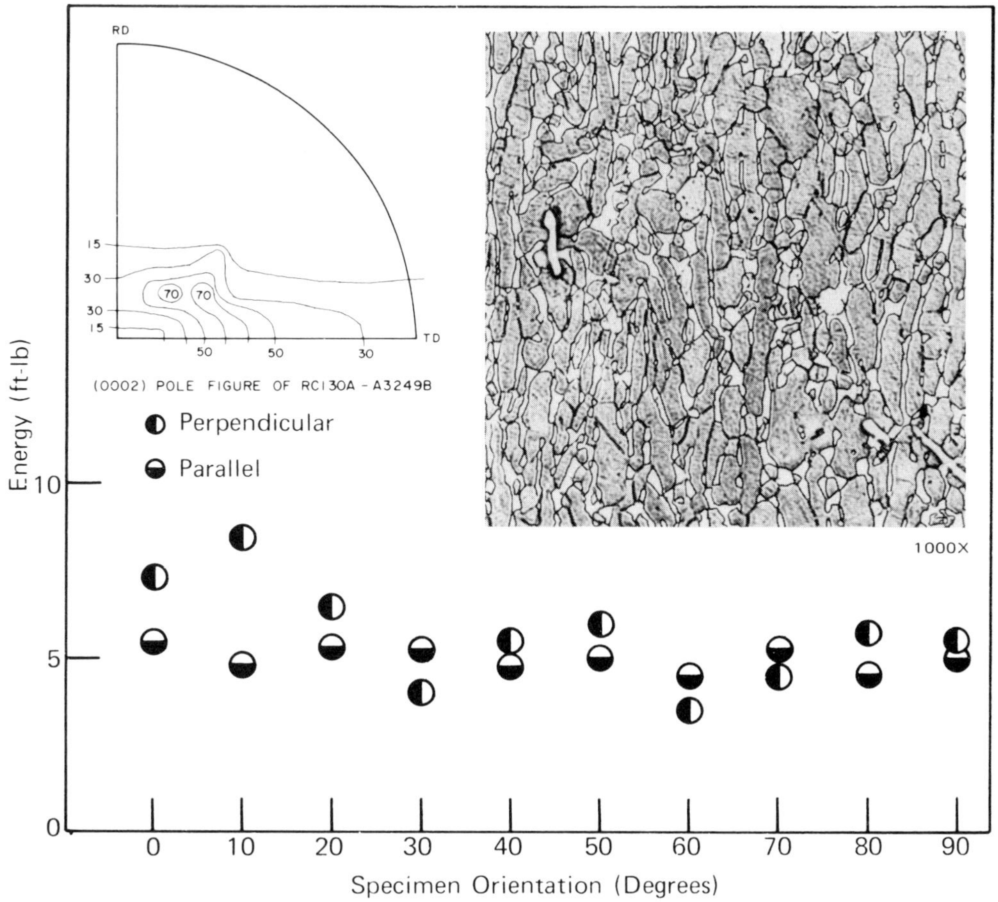

Fig. 5. Typical test data and (0002) pole figure for Ti-8Mn-A32495.

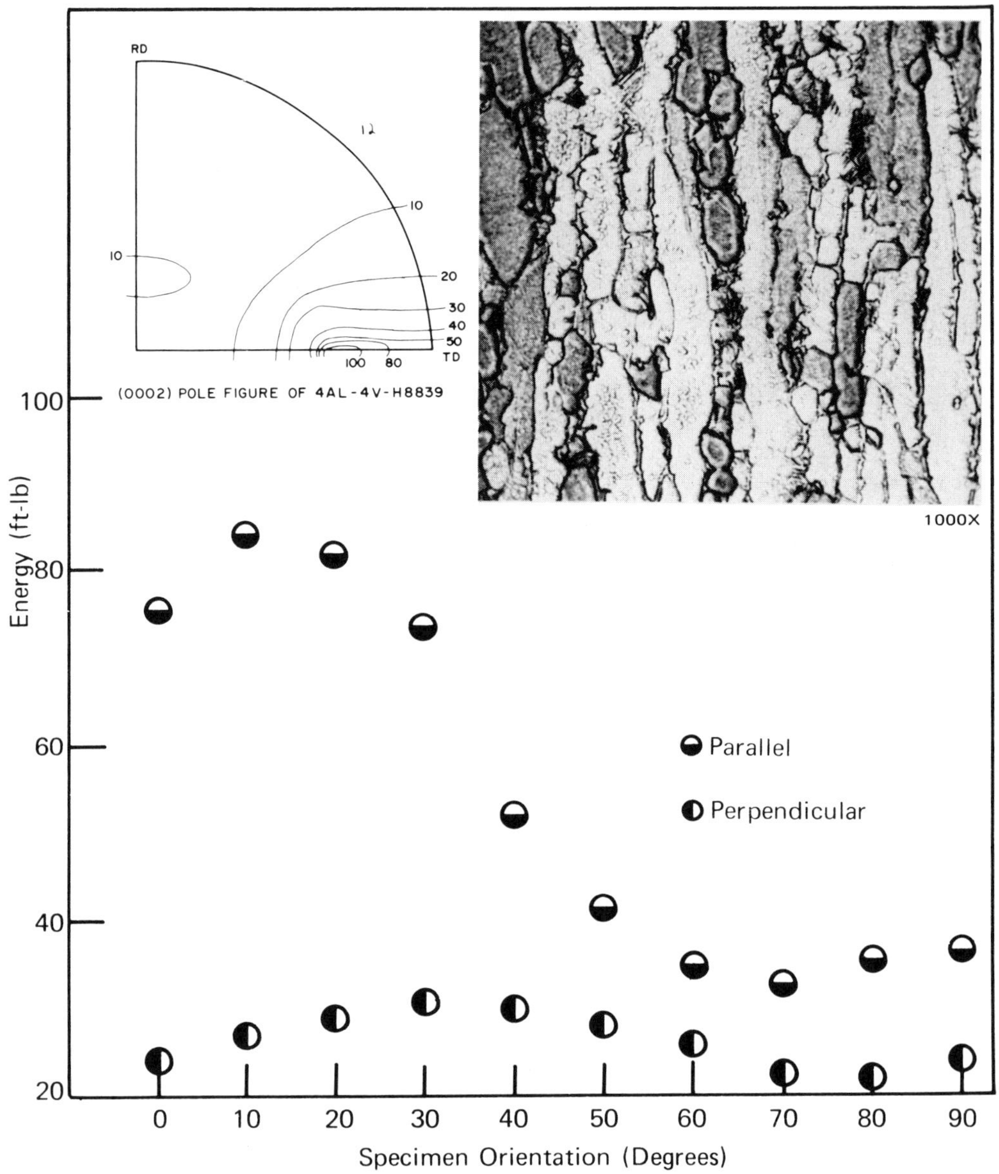

Fig. 6. Typical test data and (0002) pole figure for Ti-4Al-4V-H8839.

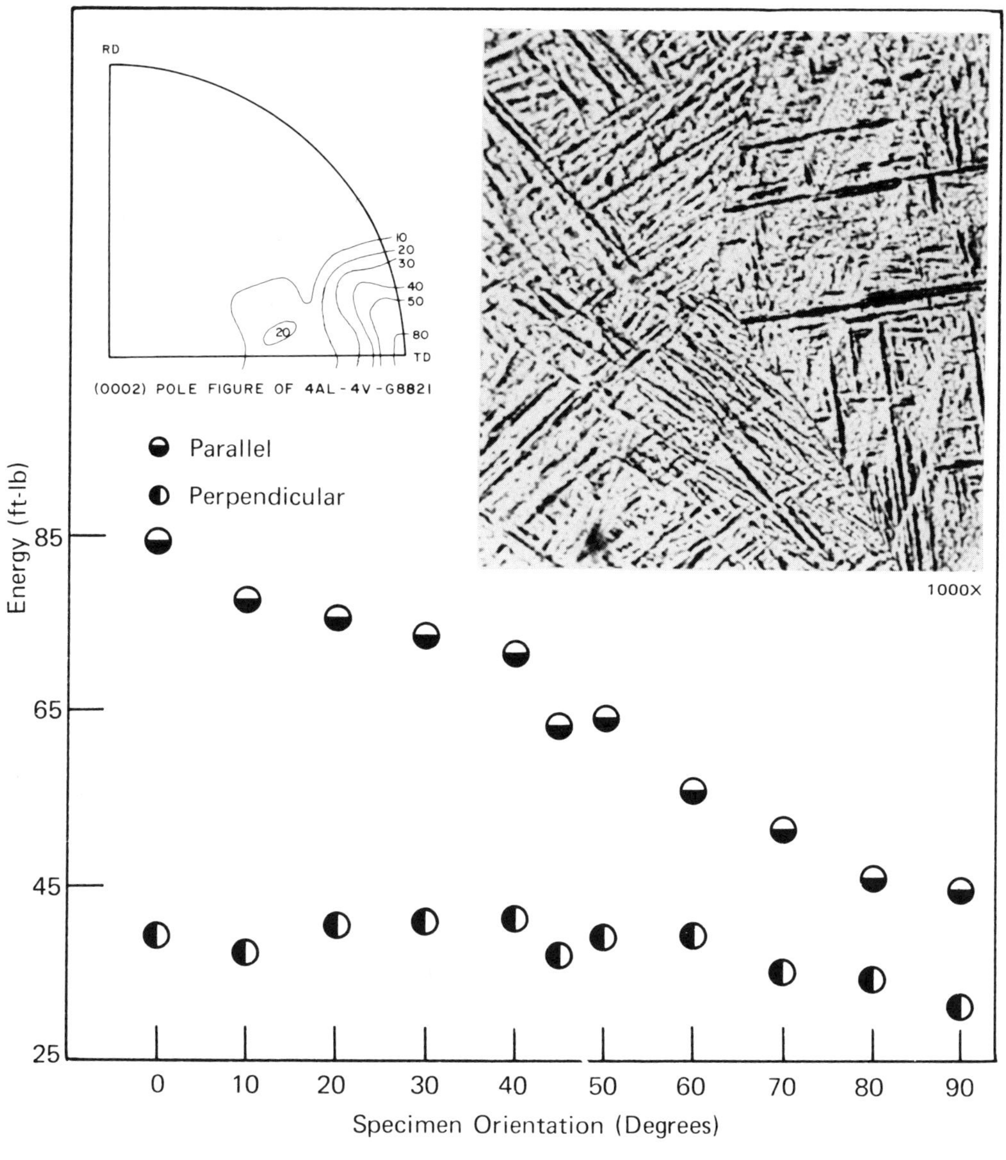

Fig. 7. Typical test data and (0002) pole figure for Ti-4Al-4V-G8821.

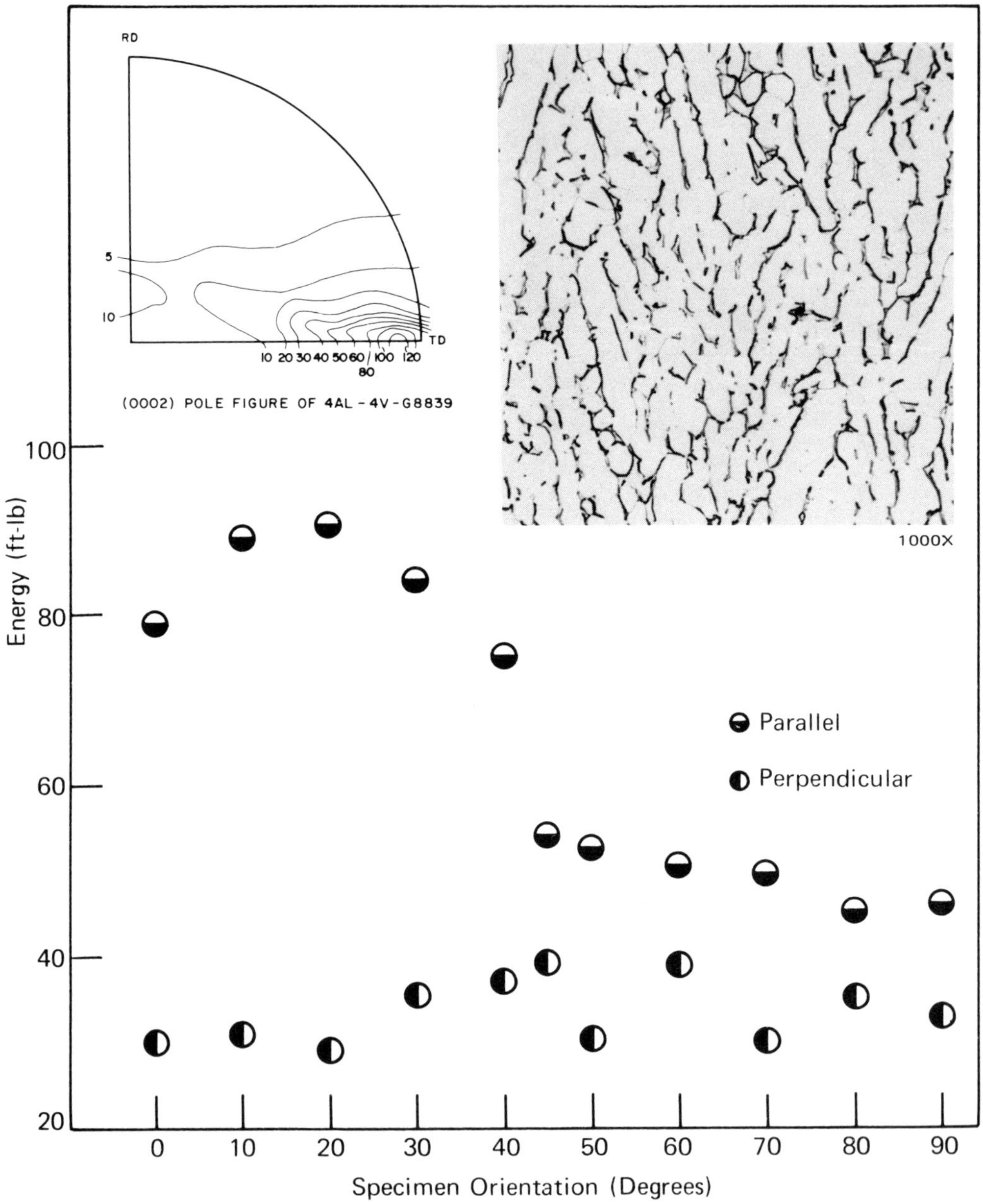

Fig. 8. Typical test data and (0002) pole figure for Ti-4Al-4V-G8839.

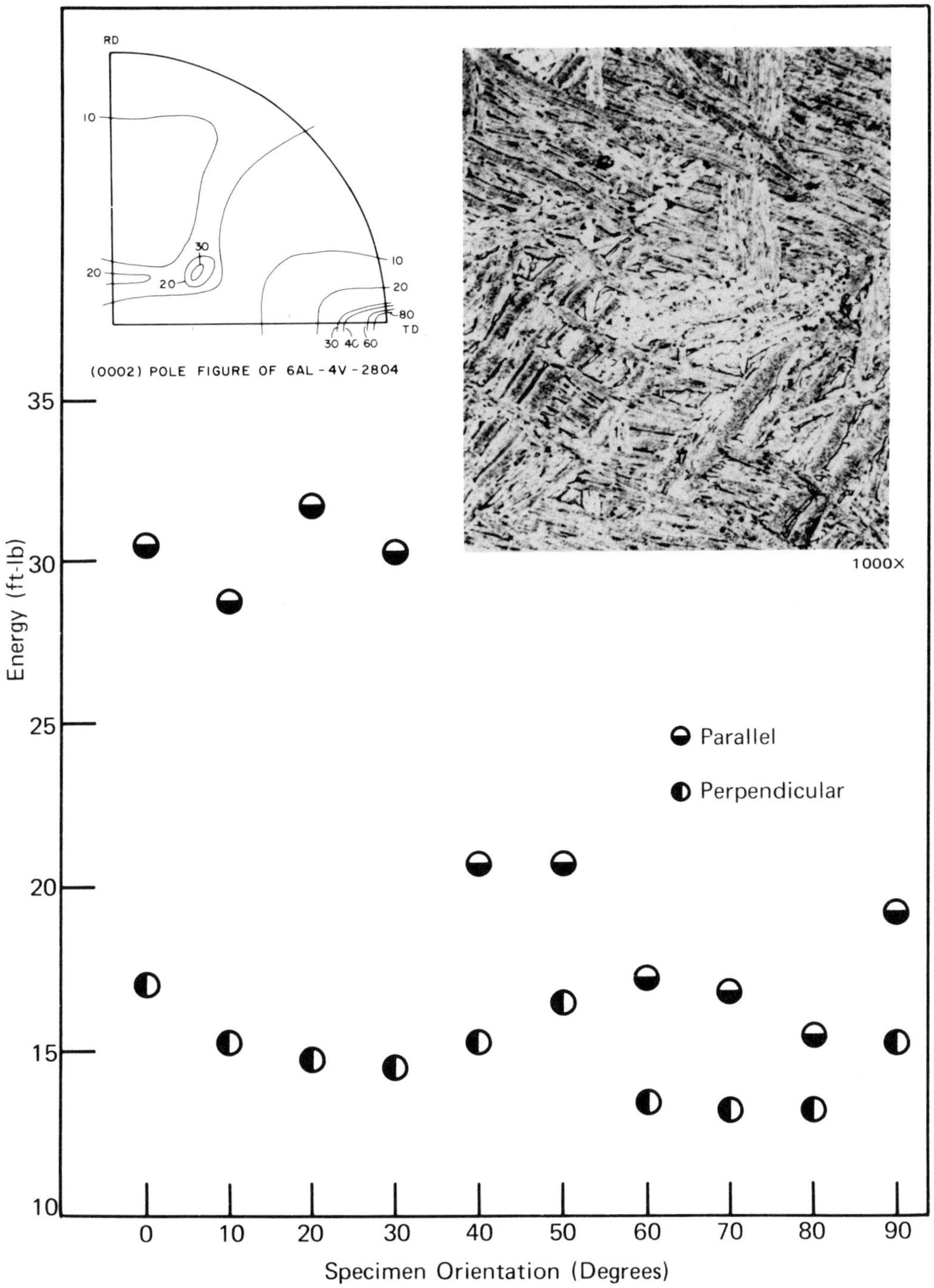

Fig. 9. Typical test data and (0002) pole figure for Ti-6Al-4V-M2804.

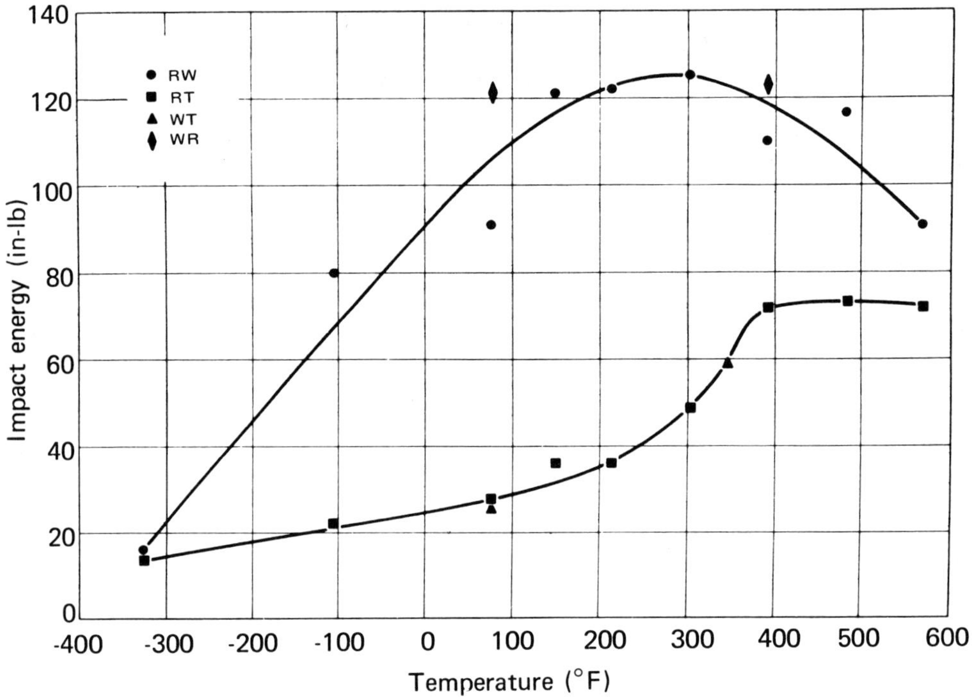

Fig. 10. Impact energy curves for Zircalloy-2, Reference 12.

Table II

Impact and R Values for Ti-4Al, 0.5-Inch Plate

Test Direction	Charpy V-Notch Impact Energy Ft - Lbs Edge Notch	Face Notch	R
L	RW 63.5	RT 17.5	5.1
T	WR 67.0	WT 17.3	9.5

thus the RW or WR orientation specimen would have a tough texture orientation. The hypothesis is that for the soft orientation plastic flow occurs and high energy is absorbed. For the hard orientation higher levels of strength are required to induce flow and then fracture by twinning occurs at lower strains and therefore is a less tough condition. As the temperature is raised, the hard orientation becomes tougher due to the temperature dependence of slip and twinning modes.

In order to verify the effect of texture on the transition temperature, a series of specimens were cut from Ti-75A, heat M-230. This plate had a texture with a basal pole intensity near the plate normal with a slight transverse spread. The tough specimen orientation would then correspond to the through-thickness notch orientation. From examination of data at several test temperatures, it can be seen that the transition temperature is lower and the toughness is higher for the through thickness notch orientation (Figure 11). These results are in agreement with the results for Zircalloy 2.

A better understanding of the relationship between plastic deformation and fracture can be had from studying Figure 12 (14, 15). At high temperatures the critical resolved stress for $(10\bar{1}0)[11\bar{2}0]$ slip is low and a specimen in a Pa orientation (Figure 12a) is favorably orientated for this mode of deformation. As the temperature is lowered, the critical resolved stress for $(10\bar{1}0)$ $[11\bar{2}0]$ slip rises until $(11\bar{2}2)$ twinning begins, then microcracks are formed at the twin-matrix interface and brittle fracture ensues. For the Pc specimen orientation (Figure 12b) a similar behavior is apparent; however, due to the plane strain constraints at the base of the notch, a higher temperature is needed for plastic flow and high toughness. Predictions are for large compressive deformations at the back side of the specimen. Finally, the Ba orientation specimen is not favorably oriented for slip on any of the known slip systems since the stress is perpendicular to the $[11\bar{2}0]$ direction. In this case $(10\bar{1}2)$ twinning probably occurs, and the reoriented matrix would now be favorably oriented for $(11\bar{2}2)$ twinning and susceptible to fracture (Figure 12c). The above comments are, of course, highly speculative and required further research. This discussion however, does provide a simple model for the description of texture

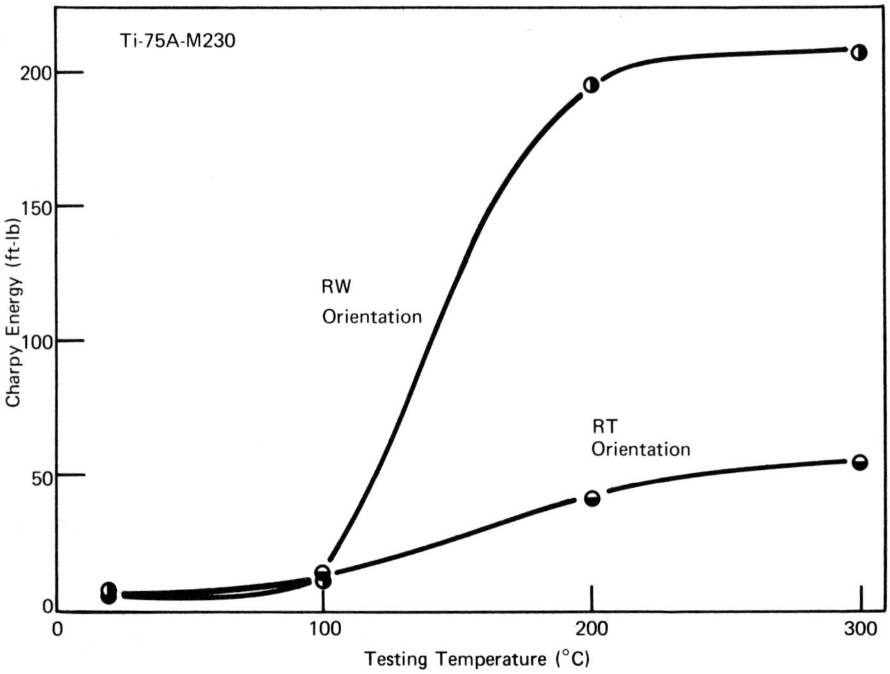

Fig. 11. Effect of texture on the transition temperature.

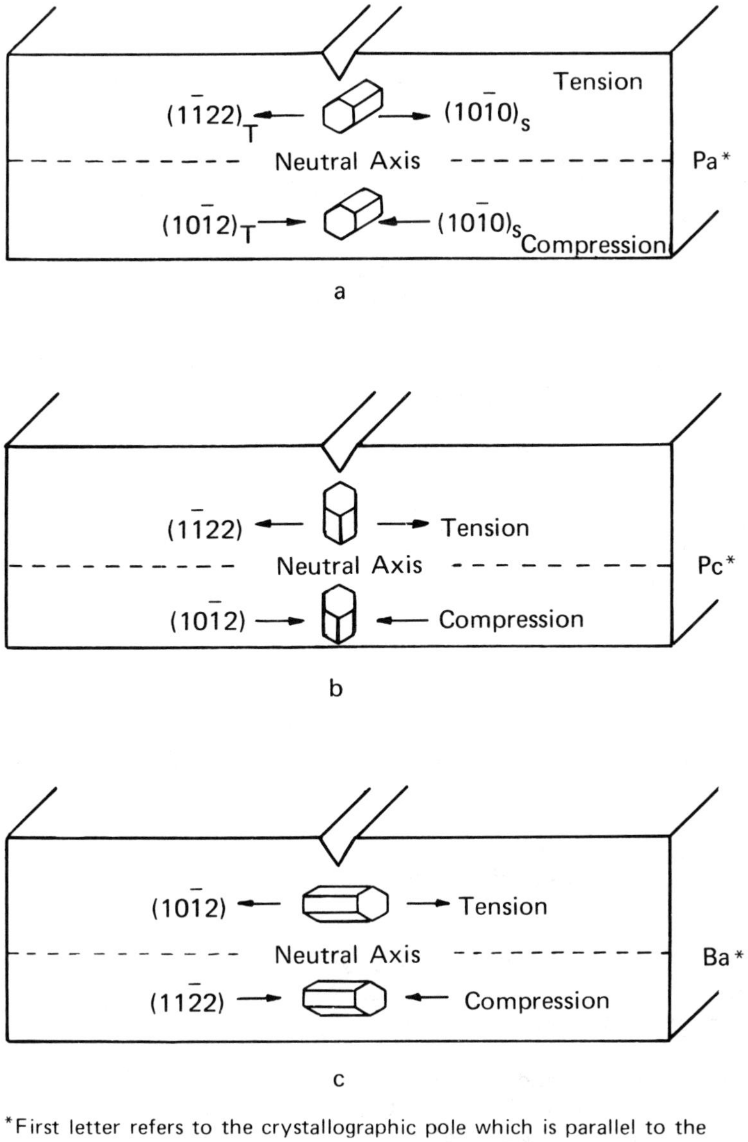

*First letter refers to the crystallographic pole which is parallel to the specimen axis, P = Prism, B = Basal. Second letter refers to the crack propagation direction, i.e., a and c as crystallographic directions.

Fig. 12. Schematic of relationship between deformation and texture.

and its influence on toughness and this beginning of understanding should allow the utilization of textures to produce a high toughness material.

Conclusions

1. A significant toughness variation has been found in titanium and titanium alloy plate as a function of specimen and notch orientation.

2. This toughness variation can be related to the texture of plate with the normal fibering effect being small or absent.

3. This toughness variation can be explained on a simplified crystallographic slip basis where plastic flow predominates.

4. For brittle fracture, the effect of crystallographic texture was not evident.

5. The texture has an influence on the toughness transition with the lower transition temperatures being tied to a specific texture orientation.

References

1. Hill, R. A., "Theory of Yielding and Plastic Flow in Anisotropic Metals," Proceedings of the Royal Society, London, Vol. 193, 1948, p. 281.

2. Backofen, W. A., Hosford, W. F., Jr., and Burke, J. J., "Texture Hardening," Transactions of American Society Metals, Vol. 55, 1962, p. 865.

3. Lopata, S. L. and Kula, E. B., "A Reflection Method for Pole Figure Determinations," Transactions of the Metals Metallurgical Society of AIME, Vol. 224, 1962, p. 865.

4. Dillamore, I. L. and Roberts, W. T., "Preferred Orientation in Wrought and Annealed Metals," Institute of Metals, Metallurgical Reviews, Vol. 10, No. 39, 1965, p.271.

5. Larson, F. R., Zarkades, A., and Avery, D. H., "Twinning and Texture Transitions in Titanium Solid Solution Alloys," Army Materials and Mechanics Research Center, AMMRC TR 71-11, June 1971.

6. Larson, F. R., "Textures in Titanium Sheet and Its Effect on Plastic Flow Properties," Army Materials Research Agency, AMRA TR 65-24, October 1965.

7. Zarkades, A. and Larson, F. R., "Experimental Determination of Texture and Mechanical Anisotropy of Tensile Properties in Commercially Pure Titanium Sheet," Army Materials and Mechanics Research Center, AMMRC TR 67-05, December 1967.

8. Zarkades, A. and Larson, F. R., "Sheet Tensile Properties of Titanium Alloys as Effected by Texture," Army Materials and Mechanics Research Center, AMMRC TR 68-03, January 1968.

9. Zarkades, A. and Larson, F. R. "Texturing in an Aged Hardened Ti-6Al-4V," Army Materials and Mechanics Research Center, AMMRC TR 69-32, December 1969.

10. English, A. S., "Influence of Mechanical Fibering on Anisotropy of Strength and Ductility," Journal of Metals, April 1965, p. 395.

11. Hatch, A. J., "Mechanical Metallurgy and Fracture Toughness of Titanium Alloys." Titanium Technical Conference, Air Force Materials Laboratory and Research Institute, Dayton, Ohio, November 1967.

12. Rittenhouse, P. L. and Picklesimer, M. L., "The Effects of Fabrication Variables on the Anisotropy of Mechanical Properties Part I," Oak Ridge National Laboratory, ORNL-2944, November 1960.

13. Rittenhouse, P. L. and Picklesimer, M. L., "The Effects of Fabrication Variables on the Anisotropy of Mechanical Properties, Part II," Oak Ridge National Laboratory, ORNL-2948, February 1961.

14. Burrier, H. I., Jr., Amateau, M. F. and Steigerwald, E. A., "The Relationship Between Plastic Deformation and Fracture in Alpha Titanium," Tech Report ASML-TR-65-239, July 1965.

15. Amateau, M. F. and Steigerwald, E. A., "The Relationship Between Plastic Deformation and Fracture in Alpha Titanium," Tech Report AF ML-TR-66-263, June 1966.

EVALUATION OF ANISOTROPY IN TITANIUM SHEET BY KNOOP HARDNESS MEASUREMENTS

R.A. Fishburn and W.T. Roberts

Department of Industrial Metallurgy
University of Birmingham
Birmingham, England

The flow stress of thin sheet material subjected to biaxial stresses can be characterised by a plane stress yield locus. It is only when information is available in this form that the effectiveness of texture strengthening under a wide range of stress situations can be assessed. The plotting of a yield locus by conventional uniaxial tension, compression and plane strain tests is time consuming, and, for thin sheets, data in the compression-compression quadrant are difficult to obtain. The possibility of using hardness measurements to derive a plane-stress yield locus therefore appears attractive.

The use of the Knoop hardness indenter for establishing the shape of the yield locus was first proposed by Wheeler and Ireland (1). Two indentations are made on each of the three principal anisotropic planes, with the indenter axes aligned along the principal directions in these planes. (Figure 1.) The K.H.N. for any indenter orientation was represented as a vector quantity on the octahedral stress plane. Subsequently, Hosford (2) and Lee, Jabara and Backofen (3) plotted the data on a plane stress diagram with principal stresses in the plane of the sheet, and this method of presentation has been used in the present work.

Although the theoretical analysis is not satisfactory for reasons outlined below, the method has given reasonable agreement with yield loci determined by alternative methods for some materials (1,3), but not for magnesium (4). A major difficulty with the Knoop hardness method is the assignment of a strain level for the yield locus. The problem of defining the stress state associated with each indentation involves an assumption about the principal strains based on the geometry of the indenter, and a curious mixture of isotropic and anisotropic plasticity theory.

Following references (2) and (3), fixed stress states are assigned to each of the six indentations, these stress states being independent of the anisotropy of the material. On the other hand, Wonsiewicz and Wilkening (W & W) (4) suggested that the strain-ratios for each impression were fixed, and using the average plastic strain ratio, $\bar{r}$, as a measure of anisotropy, with an additional assumption of planar isotropy, stress states were calculated. The W and W method does not appear to have been applied to experimental data.

Both methods of plotting the hardness data on a plane stress diagram have been used in the present contribution, the W and W method being modified to take into account differences between r_0 and r_{90}.

Results and Discussion

The materials used were commercial-purity titanium sheet (I.M.I. Titanium 130) and Ti-6 Al - 4 V sheet (I.M.I. Titanium 318). These materials had been processed to give a wide range of textures and anisotropic conditions, and evaluated by tensile tests, strain ratio measurements and plane-strain compression tests. This part of the work will be described elsewhere (5).

Knoop hardness indentations were made on electropolished surfaces, a load of 1 Kg being used. Each quoted hardness value is an average of at least ten impressions. The locus derived from the Knoop values is in each case compared with the locus derived from the measured r values, using the Hill theory of plastic anisotropy (6). Tangents to the locus derived from experimental plane strain compression values are also drawn on the diagrams (Figures 2. and 3.). Experimental and theoretical loci have been calibrated to coincide at the uniaxial tensile yield point, σ_y.

Evidence of crystallographic deformation in the surface grains adjacent to the indentations was noted, and then surface layers were removed by electropolishing in order to reveal evidence of deformation twinning below the Knoop impressions. The presence or absence of twins below the Knoop indentations is summarised in Table I. for the two extreme textural conditions.

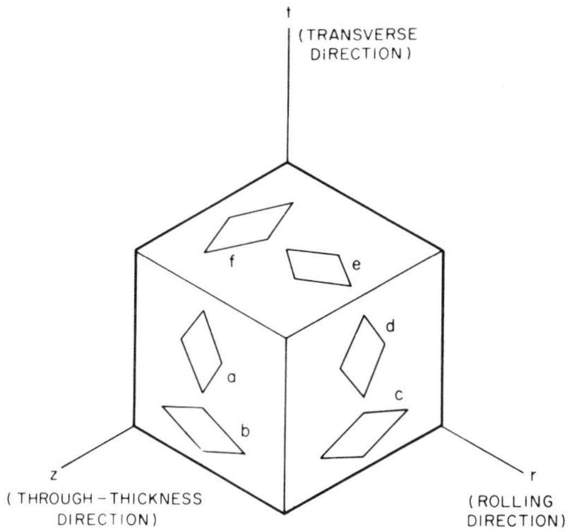

Fig. 1. Knoop indenter orientations (after Wonsiewicz and Wilkening (4)).

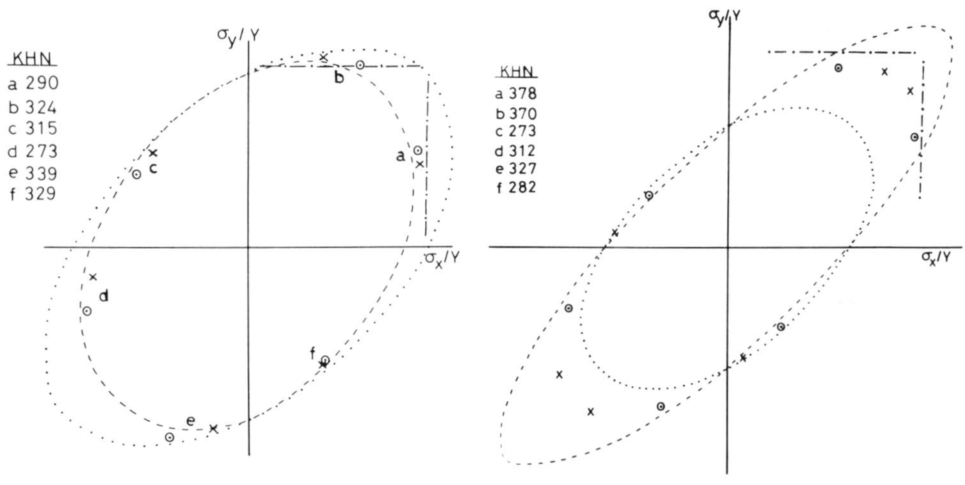

Fig. 2. Plane-stress (σ_2 = 0) yield loci for titanium 318. Left - low $\bar{r}$ sheet, r_0 = 0.4, r_{90} = 0.5; Right - high $\bar{r}$ sheet, r_0 = 4.3, r_{90} = 5.3.

ʘ	Hosford K.H.N. points
X	W & W K.H.N. points
-.-.-.	Tangent to locus based on plane-strain compression results
------	Theoretical locus based on Hill's anisotropic theory
••••••	Theoretical locus for isotropic material

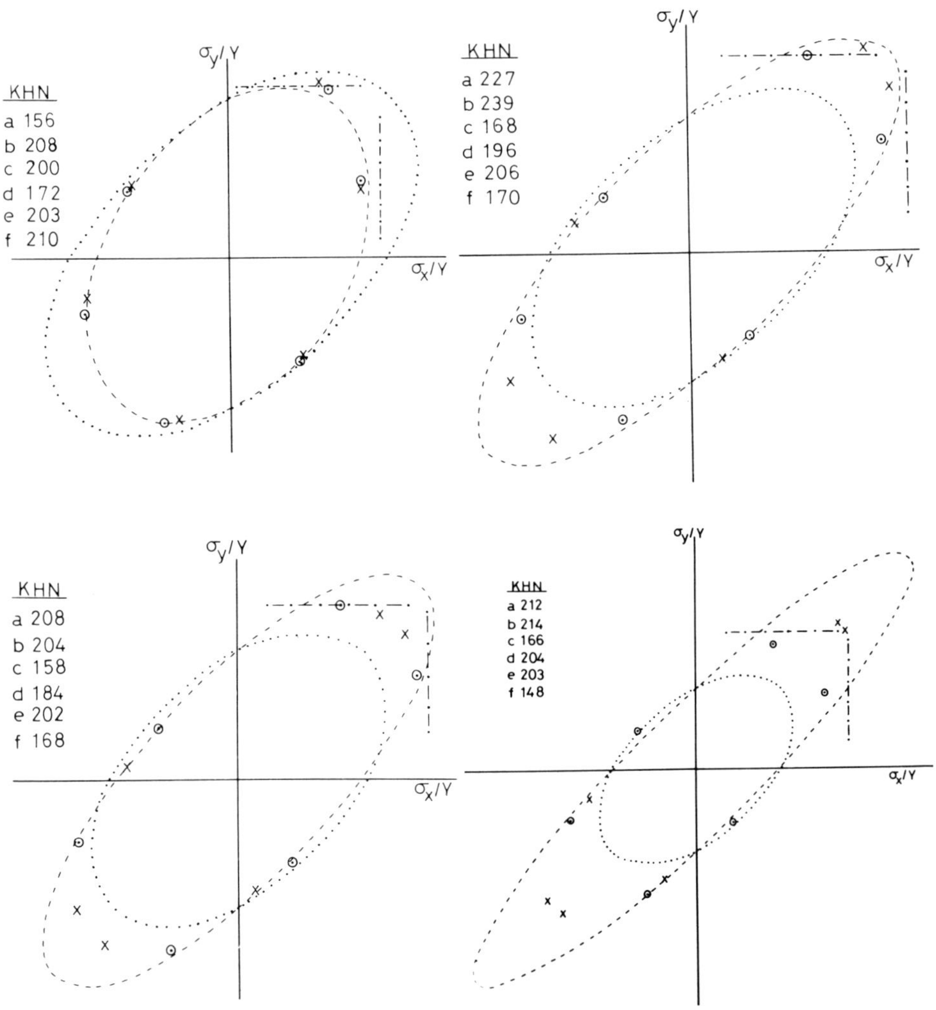

Fig. 3. Plane-stress ($\sigma_2 = 0$) yield loci for titanium 130. Top left - low $\bar{r}$ sheet, $r_0 = 0.5$, $r_{90} = 0.8$. Top right - medium-high $\bar{r}$ sheet, $r_0 = 2.8$, $r_{90} = 4.5$. Bottom left - high $\bar{r}$ sheet, $r_0 = 3.1$, $r_{90} = 6.6$. Bottom right - very high $\bar{r}$ sheet, $r_0 = 11.6$, $r_{90} = 13.7$. Symbols as for Fig. 2.

Table I. - Frequency of Deformation

Twins below Knoop indentations

Knoop Indentation	Very high $\bar{r}$ Titanium 130 ($\bar{r} \simeq 12$)	Low $\bar{r}$ Titanium 130 ($\bar{r} \simeq 0.6$)
a	Twins	Few twins
b	Twins	Twins
c	No twins	Twins
d	Twins	Few twins
e	Twins	No twins
f	No twins	Twins

The twinning behaviour of the high $\bar{r}$ material, having a well-defined texture with basal planes parallel to the surface of the sheet and a fine grain size, was reproducible. The low $\bar{r}$ material showed more variation from impression to impression, probably due to the less well-defined texture and the larger grain size.

From the general shape of the twins, they appear to be of the $\{10\bar{1}2\}$ type, but the sense of shear involved is difficult to reconcile with the orientation and stress system. Although the stress system is admittedly complex for the Knoop indentations (4), twins similar in appearance have been observed under plane-strain compression (5). In particular, copious twinning was observed when plane strain indentations were made on the surface of very high $\bar{r}$ sheet, conditions comparable to Knoop impressions a and b. Further work is required to establish the identity of the twins.

Examples of metallographic observations are shown in Figure 4. Profuse sub-surface twinning was generally associated with less surface slip marking. In the case of the very high $\bar{r}$ material, crystallographic slip is particularly difficult for plane-strain indentations parallel to directions a, b, d and e, and it is these Knoop impressions that show maximum twinning. Similarly, for an idealised texture with basal plane poles parallel to the transverse direction (low $\bar{r}$ material) directions e, f, b and c are associated with unfavourable slip situations.

Some representative yield loci for Titanium 318 and Titanium 130 are shown in Figures 2. and 3. For the low $\bar{r}$ materials, there is little to choose between the two methods used for plotting the Knoop data. However, for the very high $\bar{r}$ materials, the W and W points a, b, d and e define the shape of

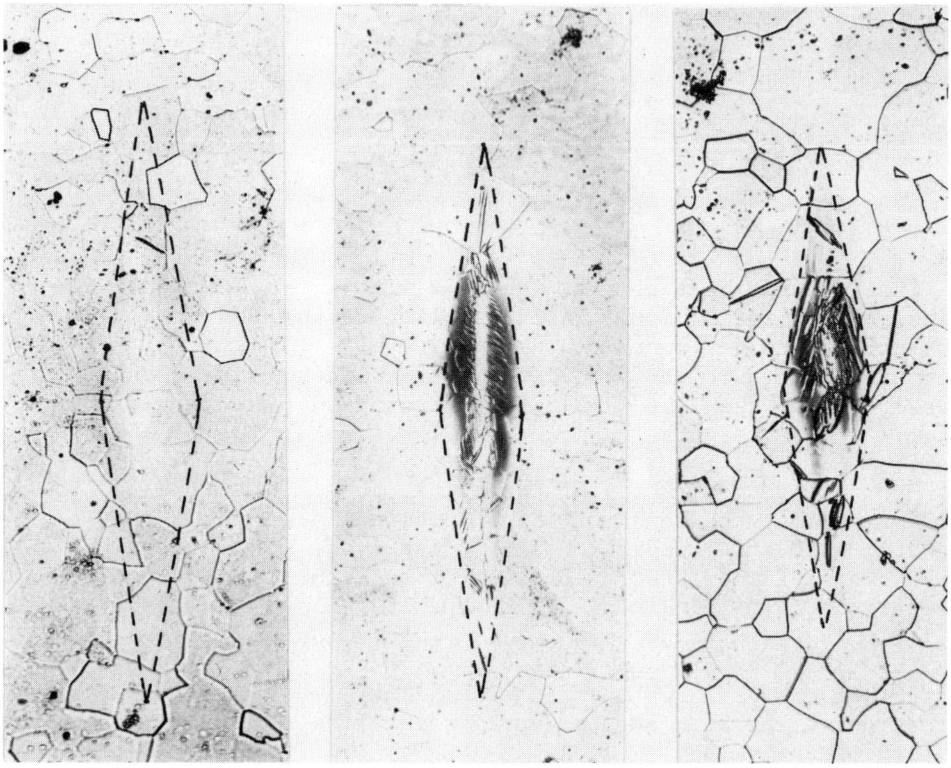

Fig. 4. Microstructure beneath Knoop indentations (X 240). Very high $\bar{r}$ titanium 130; left - impression a; centre - impression d; right - impression f.

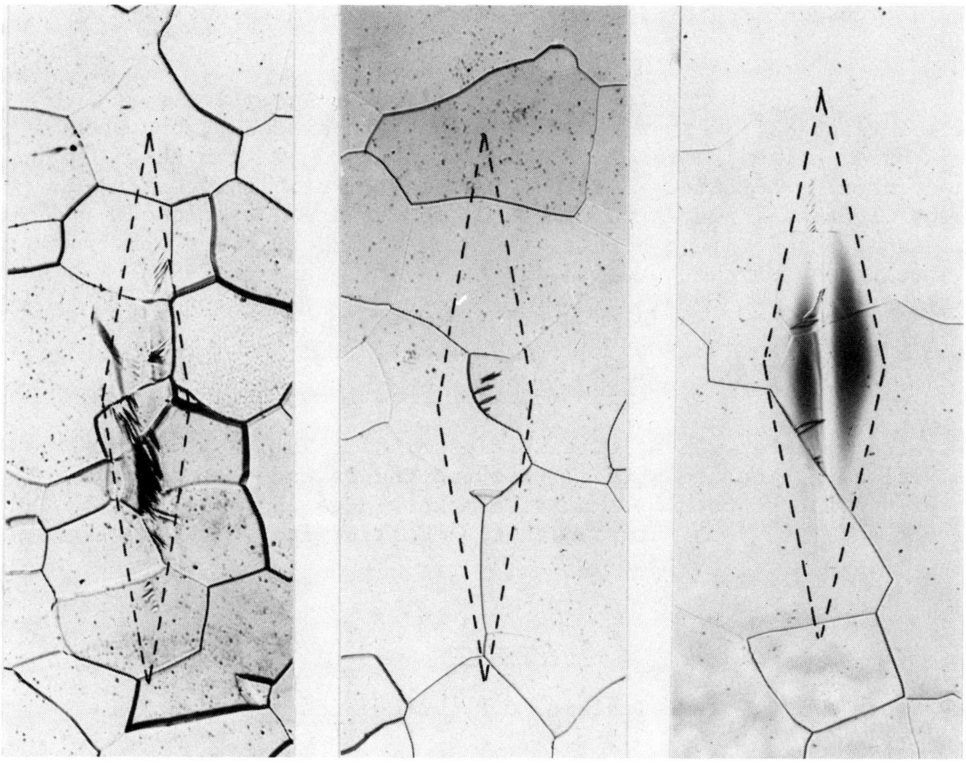

Fig. 4. (cont'd) - Low $\bar{r}$ titanium 130; left - impression a; centre - impression d; right - impression f.

the locus rather better than the Hosford, Backofen points which are comparatively insensitive to large variations in the shape of the locus.

For low and medium-high $\bar{r}$ values, the plane stress yield locus determined by Knoop hardness measurements and evidence from plane strain compression tests both agree well with the theoretical yield locus derived by the application of Hill's anisotropic theory.

For very high $\bar{r}$ values, the experimental Knoop locus again agrees reasonably well with the plane-strain evidence, but both sets of experimental results deviate considerably from the theoretical prediction. This means that, for highly textured sheet with basal planes parallel to the surface, calculated strengths based on the high r values of the sheet are a considerable over-estimate of the actual strength under balanced biaxial stresses.

Acknowledgements

The authors wish to thank Professor E.C. Rollason for the provision of laboratory facilities and the Science Research Council for financial support. They also acknowledge the able assistance provided by Mr. John Farmer with the metallurgical work.

References

1. Wheeler, R.G. and Ireland, D.R., Electrochem. Technol., Vol. 4, 1966, pp. 313-317.

2. Hosford, W.F., Metals Eng. Quarterly, Vol. 6, 1966, pp. 264-267.

3. Lee, D, Jabara, F.S. and Backofen, W.A., Trans. TMS-AIME, Vol. 239, 1967, pp. 1476-1478.

4. Wonsiewicz, B.C. and Wilkening, W.W., Trans. TMS-AIME, Vol. 245, 1969, pp. 131-1319.

5. Fishburn, R.A., Roberts, W.T. and Wilson, D.V., to be published.

6. Hill, R., Proc. Roy. Soc. A, Vol. 193, 1948, pp. 281-297.

THE TEXTURE HARDENING OF TITANIUM AND ITS ALLOY SHEETS

Atsushi Hasegawa, Takashi Nishimura & Shitoshi Ohtani
Central Research Laboratory
Kobe Steel, Ltd.
Kobe, Japan

Introduction

A strong preferred orientation in the sheet metals leads to remarkable anisotropy in mechanical properties. As the titanium tends to form sharp texture and the common slip systems in it lie in the basal plane of the hexagonal close-packed crystal, it is considered that it shows a more remarkable anisotropy. If the basal plane lies parallel to the plane of the sheet, thinning of the sheet is difficult,which requires much higher stresses for plastic deformation in the thickness direction than those for common slip. This characteristics for titanium is called texture hardening[1)2)] and there is possibility of strengthening under biaxial stress conditions. On the other hand, by applying Hill's theory[3)] for the yielding of anisotropic materials, the yield condition is expressed as follows:

$$\sigma_x{}^2 \left\{ 1 - \frac{2r_x}{1 + r_x}\alpha + \frac{r_x(1 + r_y)}{r_y(1 + r_x)}\alpha^2 \right\} = X^2 \qquad (1)$$

where x, y: principal stress axis on the rolling plane
X : uniaxial yield strength in x direction
α : stress ratio $\alpha = \sigma_y/\sigma_x$
r_x, r_y: r value in x and y direction
σ_x, σ_y: principal stress

Then for the case of planar isotropy the equation (2) is obtained from equation (1)

$$\frac{\sigma_x}{X} = \left\{ 1 + \alpha^2 - \alpha\frac{2r}{r + 1} \right\}^{-\frac{1}{2}} \qquad (2)$$

The equation (2) indicates that the r value is quantitatively related to biaxial strength and with increasing r value the biaxial strength increases. Therefore, it is considered that if high r value would be able to be obtained, there is possibility of effective application under biaxial stress. In the present investigation, in order to make clear the characteristics of titanium, the relations between texture, r value, biaxial strength and Knoop hardness number on the several titanium alloys were studied.

Test Procedure

1. Materials

The materials used in this investigation are two commercial grades of pure titanium, KS50, KS85, the α alloys of Ti-4Al and Ti-5Al-2.5Sn, and the $\alpha+\beta$ alloys of Ti-6Al-4V and Ti-8Mn. These materials were tested at the thickness of mainly 1 mm, but partially 3.5 mm. The chemical compositions of these test specimens are given in the table I.

2. Uniaxial and Balanced Biaxial Test, Measurement of the r value

The uniaxial tensile test was done by the usual method, and simultaneously in order to determine the r value, two plastic strain gauges were pasted on the test section of the tensile spesimen. One gauge was mounted parallel to the tensile direction on one side of the specimen, while the other gauge was mounted transverse to the tensile direction on the other side. The strains in the both directions were recorded on the X-Y recorder. Several records obtained by this method are shown in Fig. 2. The r value was calculated from the slope of the curve of ε_w against ε_l on the assumption of volume constancy after yielding occured.

Table I Chemical Compositions

Material	Thickness	Composition, wt%								
	(mm)	Al	V	Sn	Mn	Fe	Si	O	H	N
KS 50-1	1.0	-	-	-	-	0.037	0.006	0.104	0.0013	-
-2	3.5	-	-	-	-	0.050	0.004	0.081	0.0021	-
KS 85	3.5	-	-	-	-	0.182	0.011	0.284	0.0014	-
Ti-4Al	1.0	3.98	-	-	-	0.056	0.0093	0.070	0.0010	-
Ti-5Al-2.5Sn	1.0	4.80	-	2.47	-	0.34	-	0.108	0.0081	0.0078
Ti-6Al-4V	1.0	6.15	4.31	-	-	0.13	-	0.141	0.0044	0.0088
Ti-8Mn	1.0	-	-	-	7.8	0.27	-	0.104	0.0119	0.0118

The uniaxial compression was examined on the specimens of block size, 15^w mm × 15^t mm × 20^h mm, consisting of several pieces of test sheet by the binding agent to prevent a buckling during the compression test.

The through-thickness compression test was carried out which is directly related to the balanced biaxial tensile stress condition on the basis of equivalence of stress. As they were thin, many pieces were bound similar to the compression test pieces, and machined into the shape of 15 mm squares and 20 mm height. Then the test specimens were compressed in the direction of thickness. The authors have comfirmed that the effect of the binding agent on the deformation resistance is almost negligible by the other test results.

3. Pole Figures

The pole figures of (0002) plane were obtained by both transmission and reflection technique. Molybdenum K_α radiation was used with a zirconium filter. The 34 mm diameter specimens were cut from the center portion of the sheets. Then they were ground and chemically polished to the final thickness of 0.1 mm. Diffracted intensities were calculated by comparison with diffraction from a randomly oriented titanium powder sintered specimen.

4. Knoop Hardness Test

The knoop hardness test was carried out to compare with the first quardrant of the yield locus obtained by the conventional method. The determination of the yield locus by the knoop hardness measurement (the Khn-locus) is based on the method proposed by Wheeler and Ireland[4)].

Result

1. Pole Figure

The (0002) pole figures of these specimens are shown in Fig.1. The texture of commercially pure titanium KS 50-1 (1.0 mm^t) shows the common orientation which is (0001) rotated 30° toward the transverse direction about the axis in the rolling direction. The (0002) pole figure of KS 50-2 (3.5 mm^t) and KS 85 (3.5 mm^t) which were hot rolled and annealed, shows a central peak with large varying amount of spread toward the transverse direction. The textures of both Ti-4Al and Ti-5Al-2.5Sn exhibit a strong basal plane sheet texture. This texture is ideal for the strengthening under biaxial stress conditions. The texture of Ti-6Al-4V, different from the type of the Ti-4Al and Ti-5Al-2.5Sn alloys consists of many orientation components. As this texture is easy to deform in the thickness direction, high degree of biaxial strengthening is not anticipated. Further, in the case of Ti-8Mn,

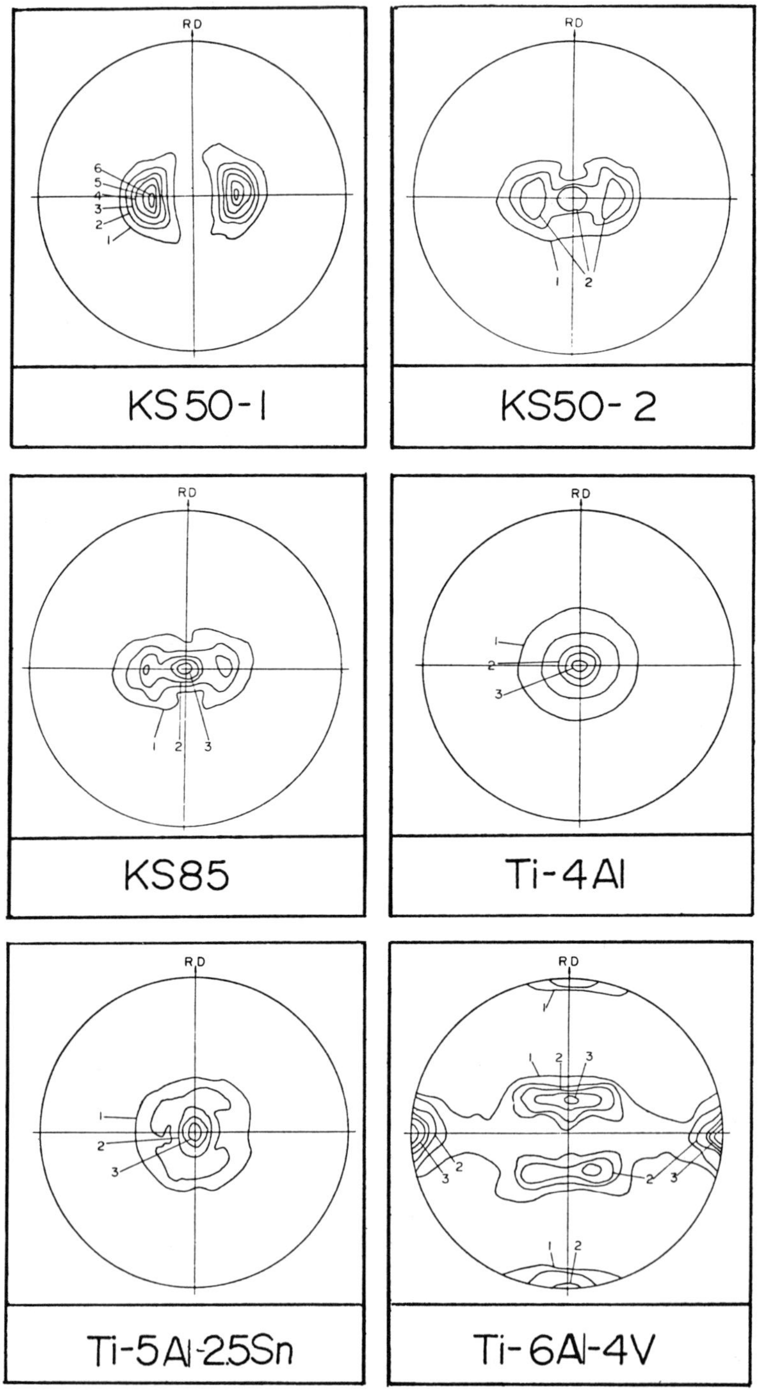

Fig. 1. Basal pole figures of titanium alloys.

it was comfirmed that the α phase was poor and the texture of the β phase consisted on (001)<110> rotated about 5° from the rolling plane. It is anticipated that the existance of the large amount of β phase reduces the effect of texture hardening characterized by the α phase.

2. r value

The r values of these specimens are shown in the table II, which contains the r values both in the rolling direction (r_x) and transverse direction (r_y). The recorded curves obtained for KS 50 and Ti-4Al are shown in Fig.2.

The r values of the Ti-4Al were very high, r_x=4.25, r_y=4.89. These values are closely related to the presence of the strong basal plane sheet texture. But the r values of Ti-5Al-2.5Sn, which showed a centralized type of pole figure similar to the Ti-4Al, were not so much high value of r_x=2.38 and r_y=2.48. This difference between these two alloys indicates that the magnitude of the r value did not always depend on only the preferred orientation.It is clear that the very small anisotropy of r value for these two alloys is directly related to this centralized type of pole figures.

The r value for three commercially pure titaniums showed the relatively high values from 2.33 to 3.63. But the r value in the transverse direction is larger than that in the rolling direction and this anisotropy is somewhat large.

The Ti-6Al-4V alloy had the r value of approximately 1.70 and showed to be planar isotropic, which was rather lower than that for Ti-5Al-2.5Sn, but as to the biaxial yield strength they have almost the same level. The rather small r value of this alloy is considered to be due to the multiple orientation components, different from the single basal plane texture such as the Ti-5Al-2.5Sn. With such a texture, the deformation in the thickness direction can occur relatively easily.

The Ti-8Mn alloy showed very low r values smaller than 1. Although this alloy was predicted to be rather texture softening under the biaxial stress condition, the observed value resulted in somewhat strengthening. It is considered that the small r value is largely affected by the presence of the β phase.

3. Uniaxial and Balanced Biaxial Yield Strengths and Their Anisotropy

The uniaxial tensile and compressive yield strengths in the both rolling and transverse directions, and the balanced biaxial tensile yield strength obtained by the thickness compression test are summarized in table II, in which the observed values of the degree of the strengthening under the balanced biaxial stress, σ_{zc}/X_T are shown compared with those calculated from the r value by the equation (2). Where X_T is the uniaxial tensile strength in the rolling direction and σ_{zc} is the balanced biaxial yield strength. In Fig.3, the stress-strain curves under the various

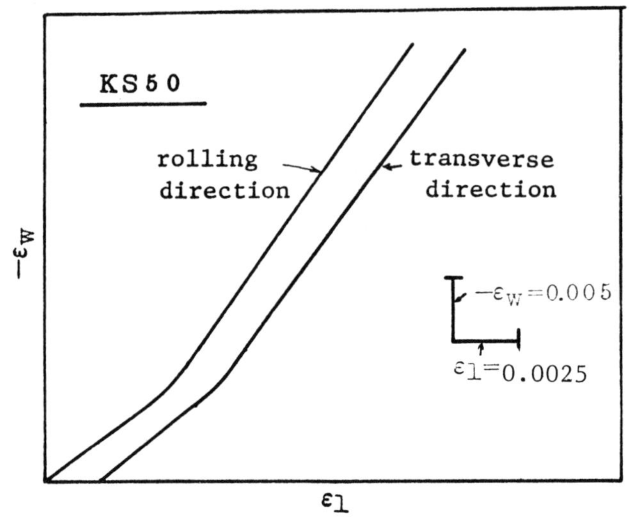

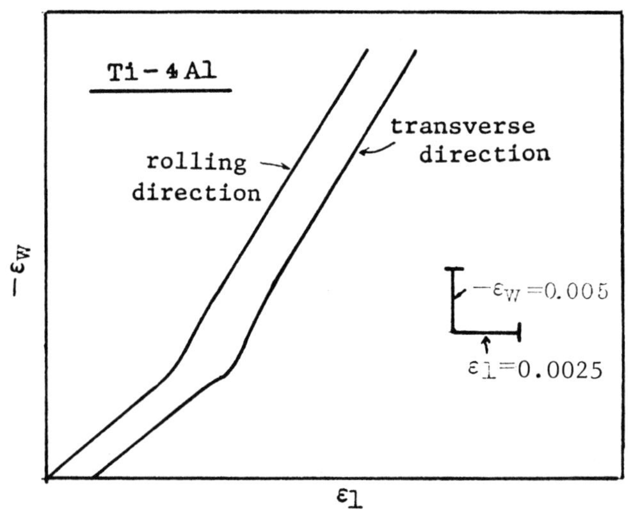

Fig. 2. Observed curves of width strain as function of length strain.

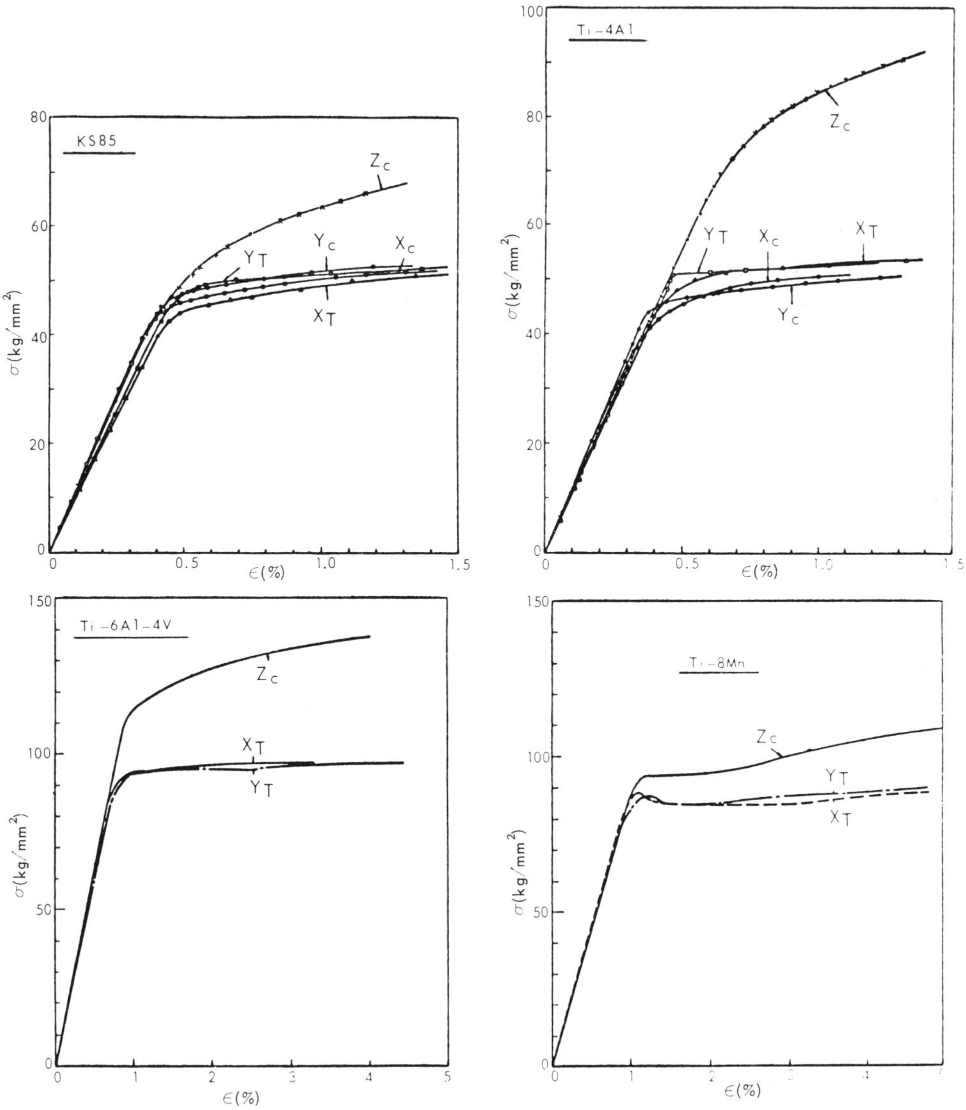

Fig. 3. Stress-strain curves under various stress condition. X_T - uniaxial tension (rolling direction); Y_T - uniaxial tension (transverse direction); X_c - uniaxial compression (rolling direction); Y_c - uniaxial compression (transverse direction); Z_c - thickness compression.

stress conditions are illustrated for KS 50-2, Ti-4Al, Ti-6Al-4V and Ti-8Mn.

For the Ti-4Al alloy which showed the strong basal plane sheet texture and high r values, the balanced biaxial yield strength significantly increased to the value of 83.4 kg/mm^2 comparing with the uniaxial yield strength. This increasing amount reached to as 64% of the uniaxial yield strength.

It is considered that this effect is directly affected by the common slips restricted on the basal plane of h.c.p. crystal and the definite preferred orientation obtained for this alloy. On the other hand, no anisotropy of the uniaxial yield strength existed and it seemed that this is related to this texture showing rotational symmetry about the sheet normal axis.

For the high strength commercially pure titanium, KS 85 (X_T=46.2 kg/mm^2, Y_T=50.1 kg/mm^2) which has the same strength level as to the Ti-4Al, the balanced biaxial tensile yield strength, σ_{zc} was 58.6 kg/mm^2 and 26% higher than uniaxial tensile yield strength, X_T. It is clear that the Ti-4Al alloy is significantly high strengthened under biaxial stresses compared with KS 85.

This difference between these two titaniums may be dependent on not only the type of texture, but also the alloying element (O and Al) added in order to strengthen the mechanical properties. On the other hand, in the case of the three titanium alloys, Ti-5Al-2.5Sn (α alloy), Ti-6Al-4V ($\alpha+\beta$ alloy) and Ti-8Mn ($\alpha+\beta$ alloy) which indicate approximately the same strength level, the differences among these alloys were examined. For the Ti-5Al-2.5Sn alloy consisting of the basal plane sheet texture similar to the Ti-4Al alloy, σ_{zc} was 115.6 kg/mm^2 and 34% higher than X_T. Although this degree of strengthening did not reach to that for the Ti-4Al (64%), this alloy will be appreciated to be able to increase under biaxial tensile stress condition. And planar anisotropy of uniaxial tensile strength for this alloy also did not exist.

For the $\alpha+\beta$ alloy, Ti-6Al-4V, the balanced biaxial tensile yield strength σ_{zc} showed 116.3 kg/mm^2 and the biaxial strength increases were 23% (X_T=94.6 kg/mm^2, Y_T=96.6 kg/mm^2).

Though the degree of strengthening of this alloy is lower than that of Ti-5Al-2.5Sn, σ_{zc} is almost the same value of the latter alloy. Although this alloy cannot be estimated to be texture hardened because of the texture with multiple orientation components, it may be considered that there is possibility of strengthening to this extent even for the commercial Ti-6Al-4V. Since the texture of this alloy is remarkably affected by the processing condition, it is sufficiently possible to increase the biaxial strength by the proper texture control.

For the Ti-8Mn, the uniaxial tensile yield strengths of 90.1 kg/mm^2 (X_T) and 87.4 kg/mm^2 (Y_T) were obtained. However, the biaxial tensile yield strength σ_{zc} was 94.4 kg/mm^2. Therefore, this alloy can be said to be almost isotropic.

As shown in Fig.2, for the alloys such as Ti-4Al, Ti-5Al-2.5Sn, the thickness compression stress-strain curve exhibits a marked increase in stress with increasing strain after the yield point.

But for this alloy, such a tendency is very little.

Since the r value of this alloy is less than 1, the softening behavior with an expression of the yield locus may occur with the increasing strain over the yield point. It is considered that such plastic properties are remarkably dependent upon the large amount of β phase.

The tendency of the biaxial strength increase for these three alloys is consistent with that of the magnitude of the r value. For the two commercially pure titanium, KS 50-1, -2, although the uniaxial yield strengths about 30 kg/mm^2 are far lower than those of these alloys, the biaxial strength increase showed 38% for the former and 44% for the latter. These increases are well correspondent to the r values obtained by these pure titaniums.

4. The Yield Locus from Knoop Hardness Number

The yield locus constructed from six knoop hardness points on the basis of the Wheeler and Ireland[4)] method are shown with a solid line in Fig.4, which contains the first quardrant of locus obtained by the conventional method shown with a dotted line. The purpose of this test was to underestand the biaxial properties, i.e., plastic anisotropy of materials, by this simple measuring method.

As shown in Fig.4, for three alloys such as Ti-5Al-2.5Sn, Ti-6Al-4V and Ti-8Mn, both yield loci in the first quardrant showed a relatively good agreement. For three pure titaniums and Ti-4Al, however, they did not agree so well. This descrepancy at the point on the path of stress ratic $\alpha=1$ reached to about 10%. These results indicate that for low r_{av} value (the r_{av} value is an average of r value, $r_{av}=\frac{1}{2}(r_x+r_y)$ up to about 2.5, the Khn-locus is a relatively well consistent with the conventional locus and is practically useful to understand the properties under the biaxial stress conditions.

The reason why the agreement grows worse with increasing r value is considered that the assumption, for example, the strain ratio of 7 : 1 when knoop hardness number is converted to the yield locus, does not hold good.

Discussion

In this investigation, the relations of various factors were discussed such as texture, r value, biaxial strength and knoop hardness number characterizing the texture hardening of titanium alloy and it was shown that there is the intimate relations among their factors. Especially the anisotropy expressed by r value exhibits well the behavior of balanced biaxial strengthening. The degree of strengthening, $(\sigma_{zc}/X_T)_{calc}$ calculated by the equation (2) and r values are listed in Table II together with the observed values. It is indicated that the observed values of (σ_{zc}/X_T) are

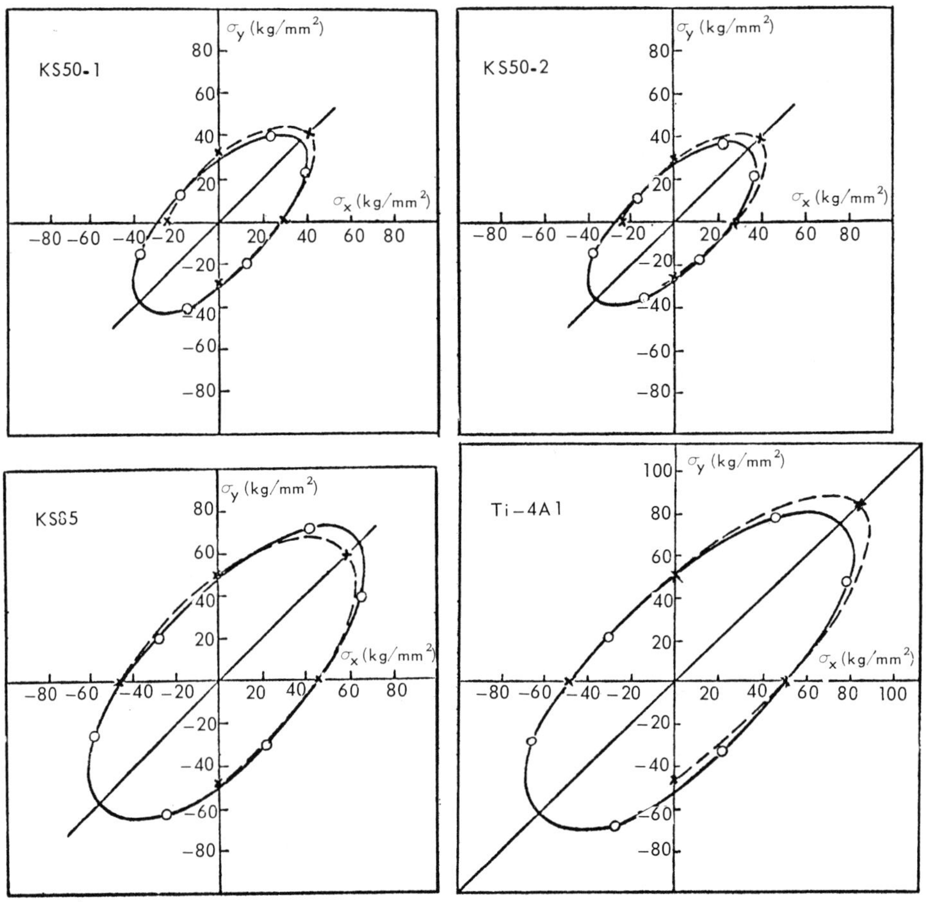

Fig. 4. Yield locus from Knoop hardness number.

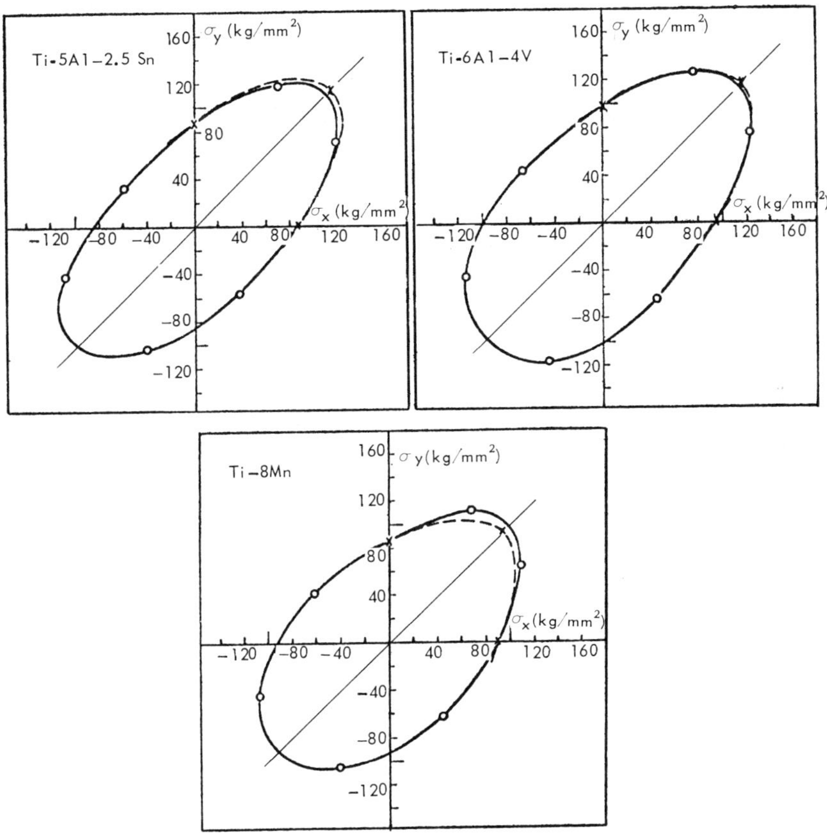
Ti-5Al-2.5 Sn
σ_y (kg/mm^2)
σ_x (kg/mm^2)
Ti-6Al-4V
σ_y (kg/mm^2)
σ_x (kg/mm^2)
Ti-8Mn
σ_y(kg/mm^2)
σ_x(kg/mm^2)

Fig. 4 (cont'd).

Table II Uniaxial and Biaxial Test Results

Alloys	Uniaxial Yield Strength				Balanced biaxial yield strength	σ_{zc}/X_T	r value (0.2%)		$(\sigma_{zc}/X_T)_{calc}$ Eqn (2)
	Tensile		Compressive						
	X_T(RD) (kg/mm^2)	Y_T(TD) (kg/mm^2)	X_C(RD) (kg/mm^2)	Y_C(TD) (kg/mm^2)	σ_{zc} (kg/mm^2)		r_x	r_y	
KS 50-1	29.5	32.9	24.4	29.4	40.8	1.38	2.33	3.50	1.42
KS 50-2	26.8	28.3	25.0	26.8	38.7	1.44	2.39	3.63	1.40
KS 85	46.2	50.1	47.1	47.8	58.6	1.26	2.55	2.94	1.38
Ti-4Al	51.0	51.1	-	-	83.4	1.64	4.25	4.89	1.68
Ti-5Al-2.5Sn	86.3	86.0	-	-	115.6	1.33	2.38	2.48	1.31
Ti-6Al-4V	94.6	96.6	-	-	116.3	1.23	1.69	1.71	1.16
Ti-8Mn	90.1	87.4	-	-	94.4	1.05	0.57	0.95	0.97

well agreed with that calculated. And these values of σ_{zc}/X_T are shown in relation to the average r value in Fig.5. Fig.5 indicates that the observed values are relatively well consistent with the prediction by Hill's theory. Therefore, the measurement of r value is the very effective method characterizing the texture hardening. Moreover, the determination of the preferred orientation is also useful.

For the Ti-5Al-2.5Sn alloy, Al and Sn are the alloying element that tend to form the basal plane sheet texture and therefore, it is relatively easy to produce the sheets with such a component. However, it is considered that even with such an orientation the degree of strengthening could reach only to a extent of 34% for this alloy comparing with that of 64% for Ti-4Al.

For the Ti-6Al-4V, however, as the texture of this alloy varies widely by the processing condition in relation to the transformation, it is considered that there is possibility of obtaining much higher biaxial strength by texture control. For the Ti-4Al, though the strength level is lower than the other alloys, the biaxial strength increase of 64% of uniaxial tensile yield strength is very attractive. This property is directly related to the centralized type of pole figure. It is desirable to increase the strength level of base materials maintaining this higher degree of strengthening.

Thus, since the yield strengths under biaxial stresses significantly increase compared with uniaxial strength, weight saving of the construction can be realized if desiged on the base of the biaxial strength, and reliability of the structure under biaxial stress is expected to increase.

Summary

1. In this paper, various factors such as r value, texture, biaxial strength and Khn-locus characterizing the texture hardening of titanium were discussed. Their results indicate that their factors are closely related with each other. The degrees of strengthening under biaxial stress showed a relatively good agreement with the Hill's theory for the anisotropy materials.

2. The two α alloys, Ti-4Al and Ti-5Al-2.5Sn, showed a strong basal plane sheet texture. For the former, the biaxial strength increase of 64% and for the latter 34% were obtained.

3. The Ti-6Al-4V alloy displayed multiple orientation components. The degree of strengthening under biaxial stress was 23%. But because of high uniaxial strength, the biaxial strength was 116.3 kg/mm^2 similar to that of the Ti-5Al-2.5Sn.

4. The Ti-8Mn alloy showed to be almost isotropic.

5. For two commercial grades of pure titanium, the biaxial strength increases were rather high. The degree of strengthening

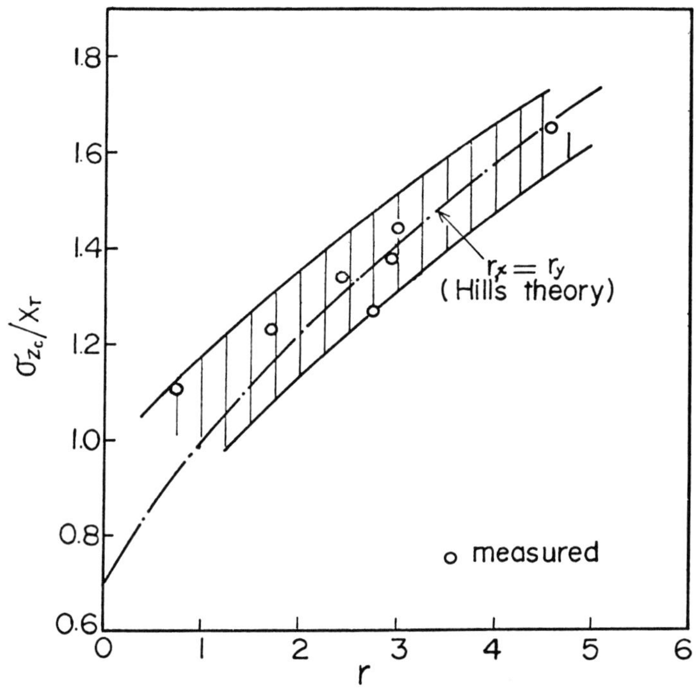

Fig. 5. The relation between the degree of strengthening σ_{zc}/X_T and r values.

under biaxial stress is about 40% for KS 50 and 26% for KS 85.

6. For the r value up to about 2.5, the shape of Khn-locus exhibited relatively a good agreement with those in first quardrant of conventional method, but for the r value more than about 2.8 they did not agree so well.

References

1. H.W.Babel, S.F. Frederick : J. Metals, 20(1968) 33
2. D. Lee, W.A. Backofen : Trans. Met. Soc., 236(1966) 1696
3. R. Hill: Proc. Ray. Soc., 193A(1948) 281
4. R.G. Wheeler, and D.R. Ireland : Electrochem. Technol., 4(1966) 313

THE RELATIONSHIP BETWEEN CRACK PROPAGATION CHARACTERISTICS AND FRACTURE TOUGHNESS IN α+β TITANIUM ALLOYS

I.W. Hall and C. Hammond

Department of Metallurgy
University of Leeds
Leeds LS2 9JT
England

Introduction

Work published over the last few years has shown that the acicular or transformed β structures obtained on solution treating and cooling α+β Titanium Alloys from above the β/α+β transsus have considerably higher plane strain fracture toughness (K_{1c}) values than the equiaxed α+β structures obtained on cooling from temperatures in the α+β field (1) (2). In practice the acicular structures may be produced as a result of forging at temperatures above or near the β/α+β transus (3). Most measurements have been made on alloys which are subsequently heat treated in the α+β field in which condition the acicular structures have inferior values of yield strength and ultimate tensile strength.

It has been suggested that the controlling factor in determining toughness is the length-thickness ratio of the acicular platelets and that layers of retained β phase at the plate boundaries may act as a crack arrestor (4). Although such arguments may be true they make the a priori assumption that the retained phase is more ductile than the α; the ductility of the β depends upon its composition, i.e. upon the segregation of β stabilizing alloying elements during the heat treatment. Furthermore, acicular structures are also obtained in the transformed β grains in α+β equiaxed structures and there seems no reason why such structures should not also show high K_{1c} values.

The object of the research programme has been to investigate the different microstructural features which give rise to different fracture toughness values and crack propagation characteristics. Three alloys were chosen for study, IMI318, 550 and 551. The compositions and transformation details are given in Table I. The presence of silicon in IMI550 and 551 confers creep

Table I

Alloy	Composition (wt.%)	$\beta/\alpha+\beta$ transus (^{o}C)	Temperature below which β does not transform on cooling (^{o}C)
IMI318	Ti-6%Al-4%V	995±15	815
IMI550	Ti-4%Al-2%Sn -4%Mo-0.47%Si	975±10	∿830
IMI551	Ti-4%Al-4%Sn -4%Mo-0.47%Si	1050±15	850

resistance and may give rise to silicide precipitation. Particular attention has been paid to the role of the retained β and it was considered that this may act as a crack arrestor by undergoing a strain induced transformation to martensite in the plastic zone ahead of a propagating crack (5). In this respect a high toughness may arise by precisely the same mechanism as occurs in Transformation Induced Plasticity (TRIP) steels (6).

Although all commercial heat treatment schedules involve ageing or stress relieving in the $\alpha+\beta$ field it was decided to restrict heat treatments to simple solution treatments and cooling from temperatures in the β and $\alpha+\beta$ fields. The microstructural changes occurring on subsequent ageing are not well understood and would give rise to complications.

Experimental Techniques

The alloys were heat treated in a flowing argon atmosphere at various temperatures in the β and $\alpha+\beta$ fields in order to obtain differing amounts of acicular α, primary (equiaxed) α and β on air cooling to room temperature. Fracture toughness tests were carried out on fatigue pre-cracked compact K_{1c} (C.K.S.) specimens loaded in tension and single edge notch (S.E.N.) specimens loaded in three point bending (Figs. 1 and 2). Plane strain conditions were obtained by setting t (thickness) $\geq 2.5\left(\frac{K_{1c}}{\sigma_y}\right)^2$ (where σ_y = yield stress) in accordance with A.S.T.M. recommended practice (7). For the present high strength alloys t = 10mm was found to be sufficient. Considerably difficulty was encountered in producing acceptably straight crack fronts particularly in the case of the C.K.S. specimens. This appears to be a general difficulty in fatigue cracking titanium alloys and was not satisfactorily corrected by careful alignment of the loading pins or by side grooving. K_{1c} values were determined by the 5% offset secant method from a load-crack opening displacement record.

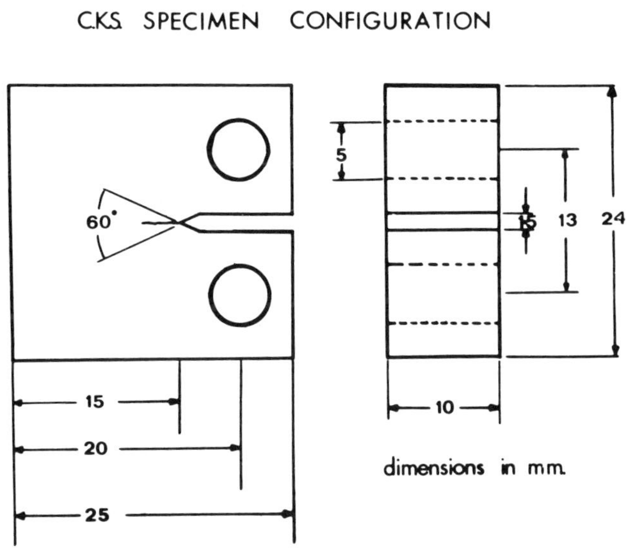

Fig. 1. Compact K_{1c} fracture toughness (C.K.S.) specimen.

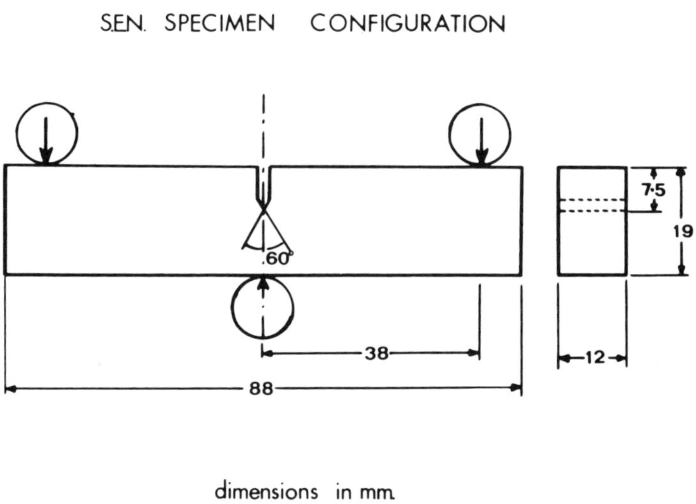

Fig. 2. Single edge notch (S.E.N.) fracture toughness specimen.

Fracture surfaces were studied using scanning electron microscopy and replica fractography. Thin foil specimens for transmission electron microscopy were prepared from 3mm discs of the heat treated material using a solution of 6% Perchloric acid 35% Butanol 59% Methanol.

Optical microscopy was used to follow fatigue crack propagation on polished and etched surfaces and tensile tests were carried out on flat 10mmx0.6mm specimens with a 30mm gauge length.

Results

(A) IMI318 (Ti-6%Al-4%V)

The results for the alloy heat treated at four different temperatures are given in Table II. It can be seen that, contrary

Table II

Heat Treatment	K_{1c}(S.E.N.) $MNm^{-3/2}$*	0.2%P.S. $MNm^{-2} \times 10^2$	U.T.S. $MNm^{-2} \times 10^2$	Microstructural Details
1050°C 1hr. Air cool	69	10.8	11.2	Plate martensite plus retained β layers (Fig. 3)
950°C 5hr. Air cool	61.5	9.8	10.0	Primary (equiaxed) α plus ∿70% acicular α (transformed β)
850°C 6hr. Air cool	46.5	8.0	10.2	ditto, plus ∿20% acicular α
750°C 6hr. Air cool	39.5	7.5	10.2	ditto, plus ∿10% untransformed β

$*1MNm^{-3/2} = 0.908 \text{ ksi} \sqrt{\text{in.}}$

to results on alloys in the aged condition, the acicular structure obtained on cooling from the β field (Fig. 3) has both greater toughness and strength than the mixed structures obtained on cooling from the α+β field (Fig. 4). Furthermore, the fracture toughness values in particular are proportional to the amount of β or transformed β in the microstructure. An important distinction between the martensite structures is that on cooling from the β field the plates occur in parallel bundles and are separated by thin (200∿600Å) layers of retained β whereas on cooling from the α+β field the plates occur in random orientations with very small amounts of retained β in isolated patches. Hence, it appears that the distribution of the β in the microstructure does not determine the crack propagation characteristics and fracture toughness.

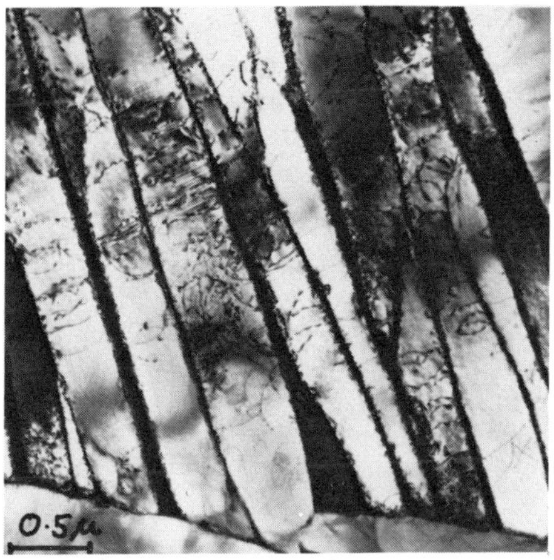

Fig. 3. IMI318 1050 1 hr air-cool. Martensite plates and retained β layers (Electron micrograph).

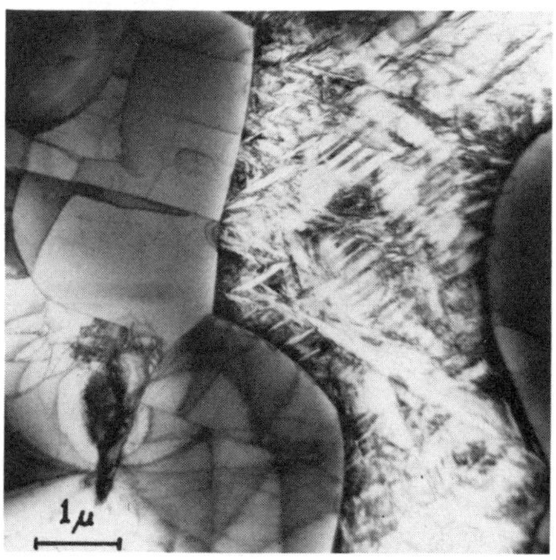

Fig. 4. IMI318 950 5 hr air-cool. Primary α and transformed β (Electron micrograph).

The variations in toughness are also reflected in the fatigue crack propagation characteristics, the crack appearing to propagate along martensite plate boundaries or perpendicular to them and branching (Fig. 5). In contrast to the fracture surface of the α+β heat treated alloy (Fig. 6) which showed a ductile dimple fracture, the β heat treated alloy (Fig. 7) consisted of regions of ductile failure and sheared facets. These facets may correspond either to prior β grain boundaries or to shear along the plate boundaries of a particularly orientated colony of martensite.

(B) IMI550 and (551) (Ti-4%Al-2%(4%)Sn-4%Mo-0.47%Si)

Although these alloys differ in composition by only 2%Sn the fracture toughness results are markedly different (Table III).

Table III

(0.2% P.S. values estimated from V.H.N. measurements)

	IMI550		IMI551	
Heat Treatment	K_{1c}(S.E.N.) $MNm^{-3/2}$	0.2%P.S. $MNm^{-2} \times 10^2$	K_{1c}(C.K.S.) $MNm^{-3/2}$	0.2%P.S. $MNm^{-2} \times 10^2$
1125°C 1hr. Air cool	82.5	~10.6	31	~13
950°C 5hr. Air cool	57	~ 9.8	31	~12.2
850°C 6hr. Air cool	46.5	~ 9.1	30.5	~11.7
750°C 6hr. Air cool	40	~ 8.4	25.5	~11.3

For the alloys solution treated at 1125°C and air cooled, K_{1c} for IMI550 was 82.5 $MNm^{-3/2}$ and for IMI551 was 31 $MNm^{3/2}$ with only a moderate difference in proof stress. The difference in fracture toughness values is so large as to suggest that the results from the C.K.S. specimens are considerably too small even though the testing procedure was valid on the basis of A.S.T.M. criteria. Work is being carried out at present to compare results obtained from the two techniques using identically heat treated specimens.

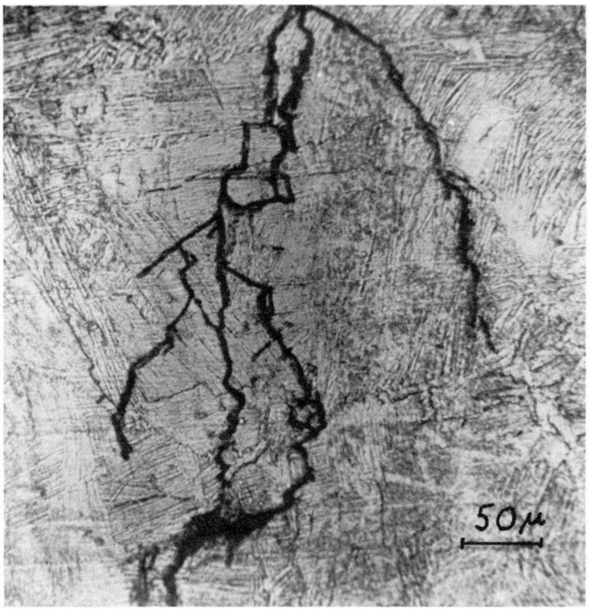

Fig. 5. IMI318 1050 1 hr air-cool. Branching fatigue crack (Optical micrograph).

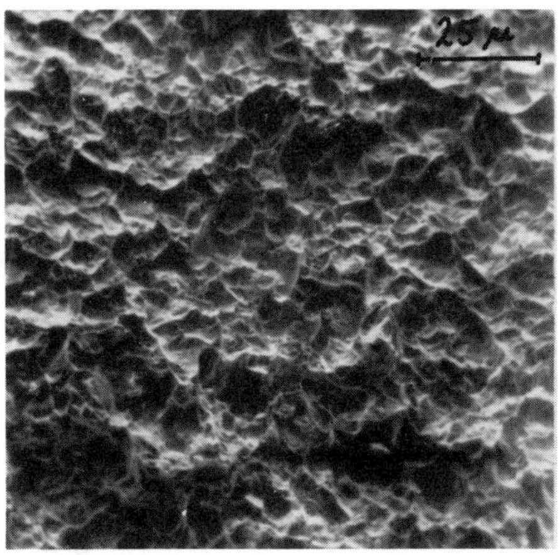

Fig. 6. IMI318 850 6 hr air-cool. Fracture surface showing ductile dimples (Scanning electron micrograph).

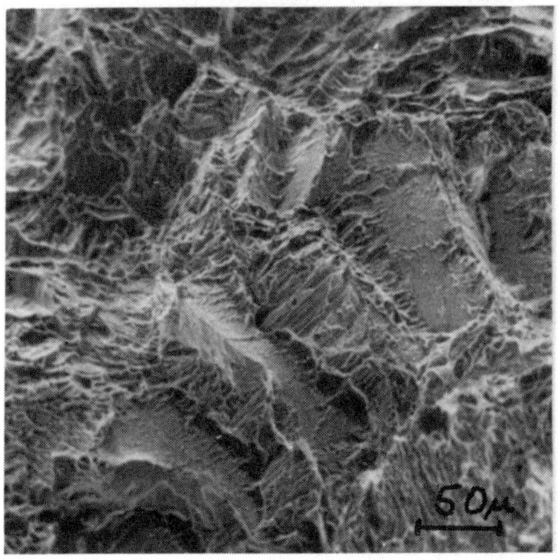

Fig. 7. IMI318 1050 1 hr air-cool. Fracture surface showing sheared facets and ductile dimples (Scanning electron micrograph).

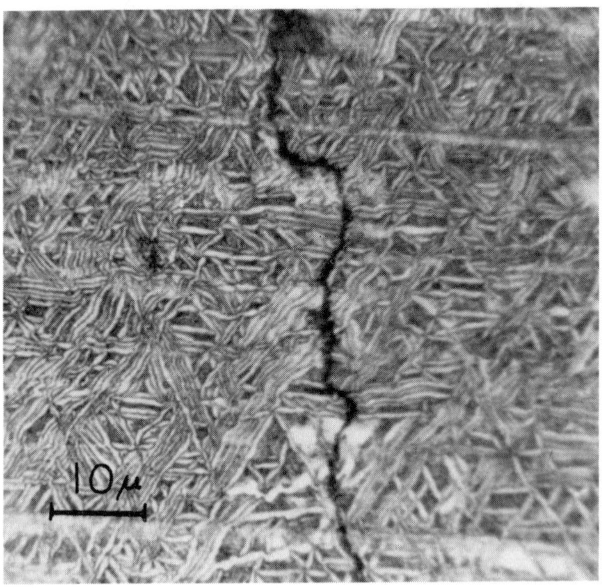

Fig. 8. IMI550 1125 1 hr air-cool. Fatigue crack propagation (Optical micrograph).

A

B

Fig. 9. IMI551 1125 1 hr air-cool. a) Dimpled intergranular facets (Scanning electron micrograph). b) Upper left intergranular facet, lower right transgranular dimples (Shadowed carbon replica).

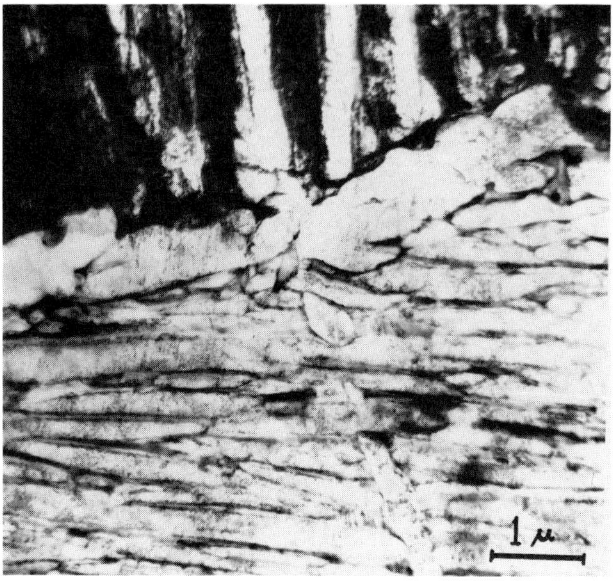

A

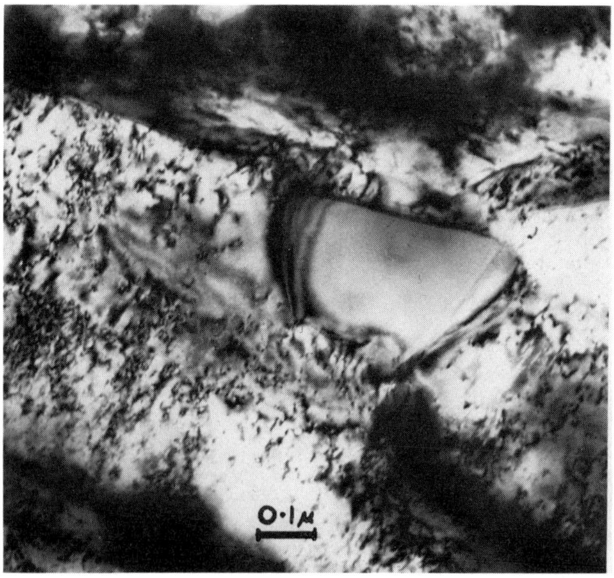

B

Fig. 10. IMI551 1125 1 hr air-cool. a) Martensite plates, grain boundary α (Electron micrograph). b) Precipitate at grain boundary (Electron micrograph).

In these alloys the fatigue crack propagates in a manner similar to that in IMI318 but without branching (Fig. 8). However, differences are observed in the fracture surfaces, in particular the IMI551 alloy exhibits dimpled intergranular facets clearly outlining the prior β grain boundaries (Fig. 9a and b) whereas the IMI550 alloy fracture surface more closely resembled that in the IMI318 alloy. The fracture surface of the IMI551 alloy is analogous to that of over-heated steels (8) which indicates that grain boundary segregation, probably silidides precipitated in the grain boundaries. Although particles have been observed at grain boundaries (Fig. 10 a and b) they have not been unambiguously identified by electron diffraction. However, scanning X-ray fluorescence analysis of the intergranular surfaces showed small Si rich regions ∿0.25μ in diameter. Lower values of K_{1c} and strength are observed for both alloys heat treated at lower temperatures in the α+β field. For the IMI550 alloy heat treated in the α+β field the load vs crack opening Displacement Record showed a change in slope which did not correspond to unstable crack propagation. A similar effect has been observed in T.R.I.P. steels (6).

Conclusions

The results on the IMI318 alloy in particular suggest that the β does not act as a crack arrester by simply diverting its path since that β which occurs as continuous layers in the martensite would be expected to give rise to a greater toughness than that which occurs as isolated patches. On the other hand any stress induced transformation of the β will occur in the plastic zone ahead of a propagating crack and the toughness increment associated with such a mechanism will be independent of the distribution of β within the plastic zone. In view of the very small proportion of retained β in the structures such a hypothesis can only be verified by the direct observation of the β phase near a propagating crack using a straining stage in the electron microscope. The other possible evidence for a T.R.I.P. mechanism - that of the deviation in the Load-Crack Opening Displacement Record - has not been explained and may be only circumstantial.

The presence of silicon appears to be deleterious to toughness when grain boundary silicide particles are formed. The problem may be overcome by heating to temperatures well above the α+β/β transus where the silicon will tend to go back into solution.

References

1. Lee, K.T., and Hubbard, R.T.J., "Relationship between Fracture Toughness, Plain Tensile Properties and Metallurgical Structure of a Titanium Alloy" Iron and Steel Institute Publication 120 1970, pp.127-129.

2. Frederick, S.F., and Hanna, W.D. "Fracture Toughness and Deformation of Titanium Alloys at Low Temperatures". Metallurgical Transactions Vol. 1, 1970, pp.347-352.

3. Coyne, J.E. "The β-Forging of Titanium Alloys" Science Technology and Application of Titanium, Pergamon Press, 1970. pp.97-110.

4. Greenfield, M.A., and Margolin, H. "The Interrelationship of Fracture Toughness and Microstructure in a Ti-5.25%Al-5.5%V-0.9%Fe-0.5%Cu Alloy". Metallurgical Transactions Vol. 2 1971, pp.841-847.

5. Crossley, F.A., and Lindberg, R.W. "Microstructural Analysis of a High Strength Martensite-Beta Titanium Alloy" 2nd International Conference on the Strength of Metals and Alloys, 1971, pp.841-850.

6. Gerbench, W.W., Hemmings, P.L., Merz, M.D. and Zackay, V.F., "Preliminary Toughness Results on a T.R.I.P. Steel". Transactions of the Metallurgical Society of A.I.M.E.Vol.61, 1968, pp.843-847.

7. Brown, W.F., and Srawley, J.A. "Plane Strain Crack Toughness Testing of High Strength Metallic Materials." American Society for Testing Materials S.T.P. 410 pp.1-65.

8. Joy, G.D., and Nutting, J. "Influence of Intermetallic Phases and Non Metallic Inclusions upon the Ductility and Fracture Behaviour of some Alloy Steels" I.S.I./B.I.S.R.A. Conference, Scarborough, 1971 published by the Iron and Steel Institute.

THE FRACTURE TOUGHNESS OF 3 Ti ALLOYS

R. Chait and T.S. DeSisto
Army Materials and Mechanics Research Center
Watertown, Massachusetts

This study investigates the influence of microstructural variables on the fracture toughness of the following heat treated Ti alloys in heavy sections: Ti-6Al-4V, Ti-6Al-6V-2Sn and Ti-8Mo-8V-2Fe-3Al. Compact tension and precracked Charpy specimens were obtained from thick-wall hollow extrusions. The extrusions were heat treated to provide a range of strength levels. Tensile specimens were obtained from the outside mid and inner wall locations to characterize strength parameters. For Ti-6Al-4V, a definite difference in fracture toughness exists between microstructures characterized by elongated or plate-like primary α and equiaxed primary α. It is shown that the equiaxed microstructure leads to a greater value of fracture toughness. Decreasing the oxygen level and increasing the amount of transformed beta matrix increases the value of fracture toughness. Of the alloys investigated, the largest value of fracture toughness was provided by Ti-6Al-6V-2Sn in the STOA condition. Beta flecks were present in the microstructure of modified Ti-6Al-6V-2Sn alloy and appeared to lower fracture toughness. The metastable beta alloy, Ti-8Mo-8V-2Fe-3Al is characterized by reduced tensile ductility and high fracture toughness. The reduced tensile ductility is associated with a large beta grain size. Increasing the amount of prior working decreases the grain size, increases tensile ductility and appears to lower fracture toughness as measured by W/A.

Introduction

The ability to make better utilization of high strength Ti alloys depends on a greater understanding of the effect microstructure has on important design parameters such as fracture toughness. This is especially true of heavy section geometries where hardenability and nonuniform degree of mechanical working can lead to varying microstructure across the section. For example, Broadwell and Heil(1) have evaluated the fracture toughness and strength variation of several high strength Ti alloys using a "Navaho" rib and web forging. To reveal the influence of microstructure in more detail, recent studies at AMMRC have focused on the heavy section mechanical properties of 3 Ti alloys: Ti-6Al-4V, Ti-6Al-6V-2Sn and Ti-8Mo-8V-2Fe-3Al.

Materials and Test Procedure

The materials for this investigation together with their respective composition and heat treatment are shown in Table I and II,respectively. Test specimens for mechanical property determination were taken from 11-1/2 in. long cylindrical hollows of the following dimensions: 8-3/4 in. OD and 3 in. wall. This configuration was achieved by extrusion at the following temperatures: Ti-6Al-4V (1750°), Ti-6Al-6V-2Sn (1675°F) and Ti-8Mo-8V-2Fe-3Al (1700°F). Tensile specimens (0.252 diameter) were taken from the outer, mid and inner wall locations for both the longitudinal and transverse orientation and were tested on a 120,000 hydraulic testing machine at a platen displacement of 0.005 in/min. For all alloys, fracture toughness measurements were performed with 3/4 in. thick compact tension specimens taken in a transverse orientation which allowed the crack to propagate toward the inner wall. With some alloys, additional measurements were obtained on specimens where crack propagation was toward the outer wall. Both orientations are shown in Fig. 1. The specimens were precracked in a 100,000 lb. MTS testing machine and tested to failure on a 120,000 lb. hydraulic testing machine. All precracking and fracture toughness testing procedures with the exception of the crack front profile conformed to ASTM Standard E 399-71. For all materials except the modified Ti-6Al-6V-2Sn, the crack front was thumbnail in appearance. Numerous attempts at AMMRC to produce straight crack fronts on Ti-6Al-4V, Ti-6Al-6V-2Sn and Ti-8Mo-8V-2Fe-3Al (all STA condition) also were not successful. Although it is felt that the data are valid and should be reported as K_{IC}, Para. 7.2.3 of E 399-71 specifies that if a difference between any two of the crack length measurements exceeds 5% of the average crack length, the test is invalid. For this reason, fracture toughness measurements are given in terms of K_Q.

Additional information on fracture energy was obtained by impacting precracked 0.394 x 0.394 in. standard Charpy specimens taken from the mid-wall location and noting the energy absorbed per unit area of the uncracked portion of the specimen. The value is given in terms of in-lb/in^2 and is represented by W/A. Precracking for these specimens was performed on a Manlabs fatigue precrack machine.

The amount of working accompanying the fabrication of the hollow cylinders noted above was approximately 30% and was limited by the capacity of the AMMRC press. It is the intent of this program to also examine the effect of increased mechanical working by extruding the above configuration at reduction ratios which correspond to approximately 83% reduction. At this writing,only the evaluation of the Ti-6Al-4V alloy (Heat B - Tables I and II) extruded at the larger reduction ratio has been completed. Limited tensile data are available for the Ti-8Mo-8V-2Fe-3Al alloy extruded at the larger reduction.

Results and Discussion

As shown in Table III, there is a wide range of strength levels resulting from the heat treatment of the 3 alloys. At the low end of the spectrum is the Ti-6Al-4V in both the STA and STOA condition. As expected, the large section size precluded any significant response to aging. In contrast, Heat B of Ti-6Al-6V-2Sn modified with 2.0% Zr and in the STA condition, exhibited yield strength close to 200 KSI at the outer wall, dropping to 181-187 KSI at the midwall. For the same STA condition, Heat C of Ti-6Al-6V-2Sn led to strength levels which were about 20-30 KSI lower than the modified composition. The drop in strength is attributable to pronounced effects of the added Zr and higher C, Cu and Fe levels. At all locations through the section, the Ti-8Mo-8V-2Fe-3Al alloy exhibited yield strength levels in the range of 174/182 KSI.

The K_Q values for each of the alloys are given in Table IV. The variation in K_Q with yield strength is shown in Fig. 2. The same plot is shown in Fig. 3 with each K_Q value adjusted for the density of the given alloy. For clarity, the average K_Q value corresponding to a given orientation (Fig. 1) is shown in Figs. 2 and 3. The yield strength value considered in Figs. 2 and 3 is the average of the outer and mid-wall for the K_Q specimen oriented for crack propagation toward the inner wall, and an average of the inner and mid-wall for K_Q specimens oriented for crack propagation toward the outer wall. The K_Q values obtained for each of the alloys are discussed in more detail below.

Ti-6Al-4V

While K_Q values for STOA condition of Heat A are relatively consistent between 44-51 KSI$\sqrt{in}$, the values for the STA condition of the same heat separate into two groups that depend on specimen

Table I. MATERIALS AND THEIR COMPOSITION

Alloy	Element (weight percent)									
	Al	V	Mo	Fe	Sn	Cu	C	O	H	N
Ti-6Al-4V										
Heat A	6.4	4.1	-	0.18	-	-	0.02	0.16	0.007	0.007
Heat B	6.3	4.2	-	0.20	-	-	0.02	0.13	0.004	0.009
Ti-6Al-6V-2Sn										
Heat A	5.9	5.5	-	0.81	2.0	0.78	0.020	0.16	0.005	0.015
Heat B*	6.0	5.9	-	0.99	2.0	0.97	0.12	0.07	0.004	0.010
Heat C	5.6	5.3	-	0.78	2.45	0.72	0.022	0.13	0.010	0.017
Ti-8Mo-8V-2Fe-3Al	2.3	8.0	8.2	1.8	-	0.006	0.022	0.16	0.007	0.018

*Modified with 2.0% Zr

Table II. HEAT TREATMENT OF MATERIALS

Alloy	Condition	Solutionizing Temperature & Time for WQ	Aging Temperature & Time for AC
Ti-6Al-4V			
Heat A	STOA	1750°F - 3 hr	1300°F - 6 hr
	STA	1750°F - 3 hr	1000°F - 4 hr
Heat B	STA	1750°F - 3 hr	1000°F - 4 hr
Ti-6Al-6V-2Sn			
Heat A	STOA	1600°F - 3 hr	1300°F - 6 hr
Heat B*	STA	1600°F - 3 hr	1000°F - 4 hr
Heat C	STA	1600°F - 3 hr	1000°F - 4 hr
Ti-8Mo-8V-2Fe-3Al	STA	1475°F - 1-1/2 hr	1000°F - 8 hr

*Modified with 2.0% Zr

Table III. SUMMARY OF TENSILE PROPERTIES

Alloy	Condition		0.2% Yield Strength, ksi			Tensile Strength ksi			Elongation %			Reduction of Area, %		
			OD	MW	ID	OD	MW	ID	OD	MW	ID	OD	MW	ID
Ti-6Al-4V														
Heat A	STOA	L	141.9	135.5	136.4	148.8	143.8	143.2	20.0	17.5	19.0	46.2	39.8	37.9
		T	137.2	139.8	136.5	147.8	151.6	149.8	16.3	17.5	16.8	36.5	41.7	39.0
Heat A	STA	L	135.9	130.9	136.0	144.6	145.0	147.5	19.5	21.0	20.0	48.1	41.6	33.0
		T	145.3	139.8	148.2	158.4	153.3	161.4	13.3	15.7	10.8	35.3	36.6	25.1
Heat B	STA	L	142.3	128.9	135.4	152.4	147.6	158.9	15.7	16.0	16.2	40.0	37.6	37.6
		T	139.5	137.4	148.5	156.4	148.4	151.9	13.1	12.5	11.2	33.1	26.2	20.8
Ti-6Al-6V-2Sn														
Heat A	STOA	L	143.3	143.2	144.7	153.9	151.9	154.8	14.5	15.0	16.5	33.8	37.1	43.8
		T	142.8	143.9	144.4	-	153.0	153.5	16.0	16.5	16.0	30.1	36.3	41.8
Heat B*	STA	L	201.7	180.9	195.3	208.9	185.6	199.5	2.2	3.8	3.5	5.6	9.2	8.1
		T	192.7	187.0	198.3	197.6	191.6	201.9	2.5	1.7	1.2	5.0	4.1	4.0
Heat C	STA	L	165.6	148.6	164.7	176.3	162.2	177.2	6.1	9.5	7.6	13.2	20.6	15.1
		T	160.5	157.5	162.5	179.0	170.7	176.5	4.4	7.9	6.5	8.4	14.6	12.4
Ti-8Mo-8V-2Fe-3Al														
	STA	L	177.5	181.3	174.2	178.9	182.1	181.4	1.1	1.3	2.0	2.4	2.1	5.4
		T	177.3	180.0	175.6	180.0	182.5	180.6	1.4	1.9	2.4	2.4	2.6	3.4

*Modified with 2.0% Zr

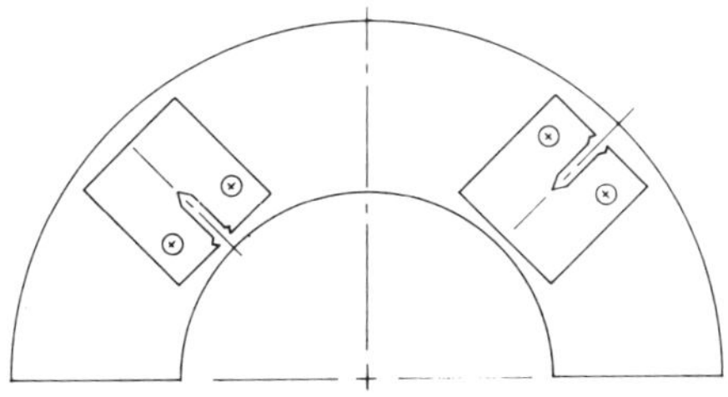

Fig. 1. Orientation of compact tension fracture toughness specimen.

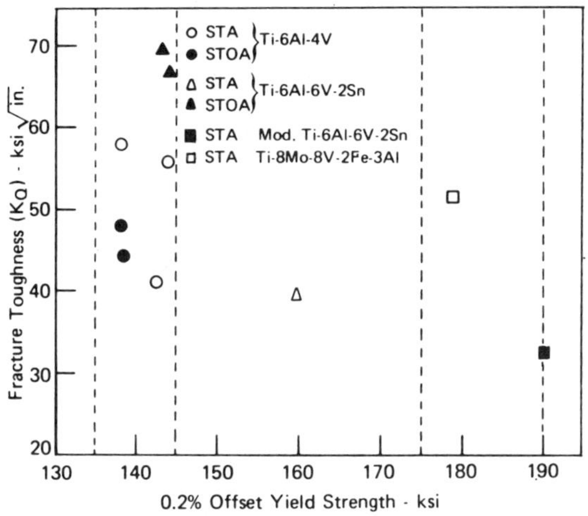

Fig. 2. Fracture toughness as a function of yield strength.

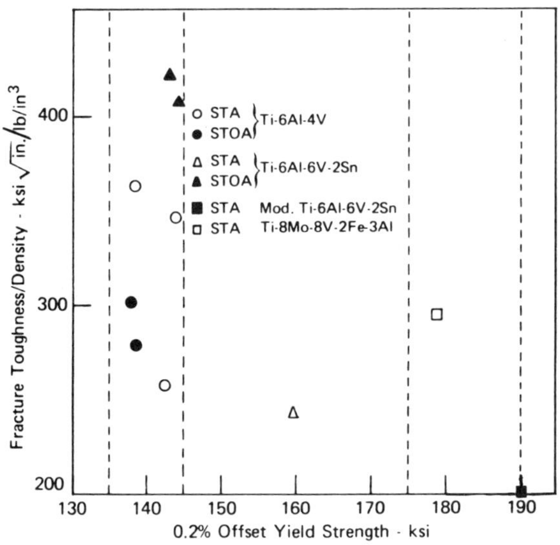

Fig. 3. Specific toughness as a function of yield strength.

orientation. The specimens with crack propagation toward the inner wall led to values in the range 41-42 KSI$\sqrt{in}$ while crack propagation in the opposite direction was characterized by significantly higher values in the range 50-56 KSI$\sqrt{in}$. Since there was a relatively small variation in tensile properties across the section, it was an opportunity to examine the possible influence of microstructure on K_Q.

Examination of the microstructure at fracture surface of the broken test specimens revealed that the higher K_Q values were associated with an equiaxed primary α (Figure 4a). The lower values of K_Q were associated with an elongated or plate-like primary α (Fig. 4b). The amount of primary α was approximately the same (32%) for both the equiaxed and plate-like microstructure as determined with a Quantimet (QTM) Image Analyzer. It is therefore evident that for this particular heat, equiaxed microstructure leads to higher values of K_Q.

Recently, Gerberich and Baker(2) have utilized relative values of crack-tip displacement and size of primary α particles to explain the fracture toughness differences between equiaxed and plate-like primary α microstructures of Ti-6Al-4V. With a duplex heat treatment, these authors were able to achieve a given plate-like microstructure that produced crack path deviation that led to increased fracture toughness values. The plate-like microstructure of the present study did not result from a duplex heat treatment and therefore could not be expected to provide the increase in K_Q noted by Gerberich and Baker. Even if it were duplicated, it is unlikely that the plate-like microstructure would produce K_Q values greater than that obtained with an equiaxed microstructure since it has been reported that an oxygen level of about 0.13% is required for plate-like α to produce increased values of fracture toughness(2). The nominal oxygen content of Heat A in the present study is listed in Table I as 0.16%, but levels averaging 0.20% were measured on the broken K_Q specimens.

As shown in Table IV, Heat B (STA condition) with a microstructure consisting of a mixture of small equiaxed and multi-grained plate-like primary α (Fig. 4c) that resulted from the increased amount of working, leads to increased K_Q values when compared to Heat A (STA condition). The W/A values for Heat B (912 in-lb/in^2) are also larger than those for Heat A (835 in-lb/in^2). In addition to the equiaxed microstructure, lower oxygen content and slightly larger amount of transformed beta matrix are noted for Heat B. Chemical analyses taken on broken K_Q specimens established the oxygen level at 0.13% compared to 0.20% for Heat A as noted above. The effect of oxygen content on fracture toughness is recognized and has caused the Ti-6Al-4V material purchased for the B-1 aircraft to be controlled at maximum oxygen content of 0.13% to insure adequate fracture toughness(3). Equally as important may be the increased amount of transformed beta matrix of Heat B (75%) compared

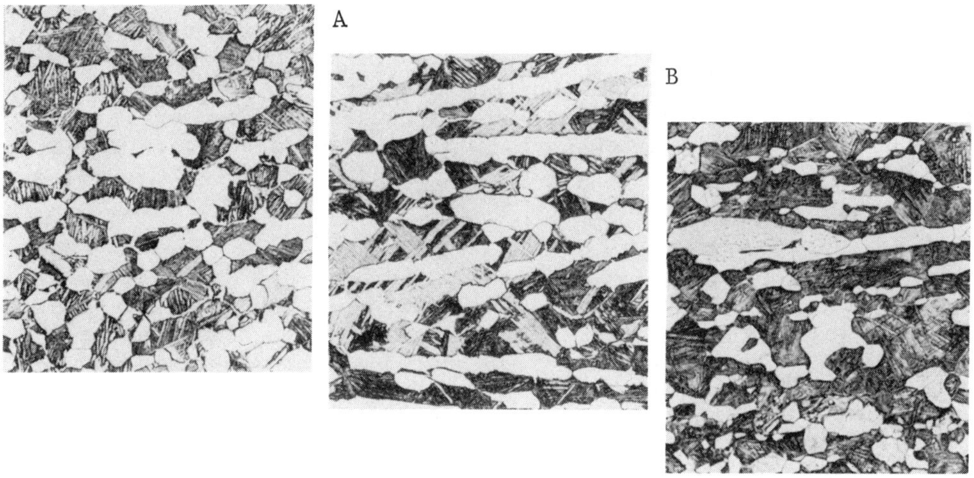

Fig. 4. Microstructure and associated fracture toughness of Ti-6Al-4V (STA). Mag. 250X.
a) Equiaxed primary α, K_Q = 55.9 ksi $\sqrt{\text{in.}}$ (Heat A).
b) Plate-like primary α, K_Q = 41.6 ksi $\sqrt{\text{in.}}$ (Heat A).
c) Equiaxed and multigrain plate-like primary α, K_Q = 59.3 ksi $\sqrt{\text{in.}}$ (Heat B).

Table IV. SUMMARY OF K_Q FRACTURE TOUGHNESS VALUES

Alloy	Condition	K_Q - ksi$\sqrt{\text{in}}$			Orientation
		Test 1	Test 2	Test 3	
Ti-6Al-4V					
Heat A	STOA	44.1	45.1	44.6	I*
		49.6	44.3	51.0	II+
Heat A	STA	41.3	41.0	41.6	I
		55.9	50.0		II
Heat B	STA	59.3	56.3	58.9	I
Ti-6Al-6V-2Sn					
Heat A	STOA	63.9	70.5	73.5	I
		68.0	67.0	66.2	II
Heat B†	STA	27.6	38.4	30.7	I
Heat C	STA	42.5	36.9		I
Ti-8Mo-8V-2Fe-3Al	STA	51.0	51.5	52.1	I

*Crack propagates toward ID of extrusion, see Fig. 1.
+Crack propagates toward OD of extrusion, see Fig. 1.
†Modified with 2.0% Zr.

to Heat A (68%), since the increased toughness of the fully transformed microstructure has been noted(4). All of these factors probably contributed to the increased toughness of Heat B. As shown in Fig. 5 for Heat B, the α (primary and in the matrix) appears to cause deviations in the crack path, thus increasing fracture toughness in the manner suggested by Gerberich and Baker (2,4).

Ti-6Al-6V-2Sn

Of the alloys considered in this study, it was possible to achieve the largest values of K_Q (63-74 KSI$\sqrt{in}$) with Ti-6Al-6V-2Sn (Heat A) in the STOA condition. The density is slightly greater than Ti-6Al-4V, but it is not enough to change the ranking on a K_Q/density basis (Figure 3). The consistency of the K_Q values stems from the strength and microstructural (equiaxed primary α) uniformity through the section. The combination of K_Q and strength values obtained for the STOA condition compare very favorably to the optimum combination of strength and toughness recently determined for various STOA heat treatments aimed at improving the toughness(5). However, the W/A value (549 in-lb/in^2), being lower than Ti-6Al-4V (STA condition), does not reflect the increased K_Q values, indicating a possible strain rate effect.

Increasing the strength level by employing the STA heat treatment (Heat C) reduced K_Q to 36-43 KSI$\sqrt{in}$. On the basis of yield strength, K_Q for the STA condition is compared to the other alloys in Figures 2 and 3.

A further increase in the strength level of Ti-6Al-6V-2Sn (STA condition) can be achieved through a modification in chemistry (Heat B - Table I). The addition of Zr together with higher C, Fe, and Cu contents lead to yield strengths of up to 200 KSI which are accompanied by lower ductility in the STA condition. For this material the K_Q is in the range 27-39 KSI $\sqrt{in}$ as shown in Table IV. The W/A values for this alloy (115 in-lb/in^2) were lower than values obtained for Heat A - STOA condition (549 in-lb/in^2) and Heat C - STA condition (300 in-lb/in^2). In view of the low oxygen level (0.07%), K_Q values might be expected to be somewhat higher. However, it has been shown that at strength levels comparable to those obtained for Heat B, both ELI and commercial grades to Ti-6Al-6V-2Sn exhibited K_{IC} in the range 29-33 KSI$\sqrt{in}$ (6).

It is interesting to note that this modified chemistry (Heat B) was also prone to the production of what have been termed "beta flecks". Apparently, these are regions of microsegregation of beta stabilizers that lead to a transformed structure after heat treatment due to a locally lower beta transus. The high Cu and Fe content which have been associated with "beta flecks"(7) are present in Heat B. The specimens used for K_Q determination were examined for the appearance of "beta flecks". Of the three specimens examined, the specimen which exhibited the lowest value of K_Q displayed "beta flecks" along the fracture surface as shown in Fig. 6. Hardness measurements made within the "beta fleck" reveal a higher hardness

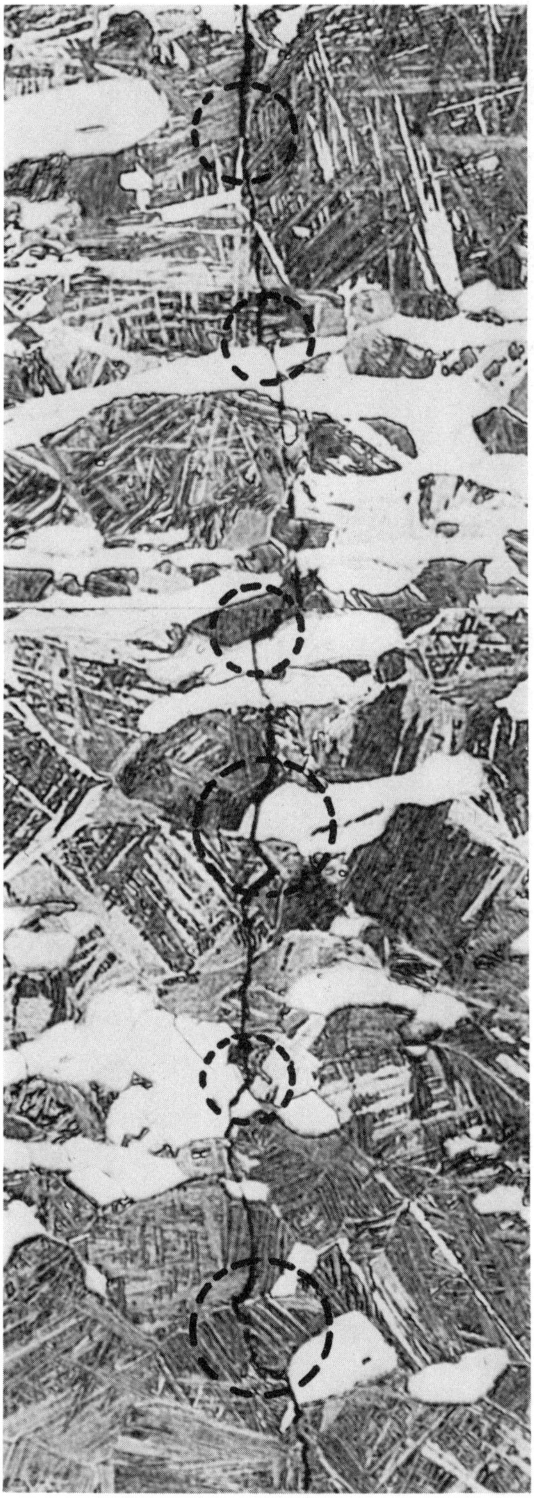

Fig. 5. Crack propagation in Ti-6Al-4V (Heat B, STA condition). Circled regions highlight locations where α has caused crack path to deviate. Mag. 500X.

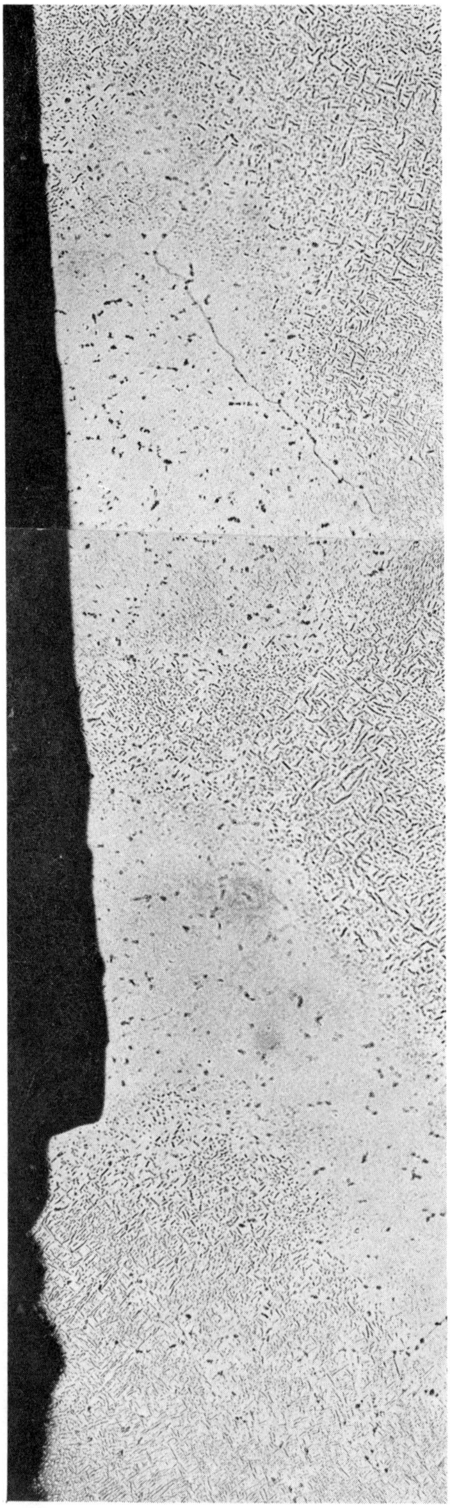

Fig. 6. The appearance in modified Ti-6Al-6V-2Sn of "Beta flecks" along fracture surface. Mag. 75X.

when compared to the hardness of the adjacent microstructure. The Knoop hardness number averaged 427 within the "beta fleck" while on either side, the average was 419. Therefore, it is not unexpected that the "beta flecks" shown detrimental to smooth specimen fatigue properties(8) also decrease fracture toughness.

Ti-8Mo-8V-3Al-2Fe

In contrast to the Ti-6Al-6V-2Sn toward the higher strength levels, Ti-8Mo-8V-3Al-2Fe provides relatively high K_Q values (Fig. 2). Its position relative to the other alloys remains essentially unchanged when density considerations are introduced (Fig. 3). The Ti-8Mo-8V-2Fe-3Al alloy possesses these values of K_Q despite rather low tensile ductilities (1-3% elongation and 2-6% RA). It has been observed that such behavior is characteristic of beta alloys with a large grain size(9). The ASTM grain size for this material is indeed coarse (Fig. 7a),being rated at approximately 0. To establish the benefits of additional working, a piece of the initial billet stock (8 in. round) was reduced approximately 78% by forging at 1900°F. The reforged material was then heat treated as before. Microstructural examination revealed a reduced ASTM grain size of 3-5. The resulting tensile properties are given in Table V. Note the increase in ductility that accompanies the reduced grain size. Limited data are also available on the tensile properties of the 3-inch wall, 11-3/4 inch O.D. hollow cylindrical extrusion that had undergone the 83% reduction at 1700°F which resulted in ASTM grain size of 3-5 (Figure 7b). As shown in Table V, the increased ductility is indicated. Therefore, as suggested by Bohanek(10) continuous hot working at sufficient reductions will reduce the grain size of large section products to a level where improved ductility is realized. However, the lower grain size does not appear to improve fracture toughness. In fact, associated with the reduced grain size obtained by the increased amount of working at 1900°F was a slightly lower fracture toughness as determined by W/A. The W/A values obtained for the larger grain size material extruded at 1700°F averaged 444 in-lb/in^2 while values averaging 330 in-lb/in^2 were obtained for the smaller grain size material processed at 1900°F. Both materials were from the same heat and were given the same aging heat treatment.

The observations emphasize the additional work that is required to detail the micromechanics of how fracture toughness of Ti-8Mo-8V-2Fe-3Al and other beta alloys is related to tensile ductility and grain size. Greenfield and Margolin(11) have shown how high fracture toughness can be associated with low tensile ductility in Ti-5.25Al-5.5V-0.9Cu-0.5Fe α/β alloy. Here, the grain boundary α that provides resistance to intergranular propagation of an existing crack also serves as a site for nucleation and growth of voids formed under tensile loading. However, in coarse grained Ti-8Mo-8V-2Fe-3Al the crack propagation is transgranular with no evidence of grain boundary α as shown in Fig. 8.

Fig. 7. Ti-8Mo-8V-2Fe-3Al grain size resulting from (a) 30% reduction shown in the aged condition and (b) 83% reduction shown in ST condition. Mag. 100X.

Table V. EFFECT OF GRAIN SIZE ON TENSILE PROPERTIES OF AGED Ti-8Mo-8V-2Fe-3Al*

ASTM Grain Size No.	Hot Working		0.2% Yield Strength, ksi	Tensile Strength, ksi	Elongation, %	RA %
	Method	Reduction				
3-5	Extrusion	83%	174.5+	182.5	4.0	9.3
3-5	Forging	78%	181.0	188.0	8.5	9.9
0	Extrusion	30%	177.7†	180.8	1.5	3.3

*1475°F (WQ) - 1000°F (8 hr).
+Transverse orientation.
†Represents average of OD, ID and MW locations given in Table III for longitudinal orientation.

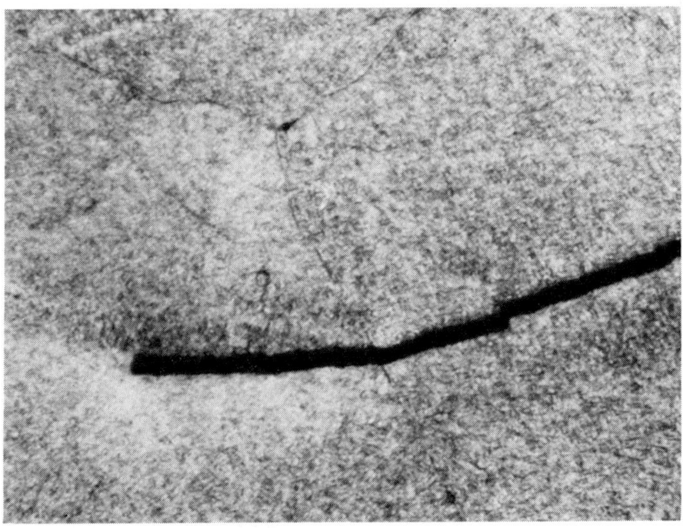

Fig. 8. Transgranular nature of crack propagation in large grain Ti-8Mo-8V-2Fe-3Al. Mag. 1000X.

Conclusions

As a result of this study which has examined the fracture toughness of three commercially available alloys (Ti-6Al-4V, Ti-6-Al-6V-2Sn and Ti-8Mo-8V-2Fe-3Al) in heavy section configuration, the following conclusions can be drawn:

1. Primary α morphology can account for the difference in fracture toughness of Ti-6Al-4V heat treated at standard solution and aging temperatures. For a given amount of primary α in Ti-6Al-4V with a commercial quantity of oxygen, equiaxed microstructure results in increased fracture toughness when compared to plate-like microstructure. Decreasing the oxygen level and increasing the amount of transformed beta matrix increase fracture toughness.

2. The Ti-6Al-6V-2Sn in the STOA conditions at approximately 143 KSI yield strength level exhibits excellent fracture toughness exceeding that of any alloy considered in the study.

3. Microsegregation in the form of "beta flecks" in Ti-6Al-6V-2Sn modified with 2.0% Zr appears to degrade fracture toughness.

4. At the high strength levels (approximately 180 KSI yield strength and above), the fracture toughness of Ti-8Mo-8V-2Fe-3Al exceeds that of modified Ti-6Al-6V-2Sn (STA condition) despite low tensile ductility. The low tensile ductility of Ti-8Mo-8V-2Fe-3Al is associated with a large beta grain size and can be improved by increasing the amount of hot work prior to heat treatment. However, decreasing beta grain size appears to lower fracture toughness as measured by W/A.

References

1. Broadwell, R.G., and Coyne, J.E., "The Fracture Toughness-Tensile Property Relationships of Deep Hardenable Titanium Alloys", Spring Meeting AIME, Las Vegas, Nevada, May 11-14, 1970.

2. Gerberich, W.A., and Baker, G.S., "Toughness of Two-Phase 6Al-4V Titanium Microstructures", ASTM STP 432, American Society for Testing and Materials, 1968, pp. 80-99.

3. Hieronymus, W.S., "B-1 Designed to New Standards", Aviation Week and Space Technology, July 26, 1971, pp. 42-45; see also Wood, R.A., "Titanium and Titanium Alloys", DMIC Review of Recent Developments, August 29, 1971.

4. Frederick, S.F., and Hanna, W.D., "Fracture Toughness and Deformation of Titanium Alloys at Low Temperatures", Metallurgical Transactions, Vol. 1, February 1970, pp. 347-352.

5. Hall, J.A., and Pierce, C.M., "Improved Properties of Ti-6Al-6V-2Sn Through Microstructure Modification", AFML-TR-76-312, February 1971.

6. DeSisto, T.S., and Hickey, C.F., Jr., "Low Temperature Mechanical Properties and Fracture Toughness of Ti-6Al-6V-2Sn", Army Materials and Mechanics Research Center, AMRA-TR-65-17, July 1965.

7. Heil, W.H., Titanium Metals Corporation of America, Private Communication.

8. Harmsworth, C.L., "Low Cycle Fatigue Evaluation of Ti-6Al-6V-2Sn and 300-M Steel for Landing Gear Applications", AFML-TR-69-48, June 1969.

9. Heil, W.H., "New Development in High Strength Titanium Alloys", New York University Titanium Metallurgy Course, September 8-11, 1969.

10. Bohanek, E., "Evaluation of Several Commercial Heats of Ti-8Mo-8V-2Fe-3Al", Titanium Metals Corporation of America, Technical Report No. 45, August 20, 1970.

11. Greenfield, M.A., and Margolin, H., "The Interrelationship of Fracture Toughness and Microstructure in a Ti-5.25Al-5.5V-0.9Fe-0.5Cu Alloy", Metallurgical Transactions, Vol. 2, March 1971, pp. 841-847.

A CHARACTERIZATION OF THE FRACTURE RESISTANCE OF THICK-SECTION TITANIUM ALLOYS

R. W. Judy, Jr., C. N. Freed and R. J. Goode

Naval Research Laboratory, Washington, D. C. 20390

Procedures for design of structures to preclude catastrophic fracture extension must be predicated on accurate fracture resistance characterizations of structural material that can only be provided by rational test procedures. For titanium alloys, there are two rational fracture resistance test procedures that apply - K_{Ic} tests for brittle materials and Dynamic Tear (DT) tests which apply to the entire range of strength and fracture resistance.

The Ratio Analysis Diagram (RAD) is a framework for determining the significance of fracture resistance data in terms of stress level and crack size required for fracture. The RAD is derived from basic linear elastic fracture concepts involving K_{Ic} and material yield strength; indexing of the RAD is accomplished from scales of K_{Ic} and DT energy. Full-thickness K_{Ic} tests and 1-in. DT tests were conducted for several 3-in. thick titanium alloy plates to confirm the relationship between the two parameters and to locate the RAD scales for thick sections. It was shown that K_{Ic} and DT could be used for indexing to the RAD with equal accuracy. Moreover, the DT test accurately predicted that several of the titanium alloys would not satisfy standard ASTM requirements for valid test results. Therefore, the results of the very expensive K_{Ic} test can be predicted by the far less costly DT test for the full range of titanium alloys.

Introduction

Safe use of high-strength titanium alloys in structures must be predicated on a complete knowledge of both the materials' inherent resistance to fracture extension and the particular requirements of the structural application. A diagrammatical procedure, called the Ratio Analysis Diagram (RAD), for interpretations of fracture resistance properties of structural metals, has existed for several years. The RAD was developed for steels (1), aluminum alloys (2), and titanium alloys (3) in section sizes up to 3-in. thick.

The RAD combines metal quality and mechanical constraint aspects of fracture to provide analyses of conditions for fracture extension in terms of parameters measured by rational test procedures. Rational test methods are required to determine a metal's intrinsic resistance to fracture; such tests impose the condition of maximum crack-tip constraint for the section size. There are a number of test methods for measuring fracture properties; however, at present there are only two rational test methods that can be applied for titanium alloys. Linear-elastic fracture mechanics technology is based on use of the critical crack tip stress-intensity factor, K_{Ic}, to describe the fracture resistance properties of brittle materials. The quantitative aspects of the RAD are derived from linear-elastic calculations of relations between critical defect or flaw sizes and applied stress levels, which are expressed in terms of the ratio of K_{Ic} to material yield strength (σ_{ys}). Linear-elastic methods are not suitable for definition of fracture properties of semi-ductile or ductile metals, i.e., metals that require high elastic or over-yield stresses for fracture. The Dynamic Tear (DT) test was developed to characterize the fracture resistance properties of metals over the entire range of strength and fracture resistance (plane strain, elastic-plastic and plastic). In the DT test the energy required to fracture a standard specimen indicates the level of fracture resistance.

For brittle materials, values measured in K_{Ic} tests and in DT tests have a definite relationship, because both tests are rational methods which characterize a single property of the materials. Thus, both the K_{Ic} and DT energy provide an entry to the RAD for the purpose of assessment of fracture safety of structures. It must be recognized that a limit of applicability for linear-elastic principles is given in terms of the section size, or material thickness. For titanium alloys, the RAD was based on K_{Ic} and DT test results from plate materials of 1-in. thickness, with reasonable extrapolations to include thicker sections up to 3 in. It is necessary to validate these early predictions by extending the experiments to include higher values of K_{Ic} and DT energy from tests involving thicker sections. This report describes the results of K_{Ic} and DT tests on 3-in.-thick titanium alloys for the purpose of corroborating the capabilities of the RAD in this size range.

Ratio Analysis Diagram for Titanium Alloys

The RAD provides a format for analysis of the fracture resistance of titanium alloys over the entire range of strength and fracture resistance levels. As a group, titanium alloys have a continuous range of yield strength and fracture resistance levels from very low to very high. To understand the significance of fracture resistance measurements, it is necessary to divide the alloys into three general groupings - brittle, plastic, and a transition region called elastic-plastic - and to analyze each group separately.

Brittle materials are those that fracture at elastic nominal stress levels; the technology of linear-elastic fracture mechanics applies only to materials of this type. The unique characteristic of brittle materials is that unstable crack propagation follows any crack initiation event; because of this, use of these materials concerns only prevention of crack initiation. Equations relating crack size and shape, applied nominal stress level, and K_{Ic} exist for several different geometrical configurations. One of the most often used equations for a part-through surface flaw is presented in graphical form, Fig. 1, for engineering use. This chart relates critical flaw depths at different applied stress levels to the ratio of K_{Ic}/σ_{ys}. It is important to note that in this and all other fracture mechanics calculations, critical flaw size is related directly to this ratio. Therefore, the K_{Ic}/σ_{ys} ratio provides an index of merit with respect to fracture properties.

It must also be cautioned that the diagram of Fig. 1 does not apply directly to all cases - only that of brittle materials. In strict terms, brittle materials are those that meet the criterion $B \geq 2.5\ (K_{Ic}/\sigma_{ys})^2$; this specification was established by the ASTM E-24 Committee on fracture as the upper-limit of application of linear-elastic fracture mechanics (5). A reference limit, called the Plane Strain Limit, separating brittle

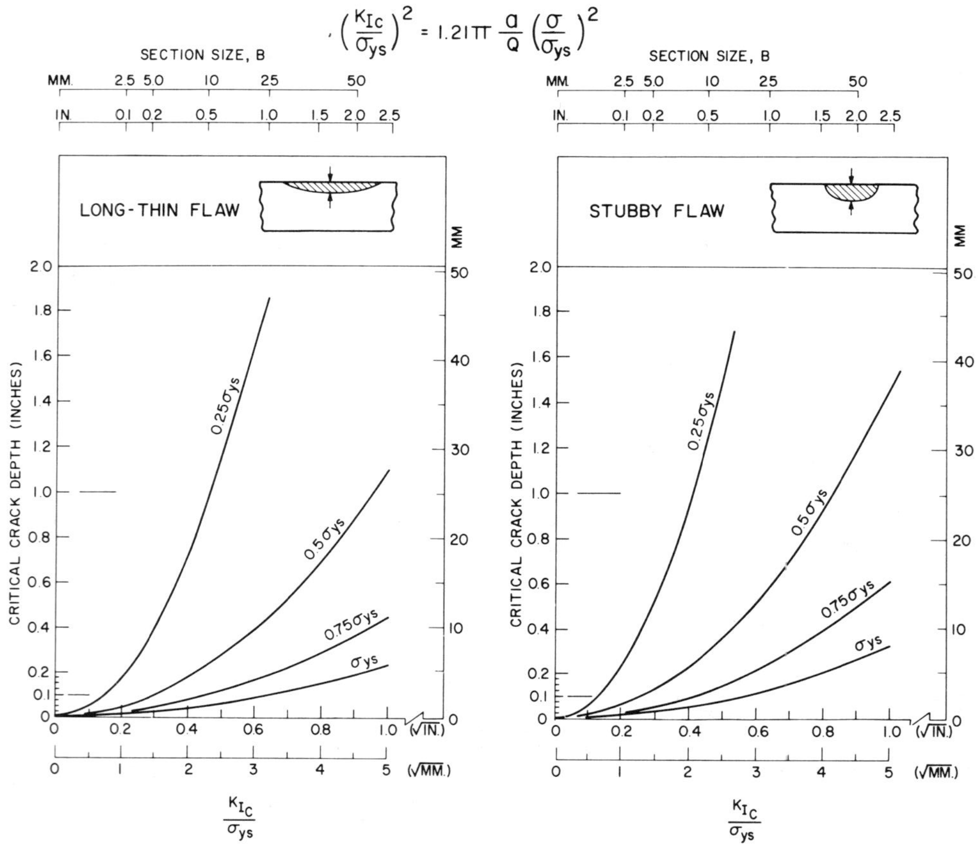

Fig. 1. Surface flaw equation in engineering diagram form for determination of critical flaw sizes for fracture of brittle metals. Note that critical flaw size depends on the ratio K_{Ic}/σ_{ys}.

materials from elastic-plastic materials, is established by this specification as a function of material thickness.

The elastic-plastic case is that of fracture at medium-to-high elastic stress levels with sufficient crack-tip plasticity to preclude use of strictly elastic analysis procedures. Elastic-plastic criteria involve through-thickness cracks that are large compared to the thickness. A through-thickness crack is illustrated in Fig. 2, along with the appropriate equation for linear-elastic flaw size calculations and an engineering diagram depicting the use of the equation for thick sections. Figure 2 shows that on an increasing fracture resistance (K_{Ic}/σ_{ys}) scale, the elastic-plastic regime is the transition between the plane strain state and the plastic fracture state. The elastic-plastic regime for any given thickness of material is bounded by the plane strain limit and by the general yield limit, which is given by $B \geq 1.0\ (K_{Ic}/\sigma_{ys})^2$ (6) where K_{Ic} is referenced by other test methods. The general-yield limit as defined is the ratio value where nominal stresses in excess of yield are necessary to cause fracture extension for flaws of reasonable size, say of length 3 times thickness (3T). The important point is that the fracture stress rises dramatically from 0.3 σ_{ys} at the plane strain limit to σ_{ys} in the narrow elastic-plastic range. This transition effect is relatively insensitive to material thickness factors and is independent of changes in flaw length, i.e., flaw sizes in the range of 2T to 6T do not significantly alter the intercepts on the plane strain limit and yield strength lines.

The third group is ductile or plastic materials, which are those that require nominal stresses over yield to cause fracture extension. The basic characterization tool for plastic materials is the fracture extension resistance curve (R curve) defined by DT test methods (7). No general relations of critical defect sizes and applied stress levels exist because the plastic fracture case is geometry dependent. For each geometrical configuration, different failure criteria and, accordingly, different materials requirements exist and problems in design with plastic materials must be solved on a case by case basis.

In the RAD, Fig. 3, all these factors are combined into a single "plotting board" for interpretation of fracture principles from laboratory test results. The RAD framework is formed from the scales of yield strength vs K_{Ic} and DT energy. The most prominent features of the RAD are the limit lines and the system of lines of constant K_{Ic}/σ_{ys} ratio. The Technological Limit (TL) line

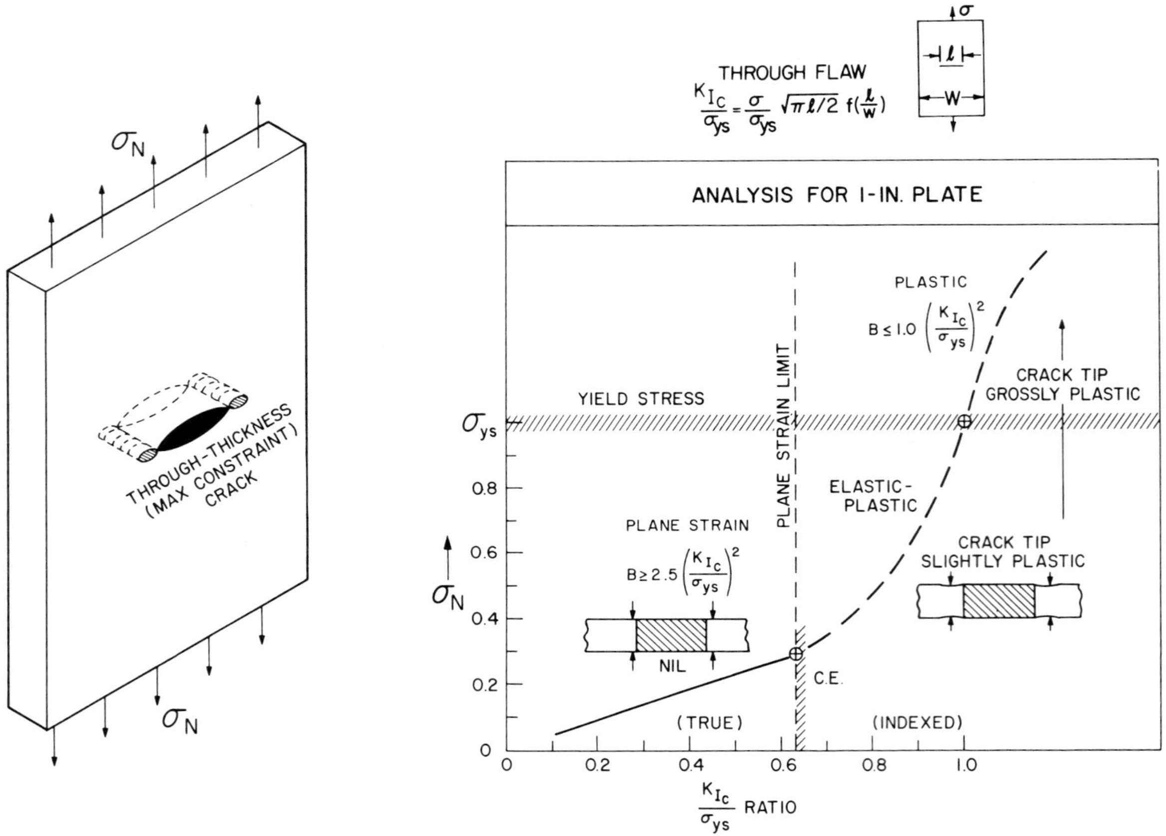

Fig. 2. Illustration of significance of elastic-plastic transition from plane-strain to plastic fracture states. Fracture stress for a 3T through flaw is given by the curve. The solid part is given by the equation of the top, while the dashed part connects the plane strain limit to the yield point.

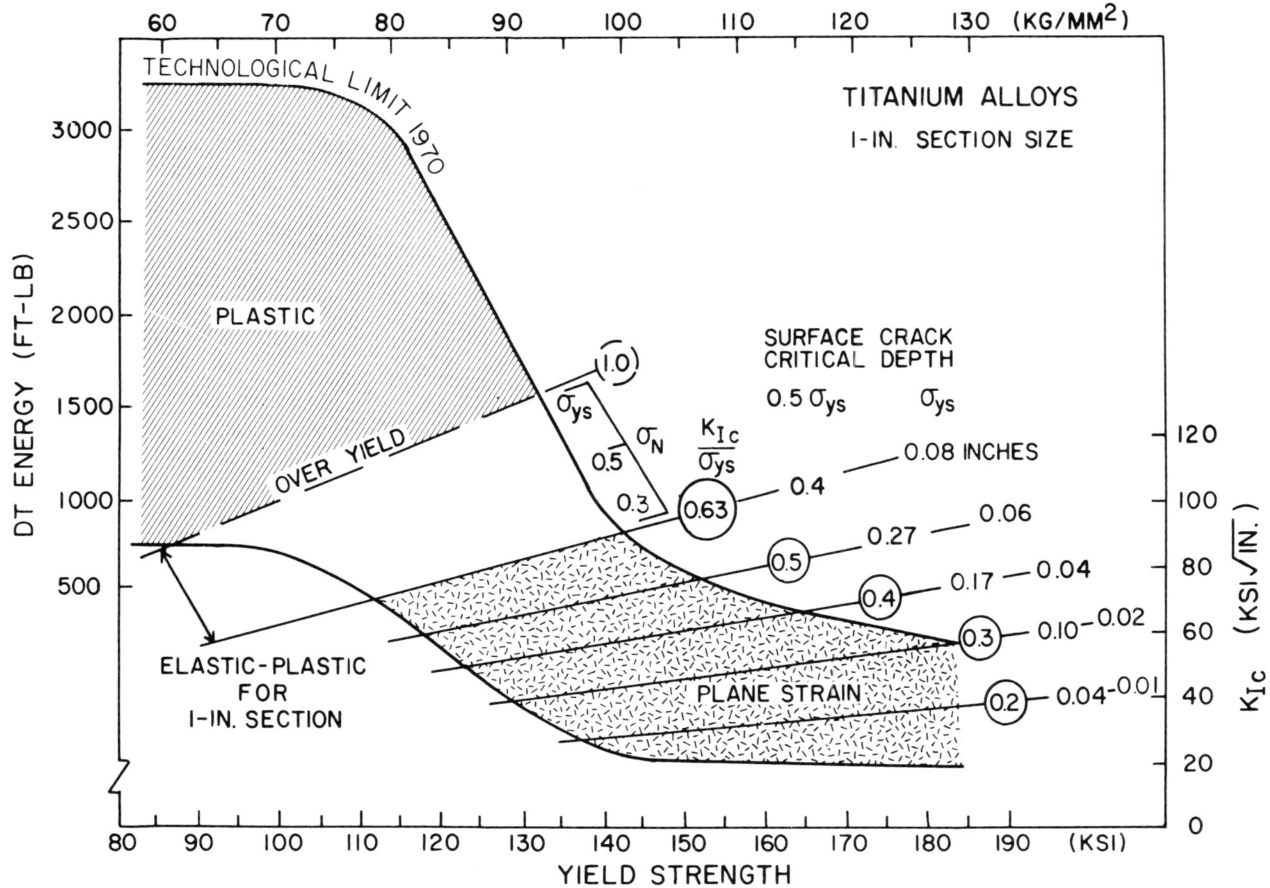

Fig. 3. Framework of the ratio analysis diagram for titanium alloys for 1 in. thickness.

represents the highest values of fracture resistance measured to date by either K_{Ic} tests or DT tests over the entire yield strength range; the Lower Bound represents the lowest levels of fracture resistance. A reference to the critical flaw size charts, Figs. 1 and 2, is provided by the system of ratio lines constructed from the scales of K_{Ic} and σ_{ys}. As an example, critical sizes for a long, thin surface flaw for half-yield and full-yield loading conditions are shown on the RAD for each ratio line.

The ratio lines also serve to divide the diagram into regions of expected plastic, brittle, and elastic-plastic fracture behavior for given material thicknesses. The separations are determined according to thickness as shown for 1-in. section size in Fig. 3. The critical edge between brittle behavior and elastic-plastic behavior is the plane strain limit. The boundary between the elastic-plastic and ductile regimes is the general-yield limit. The division of the RAD into three regions provides an engineering index of the fracture state and thereby serves to indicate the type of more detailed design approach required for each case.

The most important feature of the RAD is that both DT and K_{Ic} data can be indexed to the system of ratio lines. There is a definite advantage to using the DT test for indexing to the RAD; compared to K_{Ic} tests, DT tests are relatively inexpensive in the costs of specimen preparation, test procedures, and in the amounts of materials required. An additional advantage to using DT tests is that the entire range of materials from brittle to ductile can be characterized by DT test procedures, while use of K_{Ic} is limited to the brittle materials.

Materials and Test Procedures

The wide variety of near alpha and alpha-beta titanium alloys in 3-in. rolled plate form listed in Tables 1 and 2 were included in this investigation. K_{Ic} tests, 1-in.-thick DT tests (8), and tensile tests were conducted for each alloy. Standard .505 in. dia. tensile and standard 1-in. DT test specimens were machined from surface and plate center locations, while K_{Ic} test specimens included the full plate thickness. Dimensions of the K_{Ic} and DT specimens are shown in Fig. 4. All specimens were oriented in the "WR" or weak fracture direction with respect to anisotropy due to rolling. DT tests were conducted at 30°F, and all other tests were at room temperature.

TABLE I

Mechanical Properties of Three-Inch Thick Titanium Alloy Plates - Plane Strain

Nominal Composition	FRACTURE RESISTANCE 1-in. DT Energy (ft-lbs) Surface	Center	Avg.	K_{Ic} Avg. (ksi√in.)	K_{Ic} Range (ksi√in.)	No. of Specimens	TENSILE YIELD STRENGTH (ksi) Surface	Center	Avg.
Ti-6Al-4V-2Mo	472	714	593	69	66-71	3	151.7	139.9	147.8
Ti-6Al-4V	587	699	643	89	88-91	3	136.1	130.4	134.2
Ti-6Al-4V	986	1120	1053	100	96-103	4	112.7	112.6	112.7
Ti-6Al-4V	836	1205	1020	101	99-105	4	125.6	122.2	124.4
Ti-6Al-2Mo	1166	1320	1243	110	107-114	2	113.3	115.0	113.9
Ti-6Al-4V-1Mo	1033	1349	1191	117	111-118	4	113.1	112.1	112.8
Ti-5Al-2V-2Mo-2Sn	1250	1330	1267	117	116-118	4	115.4	111.3	114.0
Ti-6Al-2Cb-1Ta-1.2Mo	1352	1534	1443	123	115-131	3	114.8	114.5	114.7

TABLE 2

Mechanical Properties of Three-Inch Thick Titanium Alloys Elastic-Plastic or Plastic

Nominal Composition	Avg. 1-In. DT Energy (ft-lb)	Avg. Yield Strength (ksi)
Ti-6Al-2Mo	1268	109.5
Ti-7Al-1Mo-1V	1455	107.7
Ti-6Al-2Mo	1455	119.3
Ti-6Al-2Cb-1Ta-0.8Mo	1517	107.4
Ti-6Al-2Cb-1Ta-1.2Mo	1706	107.6
Ti-6Al-2Mo	1706	108.1
Ti-7Al-2.5Mo	2303	109.4

Specimen preparation procedures, test procedures, and test record analyses for the K_{Ic} tests conformed to ASTM Specification E-399-70T. In many cases, materials in the mill-annealed condition had fracture resistance levels sufficiently high that the thickness specification ($B \geq 2.5\ [K_{Ic}/\sigma_{ys}]^2$) could not be satisfied; the results of these tests are not reported. To overcome this, it was necessary to heat-treat the plates to higher yield strength levels so that valid test results could be obtained. In these cases all specimens of a given material were heat-treated at the same time.

Discussion of Results

The results of the tests are presented in Tables 1 and 2. Table 1 contains the results of K_{Ic}, DT, and tensile tests for materials where K_{Ic} could be measured. Table 2 contains DT and tensile test results for those alloys which failed the thickness requirement. The characteristic K_{Ic} values reported in Table 1 were determined by averaging the results of 2-4 specimens; the range of K_{Ic} values is also indicated. Tensile and DT test results in Tables 1 and 2 reflect an average of two specimens at each location. Some significant gradients in fracture resistance properties through the thickness are apparent from DT test results, and gradients in yield strength are also evident to a lesser degree.

The relation of DT energy and K_{Ic} is presented in Fig. 5. A slight downward adjustment of the locus from that extrapolated from 1-in. plate (4) studies was necessary. Previously determined 1-in. plate data are included in Fig. 5 to give a full range of values. The line drawn through the points was determined by a least-squares procedure. This line determines the matching of the DT and K_{Ic} scales on the RAD; the non-linearity of the curve is reflected in the location of the scales on the RAD. The fit of the data in Fig. 5 is very good when all the variables are considered. Variability of properties from one point to another in a given plate and the observed gradients in fracture properties through the thickness, as well as normal experimental variations, can account for much scatter.

The significance of the relation between DT energy and K_{Ic} can only be realized by comparing these two indices of fracture resistance on the RAD (Fig. 6). The plane strain limit and general yield limit were determined for 3-in. section size in this figure. The closeness of K_{Ic} and DT values (circles vs triangles) plotted at the average yield strength for each material underscores the accuracy of either method for indexing to the RAD. Note that both values give essentially the same indication of fracture resistance in all cases, as derived from the ratio lines. The flaw sizes that are expected to be critical for given stress levels can be determined for each material by reference to Fig. 1.

The primary difference between the two test methods is a matter of cost. A K_{Ic} test costs approximately $100 or more; a DT test might cost $10. An even more significant factor emerges from the DT data for materials where K_{Ic} could not be measured in a valid manner. The "x" symbols in Fig. 6 represent average DT energy values for seven materials for which K_{Ic} tests were conducted but the fracture resistance was high enough that the ASTM specifications could not be satisfied, i.e., $B < 2.5\ (K_{Ic}/\sigma_{ys})^2$. (See Table 2.) Such values, termed "K_Q", are not reported because, as invalid data, they have no capability for defining the fracture resistance of the test materials. In Fig. 6 the data points predict that five of the seven tests would definitely be invalid and that the other two would be very near the plane strain limit line for 3-in. plate ($K_{Ic}/\sigma_{ys} = 1.1$). It should

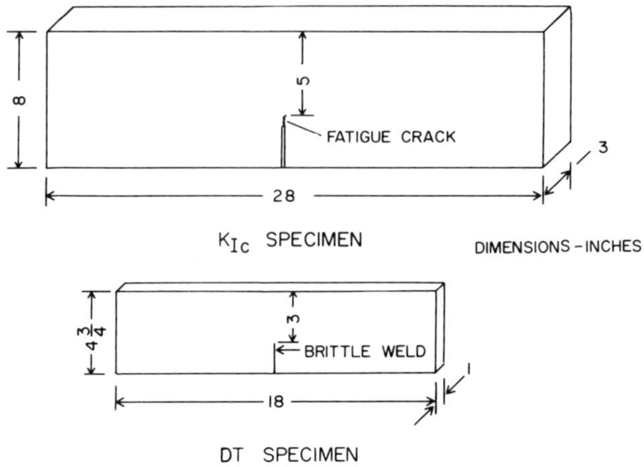

Fig. 4. Schematic of K_{Ic} and dynamic tear test specimens.

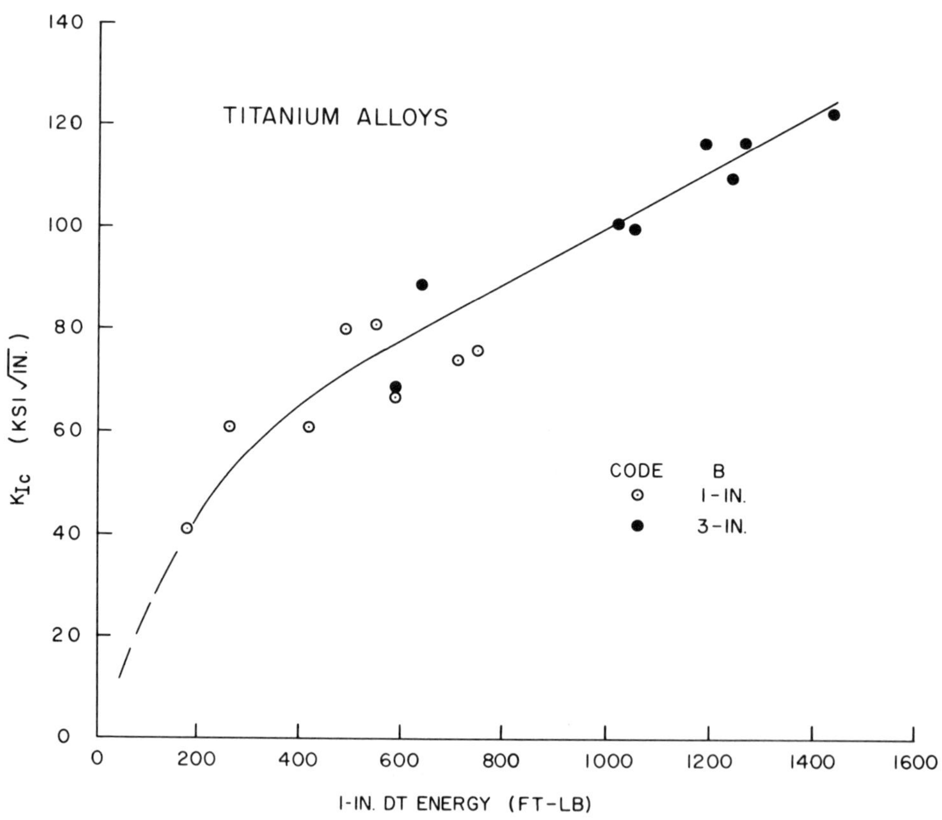

Fig. 5. Experimental correlation of DT energy and K_{Ic} for 1-in. and and 3-in. section sizes.

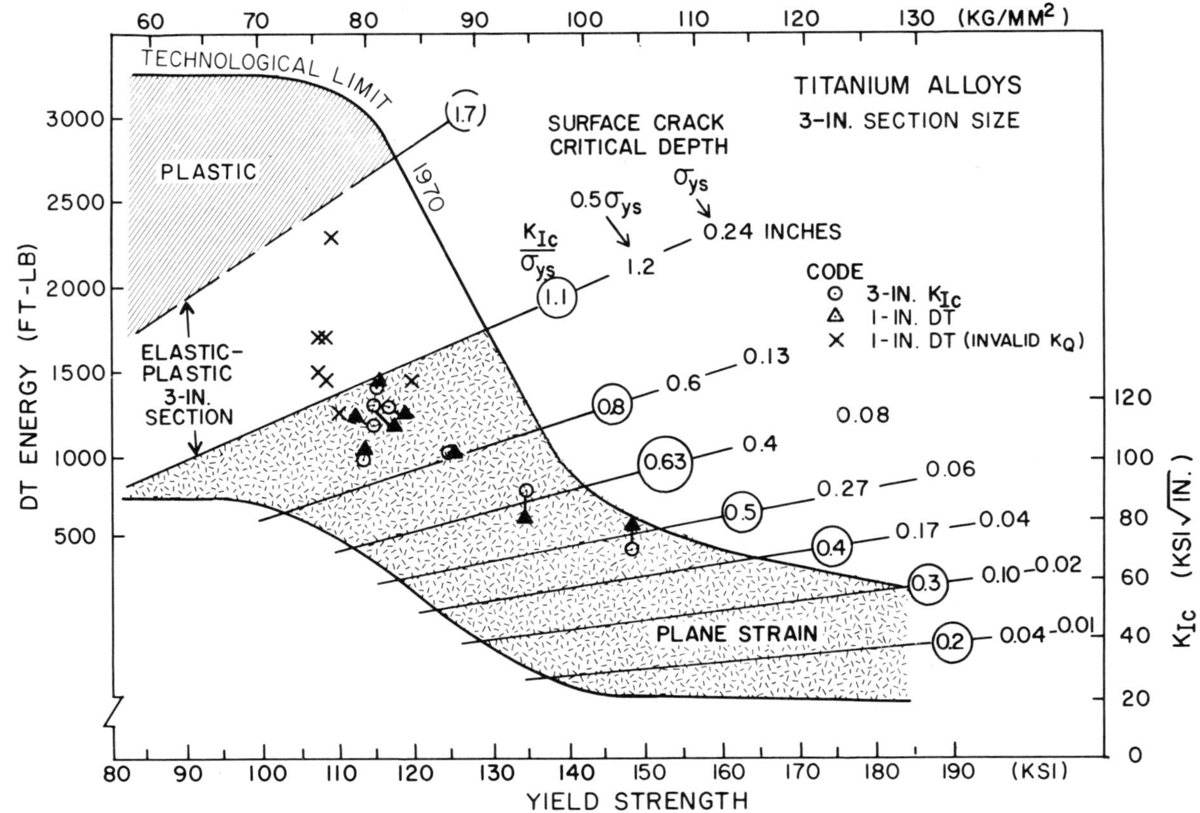

Fig. 6. Illustrating that K_{Ic} and DT energy give the same indication of fracture resistance on the RAD. Points coded "X" are 1 in. DT energy values for materials for which K_{Ic} could not measured.

be noted that the behavior predicted for materials very near the line on either side are not greatly different, particularly for thick sections. Therefore, predictions of valid K_{Ic} properties by the two "x" points just below the ratio 1.1 line, when K_{Ic} could not be measured, are not significant discrepancies. In any event, the values fall within the scatter of the correlation plot.

Use of the titanium RAD is illustrated in Fig. 7, which summarizes data for 1-in. thick titanium alloys. It has been known for several years that a reduction of interstitial oxygen content to low levels is necessary to attain plastic fracture properties in titanium alloys in heavy sections (9). The dashed line across the RAD separates low oxygen materials (0.08 wt. percent or less) from commercial grade alloys and depicts the capabilities of each type as a function of yield strength. Since oxygen is a strengthening agent, low oxygen materials are not found above 150 ksi σ_{ys} values. Conventional titanium alloys produced to commercial-practice oxygen levels and even ELI (Extra-Low Interstitial) grades occupy the two boxes in the plane strain region of the RAD. To climb from conventional materials to the plastic fracture region of the RAD, it is necessary to reduce both oxygen content and yield strength. In addition, special processing of commercially produced alloys, such as Ti-6Al-4V, is necessary to optimize fracture resistance and yield strength combinations typified by TL line quality materials. Most of the 3-in. plates of this study had low oxygen levels but lacked the special processing necessary to maximize fracture resistance. Whether the properties of 1-in. plates can be duplicated in 3-in. plate is not known at the present time.

Summary

Design of structures using titanium alloys requires a knowledge of the fracture properties of the materials in relation to the imposed loadings to preclude catastrophic failures. A systematic method for making such assessments is provided by the Ratio Analysis Diagram, which is formulated from linear elastic calculations. Both K_{Ic} and DT test data can be entered on the RAD for indexing purposes. Therefore, it is necessary to have a good knowledge of the relation between these two independent parameters.

Fracture resistance tests of 3-in. thick plate materials showed that the correlation between the DT energy and K_{Ic} is very good. Combining the data from

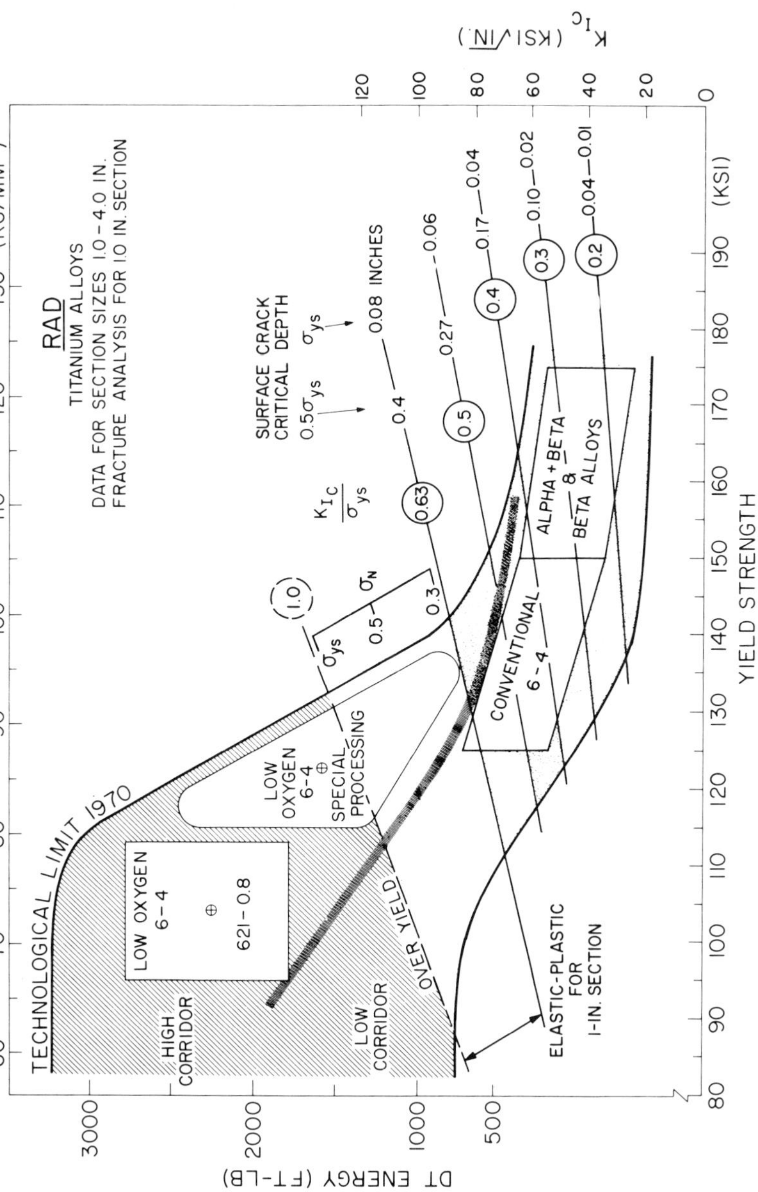

Fig. 7. RAD for titanium alloys, as prepared for trade-off analyses involving plate of 1.0 in. section size. The metallurgical zoning provides for metal selection, depending on structural requirements. For example, at 120 to 130 ksi yield strength a choice may be made for a specific alloy, featuring two levels of oxygen content. The properties are vastly different - plane strain versus plastic fracture states.

tests of 3-in. thick materials with data from earlier studies of 1-in. thick materials permitted validation of the RAD interpretations for section sizes up to 3-in. thick. Comparing the data (K_{Ic} and DT energy) on the RAD showed that they are equivalent and that either method could be used for indexing to the RAD; however, because of the much lower cost of DT tests compared to the K_{Ic} tests, DT tests are preferred for general use. Most importantly, it was shown that the DT test could be used to predict when K_{Ic} tests would be invalid, thereby saving expensive tests that would not have any value.

References

1. W. S. Pellini, "Advances in Fracture Toughness Characterization Procedures and in Quantitative Interpretations to Fracture-Safe Design for Structural Steels," NRL Report 6713, April 3, 1968; also Welding Research Council Bulletin 130, May 1968

2. R. W. Judy, Jr., R. J. Goode and C. N. Freed, "Fracture Toughness Characterization Procedures and Interpretations to Fracture Safe Design for Structural Aluminum Alloys," NRL Report 6871, Welding Research Council Bulletin 140, May 1969

3. R. J. Goode, R. W. Judy, Jr. and R. W. Huber, "Procedures for Fracture Toughness Characterization and Interpretations to Failure-Safe Design for Structural Titanium Alloys," NRL Report 6779, 5 December 1968; Welding Research Council Bulletin 134, October 1968

4. C. N. Freed and R. J. Goode, "Correlation of Two Fracture Toughness Tests for Titanium and Ferrous Alloys," NRL Report 6740, 16 January 1969

5. "Tentative Method of Test for Plane Strain Fracture Toughness of Metallic Materials," ASTM Designation E399-70T, Part 31 of ASTM Standards, July 1970

6. W. S. Pellini, "Criteria for Fracture Control Plans," NRL Report (pending publication)

7. R. W. Judy, Jr. and R. J. Goode, "Fracture Extension Resistance (R-Curve) Concepts for Fracture-Safe Design with Nonfrangible Titanium Alloys," to be presented at the 2nd International Conference on Titanium, 2-5 May 1972, Massachusetts Institute of Technology, Cambridge, Massachusetts

8. P. P. Puzak and E. A. Lange, "Standard Method for the 1-Inch Dynamic Tear Test," NRL Report 6851, 13 February 1969

9. R. J. Goode and R. W. Judy, Jr., "Fracture-Safe Design of Aluminum and Titanium Alloy Structures," NRL Report 7281, (pending publication)

PLANE STRESS FRACTURE RESISTANCE OF HIGH STRENGTH TITANIUM ALLOY SHEET

A. M. Sullivan, J. Stoop, and C. N. Freed

Naval Research Laboratory, Washington, D. C. 20390

Introduction

The demand for weight-sensitive structures requires the use of high-strength materials. Because of the favorable strength-to-weight ratio, the titanium alloys are increasingly favored. However, coupled with high strength is lowered fracture resistance. This property is a measure of the ability of a material to withstand the deleterious effects of cracks, flaws, or notches while under stress. A low fracture resistance will permit the occurrence of catastrophic fracture at nominally elastic stress levels in the presence of such discontinuities.

Recognition that materials and structures will contain cracks or flaws and that these defects will impair load-bearing capacity requires definition of the degree of impairment. To insure fail-safe structures, fracture resistance must be assessed in some form useful to the designer. Such an assessment can be provided through the basic parameters of linear-elastic fracture mechanics (LEFM) which define load and or crack length for failure according to the infinite sheet equation

$$K_c = \sigma\sqrt{\pi a} \quad (1)$$

The fact that geometric variables attendant upon the measurement of fracture resistance were minimized under conditions of plane strain (thick-section material) led initial research emphasis in this direction and a standard test method has been accepted (1). No such standard has, however, been adopted for testing thin sheet (essentially under conditions of plane stress) although the underlying principle is the same. To assist in the development of such a standard, it has seemed expedient to concentrate on the center cracked tension sheet specimen (CCT). Not only is it a natural structural prototype precursor, but the stress analysis of this specimen is well documented (2).

Experimental Parameters

Materials

The fracture resistance of four structural titanium sheet alloys has been investigated. Mechanical property data, fracture resistance values, and heat treatments are presented in Table I. All test specimens were so oriented that the path of the fracture was parallel to the rolling direction of the sheet since, if anisotropy is present, this will be the direction less resistant to fracture. It is evident that the level of fracture resistance in the weaker direction should categorize materials inasmuch as the direction of operating loads cannot be assured.

Test Procedure

The configuration of the CCT specimen is shown in Fig. 1. The central slit is produced by electric discharge (Elox) to give a .063-in.-wide slit with slit tip radii typically .003-.006 in. (designated "blunt"). To assess the effect of tip radius size on the K_c value, companion specimens had slits sharpened by a second Elox operation which extended the slit .060 in. at each tip by a fine electrode developing an extended slit width of .002 in. and a tip radius of .001 in.

When the specimen is loaded in tension, load and central crack-opening-displacement (COD) are simultaneously graphed by a Moseley X-Y recorder until failure occurs. Crack opening is measured by a displacement probe instrumented with strain gages which is positioned in a circular hole in the center of the initial slit. The COD measurement is referred to a normalized experimental calibration curve which relates the amount of crack opening to the instantaneous crack length in the specimen. Full details of this procedure have been published (3). These values, load (proportional

TABLE I

Titanium Alloy	Sheet Thickness (in.)	YS 0.2% Offset (ksi)	TS (ksi)	Elong. (%)	Re. in Area (%)	K_c Blunt (ksi$\sqrt{in}$)	K_c Sharp (ksi$\sqrt{in}$)	Heat Treatment
6Al-4V	.032	-	135.9	-	-	51.5	53.5	Solution Anneal:
	.063	151.1	156.3	1.0	2.0	74.4	64.2	1780°F, 20 Min.
	.090	158.5	168.9	3.0	-	97.6	77.5	Water Spray Quench
	.125	146.0	158.2	4.2	7.5	90.4	-	Age: 975°F, 8 hr., Aircool
4Al-3Mo-1V	.042	161.5	167.0	.5	.3	-	50.2	Solution Anneal:
	.058	153.4	174.5	.75	1.8	-	54.8	1650°F, 15 Min.
	.087	152.9	172.0	.5	1.1	64.0	62.5	Water Spray Quench
	.124	159.8	172.1	1.0	1.0	64.5	35.0	Age: 925°F, 14 hr., Aircool
16V-2.5Al	.041	181.6	190.5	1.5	3.9	56.6	52.4	Solution Anneal:
	.059	176.5	187.6	4.7	9.5	65.2	46.3	1380°F, 15 Min.
	.118	181.7	192.0	4.0	9.5	51.2	44.4	Water Spray Quench
								Age: 960°F, 4 hrs., Aircool
13V-11Cr-3Al	.040	198.1	199.4	0	2.9	55.8	34.1	Age: 900°F, 72 hrs., Aircool
	.063	207.3	220.9	2.7	6.3	41.5	38.9	
	.090	200.6	203.5	0	-	52.6	29.8	
	.125	216.5	227.0	2.2	2.1	35.7	23.2	

to gross stress, σ_G) and crack length, 2a, are required for the calculation of fracture resistance as K_c where the appropriate equation

$$K_c = \sigma_c \sqrt{\pi a_c} \; f \; 2a/W \tag{2}$$

is utilized. It has been established that, for a range of 2a/W values between 0 and .6, the following expression is accurate to within 1 percent.

$$f \; 2a/W = 1.77 \left[1 - .1 \; (2a/W) + (2a/W)^2\right] \tag{3}$$

In this technique determination of crack length from COD measurement delineates the "effective" crack length (actual crack plus plastic zone) so the plastic zone correction factor is not used in the calculation of K_c.

Geometrical Dependencies of K_c

During the early development of fracture mechanics, inasmuch as specimen dimensions were incorporated in the basic equations, it was tacitly assumed that these would be self-normalizing. Theoretically then, the K_c value could be determined using small specimens since K_c would not be geometry dependent. Implicit also was the assumption that the value obtained would reflect a plane stress situation. Soon, however, it became obvious that specimen dimensions did influence the fracture resistance value. The value of K_c was found to vary with specimen width, crack length to width ratio, sheet thickness and crack tip radius. Previous work (4, 5, 6, 7) has indicated that, since the influence of some of these dependencies can be predicted, the employment of judicious specimen design will so minimize their effect upon the fracture resistance value that its determination can be effected by routine testing procedures.

Effect of Width

Because sufficient constraint must be provided at the crack tip for general elastic conditions to prevail, the specimen width requirement will depend upon the K_c to yield stress ratio. Calculations (3) indicate a minimum such that

$$W \min = \frac{27}{2\pi} \left(\frac{K}{\sigma_{ys}}\right) \tag{4}$$

rounded values are shown in Fig. 2. Studies now in progress indicate a future downward adjustment. In any event, the specimen width of 12 in. employed for the determination of fracture resistance for the titanium alloys here discussed is more than adequate.

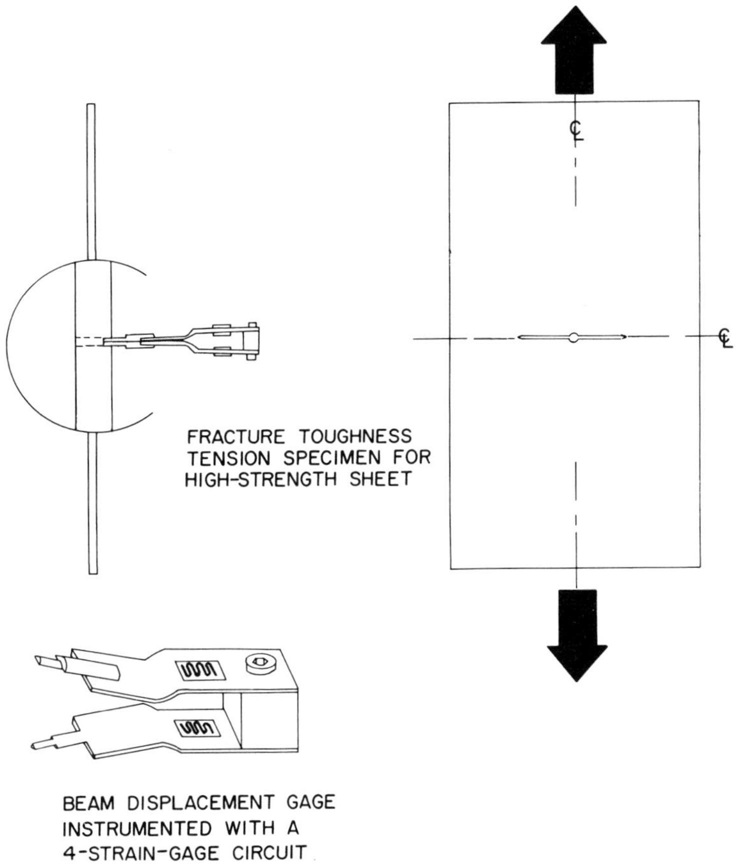

Fig. 1. Center-notched sheet tensile specimen together with beam displacement gage used for COD measurement.

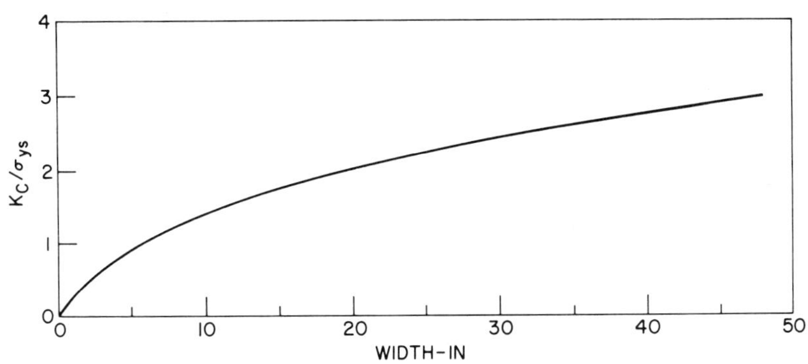

Fig. 2. Fracture resistance to yield stress ratio, K_c/σ_{ys}, plotted against specimen width estimated to produce valid data.

Effect of Relative Crack Length

Extensive investigations on the effect of crack length to width ratio in aluminum alloys (8, 9) established that, for $.15 < 2a/W < .50$, constant K_c values were obtained. This is substantiated for the titanium alloys 6-4 and 13-11-3, Fig. 3. The approximately ±10 percent scatterband is considered reasonable.

Effect of Sheet Thickness and Crack Tip Radius

A schematic diagram, Fig. 4, illustrates the anticipated dependency of K upon specimen thickness, B. At some sheet thickness specific for each alloy, a maximum value of K_c should be obtained and be accompanied by 100 percent slant fracture. As the thickness increases, lowered K_c values are accompanied by increasing amounts of flat fracture to provide the so-called "mixed-mode" appearance. Finally a minimum value is reached, K_{Ic}, and the fracture surface is entirely flat.

How closely this anticipated behavior is achieved in real materials, the titanium alloys of this investigation, is seen in Figs. 5a, b, c and d. Here K_c is plotted against sheet thickness, B. An accompanying plot also indicates the amount of slant fracture observed on the fracture surfaces. Although it is perhaps not quite correct to designate the values as K_c since they are generally "mixed-mode" fractures, this nomenclature has become customary. Curves drawn through the two sets of data points were obtained from an analytical model and will be discussed in a later section of this paper.

Some dependence upon sheet thickness is observed, but the apparent maximum K_c value is not always accompanied by 100 percent slant fracture. Values obtained from test specimens containing "sharp" elox slit tips are in general lower than those from specimens containing the "blunt" slit tips. Ideally, an extension of the initial slit by fatiguing is desirable to produce a "natural" sharp crack tip. When some crack growth precedes final separation, this tip sharpening has been considered unnecessary (10). Tests on aluminum alloys, most of which exhibited some crack growth, showed that the amount of crack growth was increased by crack tip sharpening but that the K_c values remained unchanged (11, 12). The effect of crack tip sharpening by fatigue for these titanium

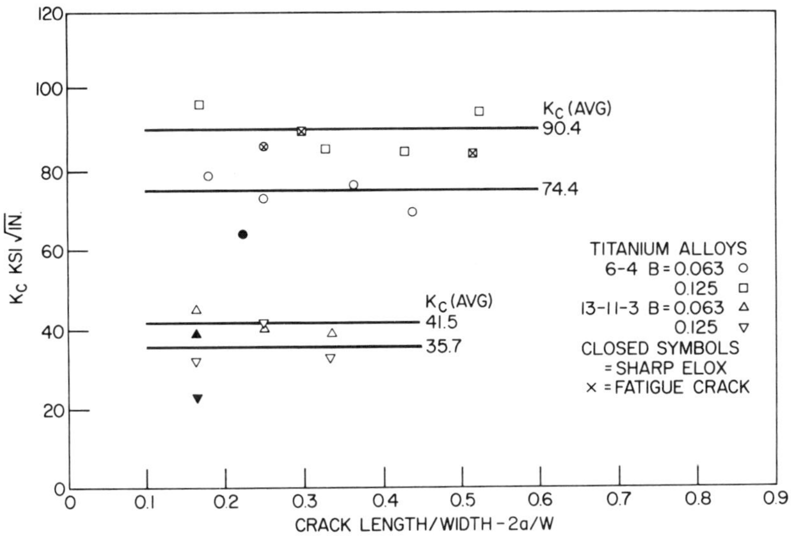

Fig. 3. Fracture resistance, K_c, plotted against crack length to width ratio, 2a/W, for titanium alloys 6-4 and 13-11-3.

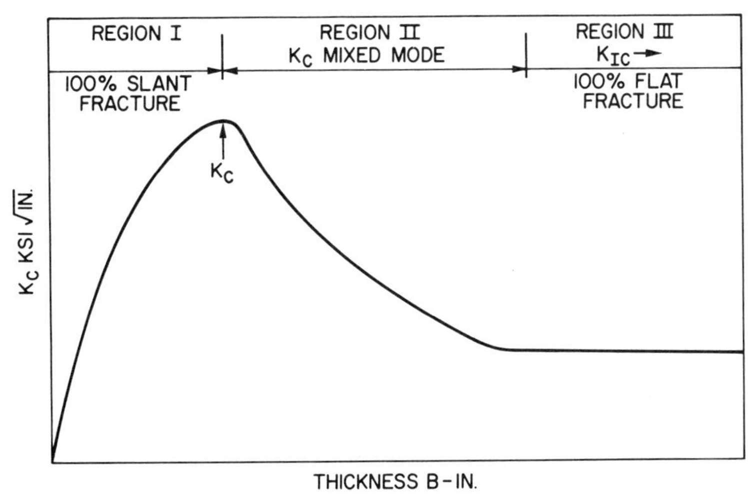

Fig. 4. Schematic figure showing postulated influence of specimen thickness, B, upon the fracture resistance, K_c.

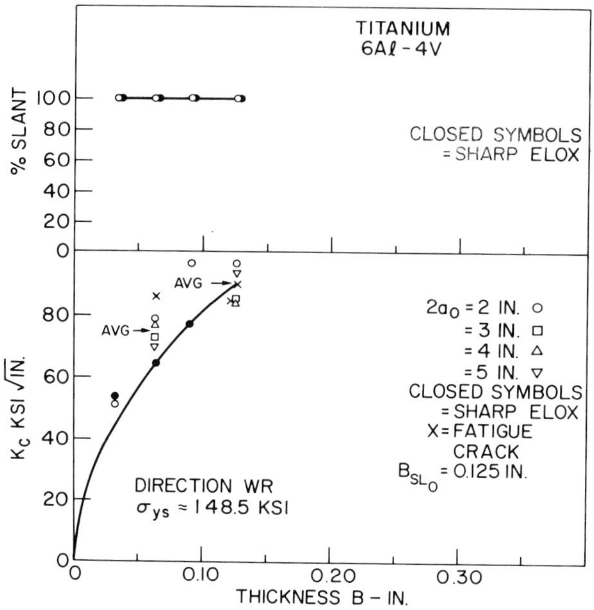

a. Titanium alloy 6-4

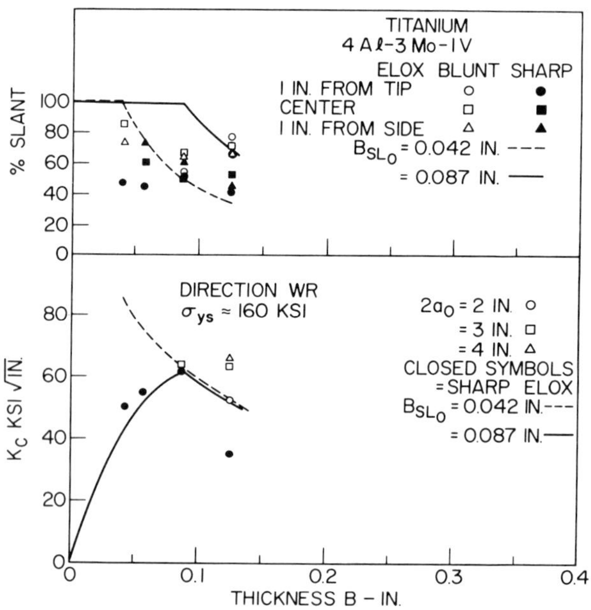

b. Titanium alloy 4-3-1.

Fig. 5. Measured K_c values (lower curve) and slant fracture percent (upper) plotted against specimen thickness, B. Curves drawn are calculated from the model of reference 13.

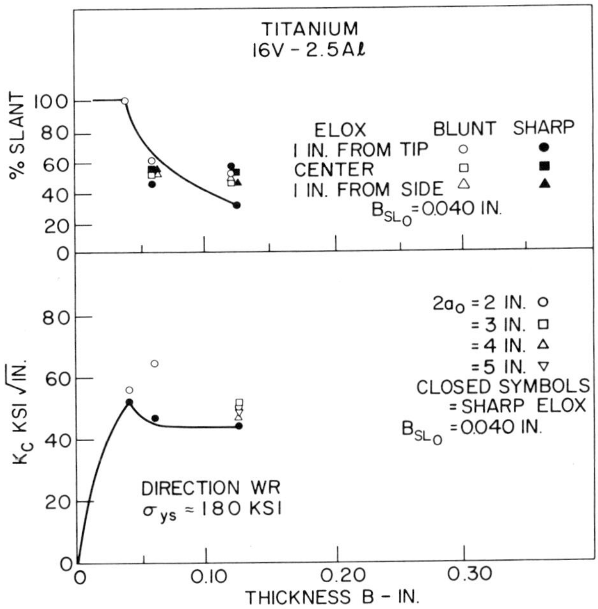

c. Titanium alloy 2.5-16.

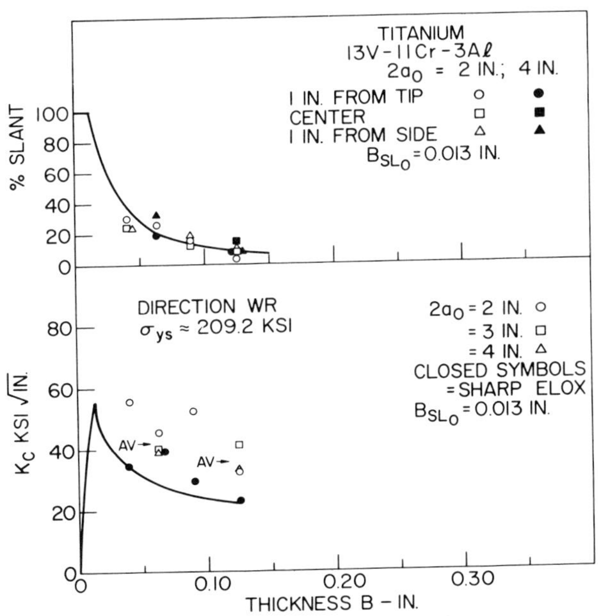

d. Titanium alloy 13-11-3.

Fig. 5. (cont'd)

alloys is the subject of a current investigation and is shown here only for titanium alloy 6-4, Fig. 5a.

Applicability of a Model for the Sheet Thickness Dependency of K_c

Since some variation in fracture resistance with sheet thickness is observed, it would obviously be a convenience if fracture resistance for all sheet thicknesses could be estimated from a limited number of actual measurements. Two quite similar models have been developed purporting to explain the effect of material thickness upon the value of K_c (13, 14). Both consider that flat fracture is a surface phenomenon whereas shear lip formation is a volume sensitive mechanism. It is further postulated that once the shear lip is fully developed, its total lip width no longer increases. Relevant equations are:

Model I

$$\mathscr{G}_c = \frac{1}{2} k_{SL} \; B_{SL} \left(\frac{B}{B_{SL}}\right); \; \left(\frac{B}{B_{SL}}\right) \leq 1 \qquad (5)$$

$$\mathscr{G}_c = \frac{1}{2} k_{SL} \; B_{SL} \left(\frac{B_{SL}}{B}\right) + k_{ff} \left(1 - \frac{B_{SL}}{B}\right); \; \frac{B}{B_{SL}} \geq 1 \qquad (6)$$

Model II

$$R = \mathscr{G}_c = \frac{1}{2} k_{SL} \; B_{SL} \left(\frac{B_{SL}}{B}\right) + k_{ff} \qquad (7)$$

where

$\mathscr{G}_c = K_c^2/E$

B_{SL} = critical (constant) shear lip thickness

B = sheet thickness

k_{SL} and k_{ff} material constants where subscripts SL and ff refer to shear lip and flat fracture, respectively, and $k_{ff} = \mathscr{G}_{Ic}$.

Examination of the equations show that Models I and II differ only in an evaluation of the contribution of the flat fracture portion. The application of Model I to the data presented here has been attempted. First, consider the plots of absolute shear lip width, B_{SL}, vs. specimen thickness, B, seen in Figs. 6a, b, c, and d. For the first three alloys, this value

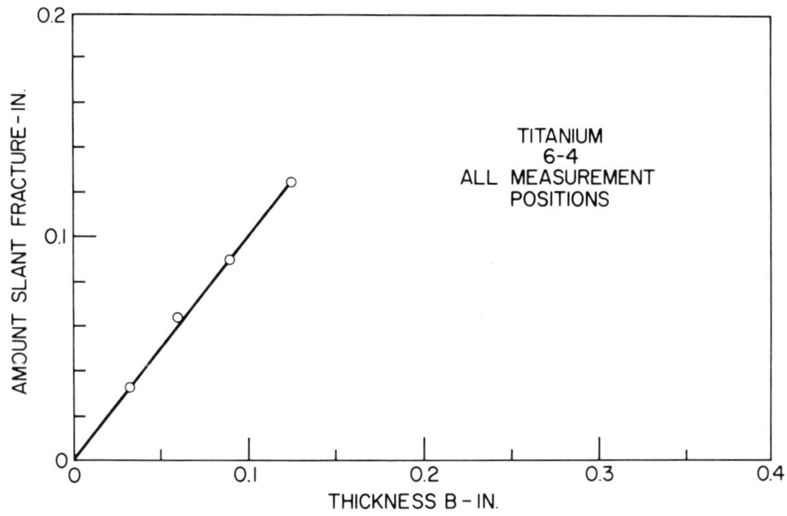

a. Titanium alloy 6-4.

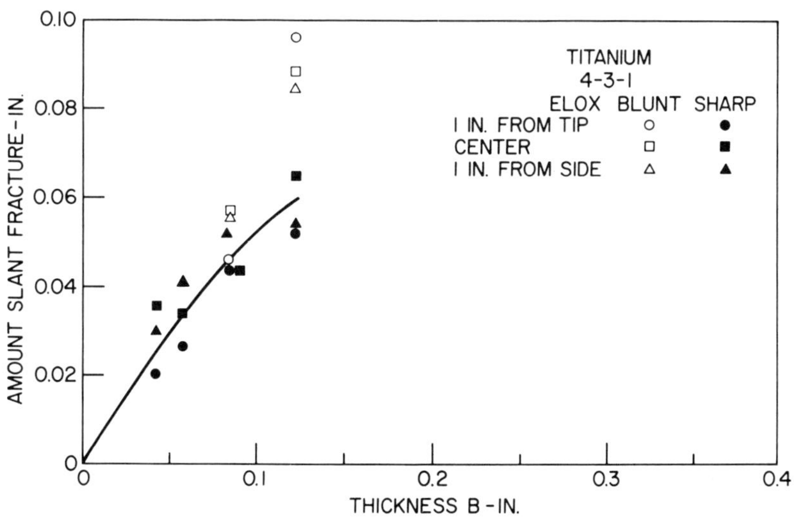

b. Titanium alloy 4-3-1.

Fig. 6. Amount of slant fracture (in.) plotted against specimen thickness, B.

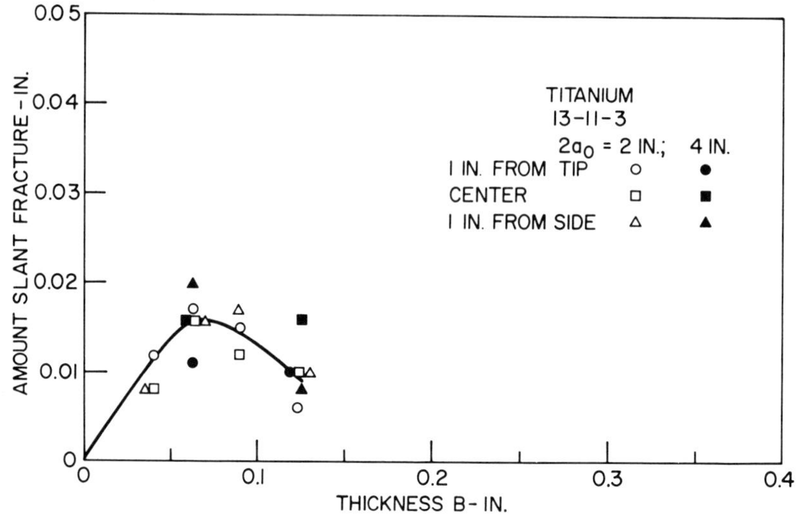

c. Titanium alloy 2.5-16

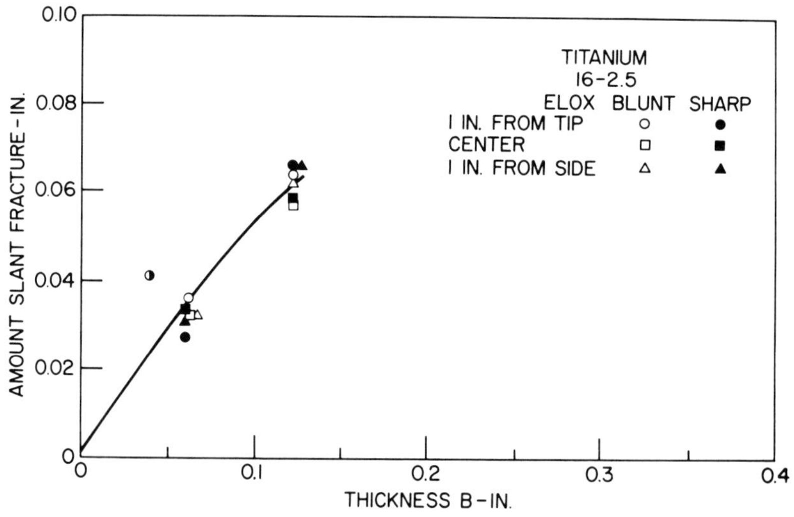

d. Titanium alloy 13-11-3.

Fig. 6. (cont'd)

continues to increase with thickness, never apparently attaining a contant value, B_{SLO}. Referring back to Figs. 6-8, however, it is noted that the percent of slant fracture is continually decreasing whereas in Fig. 5 it remains constant at 100 percent. For this reason, rather arbitrary choices of B_{SLO} were necessitated.

For Ti-6-4, Fig. 5a, since all fractures were 100 percent slant, B_{SLO} was assumed to be .125 in. and

$$k_{SL} = \frac{2\ \mathscr{G}_{max}}{B}$$

Computed values of K fit these data quite well where $\frac{B}{B_{SL}} < 1$.

Application of the model to the data of Ti 4-3-1 seen in Fig. 5b is less satisfactory. Two values of B_{SLO} were tried, but in no instance could they be fitted through the sharp elox value for B = .125 without developing a negative value for $k_{ff}(\mathscr{G}_{Ic})$ which is untenable.

Data fitted to the values of Ti 16-2.5 produce a K_c vs. B curve, Fig. 5c, which fits the sharp elox values but the percent slant vs. B curve is less satisfactory. Here B_{SLO} was taken as .040 in. since this thickness developed 100 percent slant fracture.

For Ti 13-11-3, in Fig. 5d, it is seen that the amount of slant fracture decreases with increased thickness. An average value of B_{SLO} =.0130 in. was selected for the analysis shown in Fig. 8.

Critical Crack Length at Various Levels of Operating Stress

Some generalizations can be attempted since the degree of precision sought in research is not required in the world of real structures. Loads acting on a structural member cannot be precisely defined since fabricating stresses resulting from misalignment, welding, and rivet coupling can only be estimated. Further, the alloy strength will vary slightly from sheet to sheet and even sheet thickness will fluctuate. Therefore, a reasonable estimate of K_c is a helpful design parameter. In Fig. 7 where all four alloys are compared, it is seen that average values reasonably describe three of them at least over a thickness range of .04 to .10 in.

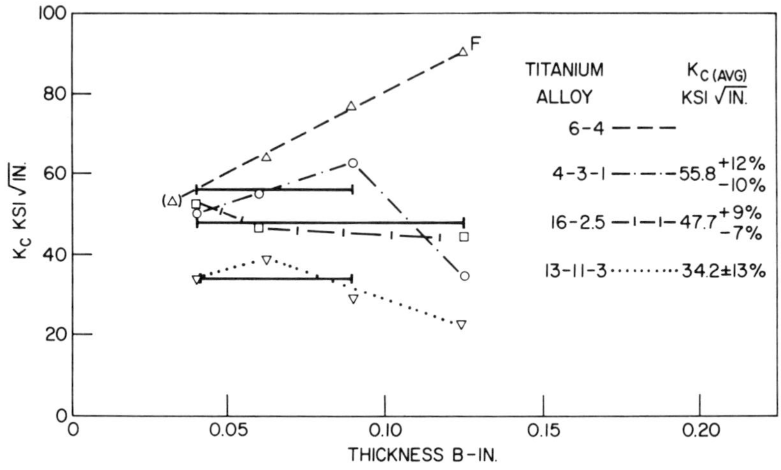

Fig. 7. Average fracture resistance, K_c, values for all four titanium alloys plotted against sheet thickness, B, for inter-comparison.

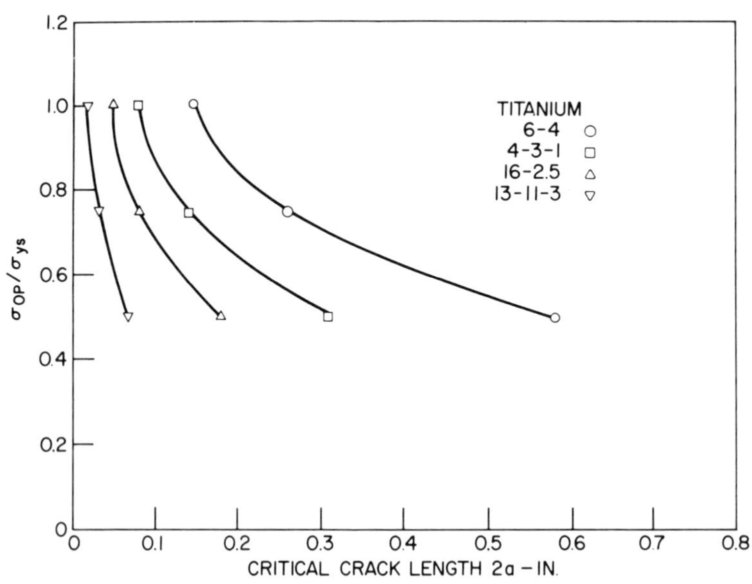

Fig. 8. Critical crack length, 2a, for instability and catastrophic fracture plotted against operating stress to yield stress ratio, σ_{op}/σ_{ys} for all four titanium alloys.

Using these average values, the crack length causing failure can be calculated at various levels of operating to yield stress ratios, where crack length,

$$2a = \frac{2}{\pi}\left(\frac{K}{\sigma_G}\right)^2$$

These values for all four titanium alloys are plotted in Fig. 8. The practicality of inspection techniques to insure against the presence of cracks of critical lengths in any structures should be a factor in material selection.

Summary

This discussion of the thin sheet fracture resistance problem with particular application to high-strength titanium alloys is intended to emphasize its scope and complexity and to suggest methods of minimizing the geometrical dependencies of K_c. The major attribute of K_c is the relationship it signifies between either the critical crack length at some predetermined operating stress level or that stress level which can be tolerated by a crack of predetermined size. The usefulness of using a single parameter to define this relationship is evident from a comparison of the fracture resistance values of four high-strength structural sheet alloys of titanium. Inconsistencies observed between the data and the model proposed indicate the need for further experimentation so that a more precise model may be developed.

References

1. "Method of Test for Plane Strain Fracture Toughness Metallic Materials," E-399-72, Annual Book of ASTM Standards, Part 31, 911, 1972.

2. Brown, W. F., Jr. and Srawley, J. E., "Plane Strain Crack Toughness Testing of High Strength Metallic Materials," ASTM STP 410, 1966.

3. Sullivan, A. M. and Freed, C. N., "The Influence of Geometric Variables on K_c Values for Two Thin Sheet Aluminum Alloys," NRL Report 7270, June 1971.

4. Carmen, C. M., Armiento, P. F., and Marcus, H., "Crack Resistance Properties of High Strength Aluminum Alloys," Proceedings of First International Conference on Fracture (Sendai, Japan) 2:995-1038, 1966.

5. Kaufman, J. G. and Holt, M., "Fracture Characteristics of Aluminum Alloys," Alcoa Research Laboratory Technical Paper No. 18, New Kensington, Pa., 1965.

6. Forman, R. G., "Experimental Program to Determine Effect of Crack Buckling and Specimen Dimensions on Fracture Toughness of Thin Sheet Materials," Air Force Flight Dynamics Laboratory Technical Report AFFOL-TR-65146, Wright Patterson AFB, Ohio, 1966.

7. Srawley, J. E. and Brown, W. F., Jr., "Fracture Toughness Testing Methods," in "Fracture Toughness Testing and Its Applications," ASTM STP 381, 133-245, 1965.

8. Freed, C. N., Sullivan, A. M. and Stoop, J., "Comparison of Plane Stress Fracture Toughness for Three Aluminum Sheet Alloys," NRL Report 7299, August 1971.

9. Sullivan, A. M. and Freed, C. N., "Plane Stress Fracture Resistance of One Sheet Steel and Two Titanium Sheet Alloys," NRL Report 7332, October 1971.

10. Brock, D., "The Residual Strength of Aluminum Alloy Sheet Specimens Containing Fatigue Cracks or Saw Cuts," National Aerospace Laboratory, (Amsterdam) Technical Report NRL-TR M 2143, 1966.

11. Sullivan, A. M., Freed, C. N. and Stoop, J., "A Comparison of R-Curves Determined from Different Specimen Types," Presented Fourth National Symposium on Fracture Mechanics, Urbana, Ill., September 1971, Publication Pending ASTM STP

12. Freed, C. N., Sullivan, A. M., and Stoop, J., "Crack Growth Characteristics of High Strength Sheet Alloys," NRL Report 7374, (Publication Pending)

13. Bluhm, J. I., "A Model for the Effect of Thickness on Fracture Toughness," ASTM Procedures, Vol. 61, 1961.

14. Krafft, J. M., Sullivan, A. M., and Boyle, R. W., "Effect of Dimensions on Fast Fracture Instability of Notched Sheet," Proceedings of Crack Propagation Symposium, Cranfield, U. K., 1961.

DISCUSSION

SECTION VIIb. FATIGUE AND FRACTURE

Rapporteur
Mr. A.C. Barber
Imperial Metal Industries (Kynoch) Ltd.
Birmingham, England

Co-Chairmen
Professor R.E. Reed-Hill
University of Florida
Gainesville, Florida

Dr. H. Kimura
National Research Institute for Metals
Tokyo, Japan

Paper VIIb-12 (I.W. Hall and C. Hammond) - The rapporteur questioned the 0.2%YS values for IMI550 and IMI551 alloys which were quoted as approximations only. Dr. Hammond replied that they were derived from Vicker's Hardness Tests. The 0.2%YS and UTS values for IMI318 were obtained from tensile specimens which had been given an identical heat treatment to the fracture toughness specimens. When questioned about the possible influence of texturing effects on the yield strength and fracture resistance, Dr. Hammond replied that texture effects were avoided by taking all specimens in the same orientation. The point claimed by Dr. Hammond is that strength and toughness are not necessarily universally related.

Paper VIIb-2 (J.I. Dickson, J.P. Owens, and A. Plumtree) - J.C. Williams said that there is usually significant scatter in fatigue test measurements and asked if sufficient measurements had been made to indicate that the reported reduction in fatigue life due to prestraining was significant and not merely due to experimental scatter. Dr. Dickson said that eight prestrained samples had been tested and all showed a 25-50% decrease in fatigue life. Moreover, the experimental scatter is usually smaller in short life fatigue where crack propagation is dominant than it is in high life fatigue where crack initiation is more important.

Paper VIIb-1 (D.E. MacDonald and W.A. Wood) - Professor H. Margolin (New York University) commented with regard to the slipless fatigue of titanium observed by the authors. At New York University we have recently been conducting studies on the determination of stress at a grain boundary. This work has shown that the grain the grain boundary stress is several times higher than the average stress and decays rapidly away from the boundary. Titanium is a highly anisotropic material and it seems reasonable to believe that such stresses would exist at some grain boundaries. If this is true, you may have slip occurring locally at grain boundaries which would initiate the crack. The crack could then propagate with slip highly localized while the grain as a whole appears elastic. Dr. MacDonald observed in reply that this narrow band of slip would not be resolvable with the scanning electron microscope. He accepted that there might be slip present that was too fine to be resolved by scanning electron microscopy, or for that matter by any other means. The problem, however, is to understand what differentiates this possible 'unobservable' slip from the slip observed which causes no cracking. Furthermore, the high anisotropy referred to, would dictate that the orientation of the narrow band of unobservable slip would change drastically at a grain boundary. This was not observed; the slipless cracks following the maximum shear direction with little regard to orientation changes at grain boundaries.

The observation that high anisotropy would restrict deformation to a narrow band was accepted by Dr. MacDonald but is was felt that this concentration of deformation can lead to internal stress levels sufficiently high to fracture grains within these bands without the benefit of slip.

As was noted for the overload model discussed in the paper, the orientation of the narrow band within the grain that absorbs the deformation is governed by the applied stress, not the local crystallography.

Paper VIIb-8 (M.J. Harrigan, A.W. Sommer, P. Reimers, and G. Alers) The rapporteur said, during his talk that it would be of value to measure the fracture properties of the textured material in the other three possible directions because work recently carried out at IMI Ltd. has indicated that K_{IC} cannot be related simply to modulus, but depends also on the orientation of both the crack propagation direction and the crack plane normal.